CATACLYSMIC VARIABLES AND LOW-MASS X-RAY BINARIES

CATACLYSMIC VARIABLES
AND
LOW-MASS X-RAY BINARIES

PROCEEDINGS OF THE 7TH NORTH AMERICAN WORKSHOP
HELD IN CAMBRIDGE, MASSACHUSETTS, U.S.A.,
JANUARY 12–15, 1983

Edited by

DONALD Q. LAMB

and

JOSEPH PATTERSON

Harvard Smithsonian Center for Astrophysics,
Cambridge, Massachusetts, U.S.A.

D. REIDEL PUBLISHING COMPANY

A MEMBER OF THE KLUWER ACADEMIC PUBLISHERS GROUP

DORDRECHT / BOSTON / LANCASTER

ASTROPHYSICS AND SPACE SCIENCE LIBRARY

A SERIES OF BOOKS ON THE RECENT DEVELOPMENTS
OF SPACE SCIENCE AND OF GENERAL GEOPHYSICS AND ASTROPHYSICS
PUBLISHED IN CONNECTION WITH THE JOURNAL
SPACE SCIENCE REVIEWS

VOLUME 113
PROCEEDINGS

Library of Congress Cataloging in Publication Data

North American Workshop (7th: 1983: Cambridge, Mass.)
   Cataclysmic variables and low-mass X-ray binaries.

   (Astrophysics and space science library; v. 113)
   Includes indexes.
   1. Cataclysmic variable stars–Congresses. 2. Stars, Double–
Congresses. I. Lamb, Donald Q. II. Patterson, Joseph. III. Title.
IV. Series.
QB835.N67   1983   523.8'446   84–29820

ISBN-13: 978-94-010-8849-7   e-ISBN-13: 978-94-009-5319-2
DOI: 10.1007.978-94-009-5319-2

Published by D. Reidel Publishing Company,
P.O. Box 17, 3300 AA Dordrecht, Holland.

Sold and distributed in the U.S.A. and Canada
by Kluwer Academic Publishers,
190 Old Derby Street, Hingham, MA 02043, U.S.A.

In all other countries, sold and distributed
by Kluwer Academic Publishers Group,
P.O. Box 322, 3300 AH Dordrecht, Holland.

TABLE OF CONTENTS

PREFACE

Since 1976 a meeting devoted to recent research on cataclysmic variables ("CV workshop") has been held annually somewhere in North America. Many of the meetings have been held – following a custom older than anyone reading this book – in locations with well-known recreational potential (e.g. Santa Cruz, CA; Boulder, CO). We thought hard about this custom while contemplating the possibility of organizing a meeting in Massachusetts in the middle of winter. Nobody wants their meeting to go down in history as the smallest and dullest, and it would surely be the coldest. But on occasion, meeting organizers have defied custom and scheduled meetings for less-than-trendy places, and gotten away with it (Urbana, IL and Rochester, NY must be reckoned as examples of this). Encouraged by the spatial and temporal proximity of the American Astronomical Society meeting (Boston, January 9–12), we thought we might get away with it again, and so came to organize a meeting for January 12–15, 1983, in Cambridge, MA.

There was another reason for a meeting at this time and place, we loftily proclaimed in early mailings. No one doubts that the CV's are closely related to the low-mass X-ray binaries ("LMXB's"), in which the accreting star is usually, or perhaps always, more compact than a white dwarf. Many of the general characteristics of LMXB's sound pretty familiar to any student of CV's: orbital periods in the range 0.02 – 2.0 days, lobe-filling donor stars with masses in the range 0.05 – 1.5 $M_\odot$, accretion rates in the range $10^{-11} - 10^{-8}$ $M_\odot$ yr^{-1}, lots of irregular variability in the light curves, outburst events probably due to unstable accretion flow through a viscous disk, etc., etc. Despite all the similarities, research programs on these two classes of accreting binaries have proceeded more or less independently, and have focused on different topics. For example, both observational and theoretical studies of CV's have taught us far more about accretion disks, while the effects of radiation transfer near the compact star have been much more extensively explored in the literature on X-ray binaries. What could be more stimulating than to bring these groups together for a joint conference? What better time than after an AAS meeting, attended by nearly 10^3 astronomers? And what more natural place to do it than Cambridge, MA, which a long line of successful satellite missions (Uhuru, SAS-3, HEAO-1, HEAO-2) has made into a world center for the study of X-ray binaries?

Anyway, that was the idea. Did it work? Well, we had a well-attended and highly successful conference. We heard a number of excellent review talks on various aspects of CV's and LMXB's, all of which are contained in these proceedings. We debated many of the critical scientific issues, and managed to collect most of the debate into this

volume. The coffee arrived on time, the microphones and recording
equipment worked, and the snowstorm held off until the last day. But
we are inclined to think that the basic <u>goal</u> of the conference, a synthesis of the two fields, didn't really quite happen. The sessions on
LMXB's were well attended by the local contingents from MIT and Harvard,
but these forces dwindled when the focus shifted to CV's, with their
pitifully low X-ray luminosities. The CV pundits had come from
thousands of miles away and therefore did not actually walk out when
outlandish numbers like 10^{38} ergs s^{-1} started flying through the air,
but their facial expressions took on the glaze of the morning donuts.
Often experts spoke to other experts on their familiar topics, while
others waited patiently for their specialized subject to come up for
discussion. To a certain extent this was inevitable and desirable, but
we were <u>disappointed</u> that there was not more of the cross-fertilization
of ideas that we had hoped for. Better things might have happened in
the corridors. But while listening to the tapes (of the formal sessions, not the corridor conversations!), we identified few exchanges in
which the CV's met the LMXB's to their mutual enlightenment. On at
least two occasions, questioners who proposed such a comparison were
actually scolded for comparing "apples and oranges". So it seems pretty
clear that the time for such a synthesis has not yet come, that the
specialists are not yet ready to recognize the unifying themes that will
surely characterize some future phase of research on close binaries containing compact stars.

But it's worth striving for. We will keep promoting the idea of
joint conferences, and keep coaxing the experts to attend each other's
meetings. It seems certain that we can all learn important things from
the orchard-keeper across the road, and probable that the most significant obstacles will disappear after a few more evenings of beer and
pizza, and a few more of John Faulkner's jokes.

In editing the discussions, we frequently found it desirable, in
the interests of clarity, to deviate from the literal transcript. In
theory, the more egregious episodes of confusion and excess verbiage
have been "cleaned up" or removed. But we also tried hard to retain the
spirit and flavor of the spoken word. Of course, we take full responsi-
bility for the results.

We are grateful to Phyllis de Fano for making everything happen,
both at the workshop and in the preparation of the proceedings. We
thank Lesley Pew, who helped us through the final, tortuous stages of
editing the proceedings.

Don Q. Lamb
Joe Patterson
January 1983

LIST OF PARTICIPANTS

Timothy Bastian	Univ. Colorado, Boulder, CO, USA
Paul Barrett	Louisiana State Univ., Baton Rouge, LA, USA
Emilia Belserene	M. Mitchell Obs., Nantucket, MA, USA
William Blair	CfA, Cambridge, MA, USA
Michael F. Bode	Los Alamos National Lab., Los Alamos, NM, USA
Forrest Boley	Dartmouth, Hanover, NH, USA
Howard Bond	Louisiana State Univ., Baton Rouge, LA, USA
J.M. Bonnet-Bidaud	CEN Saclav, FRANCE
Jim Brainerd	CfA, Cambridge, MA, USA
Bill Cabot	CfA, Cambridge, MA, USA
Brad W. Carroll	Univ. of Rochester, Rochester, NY, USA
Ganesar Chanmugam	Louisiana State Univ., Baton Rouge, LA, USA
John Clarke	Univ. of California, Berkeley, CA, USA
Larry Coleman	Univ. of Arkansas, Little Rock, AR, USA
Lynn Cominsky	Univ. of California, Berkeley, CA, USA
Guiseppina Fabbiano	CfA, Cambridge, MA, USA
John Faulkner	Lick Observatory, UCSC, Santa Cruz, CA, USA
Susan Feingold	Technion University, Haifa, ISRAEL
Bob Garrison	David Dunlap Obs., Ontario, CANADA
Albert D. Grauer	Univ. of Arkansas, Little Rock, AR, USA
Jonathan E. Grindlay	CfA, Cambridge, MA, USA
Ed Guinan	CfA, Cambridge, MA, USA
Barbara Hassall	Institute of Astronomy, Cambridge, UK
Paul Hertz	CfA, Cambridge, MA, USA
Keith D. Horne	Caltech, Pasadena, CA, USA
David Hummer	JILA, Univ. of Colorado, Boulder, CO, USA
Icko Iben	Univ. of Illinois, Urbana, IL, USA
James N. Imamura	Univ. of Virginia, Charlottesville, VA, USA
Kenneth B. Jensen	NASA-GSFC, Greenbelt, MD, USA
Paul Joss	MIT, Cambridge, MA, USA
Stephen M. Kahn	Columbia Univ., New York, NY, USA
Tim Kallman	MIT, Cambridge, MA, USA
Jonathan Katz	Washington Univ., St. Louis, MO, USA
Richard Kelley	MIT, Cambridge, MA, USA
Scott J. Kenyon	Univ. of Illinois, Urbana, IL, USA
Alan Kiplinger	NASA-GSFC, Greenbelt, MD, USA
Julian Krolik	CfA, Cambridge, MA, USA
Don Q. Lamb	CfA, Cambridge, MA, USA
Frederick K. Lamb	Univ. of Illinois, Urbana, IL, USA
Stephen Langer	Univ. of Illinois, Urbana, IL, USA
Alan M. Levine	MIT, Cambridge, MA, USA
Walter H.G. Lewin	MIT, Cambridge, MA, USA
James Liebert	Univ. of Arizona, Tucson, AZ, USA
Rich London	Livermore Lab., Livermore, CA, USA

Vincent J. Mantle   Oxford Univ., Oxford, UK
Masaru Matsuoka   Inst. of Space and Astron. Sci., Tokyo, JAPAN
Janet Mattei   AAVSO, Cambridge, MA, USA
Jon Magnuson   Dartmouth College, Hanover, NH, USA
Herman Marshall   CfA, Cambridge, MA, USA
Jeffery McClintock   MIT, Cambridge, MA, USA
Brian McLean   Queens Univ., Kingston, NY, USA
Ed Morgan   MIT, Cambridge, MA, USA
Martine Mouchet   ESO, Munich, FRG
Lorne Nelson   Queens Univ., Kingston, Ontario, CANADA
John A. Nousek   Penn. State Univ., Univ. Park, PA, USA
Bohdan Paczynski   Princeton Univ., Princeton, NJ, USA
Robert Panek   NASA-GSFC, Greenbelt, MD, USA
Jan van Paradijs   Univ. of Amsterdam, THE NETHERLANDS
Joseph Patterson   CfA, Cambridge, MA, USA
Bill Priedhorsky   Los Alamos National Lab., Los Alamos, NM, USA
Saul Rappaport   MIT, Cambridge, MA, USA
John Raymond   CfA, Cambridge, MA, USA
Robert Rosner   CfA, Cambridge, MA, USA
George Rybicki   CfA, Cambridge, MA, USA
Bradley Schaefer   MIT, Cambridge, MA, USA
Gary Schmidt   Univ. of Arizona, Tucson, AZ, USA
Dan Schwartz   CfA, Cambridge, MA, USA
Allen W. Shafter   Univ. of California, Los Angeles, CA, USA
Giora Shaviv   Technion, Haifa, ISRAEL
Edward M. Sion   Villanova Univ., Villanova, PA, USA
Jozef Smak   JILA, Univ. of Colorado, Boulder, CO, USA
Warren Sparks   Los Alamos National Lab., Los Alamos, NM, USA
Steven Stahler   CfA, Cambridge, MA, USA
Sumner Starrfied   Arizona State Univ., Tempe, AZ, USA
(H.S.) Peter Stockman   Univ. of Arizona, Tucson, AZ, USA
Richard Stover   Lick Observatory, UCSC, Santa Cruz, CA, USA
Jean Swank   NASA-GSFC, Greenbelt, MD, USA
Paula Szkody   Univ. of Washington, Seattle, WA, USA
Santiago Tapia   Univ. of Arizona, Tucson, AZ, USA
John R. Thorstensen   Dartmouth College, Hanover, NH, USA
Kenneth C. Turner   Cornell-Arricebo Obs., Arecibo, PR
Alexander V. Tutukov   Univ. of Illinois, Urbana, IL, USA
Hugh Van Horn   Univ. of Rochester, Rochester, NY, USA
Frank Verbunt   Institute of Astronomy, Cambridge, UK
Richard A. Wade   Institute of Astronomy, Cambridge, UK
Brian Warner   Univ. of Capetown, Capetown, SOUTH AFRICA
Ronald F. Webbink   Univ. of Illinois, Urbana, IL, USA
J. Craig Wheeler   Univ. Texas, Austin, TX, USA
Nicholas E. White   ESA-ESTEC, Noordwijk, THE NETHERLANDS
Lee Anne Willson   Iowa State Univ., Ames, IA, USA
Kent Wood   Naval Research Lab., Washington, DC, USA

EVOLUTION OF CATACLYSMIC BINARIES

B. Paczynski[1,2]
Princeton University Observatory

ABSTRACT

A standard scenario for the evolutionary origin of cataclysmic binaries is presented. In this scenario the initial periods of the ances-
tor binaries was between a few months and a few years, and the initial
mass ratio was very different from unity. The period was reduced, and
a lot of mass and angular momentum was lost during the common envelope
phase of evolution. If this scenario is correct then there should be
a lot of short period binaries made of pairs of white dwarfs, which
were produced by the systems with initial mass ratios close to unity.
About 10% of all "single" white dwarfs could be short period binaries,
or could have passed through such a phase some time in the past. Such
binaries should evolve due to angular momentum losses caused by gravitational radiation and could produce R CrB stars, AM CVn cataclysmic
binaries, DB white dwarfs, and Type I Supernovae. Inconsistencies in
the standard evolutionary scenario are discussed. It is possible that
a binary nature of a star is the necessary condition for a formation
of a high surface brightness planetary nebula.

1. INTRODUCTION

Cataclysmic binaries are characterized by their eruptive behaviour
stimulated by the mass transfer between the components. The secondary
component fills its Roche lobe, and in most cases it is the less massive of the two stars. Matter flows from the secondary through the
vicinity of the inner Lagrangian point (L1) towards the compact primary component, in most cases forming an accretion disk around it. In
a classical cataclysmic binary (CB) the primary is a white dwarf. Low
mass X-ray binaries with neutron star primaries have many properties

---

[1] On leave from N. Copernicus Astronomical Center, Polish Academy of
Sciences, Warsaw.

[2] A long-term member, The Institute for Advanced Study, Princeton,
New Jersey.

D. Q. Lamb and J. Patterson (eds.), Cataclysmic Variables and Low-Mass X-Ray Binaries, 1–14.
© 1985 by D. Reidel Publishing Company.

of CB and I shall discuss them to some extent as well.   In most known
cases the secondary is a main sequence star, but in some   systems it
may be a giant (T CrB), a subgiant (Cyg X-2 = V1341 Cyg),   a hydrogen
degenerate dwarf (WZ Sge), or a helium degenerate dwarf (HZ 29 = AM
CVn, G61-29 = GP Com).   In a typical system almost all visible   light
comes from the accretion disk.   In some systems the primary white
dwarf has so strong a magnetic field that the rotation of the primary
is synchronized with the binary motion, and accretion proceeds through
a column rather than a disk (polars = AM Her type stars).   No cases
are known with a magnetized neutron star corotating with a binary.

Many types of CB are known: novae, recurrent novae, Z Cam stars, dwarf
novae, SU UMa stars, polars, DQ Her stars, bulge X-ray binaries, X-ray
bursters, X-ray transients, and possibly others.

2. ORIGIN AND EVOLUTION OF CATACLYSMIC BINARIES (A Standard Scenario)

It is believed that initially a system which is going to evolve to become a CB has a binary period of a few months or a few years, so that
the two stars are very separated. The component that is initially the
more massive of the two, the primary, proceeds faster with its nuclear
evolution and evolves all the way to a red giant or a red supergiant
stage before overflowing its Roche lobe.   Therefore it may develop a
massive degenerate core, or even a massive nondegenerate core if the
initial mass of the primary was sufficiently large.   Let us concentrate on the first case.   As soon as the primary expands sufficiently
to overflow its Roche lobe a rapid mass transfer begins, and leads to
a buildup of an extended envelope around the accreting secondary,
which is a low mass main sequence star.   The binary evolves to a contact configuration and later a common envelope engulfs the two compact
cores: a degenerate primary and a main sequence secondary.   The two
cores transfer angular momentum to a common envelope and gradually
spiral in with ever decreasing period and separation. Finally, either
the two cores merge and we are left with a single star, or the envelope is lost and a short period binary is left (Ostriker 1973, Eggleton 1976, Paczynski 1976, Ritter 1976, Taam, Bodenheimer and Ostriker
1978, Meyer and Meyer-Hofmeister 1979, Livio, Saltzman and Shaviv
1979, Livio 1982).   Here we are interested in the second scenario
only, and there are binaries known that have recently emerged from a
common envelope stage and are now nuclei of planetary nebulae (UUSge,
V477 Lyr = Abell 46, VW Pyx = K1-2, cf. Patterson 1983 for references,
Abell 41 -Grauer and Bond 1983).   A somewhat older system of the same
type is LB 3459 = AA Dor, with the planetary nebula gone by now and
the hot primary evolving to become a white dwarf.   Later on the primary becomes a white dwarf, while the main sequence secondary is
smaller than its Roche lobe and the system is detached (V471 Tau, HZ
9, GK Vir = PG 1413+01, UX CVn = HZ 22, BE UMa = PG 1155+492; cf. Patterson 1983 for references).   Much later still, either as a result of
angular momentum loss from the system, or as a result of nuclear evolution of the secondary, that star fills its Roche lobe, the system
becomes semidetached, and the mass transfer begins giving rise to cat-

aclysmic activity.   The observations indicate that white dwarfs in CB are on average more massive than single white dwarfs. This is entirely consistent with the present scenario (Wai-Yuen Law and Ritter 1983).

A similar evolutionary pattern may produce short period binaries with neutron star primaries.   In this case the initial mass of the primary should be large enough, above 7 solar masses, so that a nondegnerate helium core may be formed.   The first violent mass transfer and the first common envelope stage proceed with the primary being a nondegenerate helium core with a mass exceeding the Chandrasekhar limit.   A short period binary emerging from the common envelope phase should live through a supernova explosion of the primary. We are left with a detached binary made of a neutron star and a low mass main sequence star, with a short orbital period. Now we have to wait for the binary to become semidetached, which requires either a substantial loss of angular momentum, or expansion of the secondary component due to its nuclear evolution, just like it was the case in a binary with a white dwarf primary.

There are other ways in which CB can form.   All kinds of short period binaries (main sequence star plus a neutron star, or a white dwarf, or another main sequence star) may form by tidal capture in dense stellar systems, like globular clusters (Lightman and Grindlay 1982, and references therein).   It is not clear how important the captures may be in the galactic disk. Perhaps some short period binaries with neutron star primaries were produced when accretion drove a white dwarf over the Chandrasekhar limit (Ergma and Tutukov 1976, Nomoto et al. 1979, Labay et al. 1983).

Let us suppose that finally we have a short period semidetached binary with the mass transfer going on.   If the rate of mass transfer is low and the outer parts of accretion disk are cool then disk accretion becomes unstable and gives rise to a dwarf nova activity and to some X-ray transients (Smak 1971, 1982; Osaki 1974; Hoshi 1979, 1981, 1982; Meyer and Meyer-Hofmeister 1981, 1982; Cannizzo, Wheeler and Ghosh 1982; Faulkner and Lin 1983).   At some intermediate rate of mass transfer the disk may be either stable or not stable and a binary may look like Z Cam. At high mass transfer rates the disk is stable and a binary may be classified as nova-like. Accretion of nuclear fuel onto a compact primary may give rise to various phenomena.   If the accretion rate is very high then matter cannot be processed through nuclear burning at such rate and an extended envelope builds up. As a result a contact binary, and later a common envelope binary develops.   If the rate is still lower then nuclear burning becomes unstable and shell flashes may give rise to X-ray bursts on accreting neutron star and nova explosions on accreting white dwarfs (Lewin and Joss 1981 and references therein, Fujimoto 1982 and references therein).   This type of evolution may be terminated either because there is no secondary left to accrete from, or because the primary is driven above the Chandrasekhar limit for a white dwarf or the Oppenheimer-Volkoff limit for a neutron star, or because the two stars coalesced due to excessive loss of angular momentum. Driving a white dwarf over the Chandrasekhar limit, or in general inducing helium or carbon ignition under degenerate conditions, is the most favored scenario for a Type I supernova (Mazurek 1973, Whelan and Iben 1973, Warner 1974, Tutukov and Ergma 1976, Fugimoto and Sugimoto 1982, Fujimoto and Taam 1982, Nomoto 1982a, b, Trimble 1982, Greggio and Renzini 1983).

CB with periods above 6 or 8 hours have sufficiently massive secondaries that their evolution may be driven by nuclear burning. CB with shorter periods have secondary components of such a low mass that their nuclear time scale is longer than the age of the Galaxy. Angular momentum must be driven away from such CB in order to make the sustained mass transfer possible. For binaries with periods between 80 minutes and 2 hours gravitational radiation seems to be the dominant sink of angular momentum (Paczynski 1981 and references therein, Paczynski and Sienkiewicz 1981, 1983, Rappaport, Joss and Webbink 1982, Tutukov, Fedorova and Yungel'son 1982). Cataclysmic binaries with periods between 3 and 6 hours need more efficient means of angular momentum loss, possibly magnetic winds (Eggleton 1976, Verbunt and Zwaan 1981, Patterson 1983, Spruit and Ritter 1983). It is not clear what drives the evolution of the four binaries known to have periods between 18 and 50 minutes. Notice that at least two of those binaries have no hydrogen, and nothing is known about hydrogen abundance in the two others.

There is a well known gap in the periods of CB between 2 and 3 hours. There were many explanations proposed, none too convincing. Recently, the most fashionable explanation requires a high rate of angular momentum loss above the gap which makes the secondary depart from the main sequence mass radius relation. Abrupt decrease of angular momentum losses to the level of gravitational radiation stops the mass transfer, makes the secondary shrink back to the main sequence radius, and makes the binary detached. Cataclysmic activity may be regenerated after a few billion years, when gravitational radiation brings the stars close enough for the mass transfer to become possible again. By this time the binary period is 2 hours (Paczynski and Sienkiewicz 1983, Patterson 1983, Spruit and Ritter 1983).

3. PROBLEMS WITH CATACLYSMIC BINARIES

There are a number of problems with the "standard evolutionary scenario" as outlined in the previous chapter. It is generally believed that a common mass ratio among the binaries is 1 (Lucy and Ricco 1979, and references therein), though the nearby rather wide binaries may have almost any mass ratio (van de Kamp 1971, 1975). A common envelope evolution may lead to a CB provided the initial mass ratio is very different from 1. That is so because the initial mass of a star that gave rise to a white dwarf or a neutron star must have been much larger, while at present the compact star is more massive than the secondary filling its Roche lobe. Therefore, there must be many pro-

genitor binaries with orbital periods between a few months and a few
years and the mass ratios between 10:1 and 5:1; or so. Notice that
the secondary could not accrete much matter while passing through a
common envelope stage. It is interesting what happens to all of those
binaries with the initial mass ratio close to 1, supposedly a very
common variety. We may expect that in such systems the first phase of
mass transfer is not very violent, as the mass losing star is about as
massive as the mass accreting star. Therefore, it is likely that a
formation of a contact or common envelope binary is avoided. Following the first Case B or Case C mass transfer process we are left with
a white dwarf primary (or helium burning star evolving to become a
white dwarf or a neutron star), and a more massive secondary, which
has already a well developed hydrogen depleted (or hydrogen exhausted)
core. The secondary is going to evolve to a red giant stage. At that
time the system may look like a symbiotic star, while the secondary is
smaller than its Roche lobe (Tutukov and Yungelson 1976, Paczynski and
Rudak 1980).   Subsequent evolution of the secondary must give rise to
a violent mass transfer and a formation of a common envelope binary
with two very compact cores: degenerate, neutron, or helium burning.
It is hard to believe that all such systems coalesced. Short period
nuclei of planetary nebulae indicate that it is possible for a binary
to safely emerge from a common envelope phase. The question is then:
where are all of those post common envelope binaries which had the
initial mass ratio close to 1? These should be detached pairs of compact stars: degenerate, neutron, or burning helium. The helium burning phase is short lived and such a star is either going to explode as
a supernova and make a neutron star, or it is going to die slowly and
become a white swarf. The long lived short period binaries would be
made of neutron stars, white dwarfs and possibly black holes. One object of this type is well known: the binary radio pulsar (Taylor and
Weisberg 1982). But where are the much more numerous pairs of white
dwarfs? Existence of such pairs was suggested among others by Nather,
Robinson and Stover (1981) and Eggleton (1982), but I think it has
never been emphasized that those systems should perhaps be more numerous than cataclysmic binaries are, if the standard evolutionary scenario for the origin of CB is correct.

Let us consider now what could be the consequences of the formation of
close pairs of white dwarfs, and how those could be detected. Notice
that all those Algol-type systems that do not coalesce should produce
such pairs. For example, a well studied binary AS Eri (Popper 1973,
Refsdal, Roth and Weigert 1974) is made of a main sequence star of
about 2 solar masses, and a subgiant star of about 0.2 solar masses,
mostly in a degenerate helium core. Soon the system will become detached, and the subgiant will become a white dwarf of 0.2 solar masses. The massive component will evolve away from the main sequence
with helium core of about 0.2 solar masses, and the second mass transfer must lead to a common envelope phase. If the cores remain separated the final product will be a pair of 0.2 solar mass white
dwarfs with a very short period orbit. The system may look on the H-R
diagram as a very low mass white dwarf because of a large total area

of the two degenerate components. As there is no obvious way in which
single white dwarfs of 0.2 solar masses could be formed, any very low
mass white dwarf, if discovered, may be a good candidate for a short
period binary white dwarf system.   (Notice: the white dwarf 40 Eri B
has a mass below the helium core flash mass, it could have been produced with help from a very low mass companion spiralling in a common
envelope.)   The pairs of white dwarfs of any mass might be discovered
either because their spectra are composite, or because their radial
velocities are variable.   Little is known about variation of radial
velocities of "single" white dwarfs, but a few objects are known to
have radial velocity measurements discordant by as much as 100 km/sec
(V. Trimble, private communication). It should be a fairly simple observational project to search for radial velocity variations in white
dwarfs, as the likely periods may be a few hours, and likely amplitudes may be in excess of 100 km/sec.   At least very useful limits
could be placed on nonexistence of such systems.   About 5% of nuclei
of planetary nebulae are known to be binaries with one hot and another
cool component (Bond and Grauer, private communication to Patterson
1983).   A comparable fraction could be made of two hot components,
which will evolve to a short period pair of white dwarfs.   I expect
that of the order of 10% of all white dwarfs may turn out to be short
period pairs.

The only way a short period pair of white dwarfs could evolve is
through angular momentum loss due to gravitational radiation. Nothing
spectacular would be seen until the less massive component fills its
Roche lobe at a binary period of about one minute.   Gravitational radiation time scale is only a few thousand years now, and a mass transfer must become very violent (Dearborn and Paczynski 1978, unpublished).   It is difficult to give any firm predictions, but a few possibilities are likely.   It is possible that rapid increase of a white
dwarf mass may give rise to a Type I Supernova due to violent helium
or carbon ignition.   It is possible that helium shell ignition will
not be explosive and a single helium red giant, an R CrB star, will be
formed (Paczynski and Trimble 1973).   Subsequent evolution of that
star may also give rise to a Type I Supernova, or a DB white dwarf.
Finally, if the masses are very low, and the mass transfer relatively
slow and not catastrophic, a helium rich mass transferring CB may be
formed, like AM CVn (HZ 29) or GP Com (G61-29).   These may also evolve
to become DB white dwarfs (Nather, Robinson and Stover 1981).

Let us consider now not a formation, but the evolution of a normal,
hydrogen rich CB.   Various problems arise.   It is not really known
whether the white dwarf mass increases as a result of mass transfer,
or is it in fact decreasing due to mixing of inner, helium and carbon
rich layers, which are subsequently lost in nova explosions (Gallagher
and Starrfield 1978, Kippenhahn and Thomas 1978).   Theoretical nova
explosions are possible only at rather low mass transfer rates, less
than 10^{-10} solar masses per year (Fujimoto 1982a, b, MacDonald 1982).
Observations indicate rather high mass transfer rates, above 10^{-9}
solar masses per year (Patterson 1983), which should lead to nonvio-

lent flashes. It is possible that interaction of slowly expanding envelope of flashing white dwarf with orbiting low mass main sequence star may produce a slow nova driven by energy transfer from binary motion to envelope expansion (MacDonald 1980). However, this may be the best way to kill the binary, as MacDonald (1980) estimates that a secondary of 0.4 solar masses would have to lose 0.0025 of binary binding energy (and angular momentum) to expel 0.0001 solar mass envelope. If the system is to remain semidetached about 0.001 solar mass has to be transferred from the secondary to the primary just to compensate for the loss of the binary angular momentum. This is ten times more mass than has been ejected! This may lead to a runaway mass transfer, and a formation of a "permanent" common envelope and coalescance of the two stars.

There is another problem with attempts to increase the white dwarf mass. If we want to burn the accreted matter then we must radiate an enormous amount of energy in the far ultraviolet. Even with accretion energy to be released in the boundary layer between the disk and the white dwarf there are observational problems: the mystery of the missing boundary layer (Ferland et al. 1982). Hydrogen burning releases almost one hundred times more energy than accretion onto a white dwarf, and it is even more difficult to hide it. According to Greggio and Renzini (1983) accreting white dwarfs burning hydrogen should produce in ultraviolet about 1% of the total luminosity of a typical galaxy in order to provide a sufficient production rate of Type I Supernovae. Scenarios requiring evolution towards Type I SN by accreting hydrogen onto a white dwarf face the problem: how to hide all that energy released in hydrogen burning. I am not aware of any satisfactory explanation.

The estimates of birth rate or death rate of CB are uncertain, but various authors give a total rate that is about the same (to within a factor of 10) as the rate of Type I SN (Trimble 1982, Patterson 1983, Greggio and Renzini 1983). That requires a very high efficiency of evolution from CB to Type I SN, or even a deficiency of CB needed to account for SN I. I consider this very unlikely for the reasons outlined above. It is entirely possible that hypothetical close pairs of white dwarfs could be much more effective in producing SN I than cataclysmic binaries are, and a serious observational search for those binaries would be most welcome.

According to Patterson (1983) there is a problem with the fate of novae. There are many of them among binaries with periods above 3 hours. However, all of them cannot evolve to the short period systems as there are relatively few binaries below 2 hours. What happens to them? They may all explode as SN I, but it is surprising that almost all of them have to do this before their periods decrease below 2 hours (after crossing THE GAP one way or another). Perhaps all of them have coalesced as a result of angular momentum loss with the ejected envelopes. In any case there is no clear explanation for the fate of novae.

Recent models of novae (Fujimoto 1982a, b, MacDonald 1982) require rather large masses for accreting white dwarfs, above 1 solar mass. According to Patterson (1983) the rate of formation of nova systems is much higher than the rate for all other types of CB combined. To make the last two statements compatible requires that the majority of CB have very massive white dwarf components and that very few CB are produced with low mass white dwarfs. A standard evolutionary scenario explains why some of the CB white dwarfs may be massive (Wai-Yuen Law and Ritter 1983), but it is not clear at all why this should be the case with the majority of CB. I think it is more likely that either models of novae, or the observational statistics is incorrect. Thermonuclear models of novae cannot explain the short time interval between the eruptions of recurrent novae (MacDonald 1982). I believe it means that we do not understand thermonuclear models well enough, and makes other predictions of the models (masses of white dwarfs and accretion rates) not very trustworthy.

There is a long time controversy about the nature of emission lines in cataclysmic binaries. Are they due to photoionization of the disk atmosphere by the central UV source, or are they due to local viscous heating in the external regions of optically thin accretion disk. It is known that in DQ Her a very large fraction of line and disk emission is due to the beamed radiation from rotating white dwarf, as demonstrated by the analysis of 71-second oscillations (Warner et al. 1972, Chanan, Nelson and Margon 1978). Reprocessing of the hot star radiation by the disk is clearly seen in UX UMa from the analysis of short period oscillations and the eclipse (Nather and Robinson 1974). In dwarf novae the hot spot at the outer rim of the disk is eclipsed by the disk itself over half of a binary cycle (Smak 1971). This implies that outer parts of the disk are optically thick, and emission lines are more likely due to the central UV source. Paczynski (1965) found that the strength of emission features as measured by U-B color in the UBV system declines during 80 days following every eruption of U Gem, which would be consistent with gradual cooling of the white dwarf surface following every accretion event. If this interpretation is correct, then the observed spectrum and optical luminosity of a disk may not be due to local release of accretion energy, but rather may be strongly affected by reprocessing of the UV radiation from the accreting white dwarf. This means that accretion rates deduced from observations of disks (Patterson 1983) may be systematically overestimated. This may be the reason why the "observed" rates are too high for the nova models (Fujimoto 1982b, MacDonald 1982). At the same time, a significant reduction of those rates may spoil the agreement which there seems to be between those rates and the value of binary period at the upper boundary of the gap, which is 3 hours (Paczynski and Sienkiewicz 1983).

The absence of hydrogen rich CB with binary periods shorter than 80 minutes is compatible with the suggestion that the evolution of the short period CB is driven by angular momentum losses due to gravitational radiation (Paczynski 1981, Paczynski and Sienkiewicz 1981,

1983, Rappaport, Joss and Webbink 1982). Unfortunately, theoretical results are rather sensitive to the adopted atmospheric opacities, and to chemical composition of the star overflowing its Roche lobe. The minimum period may be shorter if low metal content and low hydrogen content are chosen (Paczynski and Sienkiewicz 1983, Sienkiewicz 1983). Therefore, it is not really possible to use the 80-minute limit as a test for existence of gravitational radiation, but the two are certainly consistent with each other.

The high mass transfer rates observed in CB with periods above 3 hours, and the presence of the lower limit to the periods of high mass transfer CB implies that something much more effective than gravitational radiation is removing angular momentum from those systems (Patterson 1983, Paczynski and Sienkiewicz 1983, Spruit and Ritter 1983). Magnetic winds are usually made responsible. These winds are supposed to remove angular momentum from the spin of the secondary, while tidal effects are keeping it in synchronism with orbital motion. In effect, the secondary is used as a tramsitter of angular momentum, and energy, from binary motion to the wind. Notice that the secondary spin angular momentum is about 1% of the orbital. It follows that the secondary has to transmit one hundred times as much angular momentum as it has at any given time in order to change the binary parameters significantly. It has to do so while the binary period is close to 3 hours, and the secondary is supposed to depart from the main sequence and thermal equilibrium (Paczynski and Sienkiewicz 1983, Spruit and Ritter 1983). It is hard to believe that the transmission process is possible on such a scale, and if it is possible it should affect the structure of the secondary, as it has to transmit 100 times its total spin on its Kelvin-Helmholtz time scale.

4. CONCLUSIONS

I believe that the standard evolutionary scenario provides a reasonably good framework for the discussion and understanding of cataclysmic binaries. It helps to bring some order into a vast amount of observational data. Nevertheless, a considerable number of important phenomena is not understood. I am convinced that if the standard scenario is basically correct then there should be a lot of short period binaries with two white dwarf components. Roughly 10% of all white dwarfs may be in such systems. The most direct verification of all these considerations would be through a search for radial velocity variation among "single" white dwarfs. Not only a discovery of such variations, but also their absence in a large number of stars would be very informative.

The relation between cataclysmic binaries and Type I Supernovae is not clear in spite of many publications on the subject. The main problems are: do white dwarfs in CB increase or decrease their mass as a result of cataclysmic activity? If they increase, why cannot we account for the large ultraviolet luminosity due to hydrogen burning? If close pairs of white dwarfs are discovered, they could be much better candidates for pre-SN I.

There seems to be some disagreement between the accretion rates required for model nova explosions, and those inferred from the observed brightness of the accretion disks. It is possible that the "observed" rates are overestimated, as a large fraction of the visual luminosity may be due to reprocessing of the ultraviolet radiation from the accreting star by the disk. It is certainly the case with DQ Her and possibly with U Gem, and many other systems. If that was the case, the "observed" rates could be scaled down to make them agree better with the models of explosions, and perhaps the "mystery of the missing boundary layer" could be solved, or at least reduced.

No systematic study was done on the secular effect of repeated short time common envelope phases through which a binary must evolve while erupting like a nova. The effect of angular momentum loss may be very important for slow novae, because the binary is likely to interact strongly with slowly expanding envelope (MacDonald 1980). In fact, the interaction may turn out to be strong enough to force a coalescence of the binary.

There is still a mystery of the gap in periods, and no fully satisfactory explanation has been offered. The plausibility of magnetic winds as the means of efficient removal of binary angular momentum must be looked at more carefully.

It is surprising how many nuclei of planetary nebulae are found to be short period binaries. Certainly not all the binary nuclei can be readily discovered. A pair of hot stars, or a hot star with a very small mass and small size companion may be difficult to detect. At the same time it is not known what process is responsible for a very high rate of mass loss needed to produce a planetary nebula. It is not known how to lose the last part of the envelope, about 10^{-3} solar masses or so, the process needed to change a star from a red giant to a blue and hot one, capable of ionizing the nebula before it disperses (cf. Schonburner 1981 and the "superwinds"). It is not known if all the low and intermediate mass stars must pass through a planetary nebula phase before they become white dwarfs. If their mass loss rate is not high enough then a planetary nebula would not be seen as its surface brightness would be too low. Perhaps a binary nature of a star is necessary for a formation of a high surface density planetary nebula? The rates at which planetary nebulae and white dwarfs form are about the same (Cahn and Kaler 1971, Weidemann 1979), but it would not be possible to notice a difference by a factor of 2 or even 3.

REFERENCES

Cahn, J. H. and Kaler, J. B.: 1971, Astrophys. J. Suppl. 22, 319.
Cannizzo, J. K., Wheeler, J. C. and Ghosh, P.: 1982, Ap. J. (Letters), 260, L83.
Chanan, G. A., Nelson, J. E. and Margon, B.: 1978, Ap. J., 226, 963.

Eggleton, P. P.: 1976, IAU Symp. No. 73, p. 209.
Eggleton, P. P.: 1982, preprint.
Ergma, E. V. and Tutukov, A. V.: 1976, Acta Astron., 26, 69.
Faulkner, J. and Lin, D.: 1983, AAS Meeting, Boston.
Ferland, G. J., Langer, S. H., MacDonald, J., Pepper, G. H., Shaviv,
     G. and Truran, J. W.: 1982, Ap. J. (Letters), 262, 153.
Fujimoto, M. Y.: 1982a, Ap. J., 257, 752.
Fujimoto, M. Y.: 1982b, Ap. J., 257, 767.
Fujimoto, M. Y. and Sugimoto, D.: 1982, Ap. J., 257, 291.
Fujimoto, M. Y. and Taam, R. E.: 1982, Ap. J., 260, 249.
Gallagher, J. S. and Starrfield, S. G.:1977, Ann. Rev. Astr. Ap., 16
     171.
Grauer, A. D. and Bond, H. E.: 1983, preprint.
Greggio, L. and Renzini, A.: 1983, Astron. Ap., in press.
Hoshi, R.: 1979, Prog. Theor. Phys., 61, 1307.
Hoshi, R.: 1981, Prog. Theor. Phys. Suppl., 70, 181.
Hoshi, R.: 1982, Prog. Theor. Phys., 67, 179.
Labay, J., Canal, R. and Isern, J.: 1983, Astron. Ap., 117, L1.
Lewin, W. G. H. and Joss, P. C.: 1981, Space Sci. Rev., 28, 3.
Lightman, A. P. and Grindlay, J. E.: 1982, Ap. J., 262, 145.
Livio, M.: 1982, Astron. Ap., 105, 37.
Livio, M., Saltzman, J. and Shaviv, G.: 1979, M.N.R.A.S., 188, 1.
Lucy, L. B. and Ricco, F.: 1979, A. J., 84, 401.
MacDonald, J.: 1980, M.N.R.A.S., 191, 933.
MacDonald, J.: 1982, preprint.
Mazurek, T. J.: 1973, Ap. Space Sci., 23, 365.
Meyer, F. and Meyer-Hofmeister, E.: 1979, Astron. Ap., 78, 167.
Meyer, F. and Meyer-Hofmeister, E.: 1981, Astron. Ap., 104, L10.
Meyer, F. and Meyer-Hofmeister, E.: 1982, Astron. Ap., 106, 34.
Nather, R. E. and Robinson, E. L.: 1974, Ap. J., 190, 637.
Nomoto, E.: 1982a, Ap. J., 253, 798.
Nomoto, K.: 1982b, Ap. J., 257, 780.
Nomoto, K., Miyaji, S., Sugimoto, D. and Yokoi, K.: 1979: IAU
     Colloquium No. 53, p. 56; Osaki, Y.: 1974, Publ. Astro. Soc.
     Japan, 26, 429.
Ostriker, J. P.: 1973, private communication.
Paczynski, B.: 1964, Acta Astron., 15, 305.
Paczynski, B.: 1976, IAU Symp. No. 73, p. 75.
Paczynski, B.: 1981, Acta Astron., 31, 1.
Paczynski, B. and Rudak, B.: 1980, Astron. Ap., 82, 349.
Paczynski, B. and Sienkiewicz, R.: 1981, Ap. J. (Letters), 248, L27.
Paczynski, B. and Sienkiewicz, R.: 1983, Ap. J., 268, in press.
Paczynski, B. and Trimble, V.: 1973, Astron. Ap., 22, 9.
Patterson, J.: 1983, Ap. J., in press.
Popper, D.: 1973, Ap. J., 185, 265.
Rappaport, S., Joss, P. C. and Webbink, R. F.: 1982, Ap. J., 254, 616.
Refsdal, S., Roth, J. L. and Weigert, A.: 1974, Astron. Ap., 36, 113.
Ritter, H.: 1976, M.N.R.A.S., 175, 279.
Schonburner, D.: 1981, Astron. Ap., 103, 119.
Sienkiewicz, R.: 1983, Acta Astron., 33, in press.
Smak, J.: 1971, Acta Astron., 21, 15.

Smak, J.: 1982, Acta Astron., 32, in press.
Spruit, H. C. and Ritter, H.: 1983, Astron. Ap., in press.
Taam, R. E., Bodenheimer, P. and Ostriker, J. P.: 1978, Ap. J., 222, 269.
Taylor, J. H. and Weisberg, J. M.: 1982, Ap. J., 253, 908.
Trimble, V.: 1982, The Observatory, 102, 133.
Tutukov, A. V. and Yungel'son, L. R.: 1976, Astrofisica, 12, 521.
Van de Kamp, P.: 1971, Ann. Rev. Astr. Ap., 9, 103.
Van de Kamp, P.: 1975, Ann. Rev. Astr. Ap., 13, 295.
Verbunt, F. and Zwaan, C.: 1981, Astron. Ap., 100, L7.
Wai-Yuen Law and Ritter, H.: 1983, Astron. Ap., in press.
Warner, B.: 1974, M.N.R.A.S., 167, 61P.
Warner, B., Peters, W. L., Hubbard, W. B. and Nather, R. E. 1972, M.N.R.A.S., 159, 321.
Weidmann, V.: 1979, IAU Colloquium No. 53, p. 206.
Whelan, J. A. J. and Iben, I,: 1973, Ap. J., 186, 1007.

DISCUSSION

Katz: Do you fully accept the conclusion that nova outbursts are hydro-dynamic explosions? Or is it possible they are actually hydrostatic shell flashes and the expulsion is produced by the clumsiness of trying to get a double core in a single z envelope? In other words, could the outburst be produced by the red star orbiting through the expanded outer layers of the burning white dwarf and tossing matter off to infinity at roughly the orbital velocity?

PACZYNSKI: The only study that I know of was done by MacDonald. His conclusion was that you can reproduce slow novae that way but not fast novae, which are best produced by thermonuclear runaways.

KATZ: What aspects of fast novae cannot be reproduced?

PACZYNSKI: The high velocity of ejection and the high CNO abundances in the ejecta. But slow novae may very well be due to a binary moving in a common envelope, as you describe.

JOSS: Several of the problems you alluded to derive from the rather high mass transfer rates that have been deduced from observation. Can we be confident that the mass transfer rates that we see represent the long-term rates that are important for evolution? Or should we enter-tain the possibility that there are fluctuations on timescales longer than the time we've been observing these systems but short compared to evolutionary timescales? If this were true, we would tend to pick out, due to observational selection effects, those systems with higher lumin-osities and therefore higher mass transfer rates.

PACZYNSKI: Well, this is a real problem. I think observations over-estimate mass transfer rates because most estimates explicitly or implicitly assume that the disk luminosity we see is due to local viscous processes, whereas in many cases we know for sure that what we see is light from the hot star <u>reprocessed</u> by the disk. The classical case is DQ Herculis, where the 71-second oscillations change their phase during eclipse. This implies that the disk luminosity oscillates with a 71-second period, which I believe is the rotation period of the white dwarf. So, clearly, much of the disk luminosity is due to the rotating beam from the white dwarf. The same is true for the emission lines, which also oscillate with a 71-second period and therefore must also come from photoionization by the rotating beam of the white dwarf. For other stars, where the evidence is less clear, I still believe that the mass transfer rates are greatly overestimated. But there are problems with this also. Why are all these "overestimates" confined to systems with orbital periods greater than 3 hours? I think the two observed cutoffs at 80 minutes and 3 hours must give information on <u>secular</u> mass transfer rates, because the minimum period comes from the transfer rate averaged over a thermal time scale, i.e., $\gtrsim 10^8$ years. These two cutoffs imply rates of $10^{-10}\,M_\odot$/year and $10^{-9}\,M_\odot$/year, respectively. I don't know what the solution is. No matter what you do, you make someone unhappy.

van PARADIJS: One of your problems had to do with the lack of evidence for heating in the light curves of cataclysmic variables. Couldn't that be due to the shielding of the secondary star by a thick accretion disk?

PACZYNSKI: Sure. But then the disk should be extremely bright, if it's going to reprocess all that light. Hydrogen-burning produces about 50 times more energy than accretion. From the recent paper by the Illinois group on the "missing boundary layer", I gather that even the pure accretion models produce somewhat more light than the observations suggest. So, one should really worry about postulating an energy source that is 50 times more powerful.

SHAVIV: Your scenario for evolution implies that the matter accreted through the disk is always of normal composition. Can you allow for the possibility of peculiar compositions?

PACZYNSKI: Well, I don't know of any observations which convincingly demonstrate an anomalous composition in the disk. Also, I don't see any evolutionary scenarios which would <u>allow</u> very peculiar compositions in the secondary.

SHAVIV: That leaves unexplained the results of Bob Williams.

PACZYNSKI: He was using models which I think are fundamentally wrong. He assumed that the lines are formed in a region which is optically thin in the continuum. This is strictly inconsistent with photometric observations which show that the hot spot is eclipsed by the disk for half the binary period. It couldn't be eclipsed if the disk were optically thin. There are many cases (DQ Her is the most obvious) where we see that the emission lines are due to photoionization from the central source . . . . Maybe I am too emotional about this. To be fair, the results are simply model-dependent, and I would rather wait for other models to evaluate this point.

STARRFIELD: There's at least one object, U Sco, that appears to be transferring nearly pure helium.

PACZYNSKI: OK, and there's also HZ 29, but I'm sure you'll agree that they are a very small minority.

PRE-CATACLYSMIC BINARIES

Howard E. Bond
Department of Physics and Astronomy,
   Louisiana State University

ABSTRACT

Cataclysmic variables may be descended from initially wide binaries that underwent catastrophic loss of mass and angular momemtum during a "common-envelope" phase. If this is the case, there should exist a class of detached "pre-cataclysmic" binaries: double nuclei of planetary nebulae, and close binaries containing hot subdwarfs or white dwarfs. The secondary stars should be lower-main-sequence objects, and the orbital periods should be short (< 1.5 days).

Ten such pre-cataclysmic binaries are now known (four central stars of planetary nebulae, two close binaries containing sdOB stars, and four containing white dwarfs). There are several observational hints that these systems lose angular momentum faster than provided by gravitational radiation, through magnetic braking driven by convective motions in the envelopes of the late-type secondaries.

Numerous binaries exist that are very similar to pre-cataclysmic binaries but have periods longer than 1.5 days. This suggests that cataclysmic variables represent only the short-period tail of the distribution of objects that have emerged from common-envelope interaction.

1. THE EVOLUTIONARY ENIGMA OF CATACLYSMIC BINARIES

The cataclysmic variables (CVs) present an intriguing evolutionary puzzle because they contain white-dwarf components in extremely close binary systems. The binary separations are much smaller than the red giants that must have formed the white dwarfs. Thus it has become increasingly clear that the progenitors of CVs can only be binaries that were wide enough for the more-massive component to grow to red-giant dimensions and develop a degenerate core before encountering its companion; only then does extensive loss of mass and orbital angular momentum lead to the substantial orbital contraction necessary to pro-

15

D. Q. Lamb and J. Patterson (eds.), Cataclysmic Variables and Low-Mass X-Ray Binaries, 15–27.
© 1985 by D. Reidel Publishing Company.

duce a CV (Vauclair 1972; Ritter 1976; Livio, Salzman, and Shaviv 1979; Webbink 1979 and references therein). (The older idea that CVs are descended from W UMa-type main-sequence contact binaries [Kraft 1967] has foundered upon Webbink's [1976] conclusion that such systems coalesce before the primary leaves the main sequence.)

Several authors (e.g. Paczynski 1976; Webbink 1979; Law and Ritter 1983) have proposed a scenario in which the expanding red giant swallows its main-sequence companion, producing a "common-envelope" binary. On a timescale of about 10^3 yr (Meyer and Meyer-Hoffmeister 1979), the secondary spirals in, transferring its orbital angular momentum to the envelope. Eventually the envelope reaches breakup velocity and is ejected, leaving a detached close binary (the hot red-giant core and the main-sequence companion), surrounded by an expanding, ionized nebula. Following dissipation of the planetary nebula, one would observe a sdO + main-sequence binary. After contraction of the hot subdwarf to a degenerate configuration, the system will become a white-dwarf/red-dwarf detached binary. Further gradual orbital decay (or evolutionary expansion of the secondary) may ultimately produce a mass-transferring CV.

In this paper, we will refer to binary nuclei of planetary nebulae, and detached close binaries containing hot subdwarfs or white dwarfs, as "pre-cataclysmic" binaries.

2. CLOSE-BINARY NUCLEI OF PLANETARY NEBULAE

A crucial prediction of the above model for the origin of CVs is the existence of close-binary central stars of planetary nebulae (PN). In fact, four such objects are now known, and they are described in this section.

2.1. UU Sagittae: the binary nucleus of Abell 63

The first close-binary nucleus of a PN to be discovered was UU Sagittae, the eclipsing central star of Abell 63. The variability of UU Sge was discovered more than 50 years ago by Hoffleit (1932), and the PN was discovered on a Palomar sky survey plate by Abell (1966), but it was another ten years before the coincidence in position was noticed (Bond 1976). Subsequent photoelectric photometry (Miller, Krzeminski, and Priedhorsky 1976; Bond, Liller, and Mannery 1978) disclosed that UU Sge is a detached eclipsing binary with an orbital period of only $11^h 10^m$. The light-curve solution (Bond et al. 1978; Budding 1981) shows that the hot primary is of sdO dimensions (about $0.4\ R_\odot$, or considerably larger than a white dwarf), while the secondary (unseen except by reflected light) has the radius of a K dwarf ($0.6\ R_\odot$). The cool star may be less massive than would be inferred from this radius, if it is still out of thermal equilibrium following its recent emergence from the common envelope.

The discovery of UU Sge prompted the writer and A. D. Grauer to undertake a search for additional close-binary nuclei of PN in an attempt to provide further support for the common-envelope scenario described above. The secondary star of UU Sge suffers considerable heating (to about 10,000 K on the hemisphere facing the primary), producing an 0.3-mag hump in the light curve. Hence similar systems should exhibit detectable light variations even if the orbital plane were insufficiently inclined for actual eclipses. Accordingly, we decided to carry out our search photometrically. The survey work to date has been conducted at the 0.9-m reflectors of Kitt Peak National Observatory and Louisiana State University Observatory with the two-star high-speed photometers described by Grauer and Bond (1981a). Nearly three dozen central stars have been monitored so far, and three new short-period binaries have been found and are described below. The close-binary incidence of about 10% is an upper limit since the program objects were biased by our preference for variability reported by other workers. However, our observational technique is not suitable for detecting longer-period binaries (say P > 1-2 days), so we cannot exclude Paczynski's (1985) suggestion that most or all PN are ejected by binary nuclei.

The short-period variability of the central star of Minkowski 1-2 reported by Drummond (1980b) was not confirmed (Grauer and Bond 1981b). Some additional candidate binary central stars have been mentioned by Acker (1976), Lutz (1977), Noskova (1980), and Livio (1982) and should be investigated.

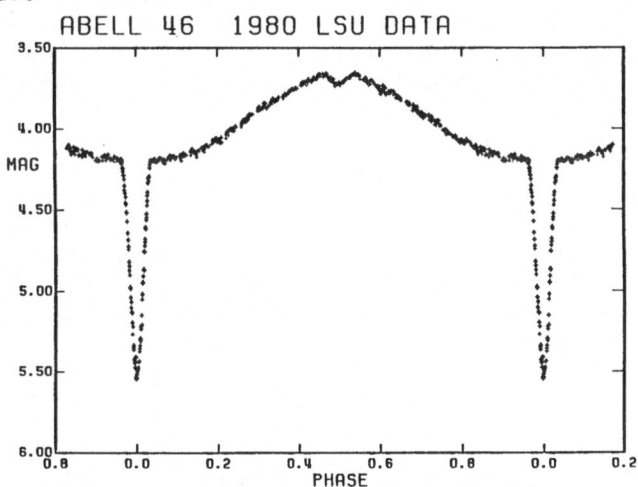

Figure 1. B light curve of the central star of Abell 46.
Magnitudes are differential with respect to a
comparison star of B = 10.7.

2.2. The eclipsing central star of Abell 46 (V477 Lyrae)

The first success of our survey was the confirmation of Abell's (1966) report that the central star of Abell 46 is variable, as recounted in Grauer and Bond (1981a, Fig. 2). Subsequent observations at LSU revealed the existence of partial eclipses (Grauer and Bond 1981a, Fig. 3); ultimately we established that the orbital period is $11^h 19^m$, and obtained the complete light curve shown in Fig. 1.

The light curve of this detached binary is remarkably similar to that of UU Sge, in that it is dominated by strong heating of the secondary star that produces the large hump centered at phase 0.5. A preliminary analysis of the light curve by L. W. Twigg yields stellar radii of about $R_1 = 0.2 R_\odot$ and $R_2 = 0.6 R_\odot$. Thus, as in the case of UU Sge, the stars have the dimensions of a pre-white dwarf and a lower-main-sequence star.

2.3. The central star of Kohoutek 1-2 (VW Pyx)

The variability of this central star was discovered on Palomar sky survey prints by Kohoutek (1964) and confirmed recently by Kohoutek and Schnur (1982). Our photometry of K 1-2 indicates that the variations are periodic, with a period near $16^h 05^m$. For several years we have been unable to obtain a complete light curve because of poor weather and the 2/3 day period, but it appears that the light variations are roughly sinusoidal, with a remarkably large amplitude of 1.5 mag. The light curve, fragmentary as it is, appears to be similar to those of the x-ray sources HZ Her (Bahcall and Bahcall 1972) and 4U 2129+47 (McClintock, London, Bond, and Grauer 1982). Thus K 1-2 might represent the birth of a neutron-star binary rather than a CV, and could be a case where common-envelope evolution directly produced a semi-detached rather than a detached binary. Further observations are clearly desirable. This object should be a prime candidate for future x-ray observations.

2.4. The central star of Abell 41

Perhaps the most remarkable result of our survey is Grauer's discovery, described elsewhere in this volume and by Grauer and Bond (1983), of the binary nature of the nucleus of Abell 41. This object is a detached binary with an orbital period of only $2^h 43^m$. The inclination is insufficient for actual eclipses, but strong heating of the secondary star produces variations of about 0.15 mag. If the mass of the sdO primary is assumed to be 0.6 $M_\odot$, the secondary has the radius of an M dwarf of 0.1-0.3 $M_\odot$ (Grauer and Bond 1983).

3. CLOSE BINARIES CONTAINING O OR B SUBDWARFS

The immediate descendants of binary nuclei of PN should resemble them very closely except for the absence of the surrounding nebula. Two systems are known that probably belong to this category.

3.1. LB 3459 = AA Doradus

This O subdwarf was found to be an eclipsing binary with a period of $6^h 16^m$ by Kilkenny, Hilditch, and Penfold (1978). Several authors

(Kilkenny, Penfold, and Hilditch 1979; Paczynski and Dearborn 1980) have discussed models in which both components of this binary are hot, highly evolved objects. However, the observed high temperature of the secondary (on the side facing the primary) is probably due entirely to heating (Paczynski 1980; Conti, Dearborn, and Massey 1981). Hence a low-mass main-sequence (or degenerate hydrogen dwarf) secondary cannot be excluded, and would require a less-complicated evolutionary history than if the secondary is another O subdwarf or a white dwarf.

The low amplitude of the radial-velocity curve of the primary star (Kilkenny, Hill, and Penfold 1981) shows that, for primary masses of $0.1-1.5$ $M_\odot$, the secondary mass is only $0.02-0.14$ $M_\odot$. Paczynski (1980), by fitting the primary to the mass-radius relation for stars with a hydrogen-burning shell surrounding a degenerate helium core, finds $M_1 = 0.36$ $M_\odot$ and $M_2 = 0.05$ $M_\odot$; similar masses have been obtained from a model-atmosphere analysis of the spectrum of the primary by Kudritzki et al. (1982). The low-mass secondary is likely to be a nearly degenerate hydrogen-rich object that was never massive enough to ignite hydrogen burning. There is no evidence for surrounding nebulosity (Hilditch and Kilkenny 1980).

3.2. HZ 22 = UX CVn

This system has an orbital period of 13^h46^m, and the primary star appears to be a B-type subdwarf (Young, Nelson, and Mielbrecht 1972). The light curve (Young and Wentworth 1982) shows no eclipses and two broad minima of slightly different depths per orbit. Although the secondary star is too faint for direct observation, it seems probable that the observed light curve could be produced by a combination of ellipticity of the primary star and a small contribution from heating of one hemisphere of the secondary. Thus the secondary could well be a low-mass main-sequence star. Greenstein (1973) deduced $T_{eff} = 28,000$ K and $\log g = 4$ for the primary star. From a fit of Greenstein's parameters to evolutionary models, Schoenberner (1978) concludes $M_1 = 0.4$ $M_\odot$. The radial-velocity amplitude of the primary is large (Greenstein 1973), and requires a secondary mass $M_2 > 0.4$ $M_\odot$.

4. DETACHED CLOSE BINARIES CONTAINING WHITE DWARFS

Once the subdwarf in a pre-cataclysmic binary has become a white dwarf, the binary is extremely inconspicuous. Nevertheless, four such systems are now known. Remarkably, two of them are members of the very nearby Hyades cluster, suggesting that binaries of this type may be common.

4.1. BD $+16^o516$ = V471 Tau

This well-studied system was shown to be an eclipsing binary with a period of 12^h30^m by Nelson and Young (1970). The components are a hot, optically faint white dwarf and a K0-K2 dwarf, and the system appears to be a member of the Hyades cluster. It is not possible to obtain radial velocities for the white dwarf, because it is much fainter in the optical region than the K star and has very broad lines. From the very well defined radial-velocity curve of the K dwarf (Young 1976) and an assumed mass of 0.7 $M_\odot$, one finds a white-dwarf mass that is also 0.7 $M_\odot$ (Young and Lanning 1975). Gilmozzi and Murdin (1983) have recently recovered the optical spectrum of the white dwarf by digital subtraction of scans taken during and out of eclipse; the spectrum is of type DA. The white dwarf is, of course, easily detectable in the ultraviolet with the IUE satellite (Guinan and Sion 1982).

4.2. PG 1413+01 = GK Vir

This white dwarf undergoes deep total eclipses that were discovered accidentally during an observation with the Palomar 5-m telescope (Green, Richstone, and Schmidt 1978). The orbital period is 8^h16^m; the primary is a hot DA, and the spectral type of the much fainter secondary is estimated to be dM3.

4.3. HZ 9

Attention was first directed to this object because of its composite spectrum (DA + dM4.5e). Spectroscopic observations revealed that the radial velocity of the H-alpha emission line from the M dwarf is variable with a 13^h32^m period (Lanning and Pesch 1981). If the spectral type of the secondary is used to estimate a mass $M_2 = 0.2 M_\odot$, the primary's mass must be greater than 0.3 $M_\odot$, on the basis of the mass function from Lanning and Pesch's radial-velocity curve. Like V471 Tau, HZ 9 is a member of the Hyades cluster.

4.4. Case 1

This DA white dwarf was discovered during an objective-prism survey with the Case Schmidt telescope (Stephenson 1960). Infrared objective-prism observations (Stephenson 1971) revealed the presence of an M2 dwarf. Spectroscopic observations by Lanning (1982) showed an H-alpha emission line from the M dwarf whose radial velocity varies with a period of 16^h01^m. For an assumed mass of the secondary star (based on the spectral type) of 0.36 $M_\odot$, the white-dwarf mass must be greater than 0.43 $M_\odot$.

Neither HZ 9 nor Case 1 is known to show eclipses or other light variability.

5. EVOLUTION TO THE CV STATE

In the above discussion, it has been pointed out that a number of detached binaries are known with subdwarf or white-dwarf primaries and either directly observed dK-M secondaries or secondaries for which the observations do not exclude unevolved stars. The secondary-star masses, where useful information exists, range from 0.05 $M_\odot$ (LB 3459) to 0.7 $M_\odot$ (V471 Tau). These objects would become CVs if their orbits could be contracted sufficiently to bring the secondary stars into contact with their Roche lobes.

One mechanism that must operate in all binaries is the emission of gravitational radiation (GR). An expression for the rate of period change, dP/dt, for two orbiting point masses due to GR has been given by Kraft, Mathews, and Greenstein (1962). We have used this expression to calculate the period evolution for two typical pre-cataclysmic binaries with masses of (M_1, M_2) = (1.0, 0.5) and (0.6, 0.2) $M_\odot$. The results are shown in Fig. 2. The ticks indicate the periods at which the 0.5 and 0.2 $M_\odot$ secondaries come into contact with their Roche lobes and initiate cataclysmic activity.

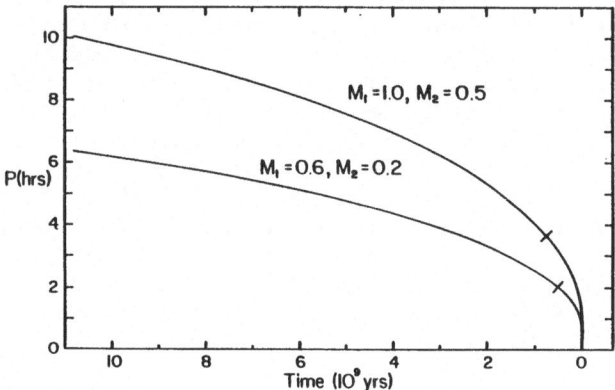

Figure 2. Orbital decay due to gravitational radiation for two detached binaries with the indicated masses (in solar units).

It is clear from Fig. 2 that pre-cataclysmic binaries with periods longer than about 10 hours are unlikely to become CVs in less than the age of the galactic disk, if GR is the only means they have for losing angular momentum. In Fig. 3 we show the orbital-period distribution for

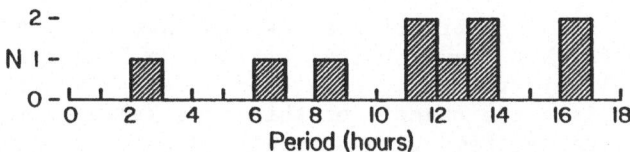

Figure 3. Distribution of orbital periods for the ten known pre-cataclysmic binaries.

the ten known pre-cataclysmic binaries discussed above. Three systems (PG 1413+01, LB 3459, and the central star of Abell 41) have periods less than 10 hours and are destined to become CVs in an astronomically short time (unless their masses are lower than those adopted in Fig. 2, as may be the case in LB 3459). What of the seven longer-period systems? We can conclude that one or both of the following must be true: (1) CVs are only the short-period tail of a much larger group of white-dwarf/red-dwarf binaries that emerge from common-envelope orbital contraction; (2) some mechanism exists that drains angular momentum from close detached binaries faster than GR (see Eggleton 1976), so that most of our pre-cataclysmics will in fact become CVs.

Statement (1) is probably true in any case (see next section). Recent discussions of the efficiency of magnetic braking in close binaries (e.g. Verbunt and Zwaan 1981; Patterson 1983) suggest that statement (2) may also be true. Indeed, there is considerable observational evidence for chromospheric activity (which should lead to a stellar wind and magnetic braking) in two of the pre-cataclysmic binaries, UU Sge and V471 Tau. Both of these objects have been detected as soft x-ray sources by Van Buren, Charles, and Mason (1980), who attribute the x-ray emission to the same mechanism as in RS CVn binaries (coronal emission from a late-star forced to rotate rapidly by tidal synchronism). A large body of optical observations exists for V471 Tau, showing its dK secondary to have many other properties seen in chromospherically active RS CVn binaries (Hall 1976). These include a light curve with a migrating wave (Cester and Pucillo 1976; Ibanoglu 1978; Rucinski 1981), strong chromospheric Ca II and Mg II emission lines (Oswalt 1979; Guinan and Sion 1982), ultraviolet flares (Beavers, Oesper, and Pierce 1979), and a variable, generally decreasing orbital period (Tumer et al. 1982).

Thus it may be that UU Sge and V471 Tau (or indeed most of the pre-cataclysmic binaries) are also destined to become CVs in less than the current age of the galactic disk. In fact, if we modify the formalism of Verbunt and Zwaan (1981) to apply to the case of magnetic braking in a detached binary, we find that systems as massive as V471 Tau can become CVs within the age of the galactic disk if they start with periods as long as roughly 1.5 days.

6. LONG-PERIOD REMNANTS OF COMMON-ENVELOPE INTERACTION

A number of detached binaries are known that contain white dwarfs or hot subdwarfs but have orbital periods so long (say P > 1.5 days) that it seems unlikely that they will ever become CVs through orbital decay. Table 1 lists the objects of this type known to the writer. Not included are several systems with separations so large that the stars have probably evolved independently, e.g. HR 3080 (sdOB + G5 III, Parsons et al. 1976), Zeta Capricorni (DA + Ba II star, Bohm-Vitense 1980), and 56 Pegasi (DA + K0 II, Schindler et al. 1982).

These systems could nevertheless become CV-like objects if the secondaries were to expand to fill their Roche lobes due to their nuclear evolution. Indeed, at least three of these systems ($-3°5357 =$

FF Aqr, 39 Cet, and HD 128220) already seem to have secondaries that
have evolved off the main sequence. A mass-transferring system formed
from a binary like HD 49798 through expansion of the secondary would
probably resemble GK Per (a long-period old nova with an evolved
secondary). HD 128220 may become a symbiotic variable or recurrent nova
(if the long period found by Wallerstein and Wolff [1966] is correct;
however, the large v sin i [100 km s^{-1}] of the G star suggests a much
shorter period).

Table 1. Systems with periods longer than 1.5 days

Name	P (days)	Spectra	References
HD 49798	1.55	sdO + (dF-K?)	Thackeray 1970; Kudritzki and Simon 1978
BE UMa	2.29	sdO + (dK?e)	Ferguson et al. 1981; Ando et al. 1982
Feige 24	4.23	DAwk + dM1-2e	Thorstensen et al. 1978
NGC 1360	8.2	sdO + ?	Mendez and Niemela 1977
-3°5357	9.2	sdO + G8 III	Dworetsky et al. 1977; Dorren et al. 1982
NGC 2346	16.0	? + A0-A5 V	See text
39 Ceti	77	DA + G5 III-IV	Simon et al. 1983
HD 128220	870	sdO + G0 III	Wallerstein and Wolff 1966

Table 1 lists the central stars of two planetary nebulae, NGC 1360
and NGC 2346; this rather clearly illustrates that binaries can emerge
from a common envelope with periods much longer than those of CVs. The
central star of NGC 2346 is a particularly unusual object. Its optical
spectrum is dominated by an A0-A5 star that is a single-lined spectroscopic binary with a period of 16.0 days (Mendez and Niemela 1981). The
A star undergoes deep eclipses, which were first observed by Kohoutek
(1982) and have been confirmed by Feibelman and Aller (1983), Marino and
Williams (1983), and the writer and A. D. Grauer. Most extraordinarily,
an examination of Harvard patrol plates has disclosed that the deep
eclipses only began in 1981 (Schaefer 1983). Mendez, Gathier, and
Niemela (1982) have suggested that a very small dust cloud is passing in
front of the system, blocking a portion of the orbit from view. An
alternative suggestion is that the unseen component of the binary is
expanding on a very fast timescale. Rapid expansion of a PN nucleus
undergoing a helium shell flash has indeed been predicted theoretically
by Iben et al. (1983), but the secondary star in NGC 2346 seems to be
expanding much faster than expected.

There is probably a large population of descendants of common-
envelope evolution beyond those listed in Table 1. Table 2 is a partial
listing of degenerate or pre-degenerate objects with main-sequence or
somewhat evolved companions that have been discovered from their
composite spectra. It would be extremely valuable to know whether any
of these systems are close binaries.

Table 2. Systems with unknown periods

Name	Spectra	References
a) Central stars of planetary nebulae		
Abell 35	? + G8 III-IV	Jacoby 1981
HD 112313	(sdO?) + G5	See text
NGC 1514	sdO + A0 III	See text
b) sdOB primaries		
HDE 283048	sdO + G	Laget et al. 1978
Feige 80	sdO + A?	Sargent and Searle 1968
+34°1543	sdB + F	Berger and Fringant 1978
+29°3070	sdOB + F	Berger and Fringant 1980
+10°2357	sdO + A	Berger and Fringant 1980
−7°5977	sdO + G	Heckathorn and Opal 1983
−11°162	sdO + G	Bond et al. 1971
c) White-dwarf primaries		
GD 325	DA + dM	Greenstein 1974, 1975
G 147−65	DC + dM	Greenstein 1976
LP 60−359 A	DA + dM	Greenstein et al. 1977
Feige 93	DA + dM	Greenstein et al. 1977

The three central stars listed in Table 2 are of particular interest. The nucleus of the PN Abell 35 is a G8 subgiant, which probably has an optically invisible hot companion that is responsible for ionizing the nebula. High-dispersion spectrograms obtained by the writer and R. E. Luck at Kitt Peak and Cerro Tololo show the G8 star to have rotationally broadened lines, suggesting synchronous rotation in a close binary; and unpublished photometric observations by Grauer show variations on a timescale of hours. Thus the central star of Abell 35 is probably a rather short-period binary.

HD 112313 is the central star of a very low-surface-brightness PN discovered by Longmore and Tritton (1980). As in the case of Abell 35, the optical spectrum is dominated by a G-type star, but ultraviolet observations with IUE have clearly revealed a hot companion (Feibelman and Kaler 1983). Recent observations obtained by the writer with the coude spectrograph of the Lick 3-m reflector reveal rotationally broadened lines, so this object is probably another short-period PN nucleus. The G star shows strong Mg II emission (Feibelman and Kaler 1983), and the Lick observations show double H-alpha emission similar to that in the chromospherically active star FK Comae (Walter and Basri 1982).

The central star of NGC 1514 has a composite optical spectrum (Kohoutek and Hekela 1967), demonstrating that two stars must be present. However, the radial velocity seems to be constant (Greenstein 1972), and both Drummond (1980a) and Bond and Grauer (unpublished) have found no light variations. Hence the central star is not a close binary (unless it happens to be viewed pole-on).

Note that the OB subdwarfs in Table 2 have companions of systematically earlier spectral types than the white dwarfs. This is simply due to the selection effect that the two components of a binary must have comparable luminosities in order for a composite spectrum to be detected. There are undoubtedly many sdO + dK-M and DA + dF-G systems where the fainter components are undetectable. Further, many of the objects in Tables 1 and 2 have rather bright apparent magnitudes (many are listed in the HD or BD), and hence such longer-period binaries are probably much more common than CVs. We conclude that common-envelope interaction usually results in binaries that are still so wide that they will never become CVs in less than the Hubble time.

The writer thanks R. P. Kraft for use of the facilities of Lick Observatory while this paper was being written.

REFERENCES

Abell, G. O. 1966, Ap. J., 144, 259.

Acker, A. 1976, Publ. Astr. Obs. Strasbourg, 4, No. 1.

Ando, H., Okazaki, A., and Nishimura, S. 1982, Publ. Astr. Soc. Japan, 34, 141.

Bahcall, J. N., and Bahcall, N. A. 1972, Ap. J. (Letters), 178, L1.

Beavers, W. I., Oesper, D. A., and Pierce, J. N. 1979, Ap. J., 230, L187.

Berger, J., and Fringant, A.-M. 1978, Astr. Ap., 64, L9.

_____. 1980, Astr. Ap., 85, 367.

Bohm-Vitense, E. 1980, Ap. J. (Letters), 239, L79.

Bond, H. E. 1976, Publ. A. S. P., 88, 192.

Bond, H. E., Liller, W., and Mannery, E. J. 1978, Ap. J., 223, 252.

Bond, H. E., Perry, C. L., and Bidelman, W. P. 1971, Publ. A. S. P., 83, 643.

Budding, E., 1981, in Photometric and Spectroscopic Binary Systems, ed. E. B. Carling and Z. Kopal (Dordrecht: Reidel), p. 405.

Cester, B., and Pucillo, M. 1976, Astr. Ap., 46, 197.

Conti, P. S., Dearborn, D., and Massey, P. 1981, M.N.R.A.S., 195, 165.

Dorren, J. D., Guinan, E. F., and Sion, E. M. 1982, in Advances in Ultraviolet Astronomy: Four Years of IUE Research, NASA Conference Publication 2238, p. 517.

Drummond, J. D. 1980a, Ph.D. Dissertation, New Mexico State University.

_____. 1980b, Astr. Ap., 88, L11.

Dworetsky, M. M., Lanning, H. H., Etzel, P. B., and Patenaude, D. J. 1977, M.N.R.A.S., 181, 13P.

Eggleton P. 1976, in IAU Symposium 73, Structure and Evolution of Close Binary Systems, ed. P. Eggleton, S. Mitton, and J. Whelan (Dordrecht: Reidel), p. 209.

Feibelman, W. A., and Aller, L. H. 1983, Ap. J., in press.

Feibelman, W. A., and Kaler, J. B. 1983, Ap. J., in press.

Ferguson, D. H., Liebert, J., Green, R. F., McGraw, J. T., and Spinrad, H. 1981, Ap. J., 251, 205.

Gilmozzi, R., and Murdin, P. 1983, M.N.R.A.S., 202, 587,

Grauer, A. D., and Bond, H. E. 1981a, Publ. A. S. P., 93, 388.

_____. 1981b, Publ. A. S. P., 93, 630.

_____. 1983, Ap. J., in press.

Green, R. F., Richstone, D. O., and Schmidt, M. 1978, Ap. J., 224, 892.

Greenstein, J. L. 1972, Ap. J., 173, 367.

_____. 1973, Astr. Ap., 23, 1.

_____. 1974, A. J., 79, 964.

_____. 1975, Ap. J. (Letters), 196, L117.

_____. 1976, Ap. J. (Letters), 207, L119.

Greenstein, J. L., Oke, J. B., Richstone, D., van Altena, W., and
    Steppe, H. 1977, Ap. J. (Letters), 218, L21.

Guinan, E. F., and Sion, E. M. 1982, Bull. A. A. S., 13, 817.

Hall, D. S. 1976, in Multiple Periodic Variable Stars, IAU Colloqium No.
    29, ed. W. S. Fitch (Dordrecht: Reidel), p. 287.

Heckathorn, H., and Opal, C. B. 1983, Bull. A. A. S., 14, 919.

Hilditch, R. W., and Kilkenny, D. 1980, M.N.R.A.S., 192, 15P.

Hoffleit, D. 1932, Harvard Bull., No. 887.

Ibanoglu, C. 1978, Ap. Space Sci., 57, 219.

Iben, I., Kaler, J. B., Truran, J. W., and Renzini, A. 1983, Ap. J.,
    264, 605.

Jacoby, G. H. 1981, Ap. J., 244, 903.

Kilkenny, D., Hilditch, R. W., and Penfold, J. E. 1978, M.N.R.A.S., 183,
    523.

Kilkenny, D., Hill, P., and Penfold, J. E. 1981, M.N.R.A.S., 194, 429.

Kilkenny, D., Penfold, J. E., and Hilditch, R. W. 1979, M.N.R.A.S.,
    187, 1.

Kohoutek, L. 1964, Bull. Astr. Inst. Czech., 15, 161.

_____. 1982, IAU Inf. Bull. Var. Stars, No. 2113.

Kohoutek, L., and Hekela, J. 1967, Bull. Astr. Inst. Czech., 18, 203.

Kohoutek, L., and Schnur, G. P. O. 1982, M.N.R.A.S., 201, 21.

Kraft, R. P. 1967, Publ. A. S. P., 79, 395.

Kraft, R. P., Mathews, J., and Greenstein, J. L. 1962, Ap. J., 136, 312.

Kudritzki, R. P., and Simon, K. P. 1978, Astr. Ap., 70, 653.

Kudritzki, R. P., Simon, K. P., Lynas-Gray, A. E., Kilkenny, D., and
    Hill, P. W. 1982, Astr. Ap., 106, 254.

Laget, M., Vuillemin, A., Parsons, S. B., Henize, K. G., and Wray, J. D.
    1978, Ap. J., 219, 165.

Lanning, H. H. 1982, Ap. J., 253, 752.

Lanning, H. H., and Pesch, P. 1981, Ap. J., 244, 280.

Law, W.-Y., and Ritter, H. 1983, Astr. Ap., in press.

Livio, M. 1982, Astr. Ap., 105, 37.

Livio, M., Salzman, J., and Shaviv, G. 1979, M.N.R.A.S., 188, 1.

Longmore, A. J., and Tritton, S. B. 1980, M.N.R.A.S., 193, 521.

Lutz, J. H. 1977, Astr. Ap., 60, 93.

Marino, B. F., and Williams, H. O. 1983, IAU Inf. Bull. Var. Stars, No.
    2266.

McClintock, J. E., London, R. A., Bond, H. E., and Grauer, A. D. 1982,
    Ap. J., 258, 245.

Mendez, R. H., Gathier, R., and Niemela, V. S. 1982, Astr. Ap., 116, L1.

Mendez, R. H., and Niemela, V. S. 1977, M.N.R.A.S., 178, 409.

_____. 1981, Ap. J., 250, 240.

Meyer, F., and Meyer-Hofmeister, E. 1979, Astr. Ap., 78, 167.

Miller, J. S., Krzeminski, W., and Priedhorsky, W. 1976, IAU Circ., No. 2974.

Nelson, B., and Young, A. 1970, Publ. A. S. P., 82, 699.

Noskova, R. I. 1980, Astr. Tsirk., No. 1128.

Oswalt, T. D. 1979, Publ. A. S. P., 91, 222.

Paczynski, B. 1976, in IAU Symposium 73, Structure and Evolution of Close Binary Systems, ed. P. Eggleton, S. Mitton, and J. Whelan (Dordrecht: Reidel), p. 75.

_____. 1980, Acta Astr., 30, 113.

_____. 1985, this volume.

Paczynski, B., and Dearborn, D. S. 1980, M.N.R.A.S., 190, 395.

Patterson, J. 1983, Bull. A. A. S., 14, 902.

Parsons, S. B., Wray, J. D., Kondo, Y., Henize, K. G., and Benedict, G. F. 1976, Ap. J., 203, 435.

Ritter, H. 1976, M.N.R.A.S., 175, 279.

Rucinski, S. M. 1981, Acta Astr., 31, 37.

Sargent, W. L. W., and Searle, L. 1968, Ap. J., 152, 443.

Schaefer, B. E. 1983, preprint.

Schindler, M., Stencel, R. E., Linsky, J. L., Basri, G., and Helfand, D. J. 1982, Ap. J., 263, 269.

Schoenberner, D. 1978, Astr. Ap., 70, 451.

Simon, T., Fekel, F., and Gibson, D. M. 1983, Bull. A. A. S., 14, 982.

Stephenson, C. B. 1960, Publ. A. S. P., 72, 387.

_____. 1971, in IAU Symposium 42, White Dwarfs, ed. W. L. Luyten (Dordrecht: Reidel), p. 61.

Thackeray, A. D. 1970, M.N.R.A.S., 150, 215.

Thorstensen, J. R., Charles, P. A., Margon, B., and Bowyer, S. 1978, Ap. J., 223, 260.

Tumer, O., Ibanoglu, C., Kurutac, M., and Tunca, Z. 1982, Ap. Space Sci., 83, 269.

Van Buren, D., Charles, P. A., and Mason, K. O. 1980, Ap. J. (Letters), 242, L105.

Vauclair, G. 1972, Astr. Ap., 17, 437.

Verbunt, F., and Zwaan, C. 1981, Astr. Ap., 100, L7.

Wallerstein, G., and Wolff, S. C. 1966, Publ. A. S. P., 78, 390.

Walter, F. M., and Basri, G. S. 1982, Ap. J., 260, 735.

Webbink, R. F. 1976, Ap. J., 209, 829.

_____. 1979, in IAU Colloquium 53, White Dwarfs and Variable Degenerate Stars, ed. H. Van Horn and V. Weidemann (Rochester: U. of Rochester Press), p. 426.

Young, A. 1976, Ap. J., 205, 182.

Young, A., and Lanning, H. H. 1975, Publ. A. S. P., 87, 461.

Young, A., Nelson, B., and Mielbrecht, R. M. 1972, Ap. J., 174, 27.

Young, A., and Wentworth, S. T. 1982, Publ. A. S. P., 94, 815.

ABELL 41, A CATACLYSMIC VARIABLE PROGENITOR

Albert D. Grauer
Department of Physics and Astronomy
University of Arkansas at Little Rock

ABSTRACT

The central star of the planetary nebula Abell 41 has been found to be a detached binary star system with a period of 2 hours 43 minutes. Photometric observations which support this conclusion were carried out at Kitt Peak National Observatory (KPNO), Cerro Tololo Inter-American Observatory (CTIO), and Louisiana State University (LSU) Observatory.

The data are consistent with a model consisting of a sdO primary of about 0.6 $M_\odot$ and a dM companion of 0.1 to 0.3 $M_\odot$. The photometric variations are most probably due to viewing different amounts of the heated hemisphere of the secondary star throughout the orbital cycle.

Flickering, night to night variations, or other evidence of mass transfer have not been observed implying that the central star of Abell 41 is the detached binary with the shortest known orbital period. Gravitational radiation is sufficient to bring the secondary into contact with its critical Roche surface in the astronomically near future initiating mass transfer and cataclysmic activity.

The existence of this star provides direct evidence for theoretical work suggesting that binary stars which have suffered catastrophic angular momentum loss through the ejection of a planetary nebula are the progenitors of the cataclysmic variables.

1. INTRODUCTION

Paczynski (1985), elsewhere in this volume, beautifully presents the current standard theory for the origin of the cataclysmic binaries. Various authors (including Vauclair, 1972; Ritter, 1976; Paczynski, 1976, 1981, 1983; Webbink, 1978; Livio, Salzman and Shaviv, 1979; and Meyer and Meyer-Hofmeister, 1979) have concluded that if the primary in a wide binary (period of 1 to 10 years) becomes a red giant and expands rapidly enough it may engulf the secondary to produce a "common envelope" binary.

D. Q. Lamb and J. Patterson (eds.), Cataclysmic Variables and Low-Mass X-Ray Binaries, 29–33.
© *1985 by D. Reidel Publishing Company.*

As the secondary spirals inward toward the red giant's degenerate core, it gradually transfers its orbital angular momentum to the red giant envelope. Eventually the envelope may be ejected, producing a planetary nebula whose nucleus is a moderately close binary containing a hot degenerate primary and a cool main sequence secondary. Such a system, with cooling will resemble V471 Tauri (the Hyades eclipsing white dwarf). Still later, it may become a cataclysmic variable when evolutionary expansion of the secondary or gravitational radiation produces a mass transferring semi-detached system.

2. OBSERVATIONS

Several years ago the writer and H. E. Bond began a search for close binary central stars of planetary nebulae. Abell 41 was first observed in April of 1981 but it was not until April of 1982 that enough observations were accumulated to provide definite clues as to its nature. At KPNO and LSU two-star photometric techniques were employed (Grauer and Bond, 1981) and at CTIO a single channel photometer was used to make the observations. Fig. 1 shows the results of observations made at CTIO using a B filter.

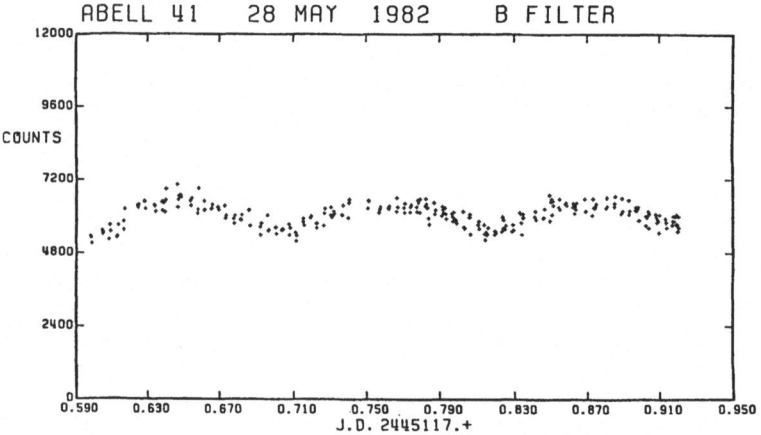

Figure 1. This B light curve for the central star of Abell 41 was obtained during an 8-hour observing run with the 0.9-m reflector at CTIO. The original 5-sec integrations have been summed into 30-sec bins, and are plotted against heliocentric Julian Date. Minima are seen to occur at 2 hour and 43 minute intervals.

Fig. 2 presents a B light curve of Abell 41, obtained with a 0.9-m reflector at KPNO on 7 nights in April of 1982 using the University of Arkansas at Little Rock two-star photometer and related techniques (Grauer and Bond, 1981). A total of 9688 5-sec integrations have been summed into 200 equally spaced phase intervals using the ephemeris:

$$\text{HJD (Min)} = 2445082.9463 + 0.1132269 \text{ E .}$$
$$\phantom{\text{HJD (Min)} = 244508}\pm 3 \pm 3$$

This equation was derived from ten well-observed minima (Grauer and Bond, 1983).

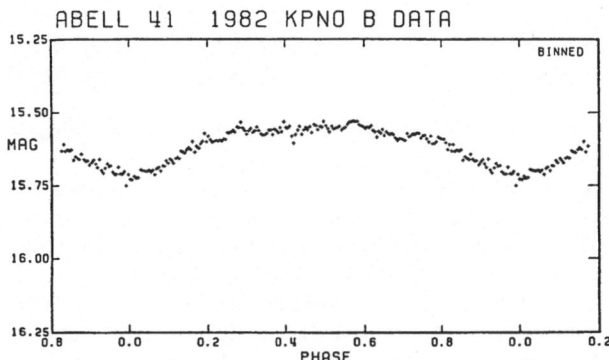

Figure 2. The mean B light curve of the central star of Abell 41.

The ultraviolet light curve of the central star of Abell 41 is nearly identical in amplitude to that in blue light. The mean magnitude and colors of the central star (V = 15.6, B-V = 0.19 and U-B = -1.14) may be slightly affected by the nebular background.

3. DISCUSSION

Upon first inspection of the data, it is not clear if the orbital period of the central star of Abell 41 is 2 hours 43 minutes (with one minimum per orbit as assumed in Fig. 2) or 5 hours 26 minutes (with two minima per orbit).

If the longer period were correct, the variations would have to be dominated by ellipticity of the primary star with reflections from the secondary being of little consequence. One would then expect sinusoidal variations rather than the distinctly flat-topped maxima which are observed.

The shorter (2 hour 43 minute) period implies that the light varia-tions are dominated by reflection from the secondary by the very hot primary and that aside from the much shorter period the system is similar to UU Sge (Bond, Liller and Mannery, 1978) and the central star of Abell 46 (Grauer and Bond, 1981 and Bond, 1985, elsewhere this volume).

The data are consistent with a detached binary system whose orbital

period is 2 hours and 43 minutes located at a distance of approximately
2 kpc. If Schoenberner's (1981) evolutionary tracks apply and the mass
of the primary is assumed to be 0.6 $M_\odot$ then for a secondary whose mass
is between 0.05 and 0.3 $M_\odot$, the separation between the two stars is
approximately 0.9 $R_\odot$. For temperatures of the primary which lie between
60,000 and 100,000 K the heated hemisphere of the secondary has a temperature of approximately 36,000 to 42,000 K. This calculation implies that
the heated hemisphere of the secondary star and the primary star have
nearly the same optical colors. This is consistent with the very nearly
equal amplitudes of the observed ultraviolet and blue light curves.

In a few tens of thousands of years the planetary nebula around
Abell 41 will dissipate and the primary star will continue to contract
until it reaches white dwarf dimensions. Depending on the mass of the
secondary, gravitational radiation will cause the secondary to come into
contact with its critical Roche surface any time between the present and
several billion years from now. The discovery that the central star of
Abell 41 is a close binary which will begin cataclysmic activity in the
astronomically near future provides solid evidence that the cataclysmic
variables are derived from wider binaries which have lost much of their
orbital angular momentum through the ejection of a planetary nebula.

A more complete discussion and presentation of the data is given
elsewhere (Grauer and Bond, 1983).

The writer would like to acknowledge support for this work from
the United States Air Force Office for Scientific Research (grant no.
82-0192, Dr. Arlo U. Landolt, principal investigator), the National
Science Foundation (grant no. Ast-82-11905), a National Aeronautics and
Space Administration grant administrated by the American Astronomical
Society, the University of Arkansas at Little Rock Office of Research
and Sponsored Programs (Mr. John Shelby, Director) and the Donaghey
Foundation of Little Rock, Arkansas.

REFERENCES

Bond, H. E. 1985, this volume.
Bond, H. E., Liller, W., and Mannery, E. J. 1978, Ap. J., 223, 252.
Grauer, A. D., and Bond, H. E. 1981, Publ. A. S. P., 93, 388.
_____. 1983, Ap. J., in press.
Livio, M., Salzman, J., and Shaviv, G. 1979, M.N.R.A.S., 188, 1.
Meyer, F., and Meyer-Hofmeister, E. 1979, Astr. Ap., 78, 167.
Paczynski, B. 1976, in IAU Symposium 73, Structure and Evolution of
      Close Binary Systems, ed. P. Eggleton, S. Mitton, and J. Whelan
      (Dordrecht: Reidel), p. 75.
_____. 1981, Acta Astr., 31, 1.
_____. 1985, this volume.
Ritter, H. 1976, M.N.R.A.S., 175, 279.
Schoenberner, D. 1981, Astr. Ap., 103, 119.
Vauclair, G. 1972, Astr. Ap., 17, 437.
Webbink, R. F. 1976, Ap. J., 209, 829.

DISCUSSION

STOCKMAN: The reflection effect in these systems seems quite strong.
Have you estimated over what range of binary inclinations these effects
can be detected?

BOND: I forget the exact results, but the effect is certainly detectable
from 90° down to about 10°.

STOCKMAN: And what percentage of observed systems show the effect?

BOND: Four out of about three dozen, but the three dozen are highly
selected because someone else has stated that they are variable. I
think the actual percentage of detectable close binaries among central
stars is around five percent.

STOCKMAN: What's the inclination for Abell 41?

GRAUER: Somewhere between 10° and 70°. Since it doesn't eclipse, it
will be very difficult to pin this number down. Also, the star is
faint and the amplitude of variation is quite small, so the inclination
is bound to be quite uncertain.

STOCKMAN: Sure. I was just thinking about the possibility that
Paczynski raised, that maybe _all_ of the central stars are binaries. It
sounds like that's improbable.

BOND: Yes, I really doubt that they could all be binaries with periods
of half a day. They could all be binaries with periods of two weeks.
One of the recently discovered systems, the central star of NGC 2346,
has a period of sixteen days.

THE BREAKDOWN OF NUCLEAR QUASI-EQUILIBRIUM IN HIGHLY COMPACT BINARIES

P.C. Joss and S. Rappaport
Department of Physics, Center for Space Research, and
Center for Theoretical Physics, Massachusetts Institute
of Technology

We have calculated the secular evolution of a highly compact binary, composed of a degenerate-dwarf primary and a low-mass main-sequence secondary, using the techniques developed by Rappaport, Joss, and Webbink (1982). Our new calculations (see Fig. 1) take into account the combined effects of (i) the gradual breakdown of quasi-equilibrium of the ^{3}He abundance in the interior of the secondary, and (ii) the progressive mixture of fresh ^{3}He into the core of the secondary due to the increasing depth of the stellar surface convection zone (see also D'Antona and Mazzitelli 1982). We find that these effects can cause the nuclear energy generation rate to pass through a sharp maximum at the time when the secondary becomes fully convective. The resultant variations in the radius of the secondary, the binary orbital period, and the mass transfer rate tend to enhance the discovery probability of systems whose orbital periods are greater than $\sim 3^h$, relative to shorter-period systems. This result may provide a natural explanation for the apparent sharp decrease in the discovery probability of a cataclysmic variable as its orbital period decreases through $\sim 3^h$ (see Fig. 2). However, the effects that we consider do not account for the relatively large number of observed cataclysmic variables with orbital periods between $\sim 80^m$ and $\sim 2^h$, so that the overall distribution of orbital periods among cataclysmic variables is not yet fully understood.

Details of our computational methods and numerical results will be presented elsewhere (Joss and Rappaport 1983).

This work was supported in part by the National Science Foundation under grant AST78-21993 and by the National Aeronautics and Space Administration under grants NSG-7643 and NGL-22-009-638 and contract NAS5-24441.

REFERENCES

Córdova, F.A., and Mason, K.O. 1983, in "Accretion Driven Stellar X-Ray Sources," ed. W.H.G. Lewin and E.P.J. van den Heuvel (Cambridge, England: Cambridge University Press).

D. Q. Lamb and J. Patterson (eds.), Cataclysmic Variables and Low-Mass X-Ray Binaries, 35–38.

D'Antona, F., and Mazzitelli, I. 1982, Ap.J., 260, 722.
Joss, P.C., and Rappaport, S. 1983, Ap.J. (Letters), in press.
Rappaport, S., Joss, P.C., and Webbink, R.F. 1982, Ap.J., 254, 616.
Ritter, H. 1982, preprint.
Webbink, R.F. 1980, private communication.

FIGURE CAPTIONS

   Figure 1. (a) Solid line, calculated rate of mass transfer, $\dot{M}$, onto the primary as a function of elapsed time, t, in our principal model of a cataclysmic variable, wherein the effects of increasing ^{3}He disequilibrium and progressive mixture of the core of the secondary by convection have been taken into account. The start of the calculated evolution corresponds to t=0. The sharp peak at t $\cong$ 8×10^9 yr has a width of $\sim$10^7 yr and is unresolved on the scale of the figure; the second peak at t $\cong$ 10×10^9 yr, which has a comparable width and is also unresolved, has a very sharp rise because of the rapid decrease in the radius of the radiative core of the secondary and the concomitant rapid mixture of ^{3}He into the core as the mass of the secondary falls below 0.4 $M_\odot$, and a very sharp decline due to the subsequent rapid cooling of the core in response to the expansion of the secondary. Dashed curve, same as the solid curve, but for a model wherein the above effects have been neglected and the burning of ^{3}He is assumed to always contribute about half the energy generation due to hydrogen burning. (b) Orbital period, P_{orb}, as a function of elapsed time for our principal model. The features denoted as 1, 2, and 3 correspond to the system coming into contact, the complete convective mixing of the secondary, and the passage of the system through its minimum orbital period, respectively. This figure is to appear in The Astrophysical Journal (Letters) (Joss and Rappaport 1983).

   Figure 2. (a) Distribution of orbital periods among known cataclysmic variables and closely related systems, for $0^h < P_{orb} < 4^h$ (data taken from Webbink 1980, Mason and Córdova 1983, and Ritter 1982). Note the short-period cutoff near 80^m and the "gap" for periods between $\sim2^h$ and $\sim3^h$. (b) Solid curve, calculated discovery probability (averaged over intervals of orbital period of width $0\overset{h}{.}01$), dp/dP_{orb}, for our principal model of a cataclysmic variable, wherein the effects of increasing ^{3}He disequilibrium and progressive mixing of the core of the secondary by convection have been taken into account. The sharpness of the maximum near $P_{orb} = 3\overset{h}{.}1$ is an artifact of our choice of a specific set of physical parameters for our model system; for an ensemble of systems with a distribution of parameter values, this peak would be somewhat broader. Dashed curve, same as the solid curve, but for a model wherein the above effects have been neglected and the burning of ^{3}He is assumed to always contribute about half the energy generation due to hydrogen burning. This figure is to appear in The Astrophysical Journal (Letters) (Joss and Rappaport 1983).

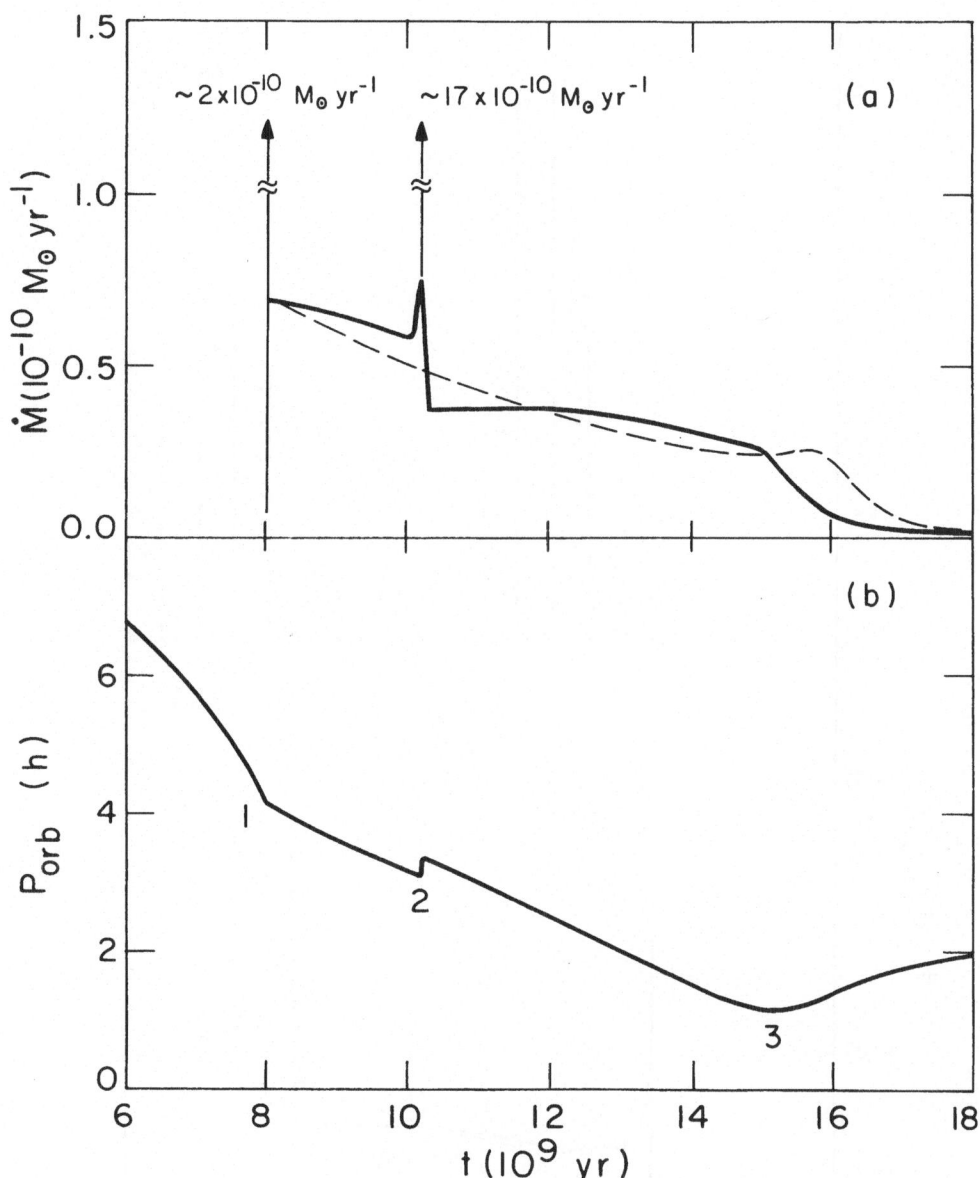

Figure 1.

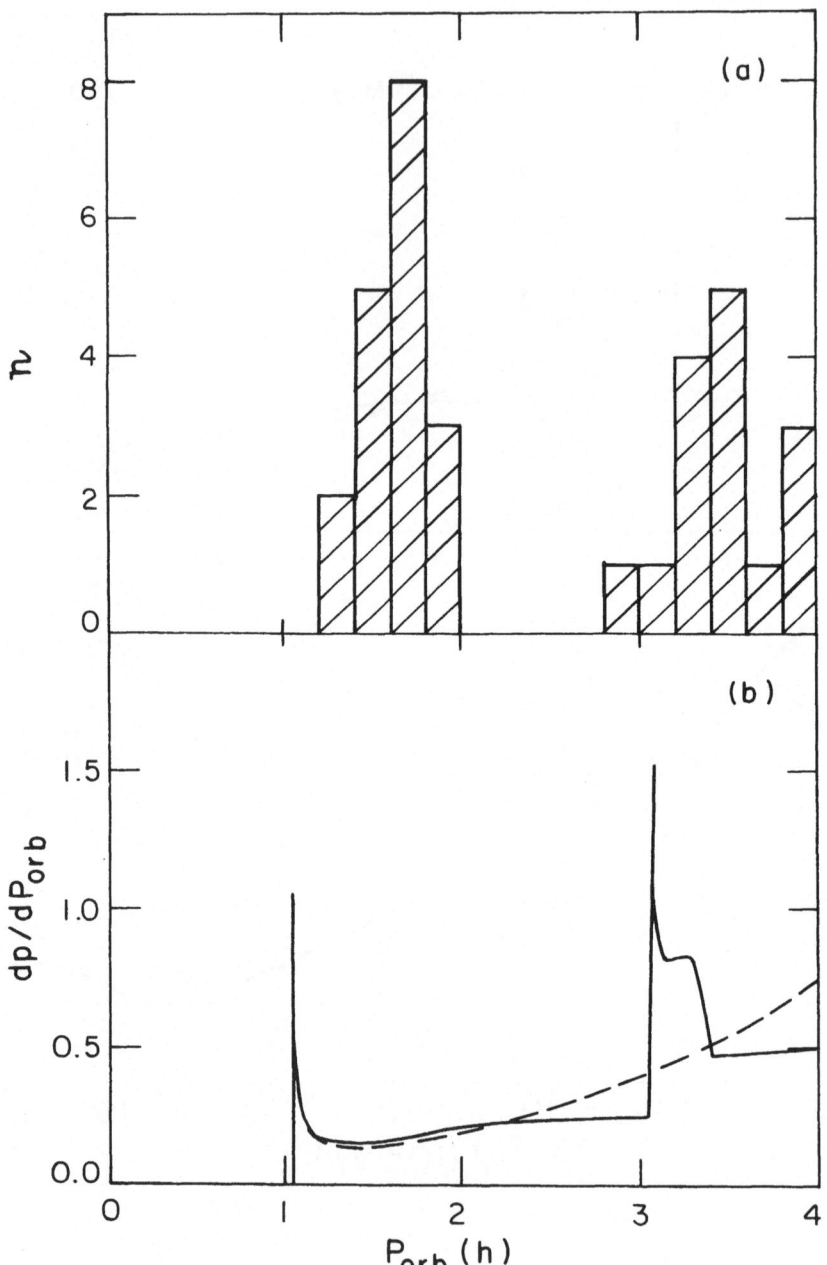

Figure 2.

A SYSTEMATIC STUDY OF MAGNETIC BRAKING IN LOW-MASS BINARIES

F. Verbunt
Institute of Astronomy
Cambridge CB3 OHA, U.K.

S. Rappaport and P.C. Joss
Department of Physics, Center for Space Research
Massachusetts Institute of Technology
Cambridge, MA 02139

Magnetic braking is probably an important process in the evolution of low-mass compact binaries, such as cataclysmic variables (CV's) and low-mass X-ray sources (Verbunt and Zwaan 1981). A simplified stellar evolution code, which describes the mass-losing star as an $n = 3/2$ polytrope, was developed previously to study the evolution of binaries with a secondary of low mass ($0.01 < M/M_{\odot} < 0.4$), when the angular momentum losses are due to gravitational radiation (Rappaport $et\ al.$ 1982). We have extended this code by using a composite polytrope model for the secondary, wherein the structure of the radiative core is described by an $n = 3$ polytrope and the convective envelope by an $n = 3/2$ polytrope. Our results are described in Rappaport $et\ al.$ 1983; this paper is a short summary.

Using our composite model we are able to study secondaries with masses up to 1 $M_{\odot}$. This is important, both because magnetic braking is likely to be more effective for G and K stars than for M stars (Verbunt and Zwaan 1981), and because K star companions have been observed in CV's and low-mass X-ray sources (Ritter 1983; Van Paradijs 1985). We use our code to analyse the effects of different magnetic braking laws on the evolution of compact binaries. We find that a) magnetic braking can result in high mass transfer rates of $10^{-8} M_{\odot}/yr$, and b) the observed period gap in CV's can be explained naturally by assuming that magnetic braking becomes less effective when the radiative core of the secondary disappears during the course of its evolution.

It may seem contradictory to talk about "evolving a polytrope". However, the runs of density and temperature in the model stellar interior are completely specified and hence the nuclear energy production can be computed. With a simple model atmosphere – a surface boundary condition and an empirical proportionality between $\rho T^{-3/2}$ at the surface of the star and $\rho T^{-3/2}$ inside the convective region – one calculates the surface luminosity. The difference between nuclear energy production and surface luminosity gives rise to a change in the entropy inside the star, which is connected with a change in the polytropic constant K of the

39

D. Q. Lamb and J. Patterson (eds.), Cataclysmic Variables and Low-Mass X-Ray Binaries, 39–43.
© *1985 by D. Reidel Publishing Company.*

convective region. This in turn gives rise to a change in the stellar
radius, and a new density/pressure structure can be calculated. This
method (outlined for simple polytropes in Rappaport *et al*. 1982) was
extended by us to be applicable to composite polytropes. Important
modifications include the self-consistent determination of changes in
the location of the interface between radiative core and convective
envelope in response to adiabatic mass loss and heat loss. To test our
code we constructed a series of thermal-equilibrium stellar models that
represents the main sequence. This main sequence corresponds well to
those found with more elaborate codes.

Next we calculated the evolution of a compact binary using
magnetic braking laws of the form

$$\tau_{mb} = 3.8 \ 10^{-30} \ M \ R_{\odot}^{\ 4} \ (\frac{R}{R_{\odot}})^{\gamma} \ \omega^3 \ \text{dyne cm} \tag{1}$$

where γ is an adjustable parameter. ($\gamma = 4$ corresponds to the braking
law as derived from the rotation of G stars in Verbunt and Zwaan 1981.)
Obviously for small γ magnetic braking remains important for stars of
smaller radius. Both theoretically (Van Ballegooijen 1981) and
observationally (Giampapa 1985; Golub 1985; Soberblom 1985) there is
evidence that magnetic activity is substantially lower in completely
convective main sequence stars than in main sequence stars with a
radiative core. Therefore we have also tested magnetic braking that has
the form of equation 1 as long as the secondary retains a sizable
radiative core, but which is set to zero as the radius of the radiative
core becomes smaller than one tenth of the stellar radius. In all our
evolutionary calculations angular momentum losses due to gravitational
radiation were also included.

Some results are shown in figure 1, where the mass transfer rate
is plotted versus the orbital period, for an initial binary with a
collapsed star of 1.2 $M_{\odot}$ and a Roche-lobe filling companion of 1 $M_{\odot}$.
The binary evolves along the curve starting in the upper right portion
of the diagram and ending in the lower left hand portion. For continuous
magnetic braking with index $\gamma = 4$ (CMB 4) both the period and the mass
transfer rate decrease with time: during an interval of 6 10^6 years
the mass transfer rate is in excess of $10^{-8} \ M_{\odot}$/yr, after 2 10^8 yr it
drops below $10^{-9} \ M_{\odot}$/yr, and after 5.4 10^9 yr the orbital period reaches
its minimum. The existence of this minimum period results from an
increase in the Kelvin time scale of the secondary to a value in excess
of the gravitational radiation time scale (which is also increasing,
but at a lower rate), which forces the secondary out of thermal
equilibrium (Paczynski and Sienkewicz 1981; Rappaport *et al*. 1982).

In model MBR 4 ($\gamma = 4$) the magnetic braking is set to zero after
$\sim 10^8$yr, when the star becomes (almost) completely convective. Up to
this point the evolution is of course exactly the same as in CMB 4; but
when the magnetic braking stops, the mass transfer rate also ceases,
and the mass losing star, trying to reach thermal equilibrium, shrinks

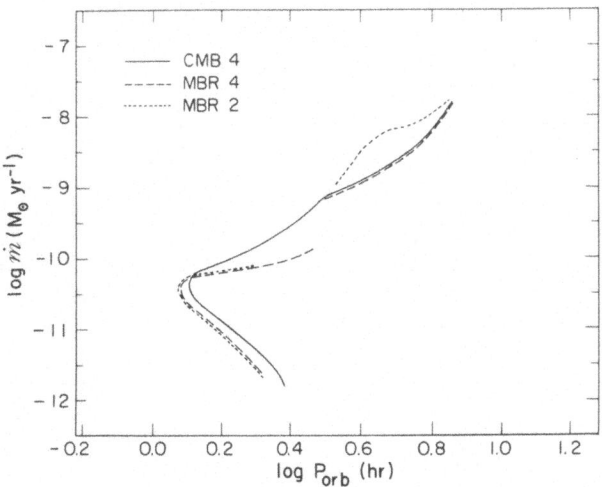

Figure 1. Mass transfer as a function of orbital period as calculated from different models with magnetic braking. CMB 4 corresponds to magnetic braking with γ = 4 throughout the evolution, MBR 4 (2) to magnetic braking with γ = 4 (2) that stops when the mass losing star becomes fully convective. (From Rappaport et al. 1983.)

within its Roche lobe. After a relatively short time gravitational radiation brings the system into contact again, and the mass transfer resumes, though at a lower rate. These same effects occur in model MBR 2 (γ = 2), but since the mass transfer rate is higher and hence the deviation from thermal equilibrium more pronounced in this model (up to the moment when the star becomes completely convective), the star shrinks well within its Roche lobe; it takes ∿8 10^8yr for gravitational radiation to bring the system into contact again, by which time the period of the binary has decreased from 3.4 to 2.0 hours.

In these models the mass of the collapsed star will exceed the Chandrasekhar limit soon after the onset of mass transfer. To be able to describe the evolution of CV's, therefore, we have calculated evolutionary sequences in which the collapsed primary has initial mass 1 $M_\odot$ and captures only half of the mass lost from the secondary, so that its mass never exceeds the Chandrasekhar limit. As expected (see Verbunt and Zwaan 1981) these evolutionary sequences are very similar to those with a more massive accreting star but with the same braking law.

With our models we can explain the high mass transfer rates of a number of low-mass X-ray sources in the galactic center region: during intervals of ∿6 - 15 10^6 yr the mass transfer rates are in excess of 10^{-8} $M_\odot$/yr for γ = 4 to 2. The braking index γ must be smaller than 4 to give results consistent with the observed luminosity function: a value of γ = 4 leads to a prediction of too many low-luminosity sources.

The general correlation between orbital period and mass transfer rate as shown in figure 1 is in reasonable agreement with observational results (Patterson 1983). Apparently, models with discontinued magnetic braking can explain the existence of the period gap (see also Spruit and Ritter 1983; Patterson 1983). From our evolutionary tracks we have calculated the discovery probability as a function of orbital period, using the prescription given in Rappaport *et al.* (1982). Some results are shown in figure 2. The correspondence of the MBR 2 model with the observed distribution (Ritter 1983) is good, in that the overall distribution is rather flat, and the location of the period gap is approximately correct. The prescription used to calculate figure 2 is not sufficiently sophisticated for us to attach any significance to the peak in the MBR 2 distribution at $P_{orb} \simeq 4$ hr.

Acknowledgements.

     This work was supported in part by the National Science Foundation under grant AST 78-21993 and by the National Aeronautics and Space Administration under contract NAS5-24441.

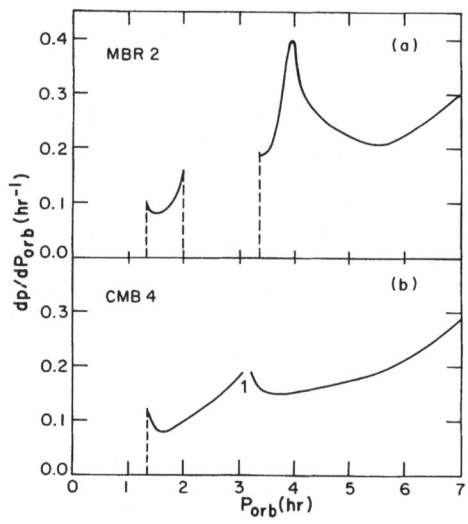

Figure 2. Observation probability as a function of orbital period as calculated from the results of two different models with magnetic braking. The designation of the models is the same as in figure 1. The hump in the MBR 2 model is not significant. (From Rappaport et al. 1983.)

References

Giampapa,M.S., 1983, in "Solar and Stellar Magnetic Fields", IAU Symp.102, Stenflo (ed.), p. 187.
Golub,L., 1983, in "Activity in Red Dwarf Stars", IAU Coll.71, Byrne,P.B. and Rodono,M. (eds.), p. 83.
Paczynski,B. and Sienkewicz,R., 1981, Astrophys.J. 248,L27.
Patterson,J., 1983, in preparation.
Rappaport,S., Joss,P.C., and Webbink,R.F., 1982, Astrophys.J. 254,616.
Rappaport,S., Verbunt,F., and Joss,P.C., 1983, Astrophys.J., submitted.
Ritter,H., 1983, "Catalogue of Cataclysmic Binaries, Low-Mass X-ray Binaries and Related Objects", MPI für Physik und Astrophysik.
Spruit,H.C. and Ritter,H., 1983, preprint.
Soderblom,D.R., 1983, in "Activity in Red Dwarf Stars", IAU Coll.71, Byrne,P.B. and Rodono,M. (eds.), P. 67.

Van Ballegooijen,A.A., 1981, Astron.Astrophys. 113,99.
Van Paradijs,J., 1983, in "Accretion-driven Stellar X-ray Sources",
    Lewin,W.G.H. and Van den Heuvel,E.P.J. (eds.).
Verbunt,F. and Zwaan,C., 1981, Astron. Astrophys. 100,L7.

DISCUSSION

WADE: I'm worried about translating masses into orbital periods. It's
my understanding that there is still a disagreement of about 25% between
theorists and observers as to what the proper mass-radius relation is
for low-mass stars. That translates into an uncertainty of 30-45 minutes in the location of these "edges".

RAPPAPORT: I'm not sure I understand. There is no assumption in these
models about any mass-radius relation. They're calculated stars.

WADE: Depending on what sources of opacity you put in the envelope of
the star, the theoretical calculation gives you a different radius.

RAPPAPORT: Oh, I see. Well, I think the uncertainties on the lower
main sequence are substantially less than 25%.

PACZYNSKI: You have a free parameter in the magnetic braking theory,
namely the power of the dependence on radius, and it is adjusted to fit
the gap.

RAPPAPORT: Yes, that's a free parameter, but it turns out to affect
primarily the width of the gap. The gap is always centered on a period
of 2.5 hours.

PACZYNSKI: Concerning this discrepancy between theoretical and observed
mass-radius relations for low mass stars, as far as I know there are
only two double-lined spectroscopic and eclipsing binaries, one at
0.2 $M_\odot$ and one at 0.6 $M_\odot$, and one of them is peculiar. All the other
measurements are much less reliable. I would really like to see a
systematic survey to find more eclipsing systems. One shouldn't be
modifying theories on the basis of data from non-eclipsing systems.

WADE: But do the theoreticians agree among themselves?

[Ed: Unintelligible muttering. Faulkner and Joss joust for a few
minutes about the assumed He^3 abundance. Extensive reference to a
figure, terminating in "Why don't we discuss this privately?"]

MAGNETIC BRAKING AND THE ORIGIN AND EVOLUTION OF CLOSE LOW-MASS BINARIES[+]

A. Tutukov [*]
Department of Astronomy
University of Illinois

ABSTRACT

Cataclysmic binaries arise from wide binaries after a common envelope phase. The frequency of formation of each type of system (short
$(P \lesssim 2^h)$ period as well as long $(P \gtrsim 3^h)$ period) is $\sim 3 \times 10^{-3}$ yr^{-1}.
The maximum semiaxis of a precataclysmic binary at the end of the com-
mon envelope stage is ~ 10 $R_\odot$ and the initial mass of the secondary is
confined to the interval $0.3 \lesssim M_2/M_\odot \lesssim 1.5$ for stars with a convective
envelope and a radiative interior. To explain the 2-3 hours gap in the
distribution of cataclysmic binaries one needs to suppose that the efficiency of angular momentum loss by an assumed magnetic stellar wind
(MSW) is limited from below. The same MSW can explain the origin and,
possibly, also the observed number of WUMa stars. Bright $(L \sim 10^2-10^4 L_\odot)$
X-ray sources may be members of binaries whose evolution is driven by
this MSW.

INTRODUCTION

The origin and the force which drives the evolution of cataclysmic
variables (CV's) has for a long time remained unclear. The low luminosity of CV's inhibits the observational study of these binaries but the
strong flashes which CV's experience from time to time enhance their
discovery rate and excite special interest in their physics and evolution. We shall discuss some aspects of the origin and evolution of
cataclysmic binaries and WUMa stars driven by an assumed magnetic stellar wind (MSW).

One process which may drive the evolution of CV's, keeping the system in the semidetached state, was pointed out by Kraft et al. (1962)
and studied by Paczynski (1967) and Faulkner (1971). It has been shown
that, for such close systems, the radiation of gravitational waves (RGW)
may remove the orbital angular momentum on a time scale shorter not only

*On leave from the Astronomical Council of the USSR Academy of Sciences,
Moscow.

D. Q. Lamb and J. Patterson (eds.), Cataclysmic Variables and Low-Mass X-Ray Binaries, 45–53.
© *1985 by D. Reidel Publishing Company.*

than the nuclear time scale of the lower mass component but shorter also
than the cosmological timescale. The interest in such evolution increased very much after observations of the binary radio pulsar PSR
1913+16 (Taylor et al. 1979) confirmed the validity of the Standard
Einstein quadrupole formula. Supposing that RGW is the single driving
force, Paczynski (1981) has explained the existence of the minimal orbital period $\sim 80^m$ for CV's.

An extensive computational exploration by Rappaport et al. (1982)
has shown that the mass exchange rate under RGW is constrained to a
value lower than 2×10^{-10} $M_\odot$/yr. But it is known that the mass exchange rate for most CV's exceeds this limit by up to one hundred times
(Tutukov, Yungelson 1979, Patterson 1982), especially for systems with
orbital periods exceeding ~ 3 hours. Practically, the upper limit on
the mass exchange rate is close to M_2/τ_{KH}, where M_2 and τ_{KH} are, respectively, the mass and the thermal timescale of the component filling
its Roche lobe. That forces us to look for another driving force of CV
evolution.

A MAGNETIC STELLAR WIND

(MSW) as another possible driving force of CV evolution was proposed by Huang (1966). In the absence of an explicit MSW theory we
can use Skumanich's law (1972) to estimate the angular momentum loss
from the binary system. The application of Skumanich's law to CV is
grounded on several suppositions. One assumption is required by the
necessity for extrapolating data on single low mass stars rotating
with velocities lower than ~ 30 km/sec to CV secondaries whose velocities are about ~ 150 km/sec, independent of the masses of the components. We also assume that the constant α in the Skumanich law (see
equation 1) does not depend on the mass of the star. The observed rotational velocity of late spectral type stars cannot help us to find
this dependence, because the main rotational deceleration of young
stars occurs primarily during the pre main sequence stage of evolution
(Kuhi 1978).

The most recent computation of low mass star formation (Stahler et
al. 1980) displays that stars with initial mass exceeding ~ 0.3 $M_\odot$ probably do not pass through a completely convective stage which can destroy
the relict magnetic field. That field stays buried in the radiative
core of the single low mass star. But, mass loss by a binary component
leads to the penetration of the base of the convective envelope into the
"magnetized" radiative core. This mechanism could also sustain the MSW
in addition to the usual rotationally induced MSW. The assumption as to
the relict nature of the MSW in CV secondaries can help us understand
why the MSW switches off for $M_2 \lesssim 0.3$ $M_\odot$: the relict magnetic field
simply becomes exhausted after the secondary becomes a completely convective star.

We now discuss the evolution of a binary driven by the MSW. Assuming Skumanich's law, $v = \alpha \, 10^{14}/t^{1/2}$, we can find

$$\frac{d\ln L}{dt} \underset{\sim}{} 9.6 \times 10^{-15} \frac{R_2^4 (M_1 + M_2)^2}{\alpha^2 a^5 M_1} \; s^{-1}, \qquad \qquad 1$$

where L is the angular momentum, R_2 and M_2 are, respectively, the radius and mass of the secondary, M_1 is mass of the primary, and a is the semi-majoraxis. All values are in solar units. We assume that the wind mass loss is negligible in comparison with the mass exchange rate. The evolution of a binary driven by this MSW is so fast that the secondary usually fills its Roche lobe before having evolved appreciably in the chemical sense; this of course is consistent with the observations.

As an additional check of equation (1), we shall apply it to nuclearly unevolved binary stars with $M_1 \underset{\sim}{<} 1.5 \; M_\odot$ and $\frac{a}{R_\odot} \underset{\sim}{<} 6 \; (M_1/M_\odot)^{1/3}$, which are probably evolving also under the influenced of MSW (Popova et al. 1982). Since $M_1 \simeq M_2$, we shall double the angular momentum loss rate. The average age of stars with $M \underset{\sim}{<} M_\odot$ is $\sim 10^{10}$ years and for stars with $M \underset{\sim}{>} M_\odot$ is equal to the MS lifetime of such stars. On integrating (1), we find that all unevolved binaries with $\frac{a}{R_\odot} \underset{\sim}{} \frac{11.5}{\alpha^{2/5}} \frac{M_1}{M_\odot}$ (0.3 $M_\odot \underset{\sim}{<} M_1 \underset{\sim}{<} M_\odot$) and with $\frac{a}{R_\odot} \underset{\sim}{<} \frac{11.5}{\alpha^{2/5}} (\frac{M_1}{M_\odot})^{2/5}$ (1 $\underset{\sim}{<} M_1/M_\odot \underset{\sim}{<} 1.5$) can lose orbital angular momentum during evolution of their components on the MS. Taking into account the distribution of unevolved binaries over semiaxis dN ≈ 0.2d log a (Popova et al. 1982), we find that about 0.06–0.4 log α of all binaries with $M \approx M_\odot$ can become detached WUMa like systems, if their previous evolution was driven by a MSW. We see that the number of WUMa systems depends sensitively on α. WUMa stars cannot occur if $\alpha \underset{\sim}{>} 1.4$. If the evolution of WUMa systems depends entirely on angular momentum loss by a MSW, we can estimate the characteristic time scale from equation (1) as $\tau_w \underset{\sim}{} \frac{3 \times 10^{+7} \alpha^2 M_1 P_{orb}^{2/3}}{(M_1 + M_2)^{1/3} M_2^{4/3}}$ years, where P_{orb} is the orbital period of the system in hours. Assuming $P_{orb} \approx 10^h$, $M_2 = 0.3$, we have $\tau_w \approx 7 \times 10^8$ yr. So now only ~ 0.07 of all semidetached binaries with $M_2 \approx M_\odot$ still exist. Therefore, the space density of those systems with $a \underset{\sim}{>} 10 \; R_\odot$ in the unit interval log a must exceed the space density of WUMa systems by approximately $50/(1-6.7 \log \alpha)$. This value may be compared with observational relative densities. Popova et al. (1982) have found that wide systems with $M_1 \approx M_\odot$ are about 3^m brighter than WUMa stars. Since the bolometric luminosities of the two classes are comparable, space densities must differ by a factor ~ 60, which agrees well with the estimate that α lies in the range $0.7 \underset{\sim}{<} \alpha \underset{\sim}{<} 1.4$. If we assume that the excess of RSCVa stars with decreasing orbital period over those with increasing period implies that orbital period decreases on a timescale $\sim 10^7$ years (Patterson 1982), then we must admit that the MSW for RSCVn stars is much more efficient in angular momentum loss than the wind just discussed. That may be due to the departure of the secondary from the main sequence in these binaries.

We now estimate the initial semimajoraxis that leads to a binary consisting of a MS star and degenerate dwarf with mass appropriate for evolving to a semidetached cataclysmic like system. Insofar as only the MS star loses its angular momentum, previous estimations becomes:

$$\frac{a}{R_0} \lesssim \frac{10}{\alpha^{2/5}} \frac{M_1}{M_\odot} \quad (0.3\ M_\odot \lesssim M_1 \lesssim M_\odot) \quad \text{and} \quad \frac{a}{R_\odot} \lesssim \frac{10}{\alpha^{2/5}} \left(\frac{M_1}{M_\odot}\right)^{2/5} \left(1 \lesssim \frac{M_1}{M_\odot} \lesssim 1.5\right).$$

The maximum semimajoraxis given by these expressions may be compared with that of binaries driven by RGW ($\sim 3\ R_\odot$). Thus the MSW is a more powerful mechanism than the RGW for creating semidetached catalysmic like systems and WUMa stars from initially more wide systems.

An evolved system with a $\lesssim 10\ R_\odot$, consisting of a degenerate carbon-oxygen dwarf and a close low mass main sequence star, is the product of previous evolution including the common envelope stage (Paczynski 1976). The common envelope is lost because of drag luminosity. Therefore we can write the energy conservation law in the form (Tutukov and Yungelson 1978)

$$\beta \frac{M_2 M_d}{a_f} = \frac{M_0^2}{a_0} , \qquad\qquad 2$$

where a_0 and a_f are initial and final values of the semimajoraxis, M_0 is the initial mass of the giant (supergiant) creating the common envelope, M_2 is the mass of the MS secondary, M_d is mass of degenerate core, and β is the efficiency of mass loss. For stars with degenerate cores, $\frac{a_0}{R_0} \simeq 10^3 \left(\frac{M_d}{M_\odot}\right)^3$, so that: $\frac{M_d}{M_\odot} = \left(\frac{a_f}{10^3 R_\odot}\right)^{1/4} \left(\frac{M_0^2}{\beta M_2}\right)^{1/4}$. To keep the secondary inside its Roche lobe, a_f must be larger than $2(M_1+M_2)^{1/3} M_2^{2/3} R_\odot$. Since $\frac{M_0^2}{\beta M_2} \gtrsim 1$ we can get a simple limitation on the mass of the dwarf: $M_d \gtrsim 0.21\ M_2^{1/6}$. That line is placed in fig. 1, together with lines for the dynamical and thermal stability of the MS secondary in conservative systems (Tutukov et al. 1982). Almost all observed CV's lie well within the region where mass exchange can proceed due to angular momentum loss by RGW and MSW or due to the nuclear evolution of a secondary with mass $M_2 \gtrsim 0.8\ M_\odot$.

Equation (2) helps us to estimate that value of M_0 required for an orbit with $a_f \gtrsim 10\ R_\odot$: $M_0 \simeq 7.8\ \beta^{1/2}$. Thus the initial mass ratio must be rather low even for initially wide (a $\sim 10^3\ R_0$) systems if only $\beta > 0.1$, as follows from similar estimates for the double radiopulsar PSR 1913 16. Now, assuming a star formation rate $dN \sim \left(\frac{M}{M_\odot}\right)^{-2.5} \left(\frac{dM}{M_\odot}\right)$ yr^{-1} and $dN \simeq 0.2$ dlog a $\simeq 0.06$, and assuming that 10 percent of all binaries have appropriate mass ratios, we find that the frequency of CV formation is $\sim \frac{0.002}{(M_d/M_\odot)^3}\ \beta^{-3/4}$ yr^{-1}. One should probably relax the assumption that M_d exceeds the solar mass, because, even for single stars, the formation of such a heavy dwarf is difficult, as follows from the distribution of red supergiants over their luminosities (Iben and Renzini 1982).

To estimate the frequency of CV formation from the observations, we shall use visual magnitudes m of the brightest CVs (during their quiescent phases between flashes), dividing them into two groups (Tutukov and Yungelson 1978). Long (ℓ) period binaries (LCV, $P \gtrsim 3^h$) have $m_\ell \sim 10^m$ and their lifetime is $\tau_\ell \simeq \dfrac{0.5\ M_\odot}{\dot{M}_\ell}$. Short (s) period binaries (SCV, $P \lesssim 2^h$) have $m_s \simeq 13^m.6$ and $\tau_s \simeq \dfrac{0.1\ M_\odot}{\dot{M}_s}$. Here, $\dot{M}$ is the exchange rate in $M_\odot$ yr^{-1}. Then, supposing that the luminosity of CV's is determined only by the accretion rate, we can estimate the distance to the closest CV's, their number in the Galaxy and the frequency of formation of both groups. The result for LCV's is: $\nu_\ell \approx 10^{-6.5-0.5\log\dot{M}_\ell}$ and $\nu_s \approx 10^{-8-0.5\log\dot{M}_s}$. If $\dot{M}_\ell \simeq 10^{-8}$ and $\dot{M}_s \simeq 10^{-11}$ (Patterson 1982). Then $\nu_\ell \sim 0.003$ yr^{-1} and $\nu_s \sim 0.003$ yr^{-1}. These semiempirical estimates agree with theoretical ones, but regretfully, uncertainties in all estimations are still too large. Numbers of both CV types in our Galaxy are $N_s \sim 3 \times 10^7$ and $N_\ell \sim 1.6 \times 10^5$. The closeness of formation rates of both CV types suggests the possibility that all LCV's evolve into SCV's. Since the lifetime of SCV's is comparable to the cosmological timescale, the number N of SCV's must be $10^{10}\ \nu_s$, which agrees with the observations if $\dot{M}_s \simeq 10^{-11}$. In 10^{10} yr a binary evolving under the exclusive influence of RGW will achieve such a mass loss rate.

The formation rate ν_n of nova systems can be estimated independently. One easily finds $\nu_n \sim N \dfrac{M_e}{M_2} \sim 6 \times 10^{-4}$, where N = 10 is the total number of nova flashes in the Galaxy per year, $M_e \sim 3 \times 10^{-5}\ M_\odot$ is the mass of the nova envelope and $M_2 = 0.5\ M_\odot$ is the secondary mass. We see that the nova formation rate is several times lower than the rate of CV formation.

DISCUSSION

We now briefly discuss the evolution of CV's driven by angular momentum loss due to a MSW. From equation (1) it follows that the mass exchange rate in semidetached systems is,

$$\dot{M} \simeq 10^{-7.6} \frac{M_2^{5/3}(M_1 + M_2)^{1/3}}{\alpha^2 M_1}\ M_\odot/yr$$ if the radius of the MS star is a

function only of its mass M_2. For stars with $0.3 \lesssim M_2 \lesssim 1$ this rate is comparable to the mass exchange rate, if mass exchange occurs on a thermal timescale. This leads to a significant deviation of the secondary from a thermal equilibrium state when the radiative core disappears. The consequent cessation of the MSW changes the timescale of evolution of the binary to that given by the RGW. The contracting sec-

ondary ceases exchanging mass for some time. This may explain the absence of CV's with orbital period between 2 and 3 hours if most of the
secondaries have $\alpha \lesssim 3$. The low mass exchange rates observed for YY Dra
(Patterson 1982) suggest that systems with $M_2 \gtrsim 0.3\ M_\odot$ and with higher
α ($\simeq 20$ for YY Dra) also exist. Such systems could evolve through the
period gap in the semidetached state, partly filling that gap. But it
is also possible that the observed low mass exchange rate is only due to
non-stationary exchange. The presence of some CVs in the 2-3 hours pe-
riod gap would be a consequence of existing LCVs with $\dot{M} \simeq 10^{-10}\ M_\odot/\text{yrs}$
due only to RGW. Thus, MSW can explain the 2-3 hour period gap only if
(for the most part) the Roche lobe filling components with $M \gtrsim 0.3\ M_\odot$
have $\alpha \lesssim 3$; this implies that mass is exchanged more rapidly than on a
thermal time scale.

     The rather high mass exchange rate driven by MSW leads to two con-
sequences. First, this high rate helps avoid strong flashes and can
supply the degenerate companion enough matter to exceed the Chandrasekhar limit and create a supernova type I event. The frequency of SNI's
is probably only several times greater than the CV formation frequency.
Such SNI events can leave no neutron star remnants; otherwise the system
must be disrupted to avoid the overproduction of low mass X-ray binaries. The second consequence of the high mass exchange rate driven by
MSW is that it can account for the observed luminosities of some low
mass X-ray sources in binaries with orbital periods 3-15 hours. Possible luminosities are 10^2-$10^4\ L_\odot$, which are appropriate for X-ray
bursters (see Fig. 2).

     I would like to thank Prof. I. Iben for his hospitality, stimula-
ting discussions, and help. I am grateful to Dr. R. Webbink and Dr. T.
Mouschovias for discussions.
[+]Supported in part by the US NSF grant AST 81-15325.
REFERENCES

Faulkner, J., 1971, Ap. J. (Letters), 170, L99.
Gray, D.F., 1982, Ap. J., 261, 253.
Iben, I., Renzini, A., 1982, Preprint IU, IAP 82-2.
Huang, S.S., 1966, Ann. d'Astroph., 29, 331.
Kraft, R.P., Mathews, J., Greenstein, J.L., 1962, Ap. J., 136, 312.
Kuhi, L.V., 1978, Moon and Planets, 19, 199.
Paczynski, B., 1967, Acta. Astron., 17, 287.
Paczynski, B. 1976, in Proc. of the IAU Symp. #73, Eds. P. Eggleton, S.
     Mitton, J. Wheelan, p. 75.
Paczynski, B. 1982, Acta. Astron., 31, 1.
Patterson, J., 1982, preprint.
Popova, E.I., Tutukov, A.V., Yungelson, L.R., 1982, Ap. Sp. Sci., 88,
     55.
Ritter, H., 1982, preprint, MPI 22.
Rappaport, S., Joss, P.C., Webbink, R., 1982, Ap. J., 254, 616.
Skumanich, A., 1972, Ap. J., 171, 565.
Stahler, S.W., Shu, F.N., Taam, R.E., 1980, Ap. J., 241, 637.
Tutukov, A.V., Yungelson, L.R., 1978, in Proc. of the IAU Sym. #83.

Tutukov, A.V., Yungleson, L.R., 1979, Acta. Astron., <u>25</u>, 665.
Tutukov, A.V., Fedorova, A.V., Yungelson, L.R., 1982, Pisma. Astron.
    Zh., <u>8</u>, 365.

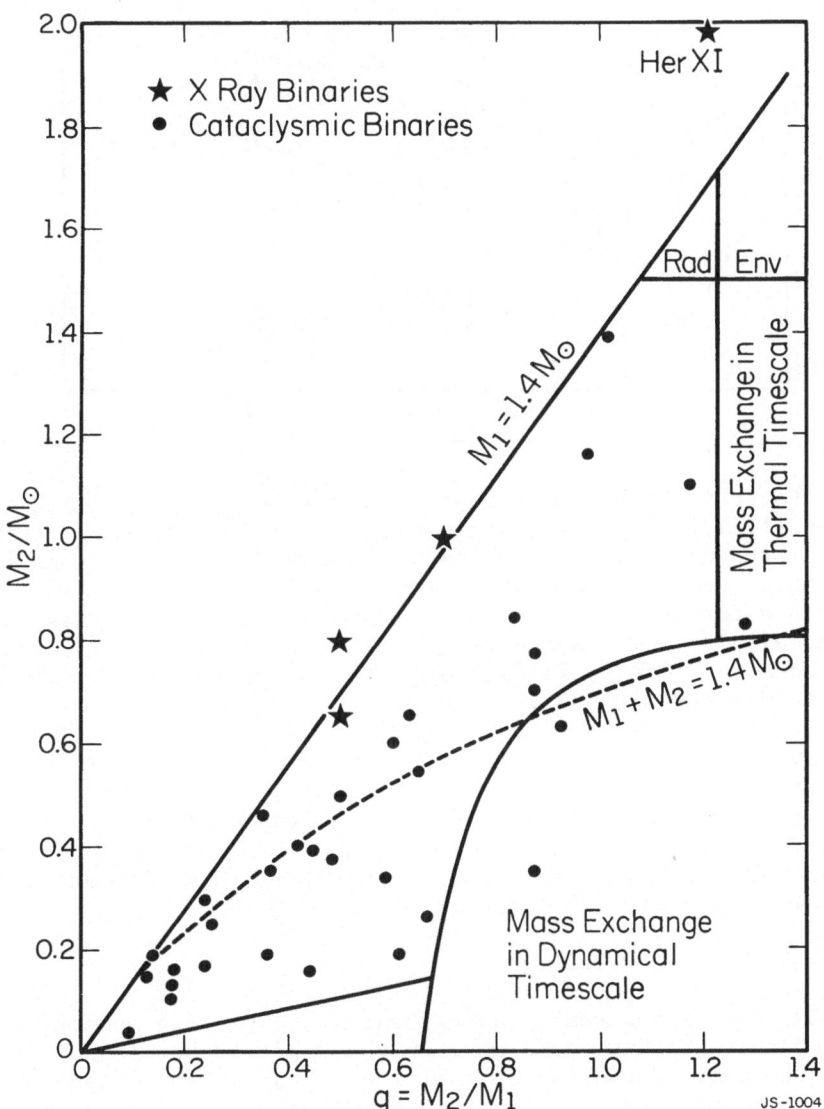

Fig. 1. Cataclysmic binaries and low mass x-ray binaries in the plane
$\frac{M_2}{M_\odot}$ - q. Borders of the region with different types of evolution is
pointed according to Tutukov et al. (1982). Systems placed above the
dashed line have mass larger than the Chandrasekhar limit. Masses of
components taken according to Ritter (1982) and Tutukov et al. (1982).

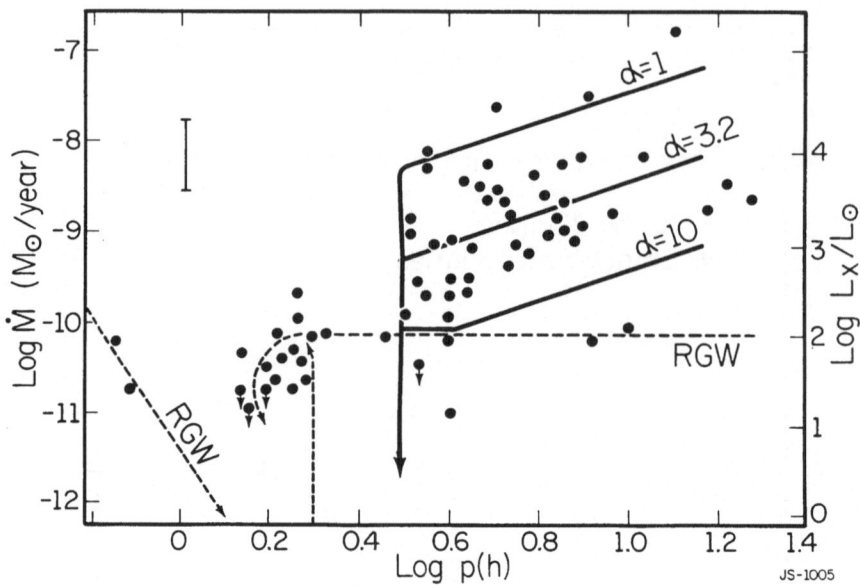

Fig. 2. Cataclysmic binaries in the plane $\dot{M}$ - log P according to Patterson (1982). The right scale shows x-ray luminosities if the accreting components are neutron stars. Theoretical estimations for the mass exchange rate in the systems driven by RGW (dashed lines) and MSW (solid lines for different values α) are shown. The left dashed line refers to a system with secondary which is a degenerate hydrogen rich low mass dwarf (Tutukov, Yungelson 1978). The error bar is shown according to Patterson (1982).

DISCUSSION

PACZYNSKI: Your rate of formation of cataclysmic binaries is lower than the rate at which Type I supernovae explode. I wonder if any of those enthusiasts who ascribe Type I SN to cataclysmic binaries are in the audience and would like to comment on this.

[Ed: Nobody takes the bait.]

PATTERSON: While not really an enthusiast, I heard a number from Bond and Grauer tonight that could make that connection promising. If you take a white dwarf formation rate of 1 per year in the galaxy and about 5% of those become cataclysmic binaries, then you form a CV every 20 years. The Type I SN rate is thought to be about 1 per 50 years. So, if 30% of the CVs possess white dwarfs that will be pushed over the Chandrasekhar limit, then everybody's happy.

PACZYNSKI: No!

TUTUKOV: We could discuss this point, but I think the cataclysmic binary formation rate is very uncertain.

RAYMOND: Regarding the extrapolation of the Skumanich law for magnetic braking, there is now evidence that both the chromospheric and X-ray emission of late-type stars doesn't just keep increasing when you go to shorter periods, but eventually levels off. This makes it difficult to extrapolate the Skumanich law as far as you want to.

TUTUKOV: Yes, I didn't have time to discuss that. I'm no big enthusiast for the Skumanich law, but it's all we have right now.

VERBUNT: A remark about the Skumanich law: what Zwaan and I did was to look directly at the history of the rotation velocity. The other quantities you mentioned, the X-ray and Ca II H&K emission luminosities, are indirect indicators of magnetic activity. I completely agree with you, I would not bet anything on the Skumanich law. But the mere fact that the X-ray luminosity decreases does not mean that the braking effect decreases, because the braking effect operates through the wind and there is no direct indicator of the wind. All we can say is that we know very little about it. We'll need another ten years, I think, before we know anything about the magnetic dynamo and the wind. Those are the two ingredients we need before we can start the discussion.

PATTERSON: Is it OK if we think about it in the intervening ten years?

PACZYNSKI: A question for the observers: do we know orbital period changes for detached main-sequence binaries in which this mechanism for period change might be relevant; V 471 Tauri, for example?

WEBBINK: There's a measured period change, but I would be very skeptical about attaching significance to it.

TUTUKOV: I remember discussing this point with Krzeminski in Warsaw. We concluded that period changes of the size expected are too small to be seen over the ~ 30 years of data.

ON THE EVOLUTIONARY STATUS OF BRIGHT, LOW-MASS X-RAY SOURCES

R.F. Webbink
Department of Astronomy, University of Illinois

S. Rappaport
Department of Physics and Center for Space Research,
Massachusetts Institute of Technology

G.J. Savonije
Astronomical Institute, University of Amsterdam

1. INTRODUCTION

Congregated near the nuclear bulge of our own Galaxy, and also that of M31, are a handful of bright ($\gtrsim 10^{37}$ ergs s^{-1}) persistent X-ray sources. These sources account for most of the total X-ray luminosity of each of these bulges, and their space distribution conforms closely to that of an old disk population (Markert *et al.* 1977; Van Speybroeck *et al.* 1979). They have (at least in our own Galaxy) very similar X-ray flux distributions (Jones 1977), and lack the X-ray pulsations frequently found in massive X-ray binaries. Presumably, they are old, low-mass binaries in which the magnetic fields of the neutron stars have decayed.

A number of possible evolutionary states in old binaries can give rise to the mass transfer rates, $\gtrsim 10^{-9}$ M$_\odot$ yr^{-1}, needed to power these sources by accretion onto neutron stars. Nearly all of them are either extremely brief phases of evolution, or else they are followed by a much longer, lower-luminosity phase which should dominate the integrated X-ray luminosity of an ensemble of such sources, contrary to observational constraints. One promising possibility, however, which does not share these potential drawbacks, is that mass transfer in these sources is driven by nuclear evolution of the donor star up the lower giant branch.

2. MODEL

As a broad rule of thumb, mass transfer driven by nuclear evolution proceeds at a rate of order

D. Q. Lamb and J. Patterson (eds.), Cataclysmic Variables and Low-Mass X-Ray Binaries, 55–59.
© *1985 by D. Reidel Publishing Company.*

$$\dot{M} \sim - M \left(\partial \ln R / \partial t \right)_M \quad ,$$

where the partial derivative is the fractional growth rate for a star
identical in structure to the donor star, but evolving at constant
mass. Among stars of roughly solar mass ($i.e.$, those of cosmological
age), one then anticipates that mass transfer rates of interest to the
bulge source problem will arise if the donor stars first fill their
Roche lobes on the lower half of the first giant branch.

We have calculated the evolution of a number of binaries of this
general type, with both metal-rich and metal-poor compositions, using a
very simple, but accurate parameterization of the structure of the
donor giant stars in terms of the masses of their degenerate helium
cores. For the sake of simplicity, we also assume conservation of
total mass and orbital angular momentum, but the evolution is qualitatively unaffected by moderate losses.

3. RESULTS

The most striking feature of mass transfer in this mode is that
the mass transfer rate is, aside from a brief initial maximum, nearly
constant throughout its duration in any one system. Furthermore, this
"plateau" transfer rate is, to a very good approximation (± 20 %), a
function only of the initial orbital period, P_o, of the binary at the
onset of mass exchange:

$$\dot{M} \simeq - 5.3 \times 10^{-10} \ M_\odot \ yr^{-1} \ (P_o/day) \quad .$$

Since P_o is typically $\gtrsim 1$ day for binaries first reaching mass exchange
after the lobe-filling star has reached the base of the giant branch,
only bright X-ray sources ($\gtrsim 10^{37}$ ergs s^{-1}) result from this model. The
complement of fainter sources which inevitably arises in other models
with decaying mass transfer rates is absent here, allowing us to understand, at least qualitatively, the dominance of bright sources in
making up the total bulge luminosity.

Our models would predict orbital periods in the range 1 - 200 days
for the bright bulge sources in our Galaxy, but more typically periods
should fall in the range from a week to a few months. If the accretion
luminosity is radiated in a pattern typical of a flat disk, the ratio
of X-ray to optical luminosities is expected to be 200 - 1000.

4. COMPARISON WITH OBSERVATIONS

Comparison of our models directly with observations is hampered by
the lack of optical identifications for most of the bright bulge
sources (Bradt and McClintock 1983). In the case of GX17+2, a normal
G star appears coincident with the X-ray position (Tarenghi and Reina
1972; Davidsen, Malina, and Bowyer 1976), but its relatively small apparent foreground reddening (Margon 1978) is inconsistent with the

marked soft X-ray cutoff observed in its X-ray spectral energy distribution (Jones 1977). Periodicities of the order of one week have been reported in the X-ray emission from this and a few other bulge sources (Ponman 1981, 1982; Matsuoka 1985), but these remain to be verified.

A more telling test of the models comes in the form of the galactic halo X-ray sources Cygnus X-2 and 2S 0921-63. Adopting the spectroscopically deduced physical parameters from the optical studies of these two sources by Cowley, Crampton, and Hutchings (1979 and 1982, respectively), we can estimate the evolutionary status (core mass and luminosity) of the mass-losing star in each of these systems, and then compare the model predictions for the mass transfer rates and spectral types of the cool components with those actually observed. For Cygnus X-2, the agreement is excellent: $L_x = 5.1$-6.0×10^{37} ergs s^{-1} (depending on composition) from the models, $L_x = 6.3 \times 10^{37}$ ergs s^{-1} observed; spectral type F2 III-IV from the models, A5-F2 III-IV observed. We do, however, find the observed orbital modulation of the optical light curve (the reflection effect) rather weaker than we would have anticipated. For 2S 0921-63, similarly good agreement can be obtained, but only if we assume the accreting star is a white dwarf. In this case, the predicted and observed X-ray luminosities agree within a factor of two, and the spectral type of the secondary falls within the broad range of those observationally allowed; little evidence of a reflection effect is either predicted or found (*cf*. Chevalier and Ilovaisky 1981, 1982).

A fuller discussion of these model systems and of their possible evolutionary origins, and a more detailed comparison with observations, will be published in the *Astrophysical Journal*, v. 270, no. 2. This work was supported in part by the National Science Foundation under grant AST 80-18859, the National Aeronautics and Space Administration under contract NAS5-24441, and the Netherlands Organization for the Advancement of Pure Research (Z.W.O.) under contract Nr. B78-183.

REFERENCES

Bradt, H.V., and McClintock, J.E.: 1983, *Ann. Rev. Astr. Astrophys.*, 21, in press.
Chevalier, C., and Ilovaisky, S.A.: 1981, *Astr. Astrophys.*, 94, L3.
Chevalier, C., and Ilovaisky, S.A.: 1982, *Astr. Astrophys.*, 112, 68.
Cowley, A.P., Crampton, D., and Hutchings, J.B.: 1979, *Astrophys. J.*, 231, 539.
Cowley, A.P., Crampton, D., and Hutchings, J.B.: 1982, *Astrophys. J.*, 256, 605.
Davidsen, A., Malina, R., and Bowyer, S.: 1976, *Astrophys. J.*, 203, 448.
Jones, C.: 1977, *Astrophys. J.*, 214, 856.
Margon, B.: 1978, *Astrophys. J.*, 219, 613.

Markert, T.H., Canizares, C.R., Clark, G.W., Hearn, D.R., Li, F.K.,
    Sprott, G.F., and Winkler, P.F.: 1977, *Astrophys. J.*, <u>218</u>, 801.
Matsuoka, M.: 1985, this volume.
Ponman, T.: 1981, *Space Sci. Rev.*, <u>30</u>, 353.
Ponman, T.: 1982, *Monthly Notices R. Astr. Soc.*, <u>200</u>, 351.
Tarenghi, M., and Reina, C.: 1972, *Nature, Phys. Sci.*, <u>240</u>, 53.
Van Speybroeck, L., Epstein, A., Forman, W., Giacconi, R., Jones, C.,
    Liller, W., and Smarr, L.: 1979, *Astrophys. J. (Letters)*, <u>234</u>,
    L45.
Webbink, R.F., Rappaport, S., and Savonije, G.J.: 1983, *Astrophys. J.*,
    <u>270</u>, no. 2.

DISCUSSION

GRINDLAY: Assuming that the G star is connected with the source, then this picture would predict variability.

WEBBINK: If the identification is correct, any model that you choose for the source imples that the G star sees virtually nothing of the X-ray source. It must be completely shielded, because it shows no variability.

GRINDLAY: I was just going to make the point that it does indeed show no variability. But there may be other ways to get around the lack of variability, as I will discuss tomorrow. I would like to make another comment, though, on the period distribution that you are getting. Aren't the periods embarrassingly long? There are a few bulge sources, after all, where much shorter periods are indicated.

WEBBINK: They have lower luminosities.

GRINDLAY: But there are a couple that don't. 1735 and 1636-53 both have suggested periods in the 3-to-5 hour range and high luminosities, around 10^{37} ergs per sec.

WEBBINK: That's the minimum luminosity we get. But I don't claim that our model applies to all these sources. There are at least a half dozen other possibilities for them.

GRINDLAY: One last comment. Tomorrow, Paul Hertz will say something about the possibility of giant identifications for hitherto unidentified bulge sources. In general, we can rule out luminous K giants in most of these systems.

WEBBINK: I think these systems are going to be rather faint infrared sources: at the galactic center, magnitude 15 at K, 15.5 or fainter at 2.2 microns.

HERTZ: Josh, I don't think we could rule out the late giants anyway. I thought we only ruled out the early giants.

PACZYNSKI: There was some discussion earlier today about possible origins of low mass binaries with neutron stars in them. At least the long-period systems cannot possibly be products of cataclysmic-binary evolution, I think, and cannot be products of tidal capture either.

GRINDLAY: Yes, certainly, the stellar density is much too low for tidal capture in the galactic bulge itself.

PRIEDHORSKY: What are the odds that the systems you propose would be observed to eclipse?

WEBBINK: Well, we can do a quick estimate.

RAPPAPORT: Thirty percent.

WEBBINK: I told you it would be a quick estimate.

THORSTENSEN: I'm curious to know what the joint probability of no eclipses of all of them is.

RAPPAPORT: Three percent.

WEBBINK: OK, this is an interesting question, and one which I neglected to discuss. There is a real question in my mind as to what the X-ray emission pattern in these systems is. It's commonly assumed that it's isotropic, and one estimates reprocessing by the companion on the assumption that you have L/4π ergs/sec/steradian. What we have done here is assume that the X-ray source radiates as a flat disk, and so we have a relatively beamed pattern; it cuts down the reprocessed radiation by a factor of roughly 3. If the X-ray emission is not isotropic, then I have the suspicion that the joint probability may be telling us more about the beam pattern of the X-ray emission than about whether these objects are binaries or not.

THORSTENSEN: Two quick comments about these two systems. At Dartmouth we have obtained spectrophotometry--Jerry LaSalla has done this--around the orbit of Cyg X-2, and we see a very clean variation of spectral type due to X-ray heating. Secondly, Phil Charles and I have finally written up a three-year photometric and radial-velocity study of 0921-63; it is an extraordinarily messy system, and I would be very cautious about accepting the present conclusions about that system. In some cases, we flatly contradict the Victoria results.

LOW-MASS X-RAY BINARIES

Jeffrey E. McClintock and Saul A. Rappaport

Department of Physics and Center for Space Research
Massachusetts Institute of Technology
Cambridge, Massachusetts 02139, USA

1. INTRODUCTION

Low-mass X-ray binaries (LMXB's) are luminous X-ray sources which are composed of a late-type optical companion $(M \lesssim 1\ M_\odot)$ and a neutron star (or possibly a black hole). They are often referred to somewhat interchangeably as galactic bulge or Population II sources, or as Sco-like sources after the historical archetype of the group Sco X-1. A schematic of one of the best-studied systems, 4U1626-67, is shown superimposed (to scale) on a photograph of the sun's disk in Figure 1. This is the smallest known LMXB; the optical companion is slightly larger than the sunspot group and the bow-tie shaped accretion disk is the same size as Saturn's ring system.

Several review articles which bear on the subject of LMXB's have appeared recently; these include the work of van Paradijs (1983), Bradt and McClintock (1983), Lewin and Joss (1983), van den Heuvel (1983), and Joss and Rappaport (1984). The first two reviews listed emphasize the optical properties of LMXB's, the third concentrates on X-ray burst sources, the fourth deals with the formation and evolution of LMXB's, and the last discusses LMXB's in the context of X-ray pulsars and bursters.

Thirty-two LMXB's have been identified with optical counterparts in the Galaxy and one has been optically identified in the Large Magellanic Cloud (Bradt and McClintock 1983). In addition, there are 20 unidentified galactic sources which are generally believed to be LMXB's: 12 in the cores of globular clusters and 8 extremely bright sources lying close to the galactic plane in the vicinity of the galactic center. A number of unidentified X-ray transients and anonymous lower luminosity X-ray sources are also likely to be LMXB's; however, the total number of active LMXB's in the Galaxy can not be very large. Worrall et al. (1982) find, for example, that the total X-ray emission from the galactic disk is only a few percent of the luminosity in resolved sources (see also Hertz 1983). It is therefore unlikely that there are more than about 100 active LMXB's in the Galaxy, an important point to which we return later in discussing the formation of LMXB's. By comparison, Warner (1974) estimates that there are more than a thousand times as many CV's ($\sim 2 \times 10^5$) in the Galaxy.

61

D. Q. Lamb and J. Patterson (eds.), Cataclysmic Variables and Low-Mass X-Ray Binaries, 61–77.
© *1985 by D. Reidel Publishing Company.*

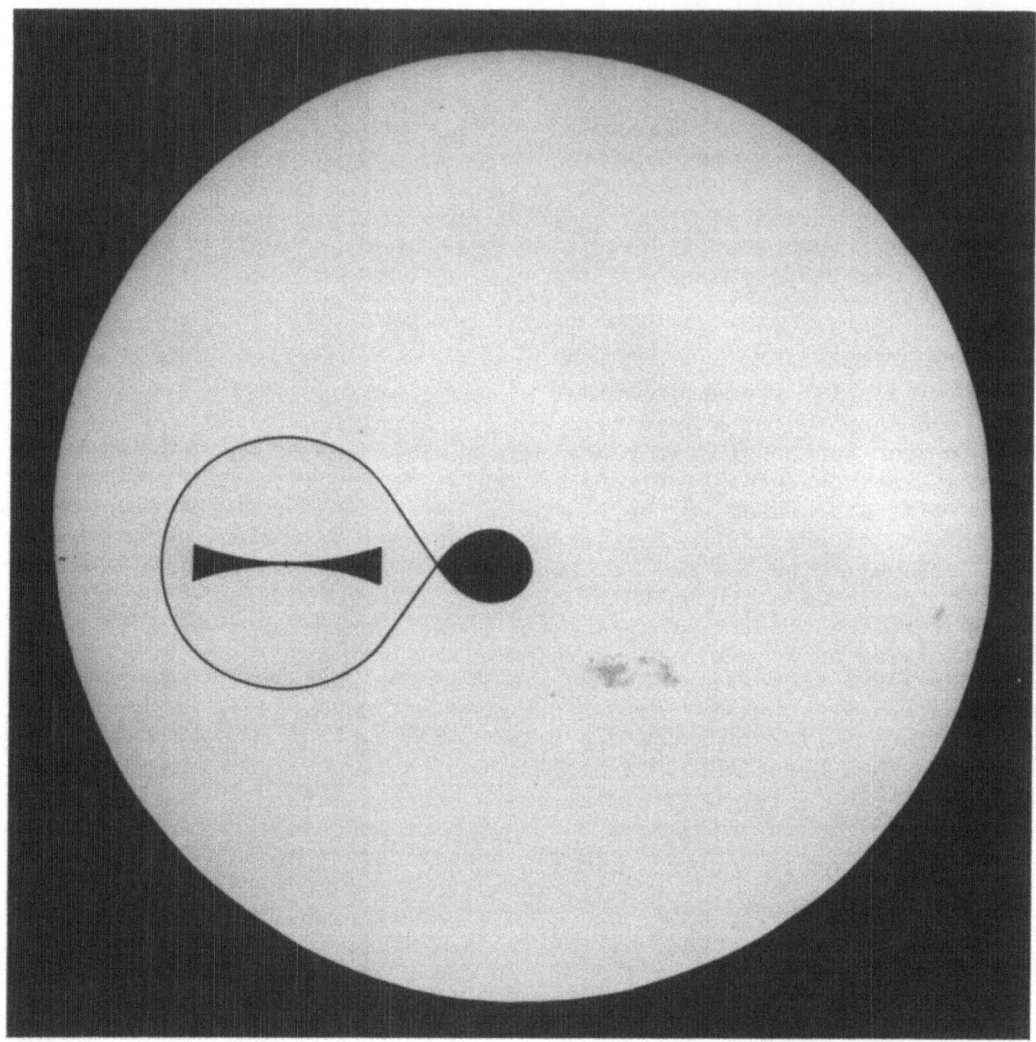

Fig. 1. The relative sizes of the LMXB 4U1626-67 and the Sun. The orbital period of 4U1626-67 is 42 min. The masses of the neutron star and the optical companion are assumed to be ~ 1 $M_\odot$ and ~ 0.1 $M_\odot$, respectively (see Section 3 and references therein). The optical companion is shown as a main-sequence dwarf filling its Roche lobe. The much larger Roche lobe of the neutron star is also shown. (Photograph of the sun by Kitt Peak National Observatory).

2. TYPICAL X-RAY AND OPTICAL PROPERTIES

In this section we list some typical X-ray and optical properties of the LMXB's. Their X-ray luminosities are in the range $\sim 10^{35}$ - 3×10^{38} erg s^{-1}. A typical luminosity of 10^{37} erg s^{-1} corresponds to a mass accretion rate of $\sim 10^{-9}$ M$_{\odot}$ yr^{-1} onto a neutron star or black hole. By comparison, CV's are $\sim 10^5$ times less luminous in X-rays (Cordova and Mason 1983). The X-ray spectra of LMXB's are approximately exponential in shape with kT $\sim$ 5 keV (Jones 1977). The spectra often exhibit a substantial low-energy cutoff at a few keV due to absorption in the interstellar medium; however, the intrinsic low-energy cutoff appears to be small ($\lesssim$ 0.5 keV, which corresponds to a neutral hydrogen column density of $\lesssim 1 \times 10^{21}$ cm^{-2}).

In most LMXB's the X-ray heated accretion disk is the dominant source of light. The average optical properties of these systems have been summarized by van Paradijs (1983). He concludes that M_v = 1.2 $\pm$ 1.0 (standard deviation), which makes them much more optically luminous than either dwarf novae, $M_v \sim$ 7.5, or quiescent novae, $M_v \sim$ 4.5 (Warner 1976). LMXB's are nevertheless fainter (B $\sim$ 18) than most of the observed CV's because the former are much more distant, typically $\sim$ 10 kpc versus $\sim$ 100 pc for CV's. Van Paradijs finds that the average colors of LMXB's, $(B-V)_o$ = 0.0 $\pm$ 0.3 (s.d.) and $(U-B)_o$ = -0.9 $\pm$ 0.2 (s.d.), are very similar to the colors of CV's (Warner 1976). The colors correspond to a constant value of spectral intensity (F_v) throughout the optical band and a blackbody temperature of $\gtrsim$ 15,000 K. Blackbody temperatures of $\sim$ 30,000 K have been deduced from far ultraviolet (1000-3000 Å) observations of four LMXB's (van Paradijs 1983).

The ratio of X-ray to optical luminosity is typically $\sim$ 1000, and the ratio of X-ray to bolometric luminosity (1000-7000 Å) is $\sim$ 100. It is therefore reasonable that the appearance of these systems at all wavelengths $\gtrsim$ 1000 Å is due to the reprocessing of X-rays in the surrounding gas; nuclear burning in the core of the secondary and viscous heating in the disk are, in general, negligible sources of energy in an active LMXB.

3. ORBITAL PERIODS

The dozen LMXB's with known orbital periods are drawn to scale in Figure 2. Three systems have evolved secondaries, Cyg X-2, 2S0921-63 and Her X-1; their orbital parameters have been estimated from optical radial velocity measurements and other data (for references see Bradt and McClintock 1983). For the remaining systems we have assumed that the secondary is a main sequence star which fills its Roche lobe and obeys the mass-radius relation $R/R_{\odot} \simeq M/M_{\odot}$. In this case the radius of the secondary can be determined from the orbital period alone: $R/R_{\odot}$ = 0.11 (P/1 hr) (Warner 1976). The further assumption that the mass of the X-ray star is $\sim$ 1.4 M$_{\odot}$ (see, e.g., Joss and Rappaport 1984) yields the binary separation.

Figure 2 contains an example of almost every class of LMXB (see also the discussion of GX339-4 in Section 4). There are the three large systems mentioned above with evolved secondaries which are located in the galactic halo. These may have been "shot out" of the galactic

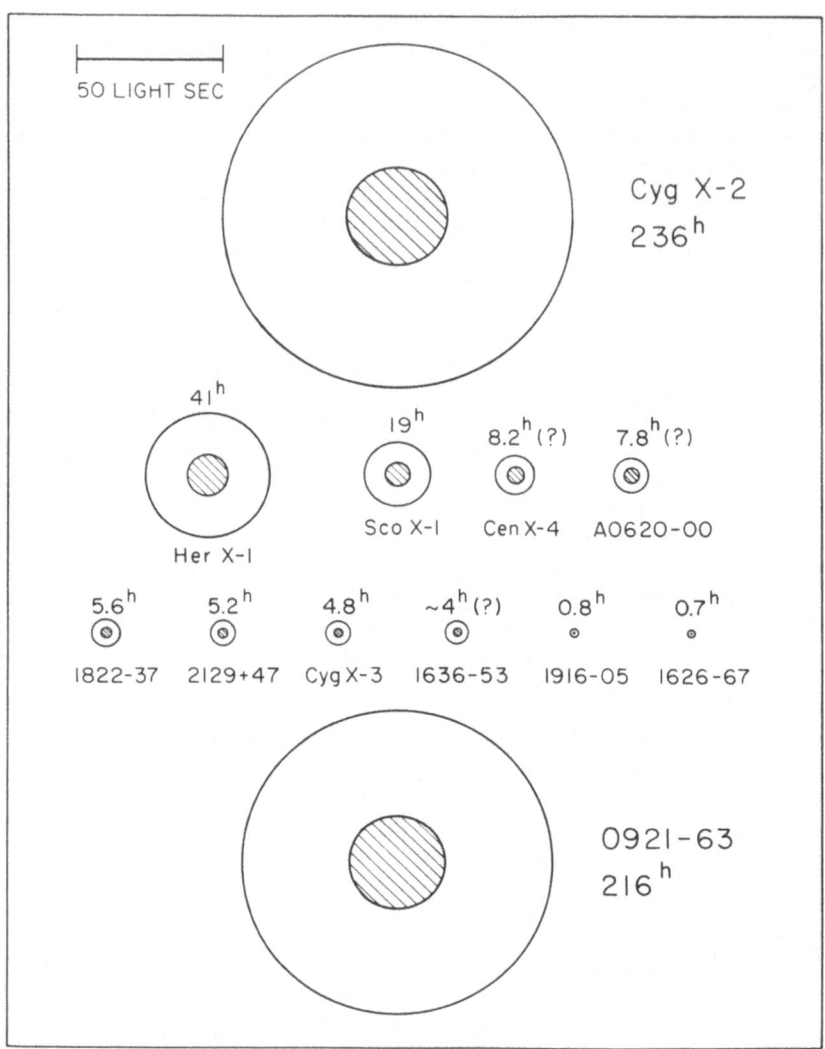

Fig. 2. Line drawing, to scale, of the orbits and companion stars (shaded regions) of twelve LMXB's (after Bradt and McClintock 1983). The geometry is based on the assumptions discussed in the text and the orbital periods which are noted in the figure. For further information see the text and the review articles listed in Section 1. Additional references not listed in the text include: 2S0921-63 (Cowley, Crampton and Hutchings 1982b; Thorstensen et al. 1983); Cen X-4 (Kaluzienski et al. 1980); A0620-00 (McClintock et al. 1983); and MXB1636-53 (Pedersen, van Paradijs, and Lewin 1981).

plane by the supernova explosions which formed their neutron stars (van den Heuvel 1983). The LMXB's in Figure 2 also include two X-ray pulsars (4U1626-67 and Her X-1); three X-ray burst sources (MXB1636-53, 4U1916-05, and Cen X-4); two systems with extended X-ray sources - accretion disk coronae (4U1822-37 and 4U2129+47); two X-ray novae (Cen X-4 and A0620-00); and the historical archetype of the LMXB's, Sco X-1. Her X-1 contains a ~ 2 $M_\odot$ optical companion and is therefore a young, stellar Population I object which cannot be older than about 6 x 10^8 years (van den Heuvel 1983). The optical companions drawn in Figure 2 encompass a wide range of spectral types. The X-ray heated face of Her X-1 can appear as early as mid-B (Crampton and Hutchings 1972) and Cyg X-2 is typically early F III (Cowley, Crampton, and Hutchings 1979). In quiescence the X-ray novae are observed to be mid-K dwarfs (Oke 1977; van Paradijs et al. 1980). As yet, the masses and corresponding spectral types of the shorter period systems can only be inferred from their orbital periods. The companion masses in the smallest systems (4U1626-67 and 4U1916-05) are < 0.1 $M_\odot$, which corresponds to a main sequence spectral type of ~ M8 or perhaps a degenerate dwarf (Joss, Avni and Rappaport 1978; Middleditch et al. 1981; Walter et al. 1982; White and Swank 1982).

A distribution of the orbital periods of LMXB's is shown superimposed on the distribution of orbital periods of CV's in Figure 3. Although the statistics are limited, it is interesting to note that two LMXB's fall below the "cutoff" at ~ 80 minutes in the orbital period distribution of CV's.

4. THE NATURE OF THE COMPACT SOURCE

The evidence that LMXB's contain neutron stars includes the following: (i) The X-ray luminosities of low-mass systems are often at or near the Eddington luminosity for spherical accretion onto a 1-2 $M_\odot$ object (~ 2 x 10^{38} erg s^{-1}). This is two orders of magnitude greater than the maximum X-ray luminosity of an accreting white dwarf (Katz 1977; Kylafis and Lamb 1979). (ii) The pulse periods of the low-mass X-ray pulsars, 4U1626-67 and Her X-1, are decreasing at a rate that is consistent with the expected accretion torques acting on a neutron star, but inconsistent with models of accretion onto white dwarfs (Lamb, Pethick, and Pines 1973; Rappaport and Joss 1977; Ghosh and Lamb 1979). (iii) The observations of X-ray bursts and their satisfying explanation in terms of the thermonuclear flash model argue strongly that the associated compact object is a neutron star (see Lewin and Joss 1983).

Are there black hole primaries among LMXB's? There are no candidates with the credentials of the massive X-ray binary Cygnus X-1 (e.g., see Bahcall 1978) or perhaps LMC X-3 (Cowley, Crampton, and Hutchings 1982a). However, the rapid (~ 10 ms) and intense X-ray and optical flaring in GX339-4 make it an intriguing candidate (Samimi et al. 1979; Motch et al. 1982). No X-ray or optical orbital period has been reported for this system. In its bright state (V ≈ 16), its periodic optical modulation is less than 3% of the steady emission on time scales from 30 sec to 4 hrs (Remillard, McClintock, and Petro; unpublished).

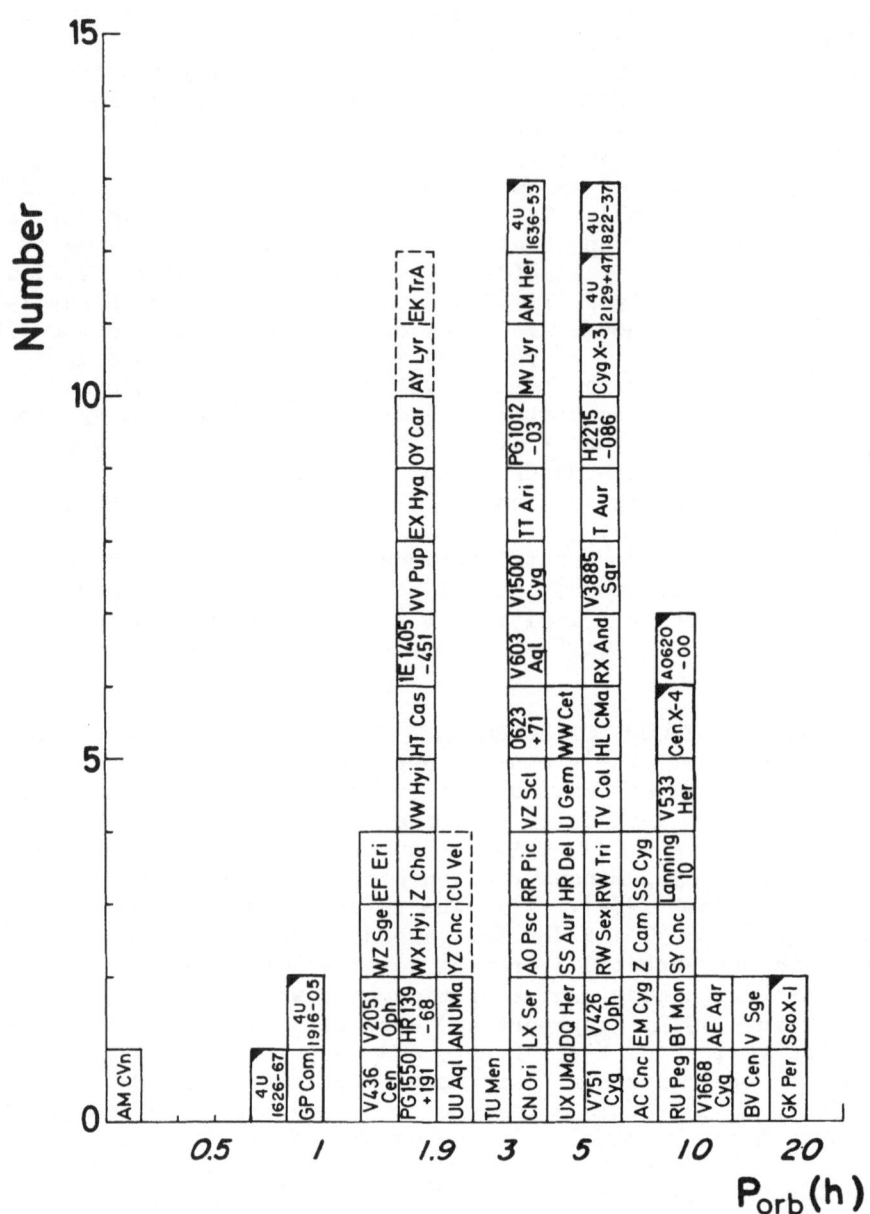

Fig. 3.   The distribution of orbital periods for LMXB's compared to the distribution for CV's with orbital periods of less than 20 hours (adapted from Ritter 1983).   The LMXB's are denoted by boxes with shaded corners.

5. ACCRETION DISKS

The dominant source of light in most LMXB's is an X-ray illuminated accretion disk. The disk spectra of three classes of LMXB's (burster, pulsar and nova) are shown in Figure 4 a-d. The spectra are characterized by a blue continuum, emission lines, notably He II λ4686 and NIII λ4640, and an absence of absorption lines. The lines are variable and sometimes double-peaked.

There are two distinct models of optical emission from X-ray heated gas which have been applied to LMXB's: (i) A stellar model in which X-rays from a compact source heat and ionize the photosphere of the optical companion (Milgrom and Salpeter 1975; London, McCray and Auer 1981). (ii) A nebular model in which a compact X-ray source embedded in a large gas cloud produces an ionization structure in the gas and an intense ultraviolet radiation field (Kallman and McCray 1982 and references therein; Halpern and Grindlay 1980). The results of these models are in qualitative agreement with optical observations of LMXB's (see Holt and McCray 1982 for a recent review). A closer correspondence between observation and theory, however, will require more realistic models which incorporate an accretion disk geometry.

An interesting feature of the spectra of MXB1735-44 and 4U1626-67 shown in Figure 4 is the absence of Balmer emission. Is the composition of the secondaries pure helium as inferred for AM CVn and G61-29 (Nather, Robinson, and Stover 1981), the CV's with 17 minute and 46 minute periods, respectively? This may be the case for 4U1626-67, but, it is definitely not the case for MXB1735-44; White, Charles, and Thorstensen (1980) have observed Balmer emission from this object.

Accretion disks in active LMXB's are qualitatively different from disks in CV's; the former are ~ 3-6 magnitudes brighter due to intense X-ray heating (van Paradijs 1983). The X-ray novae at minimum light are an important exception. In this case, the X-ray luminosity is very low ($L_x \lesssim 10^{32}$ ergs s^{-1}; Long, Helfand and Grabelsky 1981), and the optical spectrum (Figure 4e; Oke 1977) reveals a CV-type disk (M_v ~ +6) with the absorption line spectrum of a mid-K dwarf. The X-ray novae are very important objects among the shorter period LMXB's because their secondaries can be directly observed in quiescence.

The absence of eclipses among LMXB's has been long noted and was posed as a puzzle by Joss and Rappaport (1979). Excluding Her X-1, none of the ~ 20 well-studied LMXB's undergoes a total X-ray eclipse; based on simple geometric considerations the a priori probability for the lack of eclipses is very small. By comparison, 12 of 24 CV's listed by Robinson (1976), which have orbital periods and geometries comparable with LMXB's, are eclipsing systems. Milgrom (1978) proposed that the absence of X-ray eclipses in LMXB's is due to a thick accretion disk which shields the secondary from the X-ray source. One consequence of Milgrom's model is that systems with inclination angles near 90 may be unobservable as X-ray sources. The LMXB's 4U1822-37 and 4U2129+47 appear to be borderline cases. These high inclination systems contain extended X-ray sources (~ 0.5 $R_\odot$) of low luminosity. In these systems the accretion disk apparently blocks our direct view of the luminous compact source; the X-rays we observe are scattered by an extensive corona that is fed by the disk and energized by the compact source

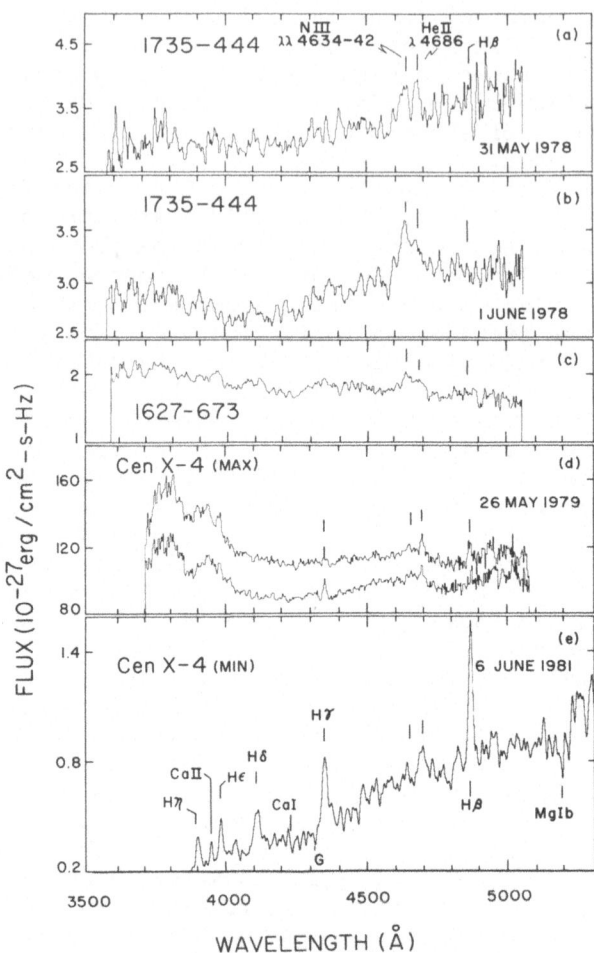

Fig. 4. Spectrophotometric data for four low-mass X-ray binaries:
(a,b) MXB1735-44 on two successive nights showing extreme variability as
well as NIII and HeII emission (Canizares, McClintock, and Grindlay
1979); (c) 4U1626-67 showing an unusually blue continuum (McClintock,
Canizares, and Grindlay, unpublished); (d) Cen X-4 on two successive
nights during outburst showing Balmer lines (labeled in (e)) and HeII in
emission (Canizares, McClintock, and Grindlay 1980); (e) Cen X-4 in
quiescence, two years after outburst showing a mid-K absorption-line
spectrum and Balmer emission (probably from an accretion disk) (Petro
and McClintock, unpublished). These data have resolutions of ≃ 12 Å
(a-d) and ≃ 7 Å (e); they were obtained with the CTIO 4-m telescope, R-C
spectrograph and SIT vidicon detector. (Figure taken from Bradt and
McClintock 1983).

(White and Holt 1982; McClintock et al. 1982). Interestingly, both
4U1822-37 and 4U2129+47 show large-amplitude optical light curves,
behavior that is atypical of LMXB's. The light curves of 4U2129+47 are
very similar to the heating light curves of HZ Her (Thorstensen et al.
1979; McClintock et al. 1981), whereas the 4U1822-37 light curves are
more complex (Mason et al. 1980).

Her X-1 is a high inclination (i > 84°; Middleditch and Nelson
1976) X-ray source which is apparently 100% modulated by the 35-day
precession of its accretion disk (see Katz et al. 1982; Levine and
Jernigan 1982 and references therein). The ratio of the inferred
precession period of the disk to the orbital period is 20. For two
massive X-ray binaries, LMC X-4 (Lang et al. 1981) and SS433 (Katz et
al. 1982), this ratio is 22 and 13, respectively. Disk precession may
be a fairly common phenomemon among the LMXB's and may explain some of
the X-ray periods proposed for LMXB's which are probably too long to be
orbital periods (e.g., 4.9 days in 4U0614+09 and 7.8 days in A0620-00).

Intense X-ray bursts, which have typical rise times of ~ 1 sec and
decay times of several seconds to minutes, are observed in about half of
the LMXB's (Lewin and Joss 1983). They are believed to be due to
thermonuclear flashes on the surface of a neutron star. Optical bursts,
which appear to be temporally smeared versions of the X-ray bursts, have
also been observed from three systems (see Lewin and Joss 1983 and
references therein). In MXB1636-53 the optical bursts suffer a fairly
constant delay of about 2 seconds (Pederson et al. 1982). The most
plausible model is that the X-rays are reprocessed promptly (< 1 sec) in
the accretion disk within ~ 1.5 lt-sec of the compact object; the delay
and the smearing of the optical bursts are due to the light travel time
across the disk. There is no evidence that reprocessing occurs on the
companion star (in which case the delay times would be variable with
orbital phase), which implies that the companion lies in the X-ray
shadow of the disk.

4U1626-67, with its 7.7 sec X-ray and optical pulsations, is
another system which provides insight into the reprocessing of X-rays
into visible light. Most of the light is emitted by the X-ray heated
disk, and much of it comes from within ~ 1 lt-sec of the neutron star
(McClintock et al. 1980). However, Middleditch et al. (1981) discovered
that a small fraction (~ 15%) of the pulsed, reprocessed light is in a
single lower-frequency sideband which is emitted by the secondary. From
this faint component of the pulsed flux (~ 25th magnitude), Middleditch
et al. determined the orbital period to be 42 minutes.

6. FORMATION

There are at least three distinct scenarios for the formation of
LMXB's that have been widely discussed (see van den Heuvel 1983 for a
comprehensive review). It seems unlikely, however, that any one of
these scenarios can account for all classes of LMXB's.

Origin from a Cataclysmic-Variable Binary. In this scenario, a
massive white dwarf in a cataclysmic variable is driven over the
Chandrasekhar mass limit as it accretes matter from its main sequence
dwarf companion. Calculations by Canal and Schatzman (1976), Canal et
al. (1980), Sugimoto and Nomoto (1980), Miyaji et al. (1980), Nomoto

(1981), and Labay et al. (1983), have indicated that C-O and O-Ne-Mg white dwarfs can undergo a "quiet collapse" to form a neutron star when they have been driven over the Chandrasekhar limit. Furthermore, it was shown that the implosion and possible subsequent mass ejection need not disrupt the binary system even for very low-mass secondaries. A possibly significant constraint in this scenario is that during the accretion phase the rate of mass transfer must be in the correct range for quiescent burning of the accreted matter to take place; in the event of nova explosions, most, if not all of the accreted matter would be ejected from the binary system (see, e.g., Nomoto 1981 and references therein).

Binary Capture in a Globular Cluster Core. The high incidence of X-ray sources in globular clusters led Clark (1975) to suggest that these were compact objects (neutron stars or black holes) that had captured a stellar companion. This was followed by proposals of specific mechanisms for binary formation, including direct collisions (Sutantyo 1975), two-body tidal capture (Fabian et al. 1975), and single-star binary encounters (Hills 1976). The high density of normal stars and presumably neutron stars leads to a sufficient rate of collisions that binary formation by, for example, two-body tidal capture can result in the numbers of observed globular cluster X-ray sources (Hills and Day 1976; Heggie 1977; Press and Teukolsky 1977; Lightman and Shapiro 1977). This type of capture mechanism, however, seems unlikely to explain the formation of LMXB's outside of globular clusters. Typical relative velocities of stars in the galactic bulge are about an order of magnitude higher than in globular clusters; the kinetic energy that would have to be dissipated in such tidal-capture collisions is therefore prohibitively high (van den Heuvel 1983). It is possible that some of the bulge sources may have been formed in globular clusters from which they were ejected, or which subsequently evaporated (Grindlay 1985; Krolik, Meiksin, and Joss 1985; van den Heuvel 1983). However, the different spatial distributions of the bulge sources and the globular clusters indicates that this scenario for the formation of LMXB's is probably not universal (see, e.g., Lewin and Joss 1983).

Common Envelope Scenario. In this picture, the more massive star in a wide binary system evolves to fill its critical potential lobe. If, at this point in the evolution, the primary has developed a 3-4 $M_\odot$ He core and the companion is a 1-2 $M_\odot$ main sequence star, the system could be a possible progenitor of an LMXB. If the companion is unable to accrete the envelope of the evolving primary which is transferred to it, a common envelope may form. This is followed by a spiral-in stage during which the unevolved star hardly captures any mass and most of the envelope is ejected from the system (Paczynski 1976; Ritter 1976; Meyer and Meyer-Hofmeister 1979; van den Heuvel 1983). The post-spiral-in system is compact with a 3-4 $M_\odot$ He star and a 1-2 $M_\odot$ main-sequence star in an orbit with a period of less than ~ 2 days. The He core of the primary evolves on a very rapid time scale, without significant expansion, and explodes leaving a neutron star remnant. The binary system may remain bound if less than about half of the total mass is ejected in the explosion. This type of evolutionary scenario is probably applicable only to the LMXB's with relatively more massive

companion stars such as Her X-1 and Sco X-1 because of the problems
associated with keeping the system bound during the explosion.

Clearly, as more LMXB's are discovered and their system parameters
determined, the constraints on progenitor systems and evolutionary
scenarios will become more significant.

Finally, we note that there are approximately equal numbers of
LMXB's and massive X-ray binaries containing early-type stars in the
Galaxy (Bradt and McClintock 1983) as well as in M31 (Long and van
Speybroeck 1983). However, as discussed by van den Heuvel (1983), the
expected lifetime of an LMXB is ~ 10^3-10^4 longer than that of a massive
X-ray binary. These facts suggest that the formation rate of LMXB's is
considerably smaller than for massive X-ray binaries. The low formation rate of LMXB's can also be appreciated in the following way. If
the lifetime of an LMXB is ~10^9 years and there are only ~ 100 in
existence (assuming the active phase is predominant), then only ~1000
have formed in the entire history of the Galaxy.

7. MASS TRANSFER MECHANISMS

The inferred mass transfer rates for many LMXB's of ~10^{-10}-10^{-8} $M_\odot$
yr^{-1} have stimulated much thinking about mass transfer mechanisms in
low-mass binaries. Three general classes of binary evolution have been
proposed: (i) orbital decay driven by the emission of gravitational
radiation (see, e.g., Paczynski and Sienkiewicz 1981; Rappaport, Joss,
and Webbink 1982); (ii) orbital decay driven by magnetic braking of the
secondary (see, e.g., Verbunt and Zwaan 1981; Rappaport, Verbunt, and
Joss 1983); and (iii) evolutionary expansion of a low-mass giant
secondary (Taam 1983a; Webbink, Rappaport, and Savonije 1983).

Detailed studies of mass transfer driven by orbital decay due to
gravitational radiation indicate that for binaries with main-sequence
secondaries, the mass transfer rates will remain close to 10^{-10} $M_\odot$ yr^{-1}
and will not exceed several times this value regardless of the other
binary system parameters (Paczynski and Sienkiewicz 1981; Rappaport,
Joss, and Webbink 1982). On the other hand, gravitational radiation
from binaries containing a neutron star and degenerate dwarf secondary
(of He or heavier elements) with mass > 0.1 $M_\odot$, can drive very large
mass transfer rates (Paczynski 1967; Lĩ et al. 1980). A serious
difficulty with this model is that such binaries will remain luminous
for only a short time and will evolve into long-lived low-luminosity
sources (Rappaport, Joss, and Webbink 1982). Such evolution leads to a
prediction of a large number of low-luminosity X-ray sources which are
not observed (Worrall et al. 1982; Hertz 1983). Finally, in this
regard, we note that a He-burning star of ~ 0.4-0.8 $M_\odot$, in orbit with a
neutron star could yield high mass transfer rates when driven by
gravitational radiation (Savonije 1983). In this model, however, the
optical companion would be far brighter than is observed for many of the
bulge X-ray sources. This model might still be valid for the extremely
X-ray luminous galactic center sources where no secure optical identifications exist and the counterparts may be heavily obscured.

Another mechanism for driving mass transfer that has recently
received much attention is "magnetic braking" (Verbunt and Zwaan 1981;
Taam 1983b; Patterson 1983; Spruit and Ritter 1983; Rappaport, Verbunt,

and Joss 1983). In this case, angular momentum is lost in a stellar
wind that is magnetically coupled to the secondary star. If synchronous
rotation with the orbital motion is enforced by the effects of tidal
interaction then these losses in the stellar wind will be extracted from
the orbital angular momentum. However, the dependence of the putative
magnetic braking on the parameters of the secondary is highly uncertain
except perhaps for G and early K stars (Skumanich 1972; Smith 1979).
Binary evolution calculations which utilize a magnetic braking law,
extrapolated to lower mass stars, indicate that mass transfer rates of
$\geq 10^{-9}$ M$_\odot$ yr^{-1} and even $\geq 10^{-8}$ M$_\odot$ yr^{-1} are easily achieved. These
calculations predict a correlation between mass transfer rate and
orbital period of the form $\dot{M} \propto P^{3\cdot3}$ (Rappaport, Verbunt, and Joss 1983;
Patterson 1983). It will be interesting to see which, if any, of the
LMXB's follow this law.

Finally, we briefly discuss a mechanism whereby mass transfer is
effected by the expansion of a low-mass ($\lesssim 1$ M$_\odot$) giant due to nuclear
evolution (Taam 1983a; Webbink, Rappaport and Savonije 1983). In this
scenario, mass is transferred stably and at very nearly constant rates
of $\sim 10^{-10}$ to 10^{-8} M$_\odot$ yr^{-1}, depending on the mass of the degenerate He
core of the secondary. The mass transfer ceases when the hydrogen-rich
envelope of the secondary has been exhausted. The model predicts
orbital periods in the range 1-200 days and X-ray to optical luminosity
ratios of $\sim$ 200-1000. This type of mass transfer mechanism may be
appropriate to sources such as Cyg X-2 and 2S0921-63 with evolved
secondaries, as well as to the unidentified bright sources in the
galactic-center region.

8. GLOBULAR CLUSTER AND BRIGHT BULGE X-RAY SOURCES

There are two independent classes of luminous X-ray sources for
which no secure optical identifications have been made: the globular
cluster sources and the brightest galactic bulge sources in our Galaxy
and in M31. Based on indirect, but compelling evidence, various workers
have concluded that these X-ray sources are LMXB's (Lewin and Joss 1983;
Grindlay 1983; van Speybroeck et al. 1979). Future optical identifications and further study of these objects may well provide significant
insight into the LMXB phenomenon. We conclude this paper by discussing
these sources.

The general properties of globular cluster X-ray sources are indistinguishable from those of LMXB's. For example, nine of the twelve
luminous globular cluster sources in our Galaxy are X-ray burst sources.
Lewin and Joss (1983) summarize the arguments that these sources are in
fact LMXB's which contain neutron stars. Grindlay (1983) has provided
an important, independent confirmation that the cluster sources are
low-mass systems by determining precise locations for eight of them (see
also, Jernigan and Clark 1979). From their locations near the cluster
centers, Grindlay (1983) deduces that M_x = 2 ± 1 M$_\odot$ (90% confidence).

Nineteen globular cluster sources have been detected in M31 at a
detection threshold of 1 x 10^{37} erg s^{-1} with an average luminosity of 5
x 10^{37} erg s^{-1} (van Speybroeck et al. 1979; Long and van Speybroeck
1983). Interestingly, the average luminosity of the 12 cluster sources
in the Galaxy is a factor of 7 smaller, and only one source (4U1820-30)
has a luminosity which exceeds the detection threshold of van Speybroeck
et al. (1979).

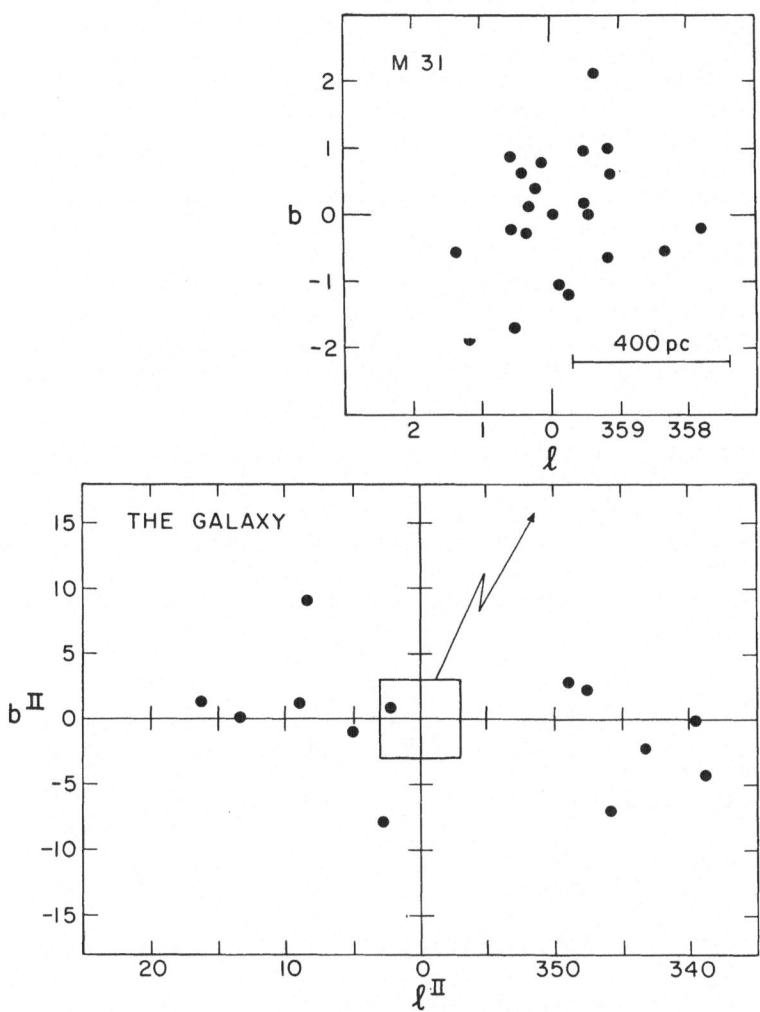

Fig. 5. A comparison of the distribution of bright X-ray sources
($L_x > 10^{37}$ erg s^{-1}) in the vicinity of the galactic center and in the
bulge of M31. Lower: The Galaxy, showing 8 sources discussed in the
text, as well as several additional luminous sources, including
4U1820-30 in the globular cluster NGC6624 and the optically identified
sources GX9+9 and MXB1735-44. A $6^{\circ} \times 6^{\circ}$ box corresponding to the size
of the M31 schematic is shown superposed on the figure; it contains only
one bright X-ray source. Upper: M31, in which the dense group of
bright bulge sources are located within about 2° of the center of the
galaxy. Simulated galactic coordinates, which would correspond to a
distance of 10 kpc from the center of M31, have been superposed on
Figure 5 of van Speybroeck et al. (1979). M31 is viewed nearly edge-on
($i \approx 77^{\circ}$).

There are 8 bright X-ray sources which are located at low galactic latitude ($|bII| \lesssim 2.5$), within 20° of the galactic center (see Fig. 5): GX17+2, GX13+1, GX9+1, GX5-1, GX3+1, GX349+2, 4U1705-44 and GX340+0. They have not been optically identified, with the possible exception of GX17+2 (Hjellming 1978; Grindlay 1981). The galactic distribution of these sources and their low-energy X-ray cutoffs imply that their typical distance is ~10 kpc (see, e.g., Webbink, Rappaport, and Savonije 1983). Consequently, they are among the most luminous sources in the Galaxy, $L_x \simeq 1\text{-}3 \times 10^{38}$ erg s^{-1}.

Van Speybroeck et al. (1979) have discovered a population of about 20 luminous ($L_x > 10^{37}$ erg s^{-1}) bulge sources in M31. Their galactic distribution, however, is dramatically different from the bulge sources in our Galaxy. As shown in Figure 5, they are contained in a volume ~ 1000 times smaller (for a discussion, see Long and van Speybroeck 1983).

Acknowledgements: We are grateful to Walter Lewin for providing us with a manuscript of "Accretion Driven Stellar X-Ray Sources" prior to publication. We also thank Paul Joss for helpful comments and Nancy Ferrari for preparing the manuscript. This work was supported by the National Science Foundation under grant AST 81-15557 and the National Aeronautics and Space Administration under contract NAS 5-24441.

REFERENCES

Bahcall, J.N.: 1978, Ann. Rev. Astron. Astrophys. 16, p. 241.

Bradt, H.V.D., and McClintock, J.E.: 1983, to appear in Ann. Rev. Astron. Astrophys. 21.

Canal, R., Isern, J., and Labay, J.: 1980, Astrophys. J. Letters 241, p. L33.

Canal, R., and Schatzman, E.: 1976, Astron. Astrophys. 46, p. 229.

Canizares, C.R., McClintock, J.E., and Grindlay, J.E.: 1979, Astrophys. J. 234, p. 556.

Canizares, C.R., McClintock, J.E., and Grindlay, J.E.: 1980, Astrophys. J. Letters 236, p. L55.

Clark, G.W.: 1975, Astrophys. J. Letters 199, p. L143.

Cordova, F.A., and Mason, K.O.: 1983, In Accretion Driven Stellar X-ray Sources   eds., W.H.G. Lewin and E.P.J. van den Heuvel, Cambridge, Cambridge University Press.

Cowley, A.P., Crampton, D., and Hutchings, J.B.: 1979, Astrophys. J. 231, p. 539.

Cowley, A.P., Crampton, D., and Hutchings, J.B.: 1982a, I.A.U.C. Circ. No. 3751.

Cowley, A.P., Crampton, D., and Hutchings, J.B.: 1982b, Astrophys. J. 256, p. 605.

Crampton, D., and Hutchings, J.B.: 1972, Astrophys. J. Letters 178, p. L65.

Fabian, A.C., Pringle, J.E., and Rees, M.J.: 1975, Mon. Not. R. Astron. Soc. 172, p. 15.

Ghosh, P., and Lamb, F.K.: 1979, Astrophys. J. 234, p. 296.

Grindlay, J.E.: 1981, I.A.U.C. Circ. No. 3620.

Grindlay, J.E.: 1985, this volume.

Halpern, J.P., and Grindlay, J.E.: 1980, Astrophys. J. 242, p. 1041.

Heggie, D.C.: 1977, Comments on Astrophys. and Space Phys. 7, p. 43.

Hertz, P.: 1983, Ph.D. thesis, Harvard University.

Hills, J.G.: 1976, Mon. Not. R. Sco. 175, p. 1P.

Hills, J.G., and Day, C.A.: 1976, Astrophys. Letters 17, p. 87.

Hjellming, R.M.: 1978, Astrophys. J. 221, p. 225.

Holt, S.S., and McCray, R.: 1982, Ann. Rev. Astron. Astrophys. 20, p. 323.

Jernigan, J.G., and Clark, G.W.: 1979, Astrophys. J. Letters 231, p. L125.

Jones, C.: 1977, Astrophys. J. 214, p. 856.

Joss, P.C., Avni, Y., and Rappaport, S.: 1978, Astrophys. J. 221, p. 645.

Joss, P.C., and Rappaport, S.: 1979, Astron. Astrophys. 71, p. 217.

Joss, P.C., and Rappaport, S.: 1984, to appear in Ann. Rev. Astron. Astrophys. 22.

Kallman, T.R., and McCray, R.: 1982, Astrophys. J. Suppl. 50, p. 263.

Kaluzienski, L.J., Holt, S.S., and Swank, J.H.: 1980, Astrophys. J. 241, p. 779.

Katz, J.I.: 1977, Astrophys. J. 215, p. 265.

Katz, J.I., Anderson, S.F., Margon, B., and Grandi, S.A.: 1982, Astrophys. J. 260, p. 780.

Krolik, J., Meiksin, A., and Joss, P.C.: 1985, this volume.

Kylafis, N.D., and Lamb, D.Q.: 1979, Astrophys. J. Letters 228, p. L105.

Labay, J., Canal, R., and Isern, J.: 1983, Astron. Astrophys. 117, p. L1.

Lamb, F.K., Pethick, C.J., and Pines, D.: 1973, Astrophys. J. 184, p. 271.

Lang, F., Levine, A., Bautz, M., Hauskins, S., Howe, S., Primini, F., Lewin, W., Baity, W., Knight, F., Rothschild, R., and Petterson, J.: 1981, Astrophys. J. Letters 246, p. L21.

Levine, A.M., and Jernigan, J.G.: 1982, Astrophys. J. 262, p. 294.

Lewin, W.H.G., and Joss, P.C.: 1983 in Accretion Driven Stellar X-ray Sources,   eds., W.H.G. Lewin and E.P.J. van den Heuvel, Cambridge, Cambridge University Press.

Li, F., Joss, P., McClintock, J., Rappaport, S., and Wright, E.: 1980, Astrophys. J. 240, p. 628.

Lightman, A.P., and Shapiro, S.L.: 1977, Astrophys. J. 211, p. 244.

London, R., McCray, R., and Auer, L.H.: 1981, Astrophys. J. 243, p. 970.

Long, K.S., Helfand, D.J., and Grabelsky, D.A.: 1981, Astrophys. J. 248, p. 925.

Long, K.S., and Van Speybroeck, L.P.: 1983, in Accretion Driven Stellar X-ray Sources   eds., W.H.G. Lewin and E.P.J. van den Heuvel, Cambridge, Cambridge University Press.

Mason, K., Middleditch, J., Nelson, J., White, N., Seitzer, P., Tuohy, I., and Hunt, L.: 1980, Astrophys, J. Letters 242, p. L109.

McClintock, J.E., Canizares, C.R., Li, F.K., and Grindlay, J.E.: 1980, Astrophys. J. Letters 235, p. L81.

McClintock, J.E., Remillard, R.A., and Margon, B.: 1981, Astrophys. J. 243, p. 900.

McClintock, J., London, R., Bond, H., and Grauer, A.: 1982, Astrophys. J. 258, p. 245.

McClintock, J., Petro, L., Remillard, R., and Ricker, G.: 1983, Astrophys. J. Letters 266, p. L27.

Meyer, F., and Meyer-Hofmeister, E.: 1979, Astron. Astrophys. 78, p. 167.

Middleditch, J., Mason, K., Nelson, J., and White, N.: 1981, Astrophys. J. 244, p. 1001.

Middleditch, J., and Nelson, J.: 1976, Astrophys. J. 208, p. 567.

Milgrom, M., and Salpeter, E.E.: 1975, Astrophys. J. 196, p. 583.

Milgrom, M.: 1978, Astron. Astrophys. 67, p. L25.

Miyaji, S., Nomoto, K., Yokoi, K., and Sugimoto, D.: 1980, Publ. Astron. Soc. Japan 32, p. 303.

Motch, C., Ilovaisky, S., and Chevalier, C.: 1982, Astron. Astrophys. 109, p. L1.

Nather, R.E., Robinson, E.L., and Stover, R.J.: 1981, Astrophys. J. 244, p. 269.

Nomoto, K.: 1981, in Fundamental Problems in the Theory of Stellar Evolution, Proc. IAU Symp. No. 93, D. Reidel Publ. Co., Dordrecht, p. 295.

Oke, J.B.: 1977, Astrophys. J. 217, p. 181.

Paczynski, B.: 1967, Acta Astr. 17, 287.

Paczynski, B.: 1976, in Structure and Evolution of Close Binary Systems, eds., P. Eggleton et al., D. Reidel, Dordrecht, p. 75.

Paczynski, B., and Sienkiewicz, R.: 1981, Astrophys. J. Letters 248, p. L27.

Patterson, J.: 1983, in preparation.

Pedersen, H., Lub, J., Inoue, H., et al.: 1982, Astrophys. J. 263, p. 325.

Pedersen, H., van Paradijs, J., and Lewin, W.H.G.: 1981, Nature 294, p. 725.

Press, W.H., and Teukolsky, S.A.: 1977, Astrophys. J. 213, p. 183.

Rappaport, S., and Joss, P.C.: 1977, Nature 266, 683.

Rappaport, S., Joss, P.C., and Webbink, R.F.: 1982, Astrophys, J. 254, p. 616.

Rappaport, S., Verbunt, F., and Joss, P.C.: 1983, submitted to Astrophys. J.

Ritter, H.: 1976, Mon. Not. R. Astron. Soc. 175, p. 279.

Ritter, H.: 1983, "Catalogue of Cataclysmic Binaries, Low-Mass X-Ray Binaries and Related Objects, First Edition", preprint No. MPA 22, Max Planck-Institut fur Physik und Astrophysik, Garching, Federal Republic of Germany.

Robinson, E.L.: 1976, Ann. Rev. Astron. Astrophys. 14, p. 119.

Samimi, J., Share, G., Wood, K., Yentis, D., Meekins, J., Evans, W., Shulman, S., Byram, E., Chubb, T., and Friedman, H.: 1979, Nature 278, p. 434.

Savonije, G.J.,: 1983, in Accretion Driven Stellar X-ray Sources, eds., W.H.G. Lewin, and E.P.J. van den Heuvel, Cambridge, Cambridge University Press.

Skumanich, A.: 1972, Astrophys. J. 171, p. 565.

Smith, M.A.: 1979, Publ. Astron. Soc. Pac. 91, p. 737.

Spruit, H.C., and Ritter, H.: 1983, preprint.

Sugimoto, D., and Nomoto, K.: 1980, Space Sci. Rev. 25, p. 155.

Sutantyo, W.: 1975, Astron. Astrophys. 44, p. 227.

Taam, R.E.: 1983a, Astrophys. J. Letters, in press.

Taam, R.E.: 1983b, Astrophys. J., in press.

Thorstensen, J., Charles, P., Bowyer, S., Briel, U., Doxsey, R.
     Griffiths, R., and Schwartz, D.A.: 1979, Astrophys. J. Letters
     233, p. L57.

Thorstensen, J.E., Charles, P.A., and Tuohy, I.R.: 1983, submitted to
     Astrophys. J.

Van den Heuvel, E.P.J.: 1983, in Accretion Driven Stellar X-ray
     Sources, ᵗ   eds., W.H.G. Lewin, and E.P.J. van den Heuvel,
     Cambridge, Cambridge University Press.

Van Paradijs, J., Verbunt, F., van der Linden, T., Pedersen, H., and
     Wamsteker, W.: 1980, Astrophys. J. Letters 241, p. L161.

Van Paradijs, J.: 1983, in Accretion Driven Stellar X-Ray Sources,
               eds., W.H.G. Lewin and E.P.J. van den Heuvel,
     Cambridge, Cambridge University Press.

Van Speybroeck, L., Epstein, A., Forman, W., Giaconni, R., Jones, C.,
     Liller, W., and Smarr, L.: 1979, Astrophys. J. Letters 234,
     p. L45.

Verbunt, F., and Zwaan, C.: 1981, Astron. Astrophys. 100, p. L7.

Walter, F., Bowyer, S., Mason, K., Clarke, J., Henry, J., Halpern, J.,
     and Grindlay, J.: 1982, Astrophys. J. Letters 253, p. L67.

Warner, B.: 1974, Mon. Not. R. Astr. Soc. 167, p. 61P.

Warner, B.: 1976, in Structure and Evolution of Close Binary Systems,
     eds., P. Eggleton, S. Mitton, and J. Whelan, D. Reidel, Dordrecht,
     p. 85.

Webbink, R.F., Rappaport, S., and Savonije, G.J.: 1983, submitted
     to Astrophys. J.

White, N.E., Charles, P.A., and Thorstensen, J.R.: 1980,
     Mon. Not. R. Astr. Soc. 193, p. 731.

White, N.E., and Holt, S.S.: 1982, Astrophys. J. 257, p. 318.

White, N.E., and Swank, J.H.: 1982, Astrophys. J. Letters 253, p. L61.

Worrall, D., Marshall, F., Boldt, E., and Swank, J.: 1982, Astrophys.
     J. 255, p. 111.

GALACTIC BULGE X-RAY BURST SOURCES FROM DISRUPTED GLOBULAR CLUSTERS ?

Jonathan E. Grindlay[1] and Paul Hertz[2]
Harvard-Smithsonian Center for Astrophysics

ABSTRACT

The origin of the bright galactic bulge x-ray sources, or GX sources, is unclear despite intensive study for the past 15 years. We suggest that the fact that many (or most) of the GX sources are x-ray burst sources (GXRBS) and are otherwise apparently identical to the luminous x-ray sources found in globular cluster cores implies that they too may have a globular cluster origin. The possibility that the compact x-ray binaries found in globulars are ejected is constrained by observations of CVs in and out of clusters. The GXRBS are instead hypothesized to have been formed by capture processes in globular clusters which have now largely been disrupted by repeated tidal stripping and shocking in the galactic plane. A statistical analysis of the 12 GXRBS which have precise positions from Einstein and/or optical (or radio) observations indicate that it is probably significant that a bright ($m_V \lesssim 19$) G or K star is found within the error circle (3" radius) in four cases. These may be surviving giants in a disrupted globular cluster core. Implications for globular cluster evolution and the GXRBS themselves are discussed.

I. INTRODUCTION

The basic nature of the luminous x-ray sources in the galactic bulge region of our Galaxy (and presumably in the bulge regions of other galaxies, e.g., M31) is reasonably clear. They are compact binary systems containing (in most cases) a neutron star with a low mass late type dwarf stellar companion in orbit around the neutron star. This basic picture was suggested by Joss and Rappaport (1979) largely on the basis of the evidence accumulated from studies of the prototype compact binary source 4U1626-67 as well as studies of x-ray bursters. Burster observations suggested that their spectra were cooling black bodies (Swank et al. 1977) and that the bursts themselves were due to thermonuclear flashes on the surface of neutron stars (Joss 1978). Seven of the eight luminous ($\gtrsim 10^{36}$erg s^{-1}) x-ray sources which are

79

D. Q. Lamb and J. Patterson (eds.), Cataclysmic Variables and Low-Mass X-Ray Binaries, 79–91.
© 1985 by D. Reidel Publishing Company.

located in globular clusters (Grindlay 1981, Hertz and Grindlay 1983a)
are x-ray bursters and are therefore also probably compact binary
sources (Lewin and Joss 1981). This conclusion was strongly supported
by the preliminary (Grindlay 1981) and final (Grindlay et al. 1983a)
determinations of the most probable mass of the globular cluster x-ray
sources from precise (~1 arcsec) measurements of their offsets from the
cluster centers: the 90% confidence limit for the source mass is in the
range 1–4 $M_\odot$. Additional direct evidence that the bursters are indeed
binaries (and not isolated neutron stars with massive disks) was
obtained with the discovery of a ~50 min binary period for the burster
4U1916–05 (Walter et al. 1982, White and Swank 1982).

Although the compact binary nature of the galactic bulge and x-ray
burst sources (GXRBS) is now clear, their origin remains poorly
understood. Despite their essentially indistinguishable x-ray
properties, models for the origin of GXRBS found in globular clusters
and those apparently in the field have been considered separately. The
globular cluster sources are generally thought to be compact binaries
formed by tidal capture in two-body interactions between neutron stars
and main sequence dwarfs in the dense cluster core (Clark 1975, Press
and Teukolsky 1977, Lightman and Grindlay 1982). Three-body exchange
collisions (Hills 1975, Hut and Verbunt 1985) between isolated neutron
stars and existing main sequence star binaries in which one of the main
sequence stars is ejected may also be significant. The fact that GXRBS
in globular clusters are much more likely to be found in clusters with
high central density suggests that the compact binary formation and/or
evolution is strongly affected by stellar encounters. However, as
discussed by Lightman and Grindlay (1982), this cannot be the case for
GXRBS apparently outside globular clusters.

The formation of GXRBS outside clusters is then the key problem.
Three major studies have been reported which attempt to account for
these sources as the result of the evolution of pre-existing binary
systems. In the first, Rappaport, Joss, and Webbink (1982) show that
compact binaries containing a neutron star (or white dwarf) and
non-degenerate companion can evolve by gravitational radiation to give
mass transfer rates appropriate to x-ray luminosities as large as
~10^{36}erg s^{-1}. The bulk of the GXRBS would not be explained, and other
mechanisms for loss of angular momentum from the system are required.
Patterson (1983) has suggested that magnetic braking may in fact be the
mechanism and has developed a scenario for the evolution of both
cataclysmic variables (CVs) and GXRBS. However this relies on a rather
uncertain extrapolation of magnetic braking laws derived originally for
~1 $M_\odot$ stars whereas the non-degenerate companion stars in GXRBS may
be only ~0.2 $M_\odot$. In the third study, Webbink, Rappaport, and Savonije
(1983) argue that at least some of the GX sources (e.g., 2S0921–63 and
Cyg X-2) may be relatively long period (~1–200 days) binaries with an
evolved giant companion. This model does not account, however, for the
presumably more typical bulge sources with much shorter periods (e.g.,
4U1916–05, with a ~50 min period as mentioned above).

In this paper we present evidence that perhaps the GXRBS were all formed in globular clusters. Those sources which do not now appear to be in globular clusters, including heavily reddened clusters (Hertz and Grindlay 1983b), may have been nevertheless born in globulars and then either ejected from collisions between the x-ray binary and other cluster stars or binaries or have had their "surrounding" clusters largely disrupted by tidal encounters with the galactic plane. Both possibilities were suggested previously (Grindlay 1981,1983) and may be supported, at some level, by recent observations which are discussed here. However, we shall concentrate on the cluster disruption picture since there are specific tests (now in progress) for this interpretation of a body of recent observational data. We also conclude (in Section II) that even if ejection or stripping of white dwarf binaries is occurring, a large primordial population of now totally disrupted clusters is required to explain the observed number of CVs in the Galaxy if they were born in globular clusters. Krolik et al. (1985) have developed the cluster ejection picture and discuss this in a companion paper.

II. OBSERVATIONAL EVIDENCE FOR EJECTION OR TIDAL STRIPPING OF BINARIES

The evidence for ejection or stripping of binaries from globular clusters is based in part on the discovery of low luminosity ($\sim 10^{32}$–10^{34}erg s^{-1}) discrete x-ray sources in the cores and haloes of globular clusters (Hertz and Grindlay 1983a). A variety of arguments suggest that these sources are white dwarf binary systems. Given the total number of dim sources detected as well as the upper limits, we have inferred that there are some ~30 accretion (or currently "active") white dwarf binaries with $L_x \gtrsim 10^{31}$erg s^{-1} in a typical globular cluster, which must then contain at least ~10^4 white dwarfs. In both clusters (ω-Cen and M22) where the longer x-ray exposure enabled detection of multiple sources, the radial offsets ranged up to 10 core radii, or nearly half the tidal radius for these clusters. This suggests the white dwarf binaries can probably be stripped from the cluster as it passes through the galactic plane. The luminosity function derived from the white dwarf sources in globulars is consistent with that for cataclysmic variables (CVs) and serendipitous x-ray sources detected by Einstein in the galactic plane (Hertz and Grindlay 1983a and 1983c). This further suggests a possible common origin for the white dwarf binaries in and out of globular clusters.

Although CVs may be stripped from globulars and thus provide at least some of the observed space density, it should be stressed that there is apparently no shortage of production mechanisms for CVs from evolution of binary stars born in the disk (Patterson 1983). This is in contrast to the origin of GXRBS, whose origin (outside globulars) is unclear as discussed above. Nevertheless, we pursue further the possibility that CVs are stripped (or ejected) from globulars since it leads to interesting constraints for the GXRBS and the origin of CVs.

The total number of white dwarf binaries which could be lost from a typical globular cluster cannot much exceed the estimate for the number now present in the cluster or the mass fraction in white dwarfs would become unreasonably large. Thus the ~200 globulars estimated (allowing for incompleteness) to be present in the entire Galaxy (Harris 1976) could populate the disk (where stripping would occur) with perhaps ~6 x 10^3 white dwarf binaries. This is much less than the total number of accreting white dwarf binaries, or CVs, estimated for the Galaxy from local measurements of the space density of CVs. Patterson (1983) estimates this space density as ~10^{-6} pc^{-3} which would suggest a total number in the Galaxy of ~5 x 10^5 for an assumed scale height of ~300 pc and galactic radius of ~15 kpc for the CV distribution. Thus if the active CV population in the Galaxy were ejected or stripped from globulars, a total cluster number of ~2 x 10^4 would be required, or a factor of ~100 more than the total number of clusters now in the Galaxy.

Interestingly enough, this factor of ~100 is just the factor by which the estimated mass density in halo subdwarfs (which are globular cluster type stars) exceeds that in the present globular cluster sample (Mihalas and Binney 1981). It has been suggested many times that perhaps the halo stars are in fact the result of a now totally stripped population of globulars (see Mihalas and Binney 1981, and Fall and Rees 1977 for references). The individual cluster stars have no way to liberate their kinetic energy when they are tidally stripped upon passage through the plane and must therefore continue with approximately the cluster velocity to populate the galactic halo. On the other hand, the compact white dwarf binaries would have binding energies comparable to or greater than their kinetic energies of motion with the cluster. Thus in their tidal interaction with objects in the galactic plane, some fraction of their kinetic energy might be "absorbed" into their orbital motion (making them less bound), and the CVs will tend to be left behind in the plane. Note that this mechanism will work only for tidal stripping by discrete objects such as giant molecular clouds rather than the galactic plane itself. Obviously a detailed calculation is required, but some such mechanism is required if the observed CVs in the plane are due to ejection or stripping from the maximum population allowed for a halo population of primordial globular clusters.

Whereas the above arguments are relevant to the origin of CVs in the plane, similar arguments might be made for the GXRBS containing neutron star binaries. That is, if the white dwarf binaries observed at $\gtrsim$ half the cluster tidal radius really are being ejected and tidally stripped from the cluster, perhaps ejection and stripping can occur for neutron star binaries. The observations are not expected to show the higher luminosity GXRBS at large offsets from the cluster center simply because the total number of sources per cluster is a factor of ~200 less than for the white dwarf binaries (Hertz and Grindlay 1983a). Furthermore, since the neutron star binaries appear to be more massive than the white dwarf binaries (as expected for ~1.4 $M_\odot$ neutron stars vs. ~0.5 $M_\odot$ white dwarfs), it is much less likely that the neutron star systems are ejected into the cluster halo where they can be stripped from the cluster. We therefore turn to the hypothesis of cluster disruption, already apparently required for the CVs if their total number is produced in globulars, as a pre-requisite for the globular cluster origin of GXRBS in the galactic bulge.

Before considering the evidence that GXRBS in the bulge are from disrupted globular clusters, we note one further constraint imposed by the CVs. That is, if the CVs are indeed from a primordial population of ~2 x 10^4 globular clusters now disrupted, then these primordial clusters must have been very deficient in their neutron star number (perhaps because of a steeper initial mass function for their stars) since the number of GXRBS from these same clusters would otherwise greatly exceed the number of luminous GXRBS now observed (unless the GXRB x-ray emission lifetimes are $\gtrsim 100$ times shorter than the corresponding values of CVs). Alternatively, the bulk of the CVs now observed (in the plane) are not from disrupted globular clusters but result from evolution of binaries in the disk.

III. OBSERVATIONAL EVIDENCE FOR CLUSTER DISRUPTION

There are many reasons to suppose that a larger number of primordial globular clusters were formed in the Galaxy and that only a fraction have survived disruption from their passages through the galactic plane. We shall return to a number of these lines of evidence in the discussion below (Sections IV and V) but summarize here only the evidence from direct observations of the field GXRBS themselves.

The evidence is based on the recently available precise positions for GXRBS from the HRI detector on the Einstein Observatory. These positions have 90% confidence radii of 3 arcsec (Grindlay et al. 1983b) and therefore greatly restrict the number of optical candidates. For (at least) two GXRBS, the x-ray error circles include relatively bright G stars and no other objects brighter than $m_v \simeq 20$. These two sources are GX17+2 (=4U1813-14) and 4U1916-05. The preliminary position for 4U1916-05 is shown in Figure 2 of Walter et al. (1982); the final position is even more centered on the bright G star ("star 2" in the figure). For GX17+2 the G star "candidate" has been known since it was first pointed out by Tarenghi and Reina (1972) as a possible counterpart for a weak and variable radio source that was presumed to be the x-ray source. Preliminary VLA studies of the radio source (Hjellming 1978) gave a ~1 x 10 arcsec error box for the radio source which included the G star. However with an x-ray position only accurate to ~30 arcsec from SAS-3, it was always possible that the radio source and/or G star were chance alignments unrelated to the x-ray source.

This is now much less likely given the Einstein HRI position and a new VLA radio position (Grindlay and Seaquist 1983). The radio position (~0.2 arcsec uncertainty) includes the G star — which is presently measured only to ~0.5 arcsec accuracy. The x-ray position (with 2 arcsec radius, from several observations) includes the radio and optical positions. Thus on positional grounds alone, there is now no doubt that the radio and x-ray sources are identical (they correspond to within 2 arcsec, and the chance alignment probability is negligible). Although the highly accurate radio position fixes the source position, the chance alignment with the G star is also small, but not totally negligible. From the star count densities given in Allen (1973) for stars within ±5° of the galactic plane, the number of $\lesssim$18th mag stars expected in a ~0.5 arcsec (radius) radio vs. optical error circle is only 1.5 x 10⁻³. However since about 15 GXRBS in the bulge have now been searched for radio counterparts (Grindlay and Seaquist 1983), the expected total number of $\lesssim$0.5 arcsec alignments with stars brighter than $m_v \simeq 18$ is ~0.02. Therefore if GX17+2 were the only GXRB with a projected close alignment on a "normal" star (which cannot itself be the optical counterpart — see discussion below), the case would not be particularly significant.

However GX17+2 is not the only case. In addition to the G stars nearly superimposed on GX17+2 and 4U1916-05, the optical counterparts of two additional GXRBS in the field, Ser X-1 (=4U1837+04) and MXB1659-29 are also nearly superimposed (within ~2 arcsec) on brighter and redder "normal" stars. The "double star" nature of Ser X-1 was first pointed out by Thorstensen et al. (1980), and the red star "companion" for MXB1659-29 was identified by us (JEG) in additional 4-m plates obtained at CTIO prior to our optical identification of the source (Doxsey et al. 1979). The spectrum of the Ser X-1 "companion", which is 2.1 arcsec NE of the UV excess x-ray counterpart and has $m_v \simeq 18$, is shown in Figure 2 of Thorstensen et al. Both Mgb(λ5175) and Hβ absorption features are apparent which suggest a spectral type of G-K. The signal to noise is not adequate in the blue (e.g., at λ4300, for the G-band) to further define the spectral or luminosity class. No direct spectrum is available for the "companion" star for MXB1659-29, which is also displaced ($\lesssim$2 arcsec south — as apparent on IV-N plates we obtained with the CTIO 4-m in August 1977 and April 1978) from the x-ray source. We identified the x-ray counterpart (Grindlay 1978, Doxsey et al. 1979) only during a period of x-ray outburst in 1978. During our May 1979 observations the uv excess object was not visible on the 4m acquisition TV ($m_v \gtrsim 21$; this also agrees with the ESO observations from June and July 1979 reported by Cominsky et al. 1983). We did obtain photometry (in May 1979) of the "companion", however, and found it to have $m_v \simeq 18.5$, B-V $\simeq 0.9$, and U-B $\gtrsim 1$ with uncertainties of $\gtrsim$0.1 mag. These colors are consistent with a moderately reddened G-K star. Thus both Ser X-1 and MXB1659-29 have G or K stars with $m_v \simeq 18-19$ within ~2 arcsec of the x-ray source.

The four GXRBS with "ordinary" G (or K) stars nearly superimposed on the x-ray source are summarized in Table 1.

Table 1 – Properties of "G–Stars" Apparently Superimposed on GXRBS

Source	ℓ,b	m_v	Offset from X–ray Source	Spectral Type
MXB1659–29	353.8,7.3	18.5	$\lesssim$2 arcsec SW	G or K
Ser X–1	36.1,4.8	18.3	2.1 arcsec NE	G or K
GX17+2	16.4,1.3	17.5	$\lesssim$0.5 arcsec	late G
4U1916–05	31.4,–8.5	16.5	$\lesssim$3 arcsec	late G

These four sources are among the 12 GXRBS in the field (i.e., not in recognizable globular clusters) for which precise positions exist from either direct optical identifications (e.g., 4U1636–53 or 4U1735–44) or Einstein HRI positions (e.g., 4U1702–42 or 4U1905+00) and have been observed to burst. The complete list of all GXRBS satisfying these criteria is taken from Bradt and McClintock (1983) and Grindlay et al. (1983b) and is, by coordinate names: 1415–32 (Cen X4), 1608–52, 1636–53, 1659–29, 1702–42, 1735–44, 1744–26, 1813–14 (GX17+2), 1837+04 (Ser X–1), 1905+00, 1908+00 (Aql X–1) and 1916–05. We now consider the chance probability that 4 (or more) of these GXRBS have "normal" stars brighter than $m_v \simeq 19$ within a 3 arcsec radius circle. (Note that although 3 arcsec is the 90% HRI error box radius, 3 of the 4 sources in Table 1 have x-ray positions within ~2 arcsec of the G star.)

The chance coincidence probability may be computed from the binomial distribution and the expected number of stars per error box. That is, for a given expected number (from star counts) of stars in a 3 arcsec radius circle with $m_v \lesssim 19$, what is the probability that 4 (or more) are observed in 12 "trials"? For each of the four fields in question, we have derived the approximate number of stars (per arcmin^2) brighter than the G star using our estimated magnitudes and either the POSS–E prints or our 4–m plates. We find that the surface density of $10^{4.7}$ stars/deg^2 given by Allen (1973) for stars with $m_v \lesssim 19$ and within $5°$ of the galactic plane is consistent with (but not smaller than) our measured values. Adopting Allen's background density, the chance coincidence (evaluated from the binomial distribution) that 4 (or more) of the 12 GXRBS have stars (of any spectral type) within 3 arcsec is then only 3.0×10^{-3}. This probability is conservative since at least 2 of the stars are definitely G stars (the other two are either G or K) and the fraction of faint G stars in the galactic plane is probably at most 30% (it is only 15% at $m_v \lesssim 10$ — Allen 1973).

The fact that the four alignments seem to be significant must of course be treated with all the caution appropriate to a posteriori statistics. However we are asking a well–defined question: What is the probability of finding, by chance, "normal" stars (i.e., not the x-ray sources themselves, since no emission lines are seen in their spectra whereas given their high x-ray luminosities, they would be dominated by x-ray heating effects if they were the actual optical counterparts) within 3 arcsec of any of the entire sample of 12 GXRBS in the field which have precise ($\lesssim$3 arcsec uncertainty) locations. The chance probability for four such alignments appears to be small.

If the alignments are indeed physical associations and not just chance projections, then tests may be devised for a given model of the physical association. We propose a model, and discuss the tests in the next section, whereby the G stars are the tracers of the cores of disrupted globular clusters in which the compact binary GXRBS were born. The G stars would then be individual giants in the surviving cluster core, which might contain only a few tens of stars in addition to the GXRB. Given the typical ratio of giants to main sequence stars in a globular cluster, we might expect only one such giant in a core stripped down to its last ≤100 stars. Alternatively, if the stripped cluster core preserves constant density or even increases its density (since the surviving core should become even more tightly bound as stars, and cluster energy, are stripped from the halo) the "G star" might in fact be the assemblage of the entire surviving stellar distribution. In the case of GX17+2, at least, where the x-ray source (and VLA source) is within ~0.5 arcsec of the G star, this might be possible. However it is not possible for MXB1659-29 or Ser X-1 where the G star is resolved from the x-ray source.

IV. POSSIBLE ORIGIN OF DISRUPTED GLOBULARS AND OBSERVATIONAL TESTS

The hypothesis that the G stars near GXRBS are tracers of disrupted globular clusters is based on a number of plausibility arguments. First, as discussed in Section I above, this would naturally account for the formation of GXRBS in the "field": like the GXRBS in globular clusters, they were also formed by tidal capture in the high stellar density environment of a globular cluster core. The indistinquishable x-ray properties of GXRBS apparently in and out of globulars is then explained. The fact that virtually all (at least 7 out of 8) of the high luminosity x-ray sources in recognizable globulars are bursters suggests that GXRBS are formed very efficiently in globular cluster cores. It is natural then to seek a link to those GXRBS apparently in the field.

The spatial distribution of the GXRBS in the field provides indirect evidence for the cluster disruption picture. That is, the 12 GXRBS with precise locations and used to derive the significance of the G star association are primarily within a band of ±35° in galactic longitude and ±10° in galactic latitude centered on the galactic center. They are a significantly flattened distribution, or "old disk" population. On the other hand the GXRBS still in globular clusters are not significantly more flattened (in z) than globular clusters in general (Lightman and Grindlay 1982), and the cluster distribution in general is spherical about the galactic center. The extended disk-like distribution of the field GXRBS in the galactic bulge is just what is expected if most of the original population of globulars which had orbits within ~20° of the galactic plane have now been disrupted. It is precisely these clusters which should be most easily disrupted by tidal effects of the galactic plane (cf. Fall and Rees 1977) and encounters with individual massive structures in the plane. In

Particular, the large population (~4000) of giant molecular clouds with masses ~10^5–10^6 $M_\odot$ now recognized to be in the plane and primarily within the ~4 kpc ring (Solomon and Edmunds 1979) should reek havoc with globular clusters in the disk. Tidal stripping and disk shocking calculations for globular clusters (e.g., Fall and Rees 1977) have not yet included this important effect.

The number of globular clusters which must have been disrupted to give rise to the GXRBS in the field may be estimated as follows. From the analysis of Lightman and Grindlay (1982), it follows that the probability of finding a GXRB in a globular cluster within the central 4 kpc of the spherical distribution of globulars is about 0.1. Thus the ~12 GXRBS in the field would require ~120 parent globular clusters to have been (largely) disrupted. The uncertainty in this number, allowing for both incompleteness in the GXRB sample (presumably small) and uncertainty in the probability of formation of a GXRB in a cluster as well as its x-ray lifetime (presumably large) is probably a factor of 3. If this number of clusters were originally present in the disk, it is possible that they were formed as a separate population in the halo collapse that may have formed the disk (Eggen, Lynden Bell and Sandage 1962). It is interesting then that the number (~120) of primordial "disk clusters" required is comparable to the number (~140) of "halo clusters" which survive today. This would suggest approximately equal numbers of halo population and disk population globulars were formed with only the halo population now surviving. These primordial "disk clusters" would presumably be metal rich. Independent evidence for a separate population of metal rich clusters in the disk has in fact been pointed out by Sandage et al. (1966). They argue that the short period RR Lyrae stars (with periods in the 0.3–0.44 day range) which populate the disk (in the galactic bulge) but not present day globular clusters could have come from a disk population of globulars which are now largely disrupted.

Alternatively, if the only clusters formed in the Galaxy were in a spherical distribution, then only those on orbits with inclinations within some critical angle of the disk plane would now be disrupted. If ~120 such disrupted clusters are needed to make the field GXRBS, then this critical angle is ~22^0. That is for ~120 disrupted clusters and ~200 surviving clusters (Harris 1976), the disrupted fraction of ~0.38 implies a "loss-cone" with this solid angle fraction, or a critical angle (galactic latitude) of $22^0 = 90^0 - \cos^{-1}(0.38)$. This estimate is in rough agreement with the flattened GXRB distribution mentioned above (with typical galactic latitudes $\lesssim 10^0$) but is probably an overestimate of the true critical angle since it assumes a fixed total number of globulars with fixed galactocentric radius R rather than the steeply decreasing distribution of clusters with R actually observed.

We now turn to observational tests of the disrupted cluster hypothesis. The first, and most obvious, is to obtain deep images of the fields to search for a clump (perhaps 10–100) of surrounding stars of much fainter magnitude around the G star. These stars would be the expected surviving main sequence stars in the cluster core. If the G stars are surviving cluster giants with absolute magnitudes $M_V \simeq 0$–1, the main sequence stars should be $\gtrsim 3$–5 magnitudes fainter. Thus for GX17+2 these stars should be at $m_V \gtrsim 21$–23 if they are of comparable spectral type. For 4U1916-05, these fainter main sequence stars should start showing up at $m_V \gtrsim 20$–22. In fact at least one such faint star within ~2 arcsec of the G star for 1916-05 has been detected by Walter et al. (1982) and in our preliminary CCD imaging studies with the MMT (Grindlay, Hertz, and Schild 1983). Deeper CCD images will hopefully be obtained in May 1983 at CTIO. We would expect the main sequence stars to be distributed in a $\lesssim 10$ arcsec (radius) region which is comparable to the original cluster core radius. However the spatial distribution is really very uncertain since simulations of the final disruption of globulars have not (to our knowledge) yet been carried out. Thus the deep images may be inconclusive.

The second obvious test, and one with minimal uncertainty, is to obtain high quality spectra of the G stars themselves to better define their spectral class (currently estimated with uncertainty of only half a spectral class from our ~15–20 Å resolution spectra from CTIO) as well as luminosity class and metallicity. The luminosity class is particularly crucial since the G stars must be giants if they are to be at the $\gtrsim 5$ kpc distances that the galactocentric distribution of GXRBS would imply. Spectra with moderately high resolution ($\lesssim 6$ Å) and good signal to noise at ~4000 Å are thus required. The traditional G-star luminosity indicator given by the CN band head at $\lambda 4215$ may not be sufficient since the stars are expected to be moderately metal poor if they are globular cluster stars. Once again, spectra are being obtained with the MMT in May 1983 in an attempt to settle this.

Finally, there is the question of interstellar reddening and the low energy cutoffs in the x-ray spectra. Of the four candidate disrupted cluster core GXRBS, only GX17+2 shows significant low energy absorption in its x-ray spectrum as derived by us from the MPC on Einstein. Best fit bremsstrahlung spectral models, which have unacceptably large χ^2 values but are nevertheless better than power law fits, have temperatures $kT \simeq 10$–20 keV and low energy cutoffs of $E_a \simeq 1.8$–2.1 keV. The corresponding (Brown and Gould 1970) column densities would then be $N_H \simeq 2.5$–4×10^{22}cm^{-2}, which in turn would imply (Gorenstein 1975) optical extinctions of $A_V \simeq 9$–18 mag. However the colors of the G star, with $(B-V) \simeq 1.3$ (Margon 1980), only allow $A_V \simeq 2$–3 mag. Two possibilities emerge if the G star is really associated with the GXRB. First, the low energy absorption is largely intrinsic to the x-ray source and the interstellar value is much less. Second, the x-ray spectrum is not really thermal bremsstrahlung (given the large χ^2 value) and the derived N_H value is therefore misleading.

Both possibilities are at least partially allowed by the data. First, variability in the low energy cutoff of GX17+2 _is_ observed — with best fit values of ~1.8 keV in some observations and 2.1 keV in others (hence the range quoted above). The 90% confidence contours in the N_H vs. kT plane for these variations do not overlap. Variations were observed both within a single ~2000 sec observation and between observations 6 months apart. The variations were "confirmed" in a model-independent way for the single (~2000 sec) observation by plotting the MPC hardness ratio of counts in the ~2-3 keV band divided by those in the ~1-2 keV band (both bands are insensitive to changes in kT) and noting smooth changes which followed those in E_a. Although the observed "base" value of $E_a \simeq 1.8$ keV is still greatly in excess of the allowed optical value, the variability suggests that at least some portion of the low energy absorption is indeed intrinsic. This should not be surprising given the large increase in low energy absorption observed (Hertz and Grindlay 1983a) for the x-ray source in the globular cluster M15: The low energy cutoff increased from $\lesssim 0.4$ keV to ~1.8 keV between two observing periods. In this picture, many of the galactic bulge sources and GXRBS which also apparently have large low energy cutoffs could have significant intrinsic absorption, though their interstellar values could also be large.

The second possible way to reconcile the x-ray absorption with the apparent optical extinction for GX17+2 is to consider multi-component x-ray spectral models. In the 1-10 keV band, or first 6 of the 8 PHA channels of the MPC, a simple blackbody model fits the GX17+2 data extremely well. The best-fit low energy absorption is small ($E_a \simeq 0.8$ KeV or $N_H \simeq 3 \times 10^{21}$) and is completely consistent with the optical value. However the corresponding temperature with $kT \simeq 2.8$ keV implies the neutron star (if it has a ~10 km radius) is radiating at very near the Eddington limit and that the x-ray burst observed with the MPC (Kahn and Grindlay 1983) greatly exceeds the Eddington limit. Furthermore a second and harder spectral component is then required to fit the ~10-15 keV data.

The critical tests for the low energy absorption vs. optical extinction towards GX17+2 are then to i) measure the <u>slope</u> of the low energy spectrum to determine whether it fits an absorption cutoff [going as $\exp(-(E/E_a)^{-8/3})$] or a blackbody cutoff [going as E^2] and to ii) measure the optical extinction vs. distance of surrounding stars close to the G star and determine whether it really is in a "window" where, despite its ~5 kpc distance and ~$1.3°$ galactic latitude, the extinction is only ~2 magnitudes. Einstein OGS spectra exist which might be sensitive enough to carry out the first analysis, and a spectrophotometric reddening study of the field will be carried out for the second test.

V.   CONCLUSIONS

We have presented evidence that the compact x-ray binaries — both lower luminosity white dwarf systems (CVs) and the higher luminosity x-ray bursters and galactic bulge sources, the so-called GXRBS, may have been formed in globular clusters. The existence of low mass white dwarf x-ray binaries at large radial offsets from the centers of globulars

suggests they may be tidally stripped from clusters. Thus at least some
of the CVs in the Galaxy may have indeed originated in globular
clusters, as also suggested by the similar shapes of the x-ray
luminosity functions (Hertz and Grindlay 1983a,c). However to populate
the disk with the total observed spatial density of CVs would require a
large primordial population of $\gtrsim 10^4$ globular clusters. It has long
been realized that a similar number of now disrupted clusters could give
rise to the Pop II stars in the galactic halo. Evidence that these halo
stars may have been stripped from globular clusters is based in part on
the existence of high velocity stellar groups (e.g., Groombridge 1830)
of cluster type stars (Oort 1965). However such a large population
($\sim 10^4$) of disrupted globulars must have been deficient in neutron stars
not to have given rise to a much larger number of GXRBS than now
observed. It is likely, then, that most CVs have evolved from binary
systems in the plane (Patterson 1983).

The existence of relatively bright G star "companions" within
~1-3 arcsec of 4 GXRBS appears to be significant. We suggest that
these may in fact be disrupted globular cluster cores in which the G
stars are surviving cluster giants and the x-ray binaries still reside
in the residual clump of cluster core stars in which they were
originally formed by tidal capture. Several key observational tests are
in progress to either further confirm or refute this hypothesis. The
GXRBS would require a sample of $\gtrsim 100$ globular clusters in the galactic
plane to have been disrupted since their formation in either a
primordial disk component (subsequent to the halo collapse) or as the
clusters in the original spherical distribution with orbits within ~20°
of the galactic plane. Such clusters would constitute ~38% of the
total and give rise to the observed number and latitude distribution of
GXRBS in the bulge. Once again, independent arguments have been made
(e.g., Sandage et al. 1966) that a non-disrupted disk population of
globulars once existed in the Galaxy. It may be that the GXRBS can
further constrain this intriguing possibility, with its fundamental
implications for the formation and evolution of the Galaxy.

This work was partially supported by NASA Contract NAS8-30751.

REFERENCES

Allen, C.W. 1973, _Astrophysical Quantities_, 3rd ed. (Athlone Press,
     London).
Brown, R. and Gould, R. 1970, _Phys.Rev. D, 1_, 2252.
Bradt, H.V. and McClintock, J. 1983, _Ann.Rev.Astron. and Astrophys.,_ in
     press.
Clark, G.W. 1975, _Ap.J.(Letters)_, 199, L143.
Cominsky, L., Ossman, W., and Lewin, W. 1983, preprint.
Doxsey, R., Grindlay, J., Griffiths, R., Bradt, H., Johnston, M., Leach,
     R., Schwartz, D., and Schwarz, J. 1979, _Ap.J.(Letters), 228_, L67.
Eggen, O., Lynden-Bell, D., and Sandage, A. 1962, _Ap.J., 136_, 748.
Fall, M. and Rees, M. 1977, _MNRAS, 181_, 37P.
Gorenstein, P. 1975, _Ap.J., 198_, 95.
Grindlay, J.E. 1978, _IAU Cir.No. 3222_.

Grindlay, J.E. 1981, in X-ray Astronomy with the Einstein Satellite,
    (R. Giacconi, ed.), Reidel, p.79.
Grindlay, J.E. 1983, in Advances in Space Research, Vol. 2, No. 9,
    pp.133-143.
Grindlay, J., Hertz, P., and Schild, R. 1983, BAAS, Vol. 14, 891.
Grindlay, J. and Seaquist, E. 1983, BAAS, Vol. 44, 888.
Grindlay, J.E., Hertz, P., Steiner, J., Murray, S., and Lightman,
    A. 1983a, in preparation.
Grindlay, J.E., et al. 1983b, in preparation.
Harris, W.E. 1976, Astron.J., 81, 1095.
Hertz, P. and Grindlay, J.E. 1983a, Ap.J., in press.
Hertz, P. and Grindlay, J.E. 1983b, in preparation.
Hertz, P. and Grindlay, J. 1983c, submitted to Ap.J.
Hills, J. 1975, Astron.J. 80, 1075.
Hjellming, R.M. 1978, Ap.J., 221, 225.
Hut, P. and Verbunt, T. 1985, these proceedings.
Joss, P. 1978, Ap.J.(Letters), 225, L123.
Joss, P. and Rappaport, S. 1979, Astron. and Astrophys., 71, 217.
Kahn, S. and Grindlay, J. 1983, submitted to Ap.J.
Krolik, J., Meiksin, A., and Joss, P. 1985, these proceedings.
Lewin, W. and Joss, P. 1981, Space Sciences Review, 28, 3.
Lightman, A. and Grindlay, J. 1982, Ap.J., 262, 145.
Margon, B. 1980, Ap.J., 219, 613.
Mihalas, D. and Binney, J. 1981, Galactic Astronomy, (Freeman, San
    Francisco).
Oort, J. 1965, in Galactic Structure, (A. Blaauw and M. Schmidt, eds.),
    h.21 (Univ. Chicago Press).
Patterson, J. 1983, Ap.J.Suppl., submitted.
Press, W. and Teukolsky, S. 1977, Ap.J., 213, 183.
Rappaport, S., Joss, P., and Webbink, R. 1982, Ap.J., 254, 616.
Sandage, A., Smith, L., and Norton, R. 1966, Ap.J., 144, 894.
Solomon, P. and Edmunds, M. 1979, Giant Molecular Clouds in the Galaxy,
    (Pergamon Press, Oxford).
Swank, J., Becker, R., Boldt, E., Holt, S., Pravdo, S., and Serlemitsos,
    P. 1977, Ap.J.(Letters), 212, L73.
Tarenghi, M. and Reina, C. 1972, Nature Phys.Sci., 240, 53.
Thorstensen, J., Charles, P., and Bowyer, S. 1980, Ap.J., 238, 964.
Walter, F., Bowyer, S., Mason, K., Clarke, J., Henry, J., Halpern, J.,
    and Grindlay, J. 1982, Ap.J.(Letters), 253, L67.
White, N. and Swank, J. 1982, Ap.J.(Letters), 253, L61.
Webbink, R., Rappaport, S., and Savonije, G. 1983, preprint.

NOTES

----

1. Alfred P. Sloan Foundation Fellow.
2. Now at the Naval Research Laboratory, Space Science Division,
    Washington, D.C. 20375.

GENERAL DISCUSSION ON LOW-MASS X-RAY BINARIES

LEWIN: Josh, you mentioned that the most convincing argument for
identifying the low-mass X-ray binaries as such was your own work in
determining the locations of the globular cluster X-ray sources. I
believe, in all fairness to a lot of people in the audience, it should
be stressed that way before the launch of Einstein it was blatantly
clear that these X-ray sources were low-mass X-ray binaries. The neutron
star character was beyond a shadow of a doubt: the radii were measured
from the X-ray bursts; the transient sources allowed us to measure the
mass of the companion star, and they were clearly seen to be low-mass
companions; periods of a few hours were measured. I believe that when
the first results from Einstein came out, important as those measurements
were, and when the rumors were spread that those objects were very
massive, most of us said, "Well, Josh will have to go back and do his
homework." As you did, and, indeed, you came out with 2 $M_\odot$. It was an
important confirmation, but don't blow it out of proportion.

GRINDLAY: Let me respond. I didn't intend to blow it out of proportion.
I thought I mentioned that there were all of these independent arguments
and that they did precede the Einstein results. I agree completely. My
point was that I think it is still fair to say that this was a quantitative estimate of the mass. All of these arguments made it clear that
these are low-mass objects. But I don't recall ever seeing a number
like 2 $\pm$ 1, or any other number, for the total mass of the system.

LEWIN: . . . OK. I want to make another point, about the picture in
which these low-mass X-ray binaries come from globular clusters. This
is an appealing idea because we have problems in forming them, as we
discussed last night. We have to explain why the distribution of these
low-mass X-ray binaries, which Jeff showed on one of his viewgraphs, is
quite different from that of globular clusters. But maybe there's a
way we can get over this. It would be nice if someone could point out
why the distributions are so different. The distribution of bulge
sources is really very pancake-shaped.

GRINDLAY: I mentioned that point on one of the viewgraphs, but there
wasn't time to talk about it. I think one possible way to disrupt these
clusters (if it's the disruption picture rather than the ejection picture)
would be for them to have a somewhat flat primordial distribution. We'd
want them to spend a lot of time in the galactic plane, where they
encounter giant molecular clouds, for example. Such an encounter will
cause a very nasty perturbation, since 10^6 $M_\odot$ in a big fuzzy cloud will
do bad things to halo stars and cluster stars. So I think to explain
the distribution of bulge sources, one might--and this is certainly less
than satisfying--invoke a primordial globular cluster distribution that
is more flattened than the surviving clusters we see today. Alternatively, and more plausible perhaps, the initially isotropic distribution
(about the galactic center) of globular cluster orbits in the proto-
galaxy might have had all orbits with low inclinations (say $\lesssim 10°$ from
the galactic equator) depleted. That is, clusters initially on these

D. Q. Lamb and J. Patterson (eds.), Cataclysmic Variables and Low-Mass X-Ray Binaries, 93–97.
© *1985 by D. Reidel Publishing Company.*

orbits may now have been largely disrupted since they spend their time
in the galactic plane. But it may have something to do with other
interesting effects in the cluster distribution; for example, metallicity
gradients which would suggest more stellar evolution and neutron star
remnants towards the bulge.

KROLIK: I would like to ask the observers whether it isn't true that
the presently observed bright globular cluster X-ray sources are in
clusters which appear not to be a fair sample of the total set of
globular clusters. Is the distribution of globular cluster X-ray
sources closer to the distribution of the galactic bulge X-ray sources
than the distribution of globular clusters as a whole?

GRINDLAY: Alan Lightman and I looked at that, you may recall, and the
conclusion we could make for the high luminosity sources was that indeed
there is a significant correlation with galactocentric radius, and in
that sense there may be a connection. But we could not demonstrate
that the X-ray globulars had a flattened distribution in scale height,
whereas the bulge sources do. So, the answer is half and half.

KATZ: I just want to inject a note of caution on the correlations. I'm
less concerned with galactocentric distance than with some correlations
that were pointed out back when it was popular to think the cluster
X-ray sources were very massive black holes, and people worried about
correlations with relaxation times. We want to be a bit careful to
consider other explanations, the most obvious of which is that the true
correlation is with the mass of the cluster. The more stars a cluster
has, the more likely it has an X-ray source. They range over quite a
bit in mass. And mass is correlated with most of these other variables,
and on top of it all, is very poorly known, except for one cluster.
Just a note of caution.

GRINDLAY: Good point. A very quick reply, if I may. The correlation
with central density does seem to be significant, and if one accepts a
constant mass/light ratio and uses the total luminosity, there does not
seem to be a striking correlation with mass.

PATTERSON: I wanted to ask Jeff McClintock if the greater incidence of
X-ray binaries in the bulge of M31 could be due to absorption in our
galaxy.

McCLINTOCK: It seems most unlikely to me. The visual extinction to the
galactic center is only supposed to be about 15-20 magnitudes. That
doesn't correspond to a very large X-ray cutoff energy, only a few keV.
But maybe.

LEWIN: This has been very carefully looked into, and your answer is
absolutely correct.

van PARADIJS: Josh, I would like to point out a possible problem with
your idea that the G stars close to these faint stars might be G giants,

the leftovers of globular clusters. If you look at the typical optical luminosities of optical counterparts of bright low-mass X-ray binaries, they're comparable to those of G giants. Now in three out of the four G giant-faint star associations, there is a luminosity difference of about 5 magnitudes. Doesn't that create a problem with having bright X-ray sources, which might be expected to produce bright accretion disks?

LEWIN: Could you rephrase the question? I don't understand it, and I would like to follow this.

GRINDLAY: I think that what Jan is saying is that if you take the G star itself as the optical counterpart of GX 17, for example, L_x/L_{opt} is a few hundred, which is already similar to that of the galactic bulge sources.

van PARADIJS: No, no, that's not the point. I thought your suggestion was that the faint star was the optical counterpart and the G star was just a G giant hanging around because they both came from the same original globular cluster.

GRINDLAY: Right.

van PARADIJS: Now the G giant has an absolute magnitude of about 0 or +1, which is about the same as the absolute magnitude that one finds for optical counterparts of bursters, bulge sources, etc. So one would expect that that faint star would be about the same magnitude as the G star we see, not 5 magnitudes fainter.

LEWIN: That's an excellent point.

GRINDLAY: If the absolute magnitude is indeed +1. . . .

van PARADIJS: Five magnitudes is a large difference.

GRINDLAY: They're not as much as 5 magnitudes.

van PARADIJS: Well, three out of four.

GRINDLAY: Three magnitudes is a more typical number. Three or four magnitudes might be OK in any case, since there may be a significant reddening effect. That is, the actual optical counterpart (= accretion disk) will be much bluer, and therefore more heavily reddened [$E(B-V) \sim$ 2-3], than the G star.

[Ed: An inaudible discussion of individual sources follows.]

LEWIN: Could that G star be not a regular G star, but something exceptional?

[Ed: Twelve people speak at once.]

GRINDLAY: No, it would have to be a giant in a cluster core, and presumably something like the individual giants we see in the core of
NGC 6624.

McCLINTOCK: Early on, it was claimed to be a G dwarf, only about 1-1.5
kpc away. What's happened? Has anyone done a careful spectral analysis?

GRINDLAY: I don't think anyone has gotten a proper high-resolution
spectrum to determine luminosity class. It's an obvious thing to do,
and we plan to do it.

THORSTENSEN: Is there any evidence from colors that says whether these
"companion" stars are at the same distance or not?

GRINDLAY: I don't think so. The only other handle on the reddening is
the low-energy cutoff seen in X-rays. In the case of GX 17, the numbers
($E_a \sim 1.5$ keV, $N_H \sim 10^{22}$ cm^{-2}, $A_V \sim 6-7$ magnitudes) might be made consistent with the G star, but would push the spectral type of the G star
to be early G and probably would make it more like a horizontal-branch
star. Again, obtaining a higher quality spectrum is obviously important.

FAULKNER: A question for Josh: I've been rather inclined to favor the
disruption picture over the ejection picture, for the latter seems to be
a case of wanting "to have your cake and throw it up too."

GRINDLAY: I agree.

FAULKNER: Because you want to form the object in the core, and yet it's
been pointed out that it's easier to eject things from the envelope.
The trouble is, the place where they naturally form is not the place
from which they're easily ejected.

GRINDLAY: And along with that is the quick remark I made that we do see
a few of what we think are low-mass systems out in the halo, where they
would belong. But the high luminosity systems are 2 $M_\odot$, and it would be
difficult to get them out where they can be plucked off.

McCLINTOCK: I would like to ask John, do you have any more comments to
make on your observations of 0921-63? The orbital period, for example?

THORSTENSEN: Yes. The period is likely to be correct, but it's not as
firm as has been advertised because it turns out that the photometric
sampling tends to preferentially give you periods near 9 days. Also, if
you do an unbiased period search with all the published and the new
velocities (except for the very latest ones from Graziella Branduardi),
it by no means unambiguously selects a 9-day period. So the 9-day period
is still unconfirmed, although it's likely to be right. Another interesting thing is that the eclipses, which have been widely touted, don't
phase very well. They seem to cluster in phase, but the phasing is not
very nice. If you demand that there be an eclipse and just force a fit,
you get about a 16-hour period, which is very, very short, considering the

size of the primary. Maybe it's just a grazing eclipse. Also, the L_x/L_{opt} ratio is very problematical, considering that the spectrum looks like that of a bulge source almost all the time.

WHEELER: I've just a comment on something McClintock said. You said with justifiable chauvinism that you probably know the accretion rate onto neutron stars better than people know it onto white dwarfs, but then you have to worry about what physics you want to conclude from that. Especially in the context of transient accretion phenomena, you want to bear in mind that the mass rate coming off the companion star is not necessarily the mass rate going through the disk, and the mass rate going through the disk--if there's mass loss--isn't the mass rate arriving at the compact object.

McCLINTOCK: Yes, I think the caution is most appropriate for a system like A0620-00. Who knows if the disk is building up for tens of years, or whether the thing is episodic, or what?

X-RAY LUMINOUS WHITE DWARF BINARIES IN GLOBULAR CLUSTERS

Paul Hertz
Harvard-Smithsonian Center for Astrophysics

An x-ray survey of galactic globular clusters has been conducted. An apparently distinct class of x-ray sources has been discovered. They are identified as accreting white dwarfs in close binaries. Implications concern the binary formation rate in globular clusters and the number of compact objects present.

THE EINSTEIN GLOBULAR CLUSTER SURVEY

We have conducted a survey of galactic globular clusters with the HRI and IPC instruments on board the Einstein X-ray Observatory [1]. A total of 71 globular clusters were observed [2]. The median luminosity limit of our survey was $\sim 10^{33.5}$ (all luminosities are in erg s^{-1} in the 0.5–4.5 keV band of the Einstein IPC); several clusters were searched down to $\sim 10^{32}$.

EINSTEIN GLOBULAR CLUSTER SURVEY

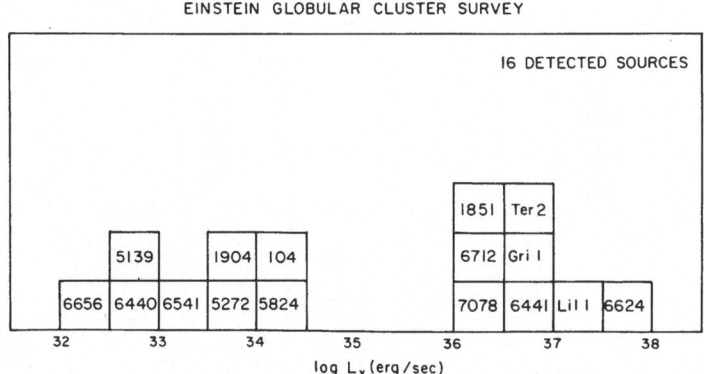

Figure 1

99

D. Q. Lamb and J. Patterson (eds.), Cataclysmic Variables and Low-Mass X-Ray Binaries, 99–102.
© *1985 by D. Reidel Publishing Company.*

Fourteen new sources were discovered in the survey; all of these sources had $L_x \lesssim 10^{34.5}$. We also observed the eight known bright sources with $L_x \gtrsim 10^{36}$. Upper limits were obtained for x-ray emission from the 55 remaining clusters. The two clusters ω Cen (NGC 5139) and M22 (NGC 6656) were observed for more than 22 kiloseconds each. Because both clusters have core radii larger than the ~ 1' resolution of the IPC, we hoped to detect multiple sources in these clusters. In fact, we detected 5 point like x-ray sources in NGC 5139 and 3 in NGC 6656. This is the first detection of more than one x-ray source in a globular cluster. We expect most clusters to have multiple sources (see below), although they were only resolved in two clusters due to selection effects.

We present a histogram of the globular cluster x-ray sources discovered in our survey in Figure 1.

THE LUMINOSITY FUNCTION

We have used the DB (detections and bounds) technique of Avni et al. [3] to construct the maximum likelihood luminosity function. We have made use of our 16 clusters detected, the 55 upper limits from Einstein data, and upper limits on x-ray emission from an additional 61 globular clusters from the Uhuru all sky survey [4]. The DB method is used because it gives a non-parametric maximum likelihood luminosity function using all available data. We present this luminosity function in Figure 2. Note the gap of 1.5 decades in luminosity between the

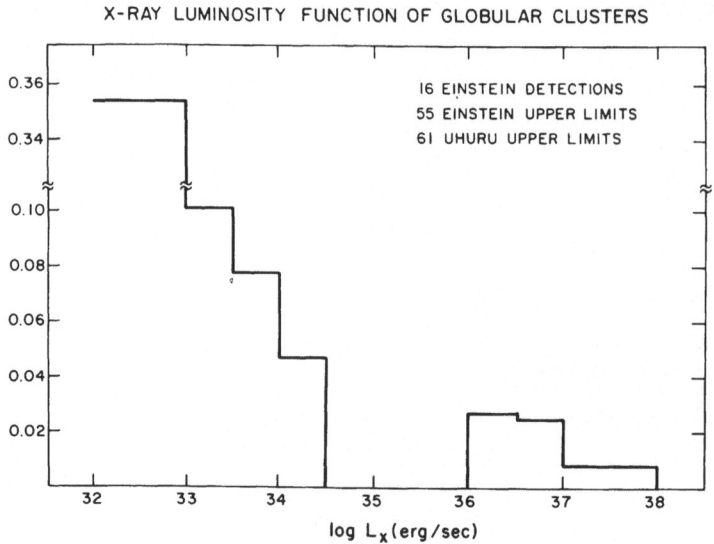

Figure 2

bright and dim sources, as well as the steeply rising function at low luminosities. Fortunately, the DB method allows us to quantify these features within the framework of likelihood analysis.

The distribution is clearly bimodal. The 90% confidence upper limit on globular cluster x-ray sources in the gap is $\lesssim$ .008 of the clusters in the galaxy per .5 decade luminosity bin. Since the .5 decade luminosity bins on either side of the gap contain .027 and .047 of the globular cluster x-ray sources, the gap is obviously significant. A Kolmogorv-Smirnov test rules out a smooth luminosity function at the 95% level of confidence.

There is no low luminosity cutoff apparent. Assuming that the luminosity function is a power law, a formal fit to the slope for $L_x \lesssim 10^{34.5}$ yields $\alpha = 1.25$ with a 90% confidence region ± 1.00. Using the DB method, we determined that the luminosity function continues to rise at low luminosities (98% level of confidence).

IDENTIFICATION OF LOW LUMINOSITY SOURCES

The bright globular cluster x-ray sources have been identified as low mass x-ray binaries, containing an accreting neutron star and a low mass secondary. The principle pieces of evidence are their x-ray similarity to optically identified low mass x-ray binaries [5], the fact that seven out of eight have been observed to burst [5], and the determination from a statistical analysis that their mass is 1-3 M_o [6].

The bimodal luminosity function indicates that there are two separate classes of globular cluster x-ray sources. We have identified the dim sources as accreting white dwarfs in compact binary systems [7]. Several reasons for this identification are:

1. The difference in the maximum luminosity of the bright sources and the dim sources is a factor of 10^3. This is easily understood as being due to the factor of $\sim 10^3$ difference in the gravitational potential of a white dwarf and a neutron star.
2. The maximum luminosity observed, $\sim 10^{34.5}$, is the maximum predicted by theoretical modeling of accretion onto a low mass white dwarf [8].
3. The distribution of projected radial offset of dim sources indicates that the ratio of the x-ray source mass to the mass of a mean globular cluster member is $q \lesssim 1.7$ (90% confidence limit [7]). This is significantly less than the $q \sim 2-5$ (90% confidence limit [6]) derived for the bright sources and is consistent with the expected mass of a binary system containing a white dwarf and a low mass companion.
4. The number of sources observed, and the maximum luminosity observed, is consistent with a single power law luminosity function describing the x-ray emission from both galactic plane cataclysmic variables [9] and dim globular cluster x-ray sources [2,7].

5. We assume that white dwarf binaries (and neutron star binaries) form in globular cluster cores via the tidal capture mechanism [10]. The total number of x-ray binaries with $L_x \gtrsim 10^{31}$ implied by the luminosity function (Figure 2), and their expected x-ray lifetime of $\sim 10^9$ years [11], is consistent with the expected capture rate in a canonical globular cluster [12] if $\sim 10\%$ of the stars in the cluster are white dwarfs. This is consistent with globular cluster luminosity functions [13].

6. Since white dwarfs and neutron stars have similar capture rates in globular cluster cores [12] and similar x-ray lifetimes, the observed ratio of bright globular cluster x-ray sources to dim globular cluster x-ray sources of 1 to 20 implies that the ratio of neutron stars to white dwarfs is also 1 to 20. This is consistent with globular cluster mass functions [14].

CONCLUSIONS

We have discovered a new class of globular cluster x-ray sources. These sources, with x-ray luminosities significantly less than the low mass x-ray binaries in globular clusters, are identified as accreting white dwarfs in binary systems. They are thus physically similar to cataclysmic variables; however their x-ray luminosity is 100 to 1000 times greater than typical cataclysmic variables.

I would like to thank my collaborator in the Globular Cluster Survey, Jonathan Grindlay, with whom this project was carried out. Figures 1 and 2 will appear in the Astrophysical Journal.

REFERENCES

[1] Giacconi,R. et al. 1979, Astrophys.J. 230, pp.540.
[2] Hertz,P. and Grindlay,J.E. 1983, submitted to Astrophys.J..
[3] Avni,Y., Soltan,A., Tananbaum,H., and Zamorani,G. 1980, Astrophys.J. 238, pp.800.
[4] Forman,W. et al. 1978, Astrophys.J.Suppl. 38, pp.357.
[5] Lewin,W.H.G. and Joss,P.C. 1981, Space Science Rev. 28, pp.3.
[6] Grindlay,J. et al. 1983, in preparation.
[7] Hertz,P. and Grindlay,J.E. 1983, Astrophys.J., in press.
[8] Kylafis,N.D. and Lamb,D.Q. 1979, Astrophys.J. 228, pp.L105.
[9] Cordova,F.A. and Mason,K.O. 1983, in 'Accretion Driven Stellar X-ray Sources', ed. W.H.G.Lewin and P.J.v.d.Heuvel (Cambridge U. Press).
[10] Fabian,A.C., Pringle,J.E., and Rees,M.J. 1975, Mon.Not.R.astr.Soc. 172, pp.15P.
[11] Lightman,A.P. and Grindlay,J.E. 1982, Astrophys.J. 262, pp.145.
[12] Press,W.H. and Teukolsky,S.A. 1977, Astrophys.J. 213, pp.183.
[13] DaCosta,G.S. and Freeman,K.C. 1976, Astrophys.J. 206, pp.128.
[14] Illingworth,G. and King,I.R. 1977, Astrophys.J. 218, pp.L109.

THREE-BODY INTERACTIONS AND CATACLYSMIC BINARIES IN GLOBULAR CLUSTERS

Piet Hut
Institute for Advanced Study, Princeton, NJ 08540, USA
Frank Verbunt
Institute of Astronomy, Cambridge, U.K.

The number density of stars in globular clusters is so large that close encounters between a single compact star and a field star occur frequently [1]. Using newly determined, accurate cross sections for gravitational three body scattering [2,3], we point out that encounters between single compact stars and binaries are of comparable importance, if the number of binaries is not too small [cf. 4]. Both processes predict the formation of a large number of cataclysmic variables [5], in analogy with earlier models for the formation of low mass bright X-ray sources (LMBX's) in globular clusters [4,6]. This explains the overabundance of cataclysmic variables in globular clusters which was indicated by the number of optical novae [7] and has been confirmed by the detection of a dozen cataclysmic variables by the EINSTEIN X-ray satellite [8]. We show that the ratio of the formation rates of cataclysmic binaries and LMBX's is appreciably smaller than the ratio of the number of white dwarfs to the number of neutron stars. A more detailed description of our work is given in [5].

The rate of binary formation through tidal capture of a compact star by a field star is given by [5, cf.6]:

$$R_{(1+1\rightarrow2)} =$$

$$6\times10^{-11} \left[\frac{N}{10^2\,\text{pc}^{-3}}\right] \left[\frac{n}{10^4\,\text{pc}^{-3}}\right] \left[\frac{M+m}{M_\odot}\right] \left[\frac{xR}{R_\odot}\right] \left[\frac{10\,\text{km}\,\text{s}^{-1}}{v_{in}}\right] \text{yr}^{-1}\,\text{pc}^{-3} \tag{1}$$

For applications to globular clusters $x \approx 3$ [6]. The rate is determined by the density of compact stars N, the density of field stars n, and - due to gravitational focussing - by the masses M and m of the compact and field stars, by their relative velocity v_{in}, and the radius R of the field star.

Three-body interactions between a binary and a compact star occur in a number of varieties (fig.1). Resonance scattering, in which the three stars form a temporarily bound three-body system with a typical life time of some hundred original binary periods, dominates over direct exchange in encounters with hard

D. Q. Lamb and J. Patterson (eds.), Cataclysmic Variables and Low-Mass X-Ray Binaries, 103–106.
© *1985 by D. Reidel Publishing Company.*

Fig. 1. Scattering between a binary (oo) and a compact star (•) can lead to a fly-by, where the binary keeps its identity; to an exchange, where one binary member joins the compact star; or to ionization, giving rise to three independently moving stars. Fly-by and exchange can either occur directly (as in the first two diagrams), or through an intermediate stage where a temporarily bound triple system is formed. This last process can only occur when the total energy is smaller than zero, and is called a resonance scattering.

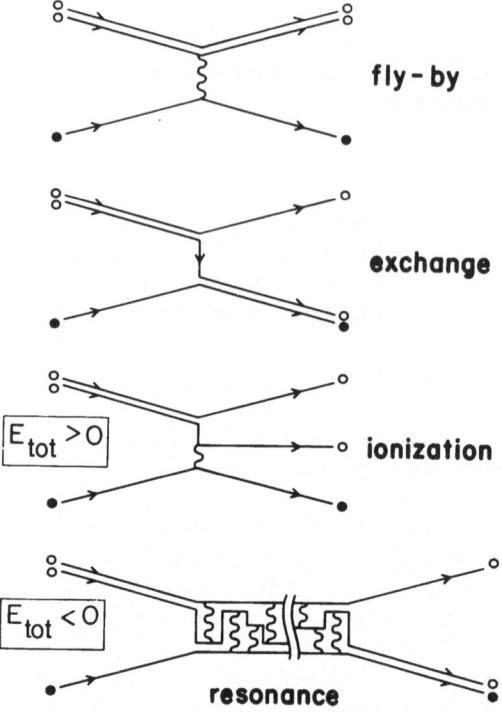

binaries [3]. The total rate of exchange scattering with a compact star is given by [3]:

$$R_{(1+2\to2+1)}(M=m) =$$

$$4.3 \times 10^{-10} \left[\frac{N}{10^2\,\mathrm{pc}^{-3}}\right] \left[\frac{n_{bin}}{10^2\,\mathrm{pc}^{-3}}\right] \left[\frac{m}{M_\odot}\right] \left[\frac{a}{1\,\mathrm{AU}}\right] \left[\frac{10\,\mathrm{km\,s}^{-1}}{v_{in}}\right] \mathrm{yr}^{-1}\,\mathrm{pc}^{-3} \qquad (2)$$

Here a the semimajor axis of the original binary and n_{bin} is the number of binaries per unit volume; This result is accurate to within 10% at the 3σ level, for an original binary orbital eccentricity of $e = 0.7$, and $v = 0.1\,v_{orb}$, where v_{orb} is the orbital velocity of the binary. As long as tidal forces are unimportant, this is an appropriate average eccentricity value since the equilibrium distribution function for binary eccentricities is $f(e) \propto e$. The dependence on v_{in} is largely determined by the gravitational focussing factor v_{in}^{-1}; the residual dependence on v_{in} is much smaller, and contributes less than a factor two over the whole range $0.001 < v_{in} / v_{orb} < 1.0$ [3].

When the compact star is heavier than the binary members, as will be the case for a neutron star coming in at a binary consisting of main sequence stars, the above exchange rate will increase significantly. For example, for a neutron star mass $M = 2m$, where again m is the mass of each of the normal stars, numerical experiments give a rate similar to Eq. (2), but a factor 2.6 larger than in the equal mass case [3] (the coefficient becomes 11×10^{-10}).

An important difference between the two-body and three-body capture processes is the larger capture cross section of a binary, with its semi- major axis as a scale length compared with a single star, where the stellar radius sets the scale. Thus for three-body interactions, a relatively wide encounter with the binary can lead to a resonance scattering resulting in exchange, which explains the difference between the rates (1) and (2). Also, instead of the number of field stars, now the number of binaries enters. This is the great unknown in globular clusters, although binaries seem to be at least somewhat underabundant in globular clusters [9]. However, even if only one percent of the stars in a globular cluster resides in binaries with the appropriate range of semi-major axes (0.02 - 1 AU) and containing a star near the turn-off point, resonance exchange is of equal importance as two-body tidal interactions. Binaries containing lighter stars can produce CV - progenitors as efficiently, but the wider orbits will cause the CV - stage to be delayed untill the star becomes a red giant. Those systems therefore will be visible for us only if the orbit after the exchange scattering has become sufficiently narrow to start mass - transfer within a Hubble-time while still on the main sequence. This increases the number of binaries needed for these lower mass range to about 10%, a number which is still significantly lower than the abundance of similar binaries in the galactic disk, and not in conflict with observations of globular clusters.

From Eq. (1) we can determine the ratio of the formation rate of CV's and the formation rate of LMBX's for two-body interactions:

$$\frac{R_{(1+1 \to 2);wd}}{R_{(1+1 \to 2);ns}} = \frac{M_{wd} + m}{M_{ns} + m} \frac{v_{in,ns}}{v_{in,wd}} \frac{N_{wd}}{N_{ns}} . \tag{3}$$

The two-body capture process favours neutron stars over white dwarfs because neutron stars have 1) larger masses and hence 2) smaller velocities, and 3) a stronger concentration to the core of the cluster, where most encounters occur. Also 4) neutron stars are older than white dwarfs, and have had more time to interact with a field star.

For the presence of single neutron stars in a globular cluster it is important that the kick velocity that a neutron star gets in the supernova event is not too high [10]. We interpret the velocity distribution of radiopulsars [11] to give a strong indication that a reasonable fraction of neutron stars is indeed born with a velocity smaller than the escape velocity of a globular cluster. The escape of a sizable fraction of neutron stars from the cluster would counteract the four abovementioned effects. As is obvious from Eq. (2) these effects operate in three-body resonance scattering as well. In a resonance scattering event the possibility exists that the compact star is the one that is thrown out. Hence the ratio of the formation rates of LMBX's and CV's is even higher for resonance scattering than for two-body tidal interactions, since 5) a massive star is less likely to be thrown out after resonance, 6) the binaries will be more centrally concentrated together with the neutron stars and therefore interact more frequently, and 7) a massive binary will have a smaller recoil velocity after resonance scattering, and is less likely to be thrown out of the cluster.

For an initial mass function $\frac{dN}{dm} \propto m^{-x}$ the ratio of white dwarfs to neutron stars is $\frac{1}{f}\left[\frac{0.8\,M_\odot}{8\,M_\odot}\right]^{1-x} = 10^{x-1}$, when white dwarfs have been formed from stars with masses between a turn-off mass of $0.8\,M_\odot$ and the minimum mass of neutron star formation of $8\,M_\odot$. Here f is the fraction of neutron stars that remain in the cluster upon formation. We then estimate the ratio of CV's to LMBX's in globular clusters as:

$$\frac{N_{bin,wd}}{N_{bin,ns}} \approx \frac{1}{f}10^{x-2} \tag{4a}$$

for two-body capture, and

$$\frac{N_{bin,wd}}{N_{bin,ns}} \approx \frac{1}{f}10^{x-2.5} \tag{4b}$$

for three-body capture.

Finally, we mention two observational consequences of the importance of three-body interactions in globular clusters. First, during a resonance scattering actual physical collisions between stars can occur (the typical distance of closest approach is of order a few percent of the original binary semimajor axis [3]). This could be one of the processes that give rise to blue stragglers. Secondly, we would expect CV's to be spread over the globular clusters, due to point 7) above, as is indeed observed both for the optically and for the X-ray discovered CV's.

REFERENCES

[1] Clark, G.W. 1975, *Ap.J.* **199**, L143.
[2] Hut, P. and Bahcall, J.N. 1983 (May), *Ap. J.* **268**.
[3] Hut, P. 1983, in preparation.
[4] Hills, J.G. 1976, *Mon. Not. R. astr. Soc.* **175**, 1p.
[5] Hut, P. and Verbunt, F. 1983, *Nature* **301**, 587.
     Hut, P. and Verbunt, F. 1983, in preparation.
[6] Fabian, A.C., Pringle, J.E. and Rees, M.J. 1975, *Mon. Not. R. astr. Soc.* **172**, 15p.
[7] Webbink, R.F. 1980, in *Close binary stars: Observations and interpretation*, IAU Symp.88, M.J.Plavec, D.M.Popper, and R.K.Ulrich (eds.), Reidel, p.561.
[8] Hertz, P. and Grindlay, J. 1983, submitted to *Ap.J.*
[9] Gunn, J.E. and Griffin, R.F. 1979, *Astron. J.* **84**, 752.
[10] Katz, J.I. 1975, *Nature* **253**, 698.
[11] Manchester, R.N. and Taylor, J.H. 1981, *Astron. J.* **86**, 1953;
     Lyne, A.G., Anderson, B. and Salter, M.J. 1982, *Mon. Not. R. astr. Soc.* **201**, 503.

THE EVOLUTION OF HIGHLY COMPACT BINARIES IN GLOBULAR CLUSTERS

Julian H. Krolik and Avery Meiksin
Harvard-Smithsonian Center for Astrophysics

Paul C. Joss
Department of Physics and Center for Space Research
Massachsetts Institute of Technology

ABSTRACT

We report on detailed calculations of the secular evolution, observational appearance, and numbers of highly compact binaries in globular clusters. Such binaries are subject to the combined influences of gravitational radiation, thermal evolution of the secondary star, gradual mass transfer from the secondary to the collapsed object primary when the system is in contact, and occasional gravitational encounters with field stars, some of which cause rapid mass transfer from the secondary to the primary. Some gravitational encounters with field stars end in direct physical collisions which can produce massive disks circling the collapsed primary. We present examples of the sorts of histories that are possible for highly compact binaries in globular clusters, and calculate the rate at which mass transfer is induced by gravitational encounters. We then describe the observational appearance of these systems, and report on calculations of their expected numbers. We also briefly discuss the relationship of highly compact binaries in globular clusters to the bright X-ray sources of the Galactic bulge.

I. INTRODUCTION

The term "highly compact binaries" is meant to encompass a broad range of close binaries in which one component is a low-mass main sequence star and the other is a collapsed stellar remnant (i.e., a degenerate dwarf, a neutron star, or a black hole). We restrict our attention today to highly compact binaries in globular clusters because we wish to explore the ways an environment of high stellar density can deflect the evolution of these binaries away from the customary path followed in the Galactic disk or elsewhere in the Galactic bulge.

Our investigation was originally motivated by the striking contrast in frequency of occurrence of highly compact binaries inside and outside of globular clusters. Katz (1975) and Clark (1975) pointed out that bright X-ray sources, most of which are now almost universally thought to be highly compact binaries in which mass is being transferred to a collapsed primary (see Joss 1981, Grindlay 1985), are more than a hundred times more common, relative to the number of neighboring ordinary stars, in globular clusters than they are in the rest of the Galactic bulge, while the ratio relative to the Galactic disk is even greater. Webbink (1980) has pointed out that classical novae, generally thought to

107

D. Q. Lamb and J. Patterson (eds.), Cataclysmic Variables and Low-Mass X-Ray Binaries, 107–116.
© *1985 by D. Reidel Publishing Company.*

be highly compact binaries in which a degenerate dwarf is the recipient of mass transfer, are substantially more common, again relative to the number of nearby ordinary stars, in globular clusters than in the Galactic disk. Tidal capture theory (Fabian, Pringle, and Rees 1975) has provided a partial answer to this question of frequency by pointing to the importance of high stellar densities in creating close binaries; we shall extend that answer here to show how the high stellar densities found in globular clusters also promote mass transfer. In fact, we shall give quantitative predictions for the total numbers of several varieties of highly compact binaries that should be found in globular clusters, including bright X-ray sources, cataclysmic variables, and a new type of system in which the main-sequence companion has been wholly or partially dissolved into a massive accretion disk surrounding the collapsed stellar remnant.

In the course of this discussion, we shall also touch on other perplexing questions sparked by the observed properties of highly compact binaries in globular clusters: The standard theory for the evolution of the bright X-ray sources, in which mass transfer is driven by gravitational radiation (Paczyński 1967; Faulkner 1971; Paczyński and Sienkiewicz 1981; Rappaport, Joss, and Webbink 1982), falls short by two orders of magnitude in its prediction of the mass transfer rate–what, then, does drive the mass transfer? Why have these systems, ostensibly close binaries, never been seen to eclipse? Why do the Galactic bulge X-ray sources, whether inside or outside of globular clusters, resemble each other so strongly, despite the disparities of stellar population?

How is it that the stellar environment in a globular cluster can strongly influence the evolution of a binary whose orbital separation is 10^{-6} times smaller than the typical interstellar separation in the cluster? The answer, in a word, is gravitational encounters. Gravitational encounters with a periastron (measured with respect to the binary center of mass) of 2×10^{12}cm or less (roughly ten or twenty times the semimajor axis of the sort of binary we are considering) can cause major changes to the binary orbit, and occur at a rate of $(6 \times 10^9 \text{yr})^{-1} \frac{n_*}{10^5 pc^{-3}}$. Thus, in clusters with stellar densities greater than 10^5pc^{-3}, almost no binary survives for a Hubble time without suffering a major gravitational encounter. This central density is exceeded in fully a quarter of the known globular clusters of our Galaxy [Peterson and King (1975) contains by far the largest collection of data, but we also use Bahcall (1976), Bahcall and Hausman (1976), and Malkan, Kleinmann, and Apt (1980); we converted the observed core surface brightnesses to stellar number density by assuming a mass-to-light ratio of $1 M_\odot / L_\odot$ and a mean stellar mass of $0.5 M_\odot$].

Many people have studied the effects on a binary's orbit due to gravitational interactions with other stars treated as point-masses (see especially Heggie 1975, Hills 1975, Bahcall and Hut 1983). They all agree that, on average, gravitational encounters between binaries and field stars whose velocity at infinity is much smaller than the binary orbital velocity tend to take energy out of the binary, decreasing its semimajor axis. Consequently, binaries created out of contact are brought into a state of mass transfer sooner than in the absence of gravitational encounters.

What has not been generally recognized is the relatively high probability of encounters in which non-point mass effects are important. Fabian, Pringle, and Rees (1975) in their study of tidal capture pointed out that in a significant fraction of captures there would be a direct physical collision between the two stars. Similar "catastrophes" can occur in encounters between a binary and a single star. Sometimes a catastrophe can occur even without a physical collision if the main-sequence companion left bound to the collapsed object after the encounter is over is in an orbit with such a small periastron distance that it substantially overhangs its critical potential lobe. The results of such catastrophes are difficult to evaluate with any confidence, but plausible arguments suggest that a massive (~ 0.1–$1 M_\odot$) disk will form around the collapsed object, whose rate of accretion may be capped only by the Eddington limit.

II. THE CALCULATION

The claims we have just made are based on extensive, detailed calculations (Krolik, Meiksin, and Joss 1983). The initial state of our model system was a binary comprising a $1 M_\odot$ collapsed primary and a $0.4 M_\odot$ main-sequence star in a circular orbit of radius 2×10^{11} cm. The evolution of this binary exclusive of gravitational scattering was traced by the same code developed by Rappaport, Joss, and Webbink (1982), in which the structure of the secondary is approximated by an $n = 3/2$ polytrope, and the orbital evolution is due to gravitational radiation and the concomitant mass transfer from secondary to primary. The thermal content of the secondary evolves due to the imbalance between heat-producing nuclear reactions and heat losses by radiation and mass transfer. Field stars with speeds at infinity of 10km s^{-1} and masses of $0.4 M_\odot$ rained down on this binary with fluxes appropriate to a stellar density of either 10^5 pc^{-3} (in one set of runs) or 3×10^5 pc^{-3} (in the other). The other free parameters of the encounters (phase angles, etc.) were chosen randomly, but the detailed trajectories were calculated exactly with a three-body integrator. A particularly pretty example of an encounter trajectory–which ends in a physical collision between the collapsed primary and the field star–is shown in Figure 1. The life of each binary was followed until either a catastrophe occurred or the elapsed time reached 2×10^{10} yr. Twenty such binary histories were constructed for each stellar density.

In conventional calculations of stellar evolution (except for the later stages of the lives of very massive stars), it is rare to find events occurring on any timescale shorter than millions of years. However, the characteristic timescales of gravitational encounters are so much shorter that it is necessary to take account of a number of unusual phenomena. Immediately after an encounter, it is likely that the new binary orbit has a finite eccentricity and that the secondary's rotation is not synchronized with the orbital period. Dissipation in the secondary can eliminate the eccentricity and non-synchronicity in a time which is very short compared to ordinary stellar evolutionary timescales but very long compared to the orbital period (Zahn 1975, 1977; Lecar, Wheeler, and McKee 1976; Hut 1982). In the meanwhile, if the periastron distance of the new orbit is so small that the secondary overhangs its critical potential lobe, there can be rapid mass transfer from the secondary to the vicinity of the primary. This transfer takes place on the orbital timescale, so there is no opportunity for dissipative processes, either in the star or in the newly-forming disk, to affect the binary orbit. However, tidal forces do exchange energy and angular momentum between the orbit and the mass-transfer stream. We have explicitly calculated the rate of exchange of energy and angular momentum per unit of transferred mass as a function of the orbital parameters so that we can follow the continuous adjustment of the orbit in response to the mass transfer. These details determine whether the secondary is completely, or only partially, consumed by this rapid mass transfer.

Figure 2 shows a typical history taken from the high-density set. At first, gravitational radiation slowly compresses the orbit, with no mass exchange taking place. Then a very close encounter occurs, substantially compressing the orbit. A short time later another encounter leaves the companion so precariously overhanging its critical potential lobe that

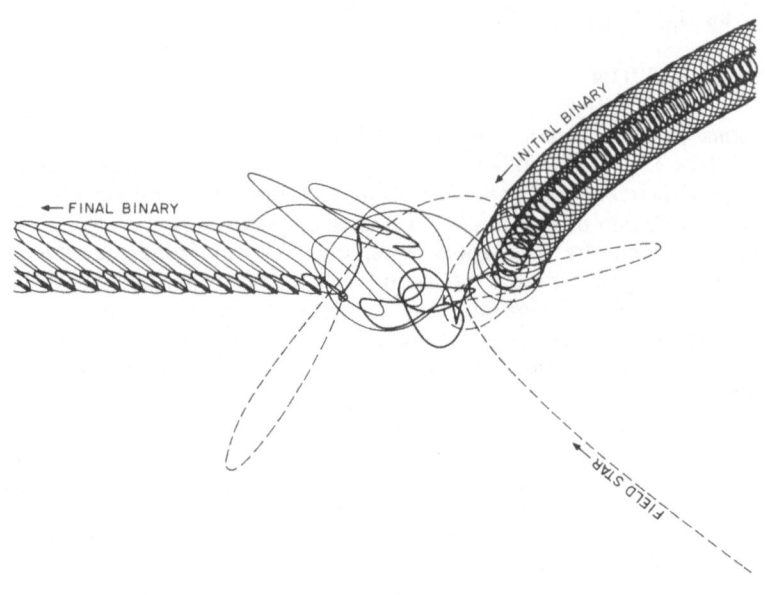

Figure 1.   The motion of the initial binary and the field star, projected onto the orbital
plane of the original binary, just prior and subsequent to the collision between
the collapsed star and the field star that terminates one of the runs in the
high-density ($n_* = 3 \times 10^5 pc^{-3}$) case. The original binary enters the figure
from the lower right, and the field star from the lower left. The collapsed
object and field star physically collide at the point marked $\otimes$. The complex
interaction among the three stars prior to the physical collision is rather
typical among encounters of this general type (see Bahcall and Hut 1982).
Following the collision, the new binary, with the original secondary star and
a new primary composed of the collapsed object and the debris of the field
star, exits at the top of the figure.

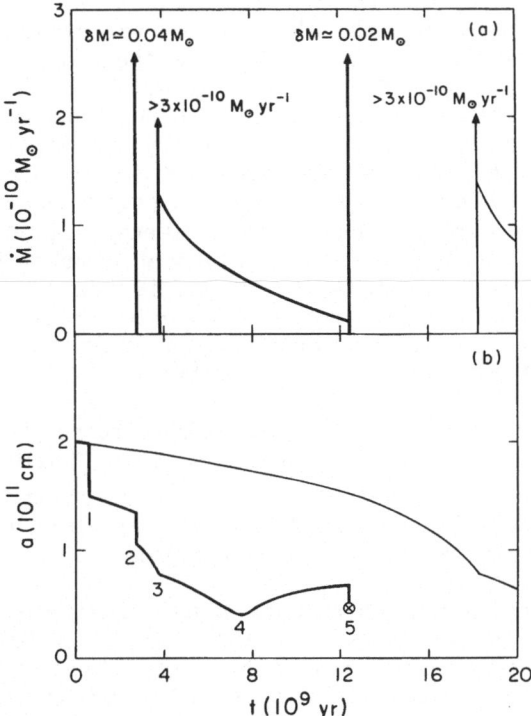

Figure 2.   Evolution of a typical highly compact binary in a globular cluster core. The case shown has $n_* = 3 \times 10^5 pc^{-3}$.

(a) <u>Heavy curve</u>–rate of mass transfer, $\dot{M}$, from the secondary to the vicinity of the primary, as a function of elapsed time, t, from the start of the binary evolution. Spikes of rapid mass transfer, due to collisions with passing field stars and entailing total quantities, δM, of transferred mass as indicated, occur at $t \simeq 2.7$ and 12.4×10^9 yr. Another, much smaller spike in the mass transfer rate occurs at $t \simeq 3.8 \times 10^9$ yr and corresponds to the onset of slow mass transfer resulting from the gradual decay of the orbit due to the emission of gravitational radiation.

<u>Light curve</u>–transfer rate for an initially identical binary, but with the effects of collisions with field stars neglected. The spike in the mass transfer rate at $t \simeq 18.2 \times 10^9$ yr corresponds to the onset of slow mass transfer driven by gravitational radiation.

(b)<u>Heavy curve</u>–semi-major axis, a, of the binary orbit. Features 1 and 2 correspond to collisions with field stars that strongly perturb a; the second of these collisions results in an episode of rapid mass transfer, as indicated in (a), but the system is again out of contact after this episode. At feature 3, the system comes into contact due to the decay of the orbit resulting from gravitational radiation, and slow mass transfer commences. At feature 4, a passes through a minimum when the mass of the secondary has dropped to $\sim 0.06 M_{\odot}$, and it falls out of thermal equilibrium. At feature 5, another collision with a field star triggers an episode of rapid mass transfer that devours the entire secondary (whose mass has dropped to $\sim 0.02 M_{\odot}$ just prior to the encounter).

<u>Light curve</u>–semi-major axis of an initially identical binary, but with the effects of collisions with field stars neglected. The gradual decline in a reflects the decay of the orbit due to gravitational radiation.

$0.04M_\odot$ is rapidly transferred to the potential lobe dominated by the primary before the secondary's stability is restored. Now the orbit is so close that evolution by gravitational radiation proceeds at a swifter pace. Shortly thereafter the system comes into contact, and steady transfer at a rate of $\sim 10^{-10} M_\odot \text{yr}^{-1}$ begins. The orbital separation shrinks as gravitational radiation removes energy and angular momentum from the system, while the secondary loses heat as nuclear reactions in its center cannot make up the losses in radiation and mass transfer from its surface. Ultimately the secondary is whittled down to such a small mass ($\sim 0.08M_\odot$) that hydrogen burning ceases and it turns degenerate. At that point, further mass transfer causes the orbit to expand, until, $\sim 4\text{x}10^9\text{yr}$ later, runaway mass transfer induced by a very close gravitational encounter completely destroys the tiny ($0.02M_\odot$) star which remains and rearranges its mass into a disk around the primary. The light curve in the figure shows what would have happened had there been no gravitational encounters: gravitational radiation would ultimately have induced slow, steady mass transfer, but only after a wait of $1.8\text{x}10^{10}\text{yr}$.

We have derived a number of quantitative results from these exploratory calculations. Most importantly, the rate of catastrophic encounters is $\simeq (2.5\text{x}10^{10}\text{yr})^{-1}\frac{n_*}{10^5 pc^{-3}}$, independent of the binary separation over the range of ~ 0.5–$2\text{x}10^{11}\text{cm}$. Weaker encounters, which merely compress the orbit, induce steady slow mass transfer (driven by gravitational radiation) at a rate approximately half that of catastrophes.

In addition to altering the binary's orbit around its own center of mass, and possibly triggering mass transfer, encounters also change the velocity of the binary center of mass itself. The field star generally receives most of the energy taken from the binary orbital motion because it is the lighter fragment, but occasionally, particularly after catastrophic encounters in which the field star escapes from the system unscathed, the (possibly former) binary's recoil speed might be sufficient to free it from the cluster. Near the low end of the cluster escape speed distribution (as given by Peterson and King 1975), at $v_{esc} = 20\text{km}$ s^{-1}, as many as a third of the catastrophes result in expulsion of the collapsed object-plus-disk fragment; near the upper end ($v_{esc} = 40\text{km s}^{-1}$), the fraction drops to about a tenth. In practise, these numbers should probably be reduced somewhat because we took no account of tidal dissipation during the encounter, which tends to reduce the final kinetic energies of the collision fragments, or may even leave them bound to one another.

III. OBSERVATIONAL CONSEQUENCES

The highly compact binaries in our study passed through a number of stages in which significant mass transfer occurred. The energy released by accretion onto the collapsed object is radiated in a different fashion in each stage. When the transfer is onto a neutron star (or a black hole of no more than a few solar masses) and is driven by gravitational radiation, the result is an X-ray source, but one having a luminosity of only $\sim 10^{35} - 10^{36}\text{erg s}^{-1}$. The bright sources in globular clusters have luminosities in the range $10^{36} - 10^{38}\text{erg s}^{-1}$, while no cluster sources with luminosities between $10^{34.5}$ and 10^{36} erg s^{-1} have been found (Hertz and Grindlay 1983). It is possible that the brightest of the systems in which mass transfer is driven only by gravitational radiation can be identified with the faintest "bright" sources, but it appears that the weaker ones are simply nowhere to be found. Consequently, some other mechanism must be present both to eliminate the fainter gravitational radiation-driven systems, and to supply the very bright observed sources.

One possibility is that the bright X-ray sources are fed from the massive disks that are the product of catastrophic encounters. The structure of such disks is almost wholly unknown, but it is likely that their cooling times are either comparable to or smaller than their angular momentum transport times, so that they settle down into a flat configuration that permits the free propagation of X-rays away from the neutron star and out into space (Krolik 1983). Because such disks are also likely to be subject to mild self-gravitating instabilities if they are sufficiently flat (i.e., sufficiently cool), the level of turbulence may be high enough to induce a significant effective viscosity and drive accretion at a substantial rate (Paczyński 1978; Kozłowski, Wiita, and Paczyński 1979). In cases where the secondary star has been completely dissolved, these collapsed object-plus-disk systems will, of course, never display eclipses, in agreement with the observed properties of both the globular cluster and Galactic bulge X-ray sources (Lewin and Joss 1983). In those cases in which a residual star remains, eclipses have only a small probability because the orbital separation following a catastrophic encounter is generally large compared to the radius of the secondary [Joss and Rappaport (1979) calculate the eclipse probability as a function of this ratio]. It is possible that Milgrom's (1978) mechanism may also play a role in accounting for the absence of eclipses when the system still possesses a secondary.

It is an open question what prevents the existence of systems in which mass is transferred onto a neutron star or solar-mass black hole in slow, steady fashion. Several proposals have been made which we simply list here without endorsement. All share the property of increasing the transfer rate at the cost of decreasing the number in existence at any given time. One such mechanism is "magnetic braking" (Verbunt and Zwaan 1981; Rappaport, Verbunt, and Joss 1983; Spruit and Ritter 1983; Taam 1983); another is the secular evolution of the secondary, which may be the remnant of a more massive star (Webbink, Rappaport, and Savonije 1983).

Four different mechanisms may make the weak (10^{32}–10^{34} erg s^{-1}) globular cluster X-ray sources seen in the Einstein Observatory survey (Hertz and Grindlay 1983). When a highly compact binary evolving by gravitational radiation losses alone reaches the stage at which the secondary turns degenerate, its mass transfer rate drops off sharply, falling below $10^{-11} M_\odot$yr^{-1} a few billion years after passing through the orbital period minimum (Rappaport, Joss, and Webbink 1982). If the collapsed component is a neutron star or a black hole of about a solar mass, X-rays would be produced at about the right rate. Alternatively, if the collapsed component is a degenerate dwarf, steady transfer onto it at a rate of $\sim 10^{-10} M_\odot$ yr^{-1} would generate an accretion luminosity of $\sim 10^{32}$ erg s^{-1}, of which some portion would come out in X-rays (Kylafis and Lamb 1979). These latter systems are, of course, identical to the usual model of a cataclysmic variable, so they can be expected to show the same range of optical variability as do other cataclysmics. Weak X-ray emission can also be obtained from the massive disks produced by catastrophic encounters. If the central object is a neutron star or black hole, the accretion rate must be very much smaller than we suggested in the previous paragraph, but there is more than enough uncertainty in the argument to allow the possibility. If the central object is a degenerate dwarf, the situation is a bit more complicated. Unlike the case in which the central object is a neutron star (Thorne and Zytkow 1977), when the cooling time is longer than the angular momentum transport time, the disk material may rearrange itself into a new, stable, nearly-spherical configuration: that of the envelope of a red giant (Krolik 1983). Consequently, weak X-ray emission can only be produced as a result of a catastrophic encounter dumping mass into a disk around a degenerate dwarf if the cooling time is short compared to the angular momentum transport time, but the angular momentum transport time is short enough to provide an adequate accretion rate.

In Section I we raised the issue of the extraordinary frequency of highly compact binaries in globular clusters. Combining the gravitational scattering rates found by Krolik, Meiksin, and Joss (1983) with those computed for tidal capture by Press and Teukolsky (1977), and using the fractional populations in globular cluster cores of degenerate dwarfs and neutron stars (as found by, e.g., Gunn and Griffin 1979, or Illingworth and King 1977), we have constructed a simple analytic theory for the number of highly compact binaries of each variety present in a globular cluster of a given core density at a given age (Krolik 1983). The only truly free parameter is the lifetime of the massive disk stage, which, in the calculation of Krolik (1983), was taken to be $\sim 10^8$ yr so as to be consistent with the assumption that massive disks around neutron stars produce X-ray sources with luminosities of 10^{37} erg s^{-1}. Sample results, arrived at by summing over all the globular clusters in the Galaxy with measured core densities, are displayed in Table 1.

Table 1. The Number of Mass-Transfer Systems

Category	PK+	HG-PK+ (predicted)	HG-PK+ (observed)
cataclysmic variables	1000	960	(750)
degenerate dwarfs-plus-disks	160	150	
very slow transfer onto neutron stars	18	17	
slow steady transfer onto neutron stars	11	10	(7)
neutron stars-plus-disks	3	3	

Notes: The abbreviation PK+ refers to a sample of all those globular clusters in the Galaxy for which we were able to find published core densities. Most of the data comes from Peterson and King (1975), but we also use Bahcall (1976), Bahcall and Hausman (1976), and Malkan, Kleinmann, and Apt (1980). HG-PK+ refers to the intersection of those globular clusters with the ones in the Einstein Observatory survey (Hertz and Grindlay 1983).

The comparison with observations is a little bit slippery because the identification of the theoretical categories with the observed is uncertain; for that reason the numbers in the "HG-PK+ (observed)" column are in parentheses. However, certain conclusions may be reached immediately. The predicted number of systems with neutron star primaries in the mass-transfer stage is roughly equal to the number of bright sources, though, as we mentioned earlier, the steady transfer ones may not actually be bright enough. The predicted number of degenerate dwarf systems is startlingly close to the inferred total number of weak sources (Hertz and Grindlay 1983). Such a close correspondence suggests that at least portions of the underlying theory may be correct.

As we mentioned earlier, encounters between binaries and field stars that result in rapid mass transfer are strongly correlated with encounters that impart high recoil velocities to the binary, or collapsed object-plus-disk, fragment. Thus, when the fragment containing the collapsed object is given a large enough velocity to escape the cluster, it is also likely to be already in, or about to enter, a stage of rapid accretion. Such systems might contribute to the population of bright Galactic bulge X-ray sources. The potential lifetimes of these systems in a high-luminosity state (up to $\sim 10^8$ yr), combined with their terminal velocities following escape from the cluster (typically ≥ 10 km s^{-1}), allow the systems to travel distances up to several kiloparsecs from their parent clusters during their lifetimes. Thus, the lack of detailed correlation between the positions of bright Galactic bulge X-ray sources and those of known globular clusters is consistent with the possibility that these sources were ejected from globular clusters.

IV. CONCLUSIONS

We have demonstrated that gravitational encounters with field stars are of considerable importance in the evolution of highly compact binaries in globular clusters. The accumulated effects of many encounters compress the orbit so that slow, steady mass transfer is begun long before gravitational radiation acting alone would have done so. More dramatically, occasional encounters result in catastrophes in which one or both of the main-sequence stars involved in the encounter are partially or wholly disrupted, much of their lost mass spread out into a disk around the collapsed object.

These massive disks are an entirely new class of accretion system. Their properties are difficult to analyze with any degree of certainty, but it is possible that, when placed around a neutron star or low-mass black hole, they may fuel the bright globular cluster X-ray sources. Their behavior when centered on a degenerate dwarf may be equally complex, but it is possible that such disks are a new source for both cataclysmic variables and red giants in globular clusters.

Enough is known about the rates of different sorts of gravitational scattering events between binaries and field stars that a comprehensive model may be assembled to predict the numbers of highly compact binaries in each evolutionary stage. This model does reasonably well when matched against the actual counts of objects in the globular clusters of our Galaxy.

Finally, it is possible that some of the bright Galactic bulge X-ray sources were created by violent gravitational encounters between highly compact binaries and field stars in globular clusters, and then expelled from their natal clusters in recoil from the same violent event that created them.

ACKNOWLEDGMENTS

This work was supported in part by the National Science Foundation under grants AST 78-21993 and PHY 80-07351, by the National Aeronautics and Space Administration under grants NSG 7643 and NGL-22-009-638, and by the University of California Lawrence Livermore Laboratory under subcontract 9744809.

REFERENCES

Bahcall, J.N., and Hut, P.: 1983, preprint.
Bahcall, N.A.: 1976, *Astrophys. J. Lett.* **204**, L83.
Bahcall, N.A., and Hausman, M.A.: 1976, *Astrophys. J. Lett.* **207**, L81.
Clark, G.W.: 1975, *Astrophys. J. Lett.* **199**, L143.
Fabian, A., Pringle, J., and Rees, M.: 1975, *M. N. R. A. S.* **172**, 15p.
Faulkner, J.: 1971, *Astrophys. J. Lett.* **170**, L99.

Grindlay, J.: 1985, this volume.
Gunn, J.E., and Griffin, R.F.: 1979, *A.J.* **84**, 752.
Heggie, D.C.: 1975, *M. N. R. A. S.* **173**, 729.
Hertz, P., and Grindlay, J.: 1983, preprint.
Hills, J.G.: 1975, *A. J.* **80**, 809.
Hut, P.: 1982, *Astron. Astrop.* **110**, 37.
Illingworth, G., and King, I.R.: 1977, *Astrophys. J. Lett.* **218**, L109.
Joss, P.C.: 1981, in *X-Ray Astronomy with the Einstein Satellite*, ed. R. Giacconi (Dordrecht: Reidel).
Joss, P.C., and Rappaport, S.A.: 1979, *Astron. Astrop.* **71**, 217.
Katz, J.I.: 1975, *Nature* **253**, 698.
Kozłowski, M., Wiita, P.J., and Paczyński, B.: 1979, *Acta Astron.* **29**, 157.
Krolik, J.H.: 1983, preprint.
Krolik, J.H., Meiksin, A., and Joss, P.C.: 1983, preprint.
Kylafis, N.D., and Lamb, D.Q.: 1979, *Astrophys. J. Lett.* **228**,
Lecar, M., Wheeler, J.C., and McKee, C.F.: 1976, *Astrophys. J.* **205**, 556.
Lewin, W.H.G., and Joss, P.C.: 1983, in *Accretion-Driven Stellar X-Ray Sources*, ed. W.H.G. Lewin and E.P.J. van den Heuvel (Cambridge, England: Cambridge University Press).
Malkan, M., Kleinmann, D.E., and Apt, J.: 1980, *Astrophys. J.* **237**, 432.
Milgrom, M.: 1978, *Astron. Astrophys.* **67**, L25.
Paczyński, B.: 1967, *Acta Astron.* **17**, 287.
Paczyński, B.: 1978, *Acta Astron.* **28**, 91.
Paczyński, B., and Sienkiewicz, R.: 1981, *Astrophys. J. Lett.* **248**, L27.
Peterson, C.J., and King, I.R.: 1975, *A. J.* **80**, 427.
Press, W.H., and Teukolsky, S.A.: 1977, *Astrophys. J.* **213**, 183.
Rappaport, S.A., Joss, P.C., and Webbink, R.: 1982, *Astrophys. J.* **254**, 616.
Rappaport, S.A., Verbunt, F., and Joss, P.C.: 1983, preprint.
Spruit, H.C., and Ritter, H.: 1983, submitted to *Astrophys. J.*.
Taam, R.: 1983, preprint.
Thorne, K.S., and Zytkow, A.: 1977, *Astrophys. J.* **212**, 832.
Verbunt, F., and Zwaan, C.: 1981, *Astron. Astrophys.* **100**, 67.
Webbink, R.: 1980, in *I.A.U. Symposium 88–Close Binary Stars: Observations and Interpretation*, eds. D.M. Popper and R.K. Ulrich (Reidel: Dordrecht).
Webbink, R., Rappaport, S.A., and Savonije, E.: 1983, preprint.
Zahn, J.-P.: 1975, *Astron. Astrop.* **41**, 329.
Zahn, J.-P.: 1977, *Astron. Astrop.* **57**, 383; erratum in **67**, 162 (1978).

GENERAL DISCUSSION ON BINARY EVOLUTION IN GLOBULAR CLUSTERS

KATZ: This question is for Julian. First, what fraction of the field
stars that collide with your binaries are expelled? And second, looking
a bit beyond what you have done, do you think it is conceivable that--
particularly if close binaries are common in these clusters--such
collisions could be a significant factor in the clusters' evolution?

KROLIK: Well, we have considered both of those. The number of field
stars expelled is considerably greater than the number of binaries, and
you can do it in a collision that isn't catastrophic. I can't recall
the numbers, but they're in the forthcoming paper, which I'll show you.
We have also worried about the effect on the globular cluster evolution,
and here I don't think we are going to differ greatly from previous
estimates made by others. The two effects are: (1) if you send off
the recoil fragment with a speed greater than the escape speed, you
simply take mass out of the cluster, and hence reduce the overall
binding energy; and (2) if you kick it out with a speed which is substantial, but less than the escape speed, that contributes to the
heating of the cluster. And in fact I wanted to comment that if the
white dwarf binary stays in the globular cluster (that is, the recoil
is less than the escape speed), then there is plenty of time for it to
share its energy with the other stars in the cluster, and it should
approach equilibrium.

VERBUNT: We considered that, but we think it gives rise to a bimodal
distribution. What you say is true: if it's not too far away from the
core, then it will go back into the core again, since it is a binary.
But it's not true if it has reached a certain distance outside the core;
then it will not be relaxed, and will stay outside. So we actually
more or less predict a bimodal distribution . . . .

KROLIK: That's a little surprising. If the orbit in the globular
cluster is reasonably eccentric, it will zip back down again. So I
think it will probably equilibrate.

LIEBERT: A question for Paul. Are these IPC sources 5σ? How was the
background subtracted?

HERTZ: The background was subtracted using a local background method
which we developed for the galactic plane survey. And the galactic
plane survey results were used to predict what the background source
rate should be for those exposures at those galactic latitudes. Only
about 3.5 out of the 16 sources can be expected to be field sources.

LIEBERT: But the background has to be higher in the globular cluster
field itself as opposed to other fields of similar galactic latitude,
doesn't it? Since every type of star along the main sequence is a
corona/chromosphere X-ray source at some level, I'm just asking if you
add together the line-of-sight contributions of all of these from the
cluster itself, that surely dominates the normal background.

D. Q. Lamb and J. Patterson (eds.), Cataclysmic Variables and Low-Mass X-Ray Binaries, 117–120.
© 1985 by D. Reidel Publishing Company.

HERTZ: Oh. For sources with luminosities as low as 10^{32} ergs/sec, that might be the case, although there haven't been many Population II stars detected, so it's really hard to make that estimate. Certainly for sources with luminosities of 10^{33} - 10^{34} ergs/sec, it's unlikely to be the case. And I might point out that for ωCen, M22, and 47 Tuc, deep IPC exposures were taken which show that there is diffuse emission associated with the cluster, but it appears to be hot gas in the cluster and not the summed emission of the stars in the cluster. That gives you an upper limit on the background for <u>giant</u> clusters, so for the more normal clusters we expect it to be even less important.

GRINDLAY: It's really very unlikely that cluster stars could contribute significantly to the background. At source luminosities of 10^{31} ergs/sec, it might begin to be important, but our survey only goes down to source luminosities of $\sim 10^{32.5}$ ergs^{-1}.

GARRISON: I can give you a new figure for the space density of cataclysmic variables as a result of our discovery of the one at m_V = 9.5 It turns out that our survey was complete to 10th magnitude, and assuming M_V = 4 or 5, we get a distance limit of about 145 pc. We then find a density of 9×10^{-5} CVs per square degree to a distance of 145 pc.

WHEELER: A comment for Frank Verbunt. Blue stragglers are not really a globular cluster phenomenon, but an open cluster phenomenon. Most globular clusters don't have them, and if you end up predicting a lot of blue stragglers in most globular clusters, ...

VERBUNT: (Ed: inaudible) ... I think there are probably 5 or 6 in the literature.

KROLIK: We have also considered the problem of blue stragglers. If the globular cluster core is reasonably dense, it's very hard to avoid making a certain number of blue stragglers because the standard tidal capture mechanism makes main-sequence binaries just as easily as it does compact object - main-sequence binaries.

WHEELER: Then how <u>do</u> your predictions compare with the observations and limits?

KROLIK: I haven't actually put in numbers yet.

VERBUNT: The reason why we got interested is that the idea of two single stars is wrong, but in the collision of a single star with a binary, you can have a higher mass than in the collision of two single field stars. And in this three-body system, you can have the possibility that all three stars coalesce. Then you can make blue stragglers. But unfortunately, from looking in the literature, it seems all these systems are in open clusters.

WHEELER: But you can use the observed limits to make sure you are not overproducing them.

GRINDLAY: Frank, can you address the Gunn & Griffin limits on binaries, and what they say about the initial fraction of binaries? Your assuming about 1% . . . .

VERBUNT: Yes, I need one percent of the cluster mass to be in binaries. I think that's consistent with the Gunn & Griffin limits.

GRINDLAY: Only barely consistent . . . .

VERBUNT: No, I think it's very easily accommodated. Also, there's a recent Canadian paper which claimed that Gunn & Griffin could not set any limits at all -- that the binary frequency in clusters need not be different from the frequency in the galactic plane. I can't judge the validity of this statement.

KATZ: I'd like to suggest a quite different scenario, and ask if there's some reason why it's wrong. It's this: the reason we have X-ray binaries in globular clusters has to do with the evolution of primordial binaries, and the reason we have so many of them is that globular clusters have evaporated 99% of their mass, and as a result the binaries, which are more massive than most of the stars, tend to remain, and are therefore now very abundant. Can we rule that out?

KROLIK: I think there's a problem with binaries that start with a moderate separation, like the binaries that Gunn & Griffin looked at. As I noted, they're right up against the edge of the separation at which they would be rapidly disrupted ... they've already got a some-what different evolution.

KATZ: The evolution of such things as binary stars is something of a mystery. Is there a stronger way of ruling out this hypothesis?

VERBUNT: Even if what you say is true, the problem may be solved by the fact that the neutron stars born in globular clusters will leave it quite often. The total number of X-ray binaries could be a factor of 10 higher if they were made in a galactic scheme (which I don't believe). If they were made the same way as in the galaxy, there might be ten times more _born_ in globular clusters than are actually observed.

KATZ: That makes it _worse_, if they're expelled on formation by recoil. The first thing you need in any of these models is that when a neutron star forms in a globular cluster, it must be formed with a _small_ recoil -- at least with a substantial probability, if not always. The question is . . . .

VERBUNT: I'm talking about recoil of the single star. We know nothing about the prior evolution of systems that will become galactic-plane sources. Now if the supernova event happens in such a source, it might be so energetic that the single neutron star, or even the binary, has a recoil velocity higher than the escape velocity from a globular cluster.

KATZ: Clearly that does not happen to all of them.

VERBUNT: It could happen in the galactic plane.

RADIATION TRANSFER IN ACCRETION DISK CORONAE

Richard A. London
University of California
Lawrence Livermore National Laboratory
Livermore, California 94550

ABSTRACT

The physics of X-ray excited accretion disk coronae is described. A numerical model for the hydrostatic inner region, including radiation transfer effects, is presented. The resulting appearance is compared to recent observations of X-ray dips in several X-ray binaries. We also discuss the dynamical effects of X-ray induced mass loss, which originates at large radius in the disk.

I. INTRODUCTION

Although the problem of external illumination of stars in X-ray binaries has been well studied during the last ten years, similar effects on accretion disks have received little attention. Recently, observational evidence for extended (size $\sim 5 \times 10^{10}$cm) highly ionized gas around compact X-ray sources, in the form of gradual, frequency independent dips in X-ray intensity has been seen in several binary sources (White et al 1981, White and Holt 1982, and McClintock et al 1982). These dips have been interpreted as eclipses by the companion stars of scattering gas, evaporated by X-rays from accretion disks. These observations have stimulated theoretical investigation of X-ray excited accretion disk coronae (ADC). Similar processes may also be important for disks around supermassive black holes in quasars.

Several effects of X-rays on accretion disks were discussed qualitatively by Shakura and Sunyaev (1973). As in the case of X-ray illuminated stars (cf. London, McCray and Auer (1981-LMA), Anderson 1981 and references therein) the effects may be broken into two categories: photospheric heating and coronal heating. Photospheric heating of disks has been discussed in papers by Cunningham (1976), Pacharintanakul and Katz (1980), and Meyer and Meyer-Hofmeister (1982). Two recent papers by Begelman, McKee and Shields (1983-BMS) and Begelman and McKee (1983-BM) discuss many of the theoretical aspects of ADC in considerable detail.

121

D. Q. Lamb and J. Patterson (eds.), Cataclysmic Variables and Low-Mass X-Ray Binaries, 121–132.
© *1985 by D. Reidel Publishing Company.*

In this paper we shall primarily describe the calculation of the structure of the inner, hydrostatic, part of an X-ray excited accretion disk corona. This requires a self-consistent calculation of the transfer of radiation from the central source through the corona, coupled with a calculation of the temperature and density of the gas. In §II we present a "pseudo 2-dimensional" method to solve the structure problem; analytic and numerical results are given in §III. We briefly discuss the effects of mass loss from the outer regions of the disk in §IV. In §V we discuss observational implications of these models.

II. MATHEMATICAL MODEL FOR X-RAY EXCITED CORONAE

We consider a thin, azimuthally symmetric disk of gas encircling a central X-ray source. We make the usual accompanying assumptions: the azimuthal velocity is nearly Keplerian and much greater than both the radial velocity and the sound speed; the gas is in hydrostatic equilibrium vertical to the rotation plane (Z-direction), and the mass in the disk is small so that the gravity is due solely to the central object. If radiation from the compact source makes its way to the surface of the disk either directly, or by scattering, it will evaporate material into a hot corona. The existence of the corona can be understood by considering the heating and cooling of the gas.

The equilibrium state of optically thin X-ray illuminated gas is a function of β--the ratio of radiation to gas pressure. (Similar parameters are used by Krolik, McKee and Tarter 1981, Kallman and McCray 1982, and references therein.) At small β, the temperature is set by a balance of atomic heating and cooling processes and is less than 10^4K, while at large β, Compton scattering fixes the temperature to a value called T_x, approximately equal to the X-ray spectral temperature (10^6-10^8K). A transition (usually very sharp) occurs at a critical value, denoted β_0, which is of order 1-10 for spectra of interest (LMA; Krolik, McKee and Tarter 1981). In regions of interest the pressure near the center of the disk is large ($\beta < \beta_0$) so that a low temperature can be maintained. As the pressure drops with height, a point is reached where β becomes larger than β_0 and the temperature rises steeply. We identify this point with the base of the corona. Knowing the radiation field here we find the base pressure: $P_0 = P_{rad}/\beta_0$.

The vertical structure of the corona at a given radius depends primarily on the ratio of T_x to the gravitational escape radius, $T_{esc} \equiv GM\mu/kR$, where G is the gravitational constant, M the mass of the X-ray star, μ the mean mass per particle in the gas, k Boltzman's constant and R the distance from the star. We define an escape radius by equating T_x to T_{esc}: $R_{esc} \equiv GM\mu/kT_x = 9.85 \times 10^{10} \overline{M}/T_{x7}$cm, for an ionized gas with 10% Helium. We denote stellar masses relative to the sun by $\overline{M}$ and we use subscript numericals to denote decimal exponents of units (e.g. $T_{x7} = T_x/10^7$). For $R < R_{esc}$, $T_x < T_{esc}$

and the corona remains gravitationally bound while for $R > R_{esc}$, $T_x > T_{esc}$ and the gas in the corona is heated to escape temperature and flows out in a wind similar to an X-ray induced stellar wind (see London and Flannery 1982 and references therein). In this paper we shall discuss primarily the bound region: $R < R_{esc}$.

Some of the elements of the theory, introduced by McClintock et al (1982), are summarized here. Perpendicular to the disk the vertical component of gravity balances the pressure gradient. For an isothermal corona at T_x and for $R \ll R_{esc}$ the density profile is Gaussian:

$$\rho(R,Z) = \rho_0(R) \, \exp[-Z^2/Z_s^2]. \tag{1}$$

The scale height is

$$Z_s = 4.51 \times 10^9 \, \overline{M}^{-1/2} T_{x_7}^{1/2} R_{10}^{3/2} \quad cm \tag{2}$$

and the base density is determined from β_0:

$$\rho_0 = [\mu \, P_{rad}(R,Z_0)/kT_x]/\beta_0 \,. \tag{3}$$

The radiation pressure is written

$$P_{rad} = \frac{4\pi J_0}{c} = \frac{L}{c 4\pi R^2} \, \overline{J}_0 \,, \tag{4}$$

where J_0 is the mean intensity of X-rays and $\overline{J}_0$ is a dimensionless value, also called the "shielding factor", which takes into account optical depth effects in the corona. In the optically thin case, $\overline{J}_0 = 1$ and the structure of the corona is completely determined. The appearance can easily be calculated by considering the scattering emissivity.

We calculate the optical depth to check for self consistency. We define two characteristic optical depths along the radial and vertical directions; respectively.

$$\tau_{\parallel} = \int_R^\infty \rho_0 \kappa_s dR = \left(\frac{L}{L_E}\right)\left(\frac{1}{\beta_0}\right)\left(\frac{R}{R_{esc}}\right)^{-1}$$

$$= 0.67 \, T_{x_7}^{-1} \, L_{37} \, \beta_0^{-1} \, R_{10}^{-1} \tag{5}$$

and

$$\tau_{\perp} = \int_0^\infty \rho(R,Z)\kappa_s dZ = \left(\frac{\pi}{2}\right)^{1/2}\left(\frac{L}{L_E}\right)\left(\frac{1}{\beta_0}\right)\left(\frac{R}{R_{esc}}\right)^{-1/2}$$

$$= 0.27 \, T_{x_7}^{-1/2} \, L_{37} \, \beta_0^{-1} \, \overline{M}^{-1/2} \, R_{10}^{-1/2} \tag{6}$$

where we have introduced the Eddington luminosity, L_E, equal to $1.47 \times 10^{38}\, \overline{M}$ ergs^{-1} for a plasma with 10% Helium. We see that for

$$R > R_{\parallel} = 6.69 \times 10^9\, T_{x_7}^{-1}\, L_{37}\, \beta_0^{-1}\ \text{cm} \tag{7}$$

the characteristic parallel optical depth is less than one, while for

$$R > R_{\perp} = 7.16 \times 10^8\, T_{x_7}^{-1}\, L_{37}^2\, \overline{M}^{-1}\, \beta_0^{-2}\ \text{cm} \tag{8}$$

the perpendicular optical depth is less than one. In these outer regions the optically thin theory holds, provided that there is no shadowing by coronal gas at smaller radii.

To consider the structure of regions interior to $R_{\parallel}$ and $R_{\perp}$, and to determine whether they shadow the outer regions, we must consider the transfer from the central source through the corona. We use a "pseudo 2-dimensional" model to do so. We assume that radiation from the compact source is attenuated by scattering along rays from the origin and thereafter remains local in radius and transfers only in the vertical direction. This approximation requires $Z_s \ll R$ and $\tau_{\perp} \ll \tau_{\parallel}$. We separate the radiation into direct and diffuse parts. For the direct mean intensity we have

$$J_{DIR} = \frac{L}{(4\pi R)^2}\, \exp[-\tau_{\parallel}(R,Z)] \tag{9}$$

with

$$\tau_{\parallel}(R,Z) = \int_0^R \rho(R',Z)\, \kappa_s\, dR' \Bigg|_{\frac{Z}{R'} = \text{constant}} \tag{10}$$

The direct radiation serves as a source for diffuse radiation, which we treat in a locally plane parallel approximation:

$$\cos\theta\, \frac{dI_{DIFF}}{dZ} = \kappa_s (J_{DIFF} + J_{DIR} - I_{DIFF}). \tag{11}$$

Here I_{DIFF} and J_{DIFF} are the specific intensity and mean intensity of diffuse radiation, and θ is the propagation angle. We assume conservative isotropic scattering, valid to first order in $kT/m_e c^2$ and $kT_x/m_e c^2$. We solve Eq. (11) by combining a formal solution for the singly scattered radiation, using the Λ-operator (e.g. Mihalas 1978), with an Eddington approximation for the multiply scattered radiation. The formal solution accurately treats the large peak in I_{DIFF} at $\theta \approx 90°$, which occurs when $\tau_{\perp} \ll 1$, while the Eddington solution handles the case $\tau_{\perp} \gg 1$. Overall the error in the mean intensity is expected to be less than 20%, as determined by comparison to test problems in Van de Hulst (1980).

The model is completed by the hydrostatic equation (giving Eq. 1 for the isothermal case) and the coronal base condition (Eqs. 3 and 4) relating density to the mean intensity. The sum of direct and diffuse radiation fields is used to determine the base density.

III. SOLUTIONS OF THE ADC MODEL

To find and present solutions, we use scaled variables, generally denoted with a super "bar" (e.g. $\bar{R}$). The radius is scaled to R_1, the value at which $\tau_1 = 1$ in the optically thin solution (Eq. 8). The height is scaled to the radius: $Y \equiv Z/R$. The mean intensity and density are scaled to their values in the optically thin model (which are radius dependent, see Eqs. 3 and 4). There remains a single nondimensional parameter characterizing the solution. We choose this to be the ratio of the thermal scale height (Z_s) to the radius at R_1. Using Eqs. (2) and (3) this ratio, defined as Y_*, is written $Y_* = \pi^{1/2} L/L_E \beta_0^{-1}$. We solve the equations numerically on a 2-dimensional grid in Y and R, typically 50 x 50 in size imposing boundary conditions that $\tau_{\parallel}(Y) = 0$ at some inner radius, R_0. The numerical technique is briefly described here. We begin at R_0 and take a step in radius to $R_0 + \Delta R$. By an iterative technique, we find a self consistent base density in this radial zone such that the scattered plus direct radiation at the base, determined from the optical depth and transfer equations is consistent with the base condition. Then we add the increments of parallel optical depth due to the region between R_0 and $R_0 + \Delta R$ to each ray and begin another radial step. Each radial step includes an iterative solution of the vertical transfer and structure and an integration of the parallel optical depth. We continue the solution outward to some specified radius. The result is the density and radiation field as functions of height and radius. A single function of radius, the scaled base radiation field, $\bar{J}_0$ (also equal to scaled base density), is sufficient to describe the solutions. We show $\bar{J}_0$ plotted versus $\bar{R}$, for five models with $Y_* = 10^{-3}$ to 10 in Fig. 1. The models all start at

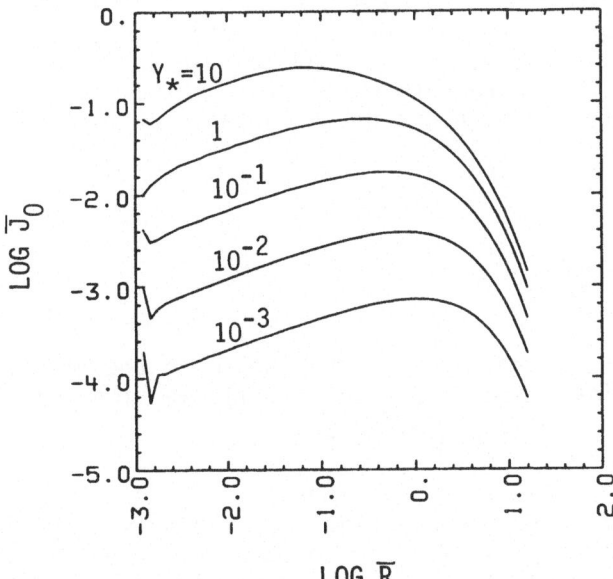

Figure 1. The dimensionless mean intensity at the base of the corona versus scaled radius. The inner boundary is at $R_0 = 1.1 \times 10^{-3}$.

$\overline{R}_0 = 10^{-3}$. The boundary conditions only effect a small region: for
$R > 2R_0$ the solution is insensitive to the exact position of the
boundary. In each case the base radiation field at $R > 2R_0$ is
completely dominated by the diffuse component. The X-rays scatter
high up in the corona where $\tau_{||} < 1$ and are redirected towards the
disk. All of the curves in Fig. 1 show a rollover near $\overline{R} \simeq 1$. The
outer regions are shadowed and cannot maintain the strong, scattering
dominated corona which exists in the inner regions. We note that an
integral equation, derived by assuming that all direct radiation
scatters at $\tau_{||} = 1$, yields an asymptotic solution valid for $\overline{R} < 10^{-2}$
(D. Eardley, private communication).

If the disk surface flares, an outer region of corona may be
excited where the disk rises above the inner shadow. Since the
flaring is probably small, for example the disk height is proportional
to $R^{9/8}$ in the Shakura and Sunyaev (1973) α-model, the location of
this outer region is very sensitive to the actual disk structure. It
is therefore difficult to be certain of whether such an outer corona
will exist.

An important effect (pointed out by BM) neglected in the previous
discussion is the influence on the corona of the radiation emitted
locally in the disk. This radiation is generally of much lower
spectral temperature than the radiation from the central source and
can cool the coronal gas via "inverse Compton" scattering. In a
two-component radiation field the Compton equilibrium temperature is

$$T_{eq} = \frac{J_0 T_X + J_D T_D}{(J_D + J_0)} \tag{12}$$

where T_D is the effective temperature of the disk and J_D is the
mean intensity of disk radiation, assumed to have a Planckian
spectrum. The radiation temperature and flux from the disk are set
by the accretion rate, independent of the details of viscosity:

$$T_D = \left[\frac{1}{\sigma_B} \frac{3}{8\pi} \frac{GM}{R^3} \dot{M} \right]^{1/4} \tag{13}$$

and

$$J_D \simeq \frac{3}{16\pi^2} \frac{GM\dot{M}}{R^3} , \tag{14}$$

where $\dot{M}$ is the accretion rate. By comparing J_D to the value of J_0
obtained ignoring disk radiation, (see Fig. 1), we find that for

$$\overline{R} \gtrsim 4.28 \times 10^{-3} \left(\frac{e}{0.1} \right)^{-3/4} T_{X_7}^{3/4} Y_*^{-2} , \tag{15}$$

where e is the efficiency of converting accreted mass to radiation,
the disk radiation is significant. In these cases we expect the corona
to be weaker. We have attempted to find new solutions of the ADC
equations with modifications made to include the cooling by the disk

radiation. When the disk radiation is large we have not been able to find continuous solutions. As we integrate from the boundary outward, a shadow develops which, when the coupled effects of thermal equilibrium and radiation transfer are accounted for, grows stronger with each radial step. A model for the structure of the inner, Compton cooler corona, based on the premise that the characteristic parallel optical depth is regulated to be of order unity, is given by BM. We have been unable to find such a solution, although it is possible that the effects of electron conduction, ignored here, would stabilize the numerical solution. The difficulty in solving the ADC equations with Compton cooling may be related to a real physical instability, in which case time variation of the inner corona would be expected for lower luminosity sources.

IV. MASS LOSS FROM THE DISK

In this section we briefly mention some of the effects of mass loss. We rely on the calculations of London and Flannery (1982) for estimates of the mass flux from the X-ray illuminated disk surface. The reader is referred to BMS for a discussion disk mass loss in greater detail, including the effects of the accompanying angular momentum loss.

The wind originates primarily from regions $R > R_{esc}$. Consider the region between R_{esc} and the disk outer radius, R_D. In the special case that the characteristic heating time (by Compton heating to temperature T_x) is much shorter than the characteristic flow time (at velocity $V_x \equiv (kT_x/\mu)^{1/2}$ to a distance $2R_D$), the transition region between the disk and corona is sharp, and the flow speed in the wind is approximately V_x. The condition on the time-scales requires

$$L > 0.18 \ T_{x_7}^{-1/2} \ L_E \ \frac{R_D}{R_{esc}} . \tag{16}$$

Imposing a momentum balance, $\rho v^2 \simeq P_0$, across the transition region the mass flux is approximated as P_0/v_x^2. The mass loss rate $(\dot{M}_w)$ is the integral of mass flux over the disk surface. Using the coronal base condition, $P_0 = P_{rad}/\beta_0$, and defining an appropriately averaged shielding factor, $<\bar{J}_0>$, we estimate the mass loss:

$$\dot{M}_w = \frac{L}{c} \ \frac{1}{\beta_0} \ \frac{1}{V_x} \ \ln(R_D/R_{esc}) \ <\bar{J}_0> \tag{17}$$

$$\simeq 9.1 \times 10^{18} \ L_{37} \ T_{x_7}^{-1/2} \ \beta_0^{-1} \ \ln(R_D/R_{esc}) \ <\bar{J}_0> \ g \ s^{-1}. \tag{18}$$

For lower luminosity the mass loss is somewhat reduced (BMS).

The dynamical effect of such a mass loss rate can be examined by comparing it to the accretion rate needed for a certain luminosity. The typical value indicated by Eq. (18) is about <u>100</u> times <u>larger</u> than the accretion rate needed to give the luminosity if the source is a neutron star. Such a situation is likely to be unstable as shown by BMS with a simple time dependent model. The following heuristic thought experiment illustrates the instability. Consider a steady flow with 100 units of mass per second injected into the disk at large radius and 99 units turned around in the wind, powered by the single unit which makes it down to the star. Now perturb the mass loss rate, say downward by 1%. This gives two units of mass not lost, which, after some time make their way to the compact star. The luminosity doubles and tries to drive twice as much mass loss. But this will completely shut off the accretion. Time variation is likely to ensue. This instability may be very common for high luminosity sources. Only if the shielding factor $\langle \overline{J_0} \rangle$ is small ($< 10^{-2}$) or the disk size small ($<< 2\ R_{esc}$) would we expect a steady flow. This may be possible, but since J_0 is very sensitive to the shape of the disk surface because of shadowing, it is difficult to be certain.

V.   OBSERVATIONAL IMPLICATIONS AND CONCLUSIONS

Of the several observable effects of X-ray excited accretion disk coronae, we discuss only the extended scattering which may explain the "dips" mentioned in §I. This effect, and several others are also discussed by BM.

A proper calculation of an X-ray light curve caused by the eclipse of the ADC requires, for each phase of the binary orbit, an integration of the specific intensity of radiation (I) emitted towards the observer, taken over the uneclipsed part of a surface surrounding the scattering gas. The intensity on this surface is computed by solving the transfer equation within the enclosed volume. If the coronal structure is known, as for example that described for the inner corona in §III, then simple solutions along rays through the corona will give the intensity. Because of several uncertainties in the coronal structure and the difficulty of performing the intensity integrations, we describe only approximate results which can be derived from the models of §III. We can estimate the apparent scattered luminosity (L_{ap}) and the brightness distribution for the two limiting orientations. When viewed face on, the apparent scattered luminosity is approximately the central source luminosity times the fraction of solid angle subtended by material which is optically thick along rays from the origin. From the numerical solutions we find, in this case, $L_{ap} \simeq 0.2\ Y_* L$ for $Y_* << 1$. The optical depths through the corona in the vertical direction are small. Therefore, the radial dependence of the outgoing intensity is approximately the same as the base mean intensity. From Fig. 1 we find that $I \propto R^{-5/3}$ for $R < R_{\perp}$; it drops more rapidly for $R > R_{\perp}$.

When viewed edge on, optical depths along the line of sight are large. The brightness can be approximated by the mean intensity at the position where the optical depth measured from infinity inward is one. We find from the solutions that this occurs near the peak in $\bar{J}_0$ (cf. Fig. 1). For $Y_* \ll 1$ the appearance is of a uniformly bright rectangle of height $2Y_* R_\perp$ and length $2R_\perp$. The apparent luminosity is $L_{ap} = 1/\pi \, Y_* \, \bar{J}_0(R = 1)L \simeq 0.04 \, Y_*^{5/3} L$.

In comparing these results to the specific observations of 4U2129+47 (White et al 1982 and McClintock et al 1982) we make the following remarks. As shown by these authors, the eclipse of a uniformly bright disk of radius 5×10^{10}cm in a binary orbit of inclination 81° fits the data well. Using the estimates for the edge-on configuration, it is difficult to match both the large size and small apparent luminosity ($\simeq 10^{34}$ erg s^{-1}) of the scattering source with an optically thick solution (see §III). An optically thin outer corona would give about the right apparent luminosity, as discussed by McClintock et al (1982). Begelman and McKee (1983) show that the eclipse of the scattering gas in the wind region, which they argue is somewhat collimated along the binary axis, gives a good fit to the 4U2129+47 with $L = 5 \times 10^{35}$. In both cases the central source must be blocked from direct view by the observer, perhaps by the disk itself.

We conclude that ADC scattering of X-rays in a system of high inclination $i \simeq 80°$ and of moderate luminosity $L \simeq 10^{36}$ erg s^{-1} can explain the observationed X-ray dips. Mass loss from disks by X-ray heating may have significant dynamical effects. Further study is needed to clarify the effect of Compton cooling on the inner corona, the two dimensional structure of the wind region, possible time dependent effects in both the hydrostatic region and the wind region, and the relationship between the corona and the underlying disk structure.

ACKNOWLEDGMENTS

I would like to thank C. McKee for many useful discussions and comments on the manuscript. Conversations with M. Begelman, J. Castor, D. Eardley, J. McClintock and P. Woodward have been helpful. This work was performed under the auspices of the U. S. Department of Energy by the Lawrence Livermore National Laboratory under contract No. W-7405-ENG-48.

REFERENCES

1. Anderson, L., 1981, Ap. J. 244, 554.
2. Begelman, M. C., McKee, C. and Shields, G., 1983, Ap. J., in press (BMS).
3. Begelman, M. C. and McKee, C., 1983, Ap. J., in press (BM).

4. Cunningham, C., 1976, Ap. J. 208, 534.
5. Kallman, T. R., and McCray, R., 1982, Ap. J. Suppl. 50, 263.
6. Krolik, J. H., McKee, C. and Tarter, C. B., 1981, Ap. J. 249, 422.
7. London, R. A., McCray, R. and Auer, L. H., 1981, Ap. J. 243, 970 (LMA).
8. London, R. A. and Flannery, B. P., 1982, Ap. J. 258, 260.
9. McClintock, J. E., London, R. A., Bond, H. E. and Grauer, A. D., 1982, Ap. J. 243, 900.
10. Meyer, F. and Meyer-Hofmeister, E., 1982, Astr. Ap. 106, 34.
11. Mihalas, D., 1978, Stellar Atmospheres 2nd Ed. Freeman: San Francisco.
12. Pacharintanakul, P. and Katz, J. I., 1980, Ap. J. 238, 985.
13. Shakura, N. I. and Sunyaev, R. A., 1973, Astr. Ap. 24, 337.
14. Van de Hulst, H. C., 1980, Multiple Light Scattering, Vol. 1, Academic Press: New York.
15. White, N. E., Becker, R. H., Boldt, E. A., Holt, S. S., Serlemitsos, P. J., and Swank, J. H., 1981, Ap. J. 247, 994.
16. White, N. E. and Holt, S. S., 1982, Ap. J. 257, 318.

DISCUSSION

SHAVIV: I am somewhat unhappy about your disk, because if the X-ray luminosity is very important, and you have a wind coming out, then $\dot{M}$ could change, and the entire structure of the disk could change. And it's not clear to me that what you got is anything more than just an iteration. If you do a second iteration to the disk structure to adjust to what has happened before, it might not converge . . . you might destroy the disk.

LONDON: Yes, I think that's a possibility: that the wind could destroy the disk.

SHAVIV: But if it destroys the disk, of course, it is very time-dependent.

LONDON: You could get a steady state established by regulation by this wind. That's another possibility. A steady state might be established by regulation of the mass going into the disk at the escape radius.

SHAVIV: Then the disk might not approach the star itself, right? Then it would start to pulsate.

LONDON: Yes, I see your point. I don't know what to say.

VERBUNT: I'm a bit worried about the overall energy considerations. Apparently 8/10 of the mass transferred never reaches the neutron star, but is thrown out again. This means we have only 2/10 left, first to give us all the observed X-rays, and second to heat the corona to 10^7 °K. I get the idea that somewhere you violate total energy conservation. Is that possible, or do I misunderstand you?

LONDON: If you have 80 units of mass going in, then this optically thin theory says that 79 units get blown out. I think the optical depths are going to have to be taken into account, because they would probably be large and the theory wouldn't apply. Nevertheless, there still is this tendency for a large wind which could disrupt the disk.

KROLIK: But isn't the point simply that the binding energy in the out-skirts of the disk is much smaller than the binding energy at the surface of the neutron star? So, if you take the energy available from accretion at the surface of the neutron star and split it equally between X-rays and wind, you can push out a pretty hefty wind.

LONDON: Yes, the binding energy is very small out there.

PACZYNSKI: What is the outflow velocity you expect for your wind?

LONDON: Well, it depends on the outer disk radius. Near the escape radius, if the luminosity is high it will be the sound speed corresponding to T_x, which is about 1000 km/sec.

PACZYNSKI: If you could make it somewhat smaller, it could be very efficient in removing angular momentum from the binary.

LONDON: Yes, you do get a smaller outflow velocity from larger radii, so if the disk has a larger radius. . . .

LANGER: Rich, I am not sure I follow this entirely, but didn't your optically thin theory imply that all the mass loss was at the inner radius of the disk? Wasn't it at the inner edge of the disk where you could boil away 80 times as much as was transferring in to the neutron star, because you need a very high radiation flux to do that? And then aren't you back to the energetics problem?

LONDON: No, the mass loss occurs at radii larger than the escape radius. It's pretty big, typically 10^{11} cm. That's much larger than the radius of a neutron star.

LANGER: OK. So you've got a factor of 10^5 more.

BLAIR: I have two questions. One, if there is a wind coming from a disk, why don't we see more P Cygni profiles? And two, in a lot of these stars like Cyg X-2 and Cen X-4, one sees peculiar line ratios in the IUE spectrum, e.g., N V is much stronger than C IV, Si IV is enhanced relative to C IV, etc. That's probably outside the scope of your model, but I wonder if you have any comments.

LONDON: OK. The first point: Why no P Cygni profiles? There are two reasons. One is that the wind is driven by thermal heating rather than radiation pressure, so it's quite a bit different. Another factor inhibiting formation of strong emission lines is the sharp transition in the disk structure between the photosphere of the disk and the corona.

This happens quite suddenly, and the temperature region where you'd expect to form emission lines is in between. There is very little gas at that temperature. Nobody has done a really detailed model of that yet, though. On the second point, I think that Tim Kallman is going to talk a little later about X-ray ionization of winds and how this can change line ratios.

HYDRODYNAMIC SIMULATIONS OF A COMBINED HYDROGEN, HELIUM THERMONUCLEAR RUNAWAY ON A 10 KM NEUTRON STAR

S. Starrfield
Department of Physics, Arizona State University
Tempe, AZ 85287 and
Theoretical Division
Los Alamos National Laboratory
Los Alamos, NM 87545

S. Kenyon and J. W. Truran
Department of Astronomy, University of Illinois
Urbana, IL 61801

W. M. Sparks
X-Division
Los Alamos National Laboratory
Los Alamos, NM 87545

ABSTRACT

We have used a Lagrangian, hydrodynamic stellar evolution computer code to evolve a thermonuclear runaway in the accreted hydrogen rich envelope of a $1.0 M_\odot$, 10 km neutron star. Our simulation produced an outburst which lasted about 2000 sec and peak effective temperature was 3 kev. The peak luminosity exceeded $2 \times 10^5 L_\odot$. A shock wave caused a precursor in the light curve which lasted 10^{-5} sec.

I INTRODUCTION

The published theoretical studies of the X-ray burst phenomena have produced simulated outbursts which are in reasonable agreement with the observations (c.f., Ayasli and Joss 1982; Taam 1980). Nevertheless, there are observed bursts which occur on much longer time scales than those modeled by the above studies and, in addition, there are the transient X-ray novae which have not, as yet, been produced by any theoretical calculation. We have, therefore, made the assumption that the longer period behavior is also the result of a thermonuclear process acting in the accreted envelope of a neutron star and proceeded to study as extreme conditions on a neutron star as possible in order to try and determine the conditions necessary for such long period outbursts.

We have used a fully implicit, Lagrangian, hydrodynamic computer code to evolve a thermonuclear runaway in the accreted hydrogen rich envelope of a $1.0 M_\odot$ neutron star with a radius of 10 km. We assume that the interior of the neutron star is cold and that the rate of accretion is low enough so that a massive envelope ($M_e \sim 10^{-11} M_\odot$) can be accreted

D. Q. Lamb and J. Patterson (eds.), Cataclysmic Variables and Low-Mass X-Ray Binaries, 133–137.
© *1985 by D. Reidel Publishing Company.*

before significant nuclear reactions are initiated at the boundary between the core and the accreted envelope (Starrfield, et al. 1982, hereafter SKST). We neglect nuclear reactions during the accretion process and our evolutionary sequence begins with a sharp composition discontinuity (hereafter, core-envelope interface: CEI).

Our computer code has been used in studies of thermonuclear runaways in the accreted hydrogen rich envelopes of white dwarfs in our successful attempts to simulate the nova outburst and is ideally suited to this task (c.f., Starrfield, Truran, and Sparks 1978). We have already described the physics used in this code and our nuclear reaction network includes the p-p chain, the $^{12}C(\alpha,\gamma)^{16}O$ reaction and all relevant neutrino loss rates (Starrfield, et al. 1982). We also assume that reacting CNO nuclei are lost to the CNO reactions at rates determined by the $^{14}O(\alpha,p)^{17}F$ and $^{15}O(\alpha,\gamma)^{19}Ne$ reactions (Wallace and Woosley 1981).

II RESULTS

The initial model used in this study had a luminosity of 0.1 $L_\odot$ and an effective temperature of 8. x 10^5 K. The temperature and the density at the CEI were 4.3 x 10^7 K and 3.1 x 10^6 gm cm^{-3}, respectively. The initial rate of energy generation was 10^{14} erg gm^{-1} sec^{-1} producing a nuclear burning time scale in the envelope of 10^2 sec to the peak of the runaway. These conditions are certainly extreme but are not unreasonable based on our intent to simulate long period outbursts.

It takes this sequence about 130 seconds to evolve to the point where the peak temperature in the shell source reaches 10^8 K. It takes about 700 seconds longer for the temperature to reach its peak value of 3.3 x 10^9 K. However, because of conductive energy losses into the interior, this does not occur at the CEI but in a zone about $10^{-12} M_\odot$ closer to the surface. The rate of energy generation exceeded 10^{21} erg gm^{-1} sec^{-1} (but not for very long).

The rapid rise of the temperature in the last stages of the thermonuclear runaway causes an overpressure of a few percent in the shell source which produces a shock wave that reaches the surface 4 x 10^{-6} sec after it is initiated. When this shock penetrates the surface layers, the luminosity climbs to 2 x $10^5 L_\odot$ and the effective temperature to 3.3 x 10^7 K (kT~3keV). Figure 1 shows the temperature history of the CEI, and the luminosity of the surface layers at the time of shock penetration is given in Figure 2. The 3 figures in this paper are reprinted from SKST.

Once the envelope has returned to equilibrium, a nuclear burning front moves both inward and outward from the original point of peak temperature. As it passes through each zone, it flashes to temperatures exceeding 10^9 K. It takes the front about 2 sec to reach the CEI and the temperature in this zone flashes to 3 x 10^9 K.

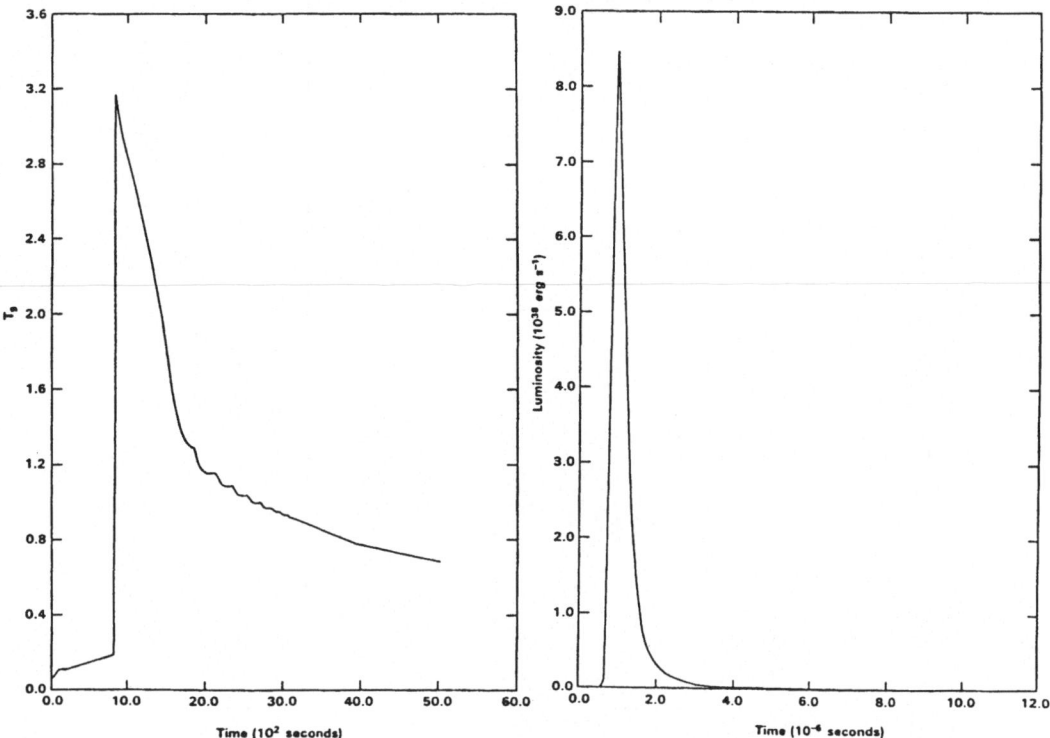

Fig. 1. The temperature of the CEI as a function of time (in units of 10^9K).

Fig. 2. The luminosity of the surface layers as a function of time when the shock wave penetrates the surface.

Up to this time the envelope has remained in hydrostatic equilibrium (except for the shock passage), but the steady heating causes the surface luminosity to reach L_{Edd} and the envelope begins to expand at a few km sec^{-1}. Shortly after it reaches a radius of 2×10^8km, the envelope becomes pulsationally unstable with the excursions in luminosity reaching factors of 2. The light curve is shown in Figure 3 where the episodes of pulsational instability appear as spikes superimposed on the steady plateau behavior.

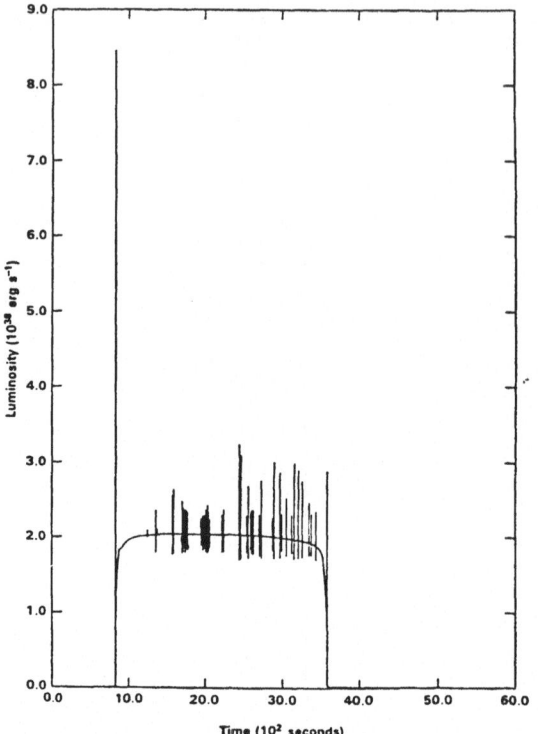

Fig. 3. The luminosity of the surface
layers as a function of time during the
entire outburst. The initial spike is
the precursor shown in Figure 2 and the
vertical bars are times of pulsational
instability during the evolution.

It takes about 2000 sec of evolution for the peak temperature in the
shell source (CEI) to drop to 10^9K and the rate of energy generation in
the same zone to 10^{12} erg gm^{-1} sec^{-1}. At the same time, because of the
large radius, the temperature has fallen to 10^6K ($kT \sim 0.1$ keV). This
value is much too soft for a normal X-ray burst.

As the hydrogen fuel is consumed and the nuclear energy production
declines, the envelope begins to collapse. This stage takes 10^2sec and
as the radius decreases, the effective temperature climbs to 2.5×10^7K
($kT = 2.2$keV). The energy emitted in the 2-10 keV range goes through an
increase followed by a rapid decrease that would appear as a burst in a
detector. Once the radius has returned to 10km, the final decline takes
only 10^3sec. By 4 hours after the decline, the luminosity is 1 $L_\odot$.

III DISCUSSION

The theoretical light curve for this simulation when folded with the
instrument response of a low energy X-ray detector will appear to that
detector as two bursts, one with an extremely short time scale,

separated by 2×10^3 sec. The peak luminosity obtained in this study, $2 \times 10^5 L_\odot$, is in close agreement with the observed values. However, the theoretical radius at maximum is certainly too large to agree with the observations (Van Paradijs 1979).

The initial conditions were chosen to represent the maximum amount of material that could be accreted by a neutron star under normal conditions. The intent was to simulate the behavior of the longer time scale outbursts. The attempt was unsuccessful and we must attribute the X-ray nova outburst to some other mechanism or combination of mechanisms. We have already proposed that the thermonuclear runaway acts as a trigger on the secondary and, in fact, causes a period of enhanced mass transfer analogous to the current hypothesis for the cause of the dwarf nova outburst.

We acknowledge partial support from the National Science Foundation through grants AST81-17177 to Arizona State University and AST80-18198 to the University of Illinois. S. Starrfield thanks G. Bell and S. Colgate for the hospitality of the Los Alamos National Laboratory and a generous allotment of computer time.

References

Ayasli, S. and Joss, P. C. 1982, Ap. J., 256, 637.

Starrfield, S., Kenyon, S., Sparks, W. M., and Truran, J. W. 1982, Ap. J., 258, 683.

Starrfield, S., Truran, J. W., and Sparks, W. M. 1978, Ap. J., 226, 186.

Taam, R. E. 1980, Ap. J., 241, 358.

Van Paradijs, J. 1979, Ap. J., 234, 609.

Wallace, R. K., and Woosley, S. 1981, Ap. J. Suppl., 45, 389.

QUASI-PERIODIC OSCILLATIONS IN GALACTIC BULGE SOURCES OBSERVED BY HAKUCHO

Masaru Matsuoka

Institute of Space and Astronautical Science
Komaba, Meguro-ku, Tokyo 153, JAPAN

X-ray sources referred to as galactic bulge sources generally show no eclipses and no periodic pulsations. However, binary periods in some of them have been discovered. Sco X-1 is the first source of this class with a binary period of 0.787 day (Gottlieb et al,1975; Cowley and Crampton,1975). The short binary periods of this class of sources are consistent with a model of Joss and Rappaport (1979) wherein the source is a neutron star in a binary system with a very low-mass companion ($\lesssim 0.5$ M$_\odot$). A much longer binary period has been discovered for Cyg X-2 (9.843 days) by optical spectroscopic observations (Cowley et al,1979). Such a long binary period is not consistent with the low-mass close binary model of Joss and Rappaport (1979); rather, it suggests that the primary is a somewhat evolved low-mass star (Cowley et at,1979).

The source 4U1627-67 is unique among low-mass X-ray sources in that it exhibits X-ray pulsations (7.7 sec) and a short orbital period of 2492 sec (Rappaport et al, 1977; Middleditch et al,1981). It also has a relatively high X-ray luminosity of about 10^{37} erg/sec (White et al,1983). These characteristics led to the low-mass close-binary system model mentioned above. Moreover, this source also a quasi-periodic oscillation of ~ 1000 sec (Li et al,1980).

The first object of this paper is to present the longer periodicities of two bulge sources (GX349+2 and 4U/MXB1636-53) suggested by the Hakucho data. The second is to present a quasi-periodicity of GX349+2 between 100-200 sec.

Long-term periodicities of the bulge sources

Hakucho has observed the bright bulge sources, GX349+2, GX340+0, GX5-1, GX9+1, GX3+1, GX17+2 and 1636-53 with its wide-field burst monitoring detectors (Kondo et al,1981). Here, we present preliminary results of our periodic analyses of two of these sources: GX349+2 and 1636-53.

Fig.1 shows one example of X-ray intensity curves in the energy range of 1-22 keV for GX349+2. These data suggest that there may be two intensity states, a small variability state (May 7-9) and a large variability state (May 1-6), which continue for several days respectively without clear regularity. Other Hakucho data support

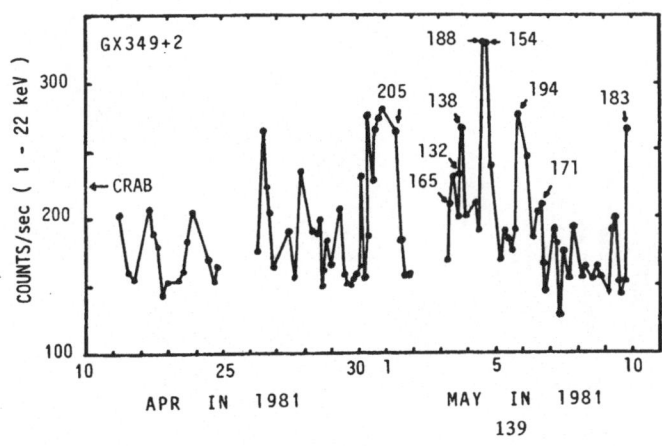

Fig.1 X-ray intensities from GX349+2 observed by FMC 2 on Hakucho. Periods (quasi-periodic pulsation) are indicated by numbers in the figure.

139

D. Q. Lamb and J. Patterson (eds.), Cataclysmic Variables and Low-Mass X-Ray Binaries, 139–142.
© 1985 by D. Reidel Publishing Company.

this two-state hypothesis.   We have also obtained a clear bimodal behavior in the hardness ratio as well as X-ray intensity for GX5-1 (Mitsuda et al,1983).   Bimodal behavior may be a common feature among bulge sources.   In order to search for periodicities we have calculated power spectra for these sources by means of the method developed by Ponman (1981).   The power spectrum of GX349+2 (Fig.2) suggests an 8.3 ± 0.7 day periodicity (90% confidence).   This period is consistent with the period of 8.71156 days obtained by Ponman (1982) from Ariel 5 data.

   Data from 1636-53 from June 14 to July 13, 1980 were also searched for periodicities.   The first aim of this was to search for the 3.8 hour period obtained from optical observation by Pedersen et al (1981).   A power spectral analysis and a standard folding method were applied to the data.   We failed to detected a periodicity at ~3.8 hours (Fig.3), but did obtain excess power at the frequencies near 0.410 day^{-1} (period ≅ 2.44 days) and at the second & third harmonics (Tsuno,1983).

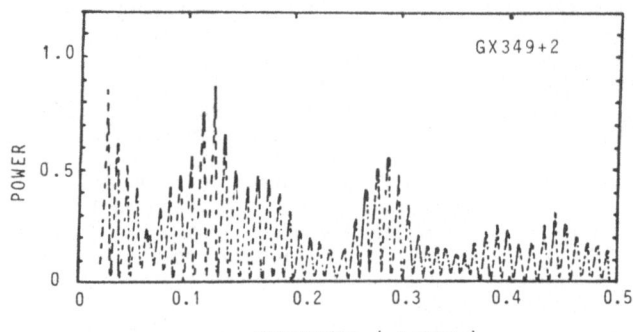

Fig.2   Power spectrum of GX349+2 from the 1980 data.

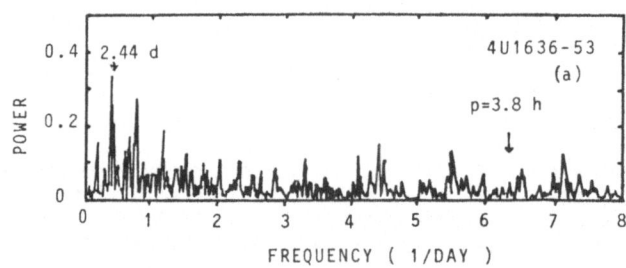

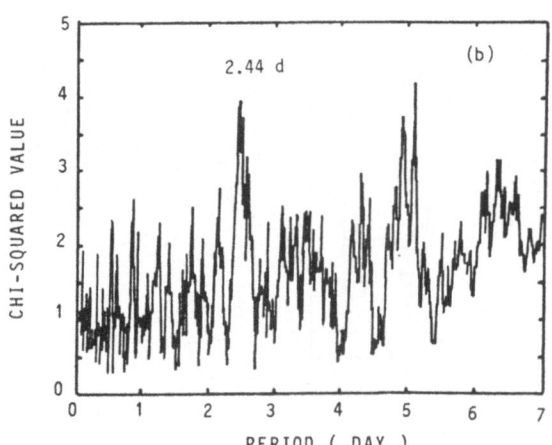

Fig.3   (a) Power spectrum and (b) reduced χ^2-value (d.o.f.=9) of 4U1636-53 from the 1980 data (Tsuno, 1983).

Reduced χ^2 values of the standard folding method in the periods of 2.44 days and 5 days were estimated to be ~4.   Thus we conclude there is some evidence for a 2.44 ± 0.02 day X-ray period in 1636-53.
     The present analyses make use of a small portion of the available Hakucho data. Further analyses are in progress.   The existence of a long-term periodicity in a bulge source is important whether it is a binary period or a quasi periodic oscilla-tion.   The evidence for both a 3.8 hour period and a 2.44 day period in 1636-53 suggests a precession period of the disk for 2.44 days (Tsuno,1983) as well as an orbital period of 3.8 hour (Pedersen et al,1981).   However, a firm conclusion regarding this interesting problem must await further analyses and further observa-tions by ASTRO-B and ASTRO-C.

Quasi-periodic oscillation of GX349+2

     The source GX349+2 was observed for about 50 days in 1980 and 1981.   Fig.4 shows one example of a short-term intensity curve for GX349+2.   For an interval of about 25 min, the data exhibit excess power at the period of 183 ± 3 sec in a power spectrum analysis.   The data when folded (Fig.4b), yield a highly significant χ^2 value (reduced χ^2 = 8 for d.o.f.= 32).   This behavior of GX349+2 often disappears and the source shows irregular fluctuations or a relatively calm steady flux. Moreover, when these pulsations appear again, a somewhat different period between 100 and 200 sec is observed.   Therefore, to search for such periods, we applied a standard folding method and a power spectral analysis to the data trains which show clear oscillations.
     In order to examine the relationship between the fluctuation and the pulsation of the X-ray intensity, we estimated the variability (Ogawara et al,1977) for each data train of ~30 min;

$$\eta = \frac{1}{\bar{x}} \sqrt{\frac{1}{n-1} \Sigma (x_i - \bar{x})^2 - \frac{1}{n} \Sigma \sigma_i^2}$$

where time bins of 48 sec were employed.   Fig.5 shows the results of this analysis. The data segments which exhibit periodicities (circles in Fig.5) lie distinctly in the high-variability and high-intensity portion of the diagram in Fig.5.   The period and the reduced χ^2 value for each segment are indicated in the figure.   These periods are also indicated in Fig.1.

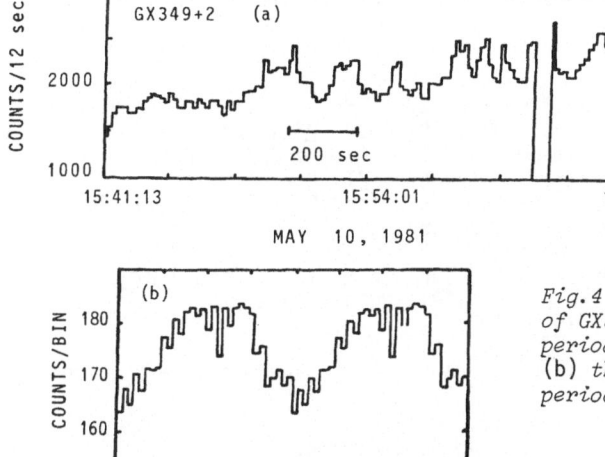

Fig.4 (a) Counting rates per 12 sec of GX349+2 when the pulsation of the period of 183 sec was excited and (b) the pulse profile folded by the period of 183 sec.

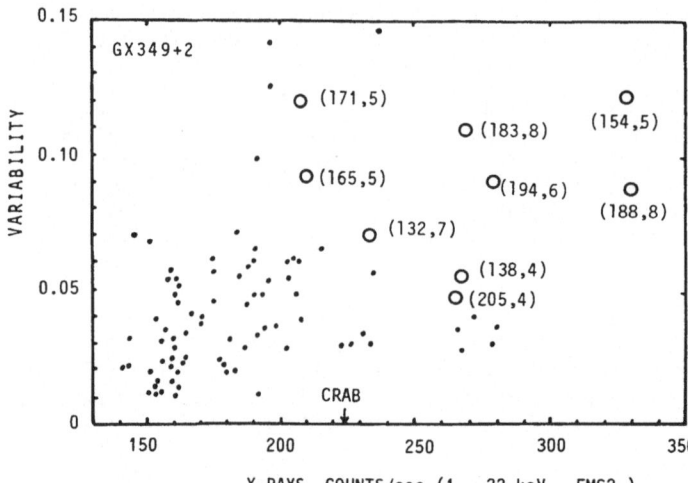

Fig.5 *Relationship between variability and X-ray intensity. All the indicated data sets were examined for periodicities with the standard folding method; those indicated by circles showed periodicities. The pulse period and reduced χ^2 values are given in parentheses.*

    In summary the oscillations of GX349+2 have the following properties: (1) the oscillations have periods between 100 and 200 sec and the pulse fraction is ~20% at maximum, (2) the same period continues only for about 30 minutes or less, (3) when the intensity was higher than the Crab intensity, the quasi-periodic oscillations were excited ~50% of the time, but no pulsations are excited for less than ~0.8 times the Crab intensity, and (4) we also point that in these data the temperature of a thermal plasma, derived from the hardness ratio, is approximately proportional to the intensity whether pulsations are excited or not.
    These quasi-periodic pulsations are reminiscent of the 1000 sec pulsation of 1627-67 (Li et al,1980) and quasi-periodic oscillations with a period of 165 ~180 sec in the optical data of Sco X-1 (Gribbin et al,1970). Further investigations of this quasi-periodic pulsation phenomenon are required to clarify the structure in the X-ray emission region of this system.

    I would like to thank K.Makishima, K.Mitsuda, K.Suzuki and K.Tsuno for the present data analysis. I am also indebted to H.Bradt for his comments and critically reading the manuscript.

References

Cowley,A.P., and Crampton,D. 1975, Ap.J.Letters 201, L65.
Cowley,A.P., Crampton,D., and Hutchings,J.B. 1979, Ap.J. 231, 539.
Gribbin,J.R., Feldman,P.A., and Plagemann,S.H. 1970, Nature 225, 1123.
Joss,P.C., and Rappaport,S. 1979, Astron. Astrophys. 71, 217.
Li,F.,Joss,P.C., McClintock,J., Rappaport,S., and Wright,E.L. 1980, Ap.J. 240,
Middleditch,J., and Nelson,J. 1976, Ap.J. 208, 567.
Mitsuda,K., and Hakucho team 1983, preprint.
Ogawara,Y., Doi,K., Matsuoka,M., Miyamoto,S., and Oda,M. 1977, Nature 270, 154.
Pedersen,H., van Paradijs,J., and Lewin,W.H. 1981, Nature 294, 725.
Ponman,J. 1981, Mon.Not.R.astr.Soc. 196, 583.
Ponman,J. 1982, Mon.Not.R.astr.Soc. 200, 351.
Rappaport,S., Markert,T., Li,F.K., Clark,G., Jernigan,G., and McClintock,J. 1977
    Ap.J.Letters 217, L29.
Tsuno,K. 1983, PhD Thesis in Osaka University.
White,N.E., Swank,J.H., and Holt,S.S. 1983, Ap.J. in press.

THE MASS TRANSFER RATE IN X1916-053: IS IT DRIVEN BY GRAVITATIONAL RADIATION?

J.H. Swank
NASA/Goddard Space Flight Center

R.E. Taam
Northwestern University

N.E. White
ESTEC

A 50 minute period for a binary system harboring an X-ray burster would allow several alternatives for the mass-giving secondary, including a H shell burning plus He degenerate core composite model. We use the burst properties of X1916-053 (4U1915-05/MXB1916-053) to argue against the He degenerate as well as the He main sequence solutions and to estimate whether for any of the other solutions the mass transfer rate could be consistent with that expected from gravitational radiation (GR). Within uncertainty of a factor of 2 the transfer rate for the composite model solution is consistent with GR, but enhancement by other mechanisms should be investigated.

I. Introduction

For low mass short period binaries the loss of angular momentum by gravitational radiation is expected to play an important role in the evolution of the system and may dominate over other angular momentum loss mechanisms. There are currently four known systems with binary periods < 1h, the two white dwarf cataclysmic variables, AM CVn (18 min.) and G61-29 (47 min.), and the two neutron star X-ray binaries, the pulsar X1627-673 (42 min.) and the burster X1916 (50 min.). The secondaries in the CVs are thought to be He white dwarfs. The nature of the companions to the neutron stars is uncertain. It is of interest to determine the mass transfer rates in these systems. For X1916-053, equating the steady, non-dip (White and Swank 1982) X-ray flux to the gravitational potential energy released by matter falling on a neutron star gives a lower limit for the mass transfer rate: $\dot{M} > 4.7 \times 10^{-10}$ $(d/10 \text{ kpc})^2$ $M_\odot$ yr^{-1}, when the mass and radius dependent gravitational red shift are taken into account. The optical identification (Walter et al. 1982) does not constrain d. However,

143

D. Q. Lamb and J. Patterson (eds.), Cataclysmic Variables and Low-Mass X-Ray Binaries, 143–146.
© *1985 by D. Reidel Publishing Company.*

since X1916-053 is a burster, a different set of arguments is
available. The observational and theoretical properties of X-ray
burst sources provide evidence both on the composition of the mass
losing star and the mass transfer rate. Here we give a brief summary
of the results of a detailed consideration (Swank, Taam and White
1983) of the evolutionary possibilities and the burst properties.

II. Possible Secondaries and $\dot{M}_{GR}$

Figure 1 shows the mass-radius constraint on a Roche lobe-filling
secondary in a 50 min. binary with a neutron star companion, together
with mass-radius relations of the degenerate dwarfs, of H and He main
sequence stars, and of stars affected by evolution with mass loss
(Paczynski and Sienkiewicz 1981; Rappaport, Joss and Webbink 1982
(RJW); Whyte and Eggleton 1981). In a somewhat evolved star the core
could collapse and the star could transit the region between a H
burning star and a white dwarf (Faulkner, Flannery and Warner 1972
(FFW)). A sequence calculated for a star put into a binary at the
point when the core begins to collapse gives a "composite" model
solution for a 50 minute binary. The mass transfer rates that would
be driven by GR alone are shown in Table 1.

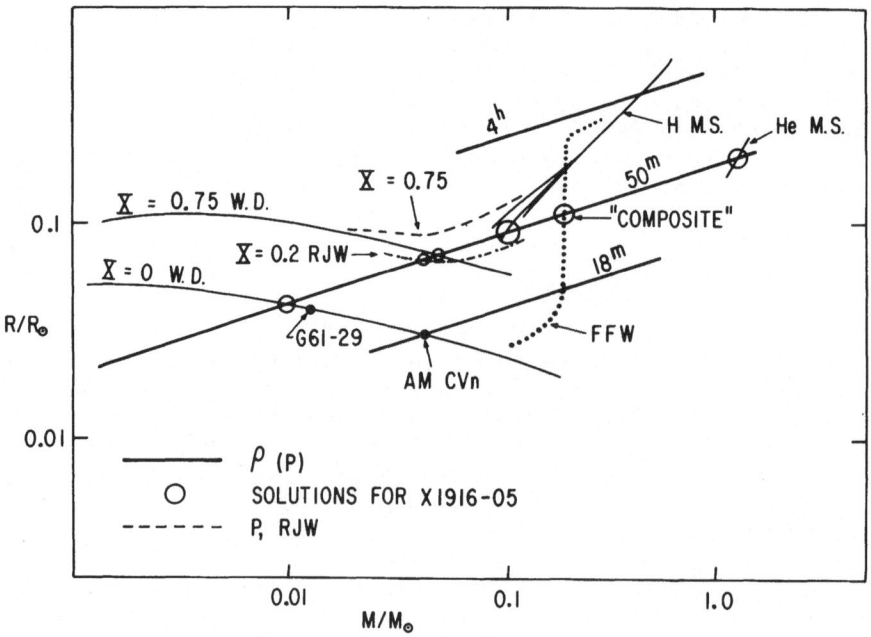

Figure 1. Radii and masses of secondaries in a 50 minute binary.
Alternative R-M relations have been suggested for the lower end of the
H main sequence (M.S.).

Table 1. Possible Secondaries and $\dot{M}_{GR}$*

Type Equilibrium	$\dot{M}_{GR}$ ($M_\odot$ yr^{-1})	Type Non-Equilibrium	$\dot{M}_{GR}$ ($M_\odot$ yr^{-1})
H burning	2×10^{-10}	H burning (X=0.2)	4×10^{-11}
He burning	6×10^{-8}	Composite	2×10^{-10}
H W.D. (X=0.75)	7×10^{-11}	*For references,	
He W.D. (X=0.)	3×10^{-12}	see Swank, Taam and White 1983.	

III. Burst Constraints

A. Implications of Thermonuclear Flash Theory

The observed burst recurrence times, $\tau_R \sim 3\text{-}6^h$, imply it is improbable that pure He is the accreting material (Taam 1981). This makes a He degenerate dwarf unlikely. For a He burning companion GR would drive too much mass transfer.

H accreting with He would not have been burned completely between bursts with $\tau_R < 10(X/0.7)(0.02/Z_{CNO})$h, so the flashes are probably of H plus He, burning together. For this type of accreting matter, $\alpha \sim 60\text{-}150$, consistent with the $\alpha = 120$ observed. Then $\tau_R \sim 3\text{-}6h$ for $\dot{M} \sim 5 \times 10^{-10}$ $M_\odot$ yr^{-1} (Ayasli and Joss 1982; Taam 1981,1982) and several times longer for substantially lower $\dot{M}$ models. Complicating effects could reduce the recurrence times for lower $\dot{M}$, but this aspect of flash theory favors $\dot{M} \sim 5 \times 10^{-10}$ $M_\odot$ yr^{-1}. The steady flux then implies d $\sim$ 10kpc.

However, flash theory also has so far predicted peak luminosities obeying the Eddington limit for the accreting material (Joss 1978; Taam 1982; Ayasli and Joss 1982) and 10 kpc implies $F_B/F_{ED} > 3$. It is not known yet whether the $\dot{M}$ or peak flux estimates are correct to better than a factor of 2.

B. Comparison to other Bursters

Although bursts from a given source vary in peak flux by $\sim$ 4, it may still be that the average is effectively a standard candle or that the maximum peak luminosity is (Lewin 1982). Values of 2 and 5×10^{38} ergs s^{-1}, for average and maximum, respectively, for the ensemble of bursters, cluster them around a distance of 9 kpc. By comparison, our observations would put X1916-053 at $\sim$ 10 kpc. Further, a subset of 6 bursters is probably within 100 pc of the galactic center (Inoue et al. 1981) and their maximum peak luminosities were $6\text{-}13\times10^{38}$ d_{10}^2 ergs s^{-1} (vs. 5×10^{38} d_{10}^2 for X1916-053).

IV. Conclusions

If the flash theory preference for $\dot{M} = 5\times10^{-10}$ $M_\odot$ yr^{-1} is correct, then $\dot{M} > 2$ $\dot{M}_{GR}$ for the composite model. If $F_B/F_{ED} \sim 1$ is correct, $\dot{M} \sim \dot{M}_{GR}$ for the composite model, but $> \dot{M}_{GR}$ for the W.D. secondary solutions. If the bursts from X1916-053 are like those from other bursters, and the galactic center distance is close to 9 kpc, the results agree with the indications of flash theory that $\dot{M} \sim 5 \times 10^{-10}$ M_θ yr. The burst properties of X1916-053 thus argue for $\dot{M} \sim 2-5 \times 10^{-10}$ M_θ yr^{-1}. Table 2 compares this with $\dot{M}_{GR}$ for the solutions considered and notes other considerations relevant to the likelihood of the secondary being of that type.

Table 2. Comparison of M and M_{GR}

Type	M/M_{GR}	Comment
Equilibrium		
H burning	$> 1-2$	solution uncertain
He burning	$\ll 1$	also inconsistent with optical id.
H W.D.(X=0.75)	$> 3-6$	
He W.D. (X=0.)	$\gg 100$	inconsistent with τ_R
Non-Equilibrium		
H burning (X=0.2)	$> 5-10$	
composite	$> 1-2$	

The composite model comes within a factor of 2 of being a self-consistent solution for which $\dot{M} \sim \dot{M}_{GR}$. But discrepancies of 3-5 are of the same order. Other mechanisms for driving mass transfer are certainly expected to contribute but which is important for very low mass companions is not clear.

Acknowledgment: RET is indebted to NSF grant AST-8109826 and JHS to a Royal Society Guest Research Fellowship for partial support of this work.

References

Ayasli, S., and Joss, P.C.: 1982, Astrophys. J. 256, p. 637.
Faulkner, J., Flannery, B.P., and Warner, B.: 1972, Astrophys. J. 175, L79.
Inoue, M. et al.: 1981, Astrophys. J. 250, L71.
Joss, P.C.: 1978, Astrophys. J. Letters 225, L123.
Lewin, W.H.G.: 1982, in Accreting Neutron Stars, ed. W. Brinkman and J. Trumper, p. 77.
Paczynski, B., and Sienkiewicz, R.: 1981, Astrophys. J. 248, p. L27.
Rappaport, S., Joss, P.C., and Webbink, R.F.: 1982, Astrophys. J. 254, p. 616.
Swank, J.H., Taam, R.E., and White, N.E.: 1983, Astrophys. J., in press.
Taam, R.E.: 1981, Astrophys. J. 247, p. 257.
Taam, R.E.: 1982, Astrophys. J. 258, p. 761.
Walter, F.M., Bowyer, S., Mason, K.V., Clark, J.T., Henry, J.P., Halpern, T., and Grindlay, J.E.: 1982, Astrophys. J. Letters 253, L67.
White, N.E., and Swank, J.H.: 1982, Astrophys. J. 253, L61.
Whyte, C.A., and Eggleton, P.: 1980, Mon.Not.R.Astron.Soc. 190, p. 801.

COS-B X-RAY OBSERVATIONS OF CYGNUS·X-3

J.M. Bonnet-Bidaud[1] and M. van der Klis[2]

[1] Section d'Astrophysique, CEA Saclay, F.
[2] Astronomical Institute, University of Amsterdam, NL.

Abstract : The X-ray source Cygnus X-3 was further observed by the X-ray detector onboard the COS-B satellite from November 1981 to February 1982. This brings to 220 days the total duration of observation of this source by COS-B. Preliminary results from the last observations are presented which show that :

- the source intensity is not periodically modulated with periods from 1 to 60d.
- the measured mean time of X-ray minimum of the 4.8h modulation is in accordance with a long-term ephemeris with a non-zero period derivative.
- short term fluctuations of the time of X-ray minima inside the observation are present but probably not strictly modulated with the period of ∿19d reported from previous observations (Bonnet-Bidaud and van der Klis 1981).

The last point does not allow to confirm the hypothesis of an apsidal motion as the cause of the short-term changes of the 4.8h period.

1. OBSERVATIONS

The COS-B satellite was pointed to Cygnus X-3 for more than three months from Nov. 5, 1981 to Feb. 18, 1982. The source was near the centre of the field of view at the near maximum sensitivity of the 80 cm^2 (2-12kev) proportional counter.

Data were recorded continuously during 25.4 s integration time every 102.4 s in the active part of the satellite orbit (24h out of every 36h). We used the **exceptional** length of this observation to test the different periods proposed for the source intensity changes (Holt et al. 1976, Molteni et al. 1980) and the phase changes of the X-ray minimum of the 4.8h modulation (Bonnet-Bidaud and van der Klis 1981).

Preliminary results are presented. Full discussion will be given elsewhere.

147

D. Q. Lamb and J. Patterson (eds.), Cataclysmic Variables and Low-Mass X-Ray Binaries, 147–150.
© 1985 by D. Reidel Publishing Company.

2. RESULTS

2.a. Intensity changes

After background subtraction, heliocentric-corrected data in each
satellite orbit were fitted to a sinusoidal curve of the form
$A_0-A_1\cos(2\pi(t-T_0)/P)$. The quantities T_0 and P were computed from
the quadratic ephemeris of van der Klis and Bonnet-Bidaud (1981).
The results for A_0 and A_1 are presented in Figure 1.
The source mean counting rate varies from 10 to 35 c/s (1 COS-B c/s
~ 2.3 to 2.7 10^{-10} erg.cm^{-2}.s^{-1} for the range of the source spectra),
while the amplitude of the 4.8h modulation varies from 2 to 12 c/s.
Figure 1 shows that the source intensity is nearly constant in an
interval of 60d from JD2444930. to JD2444990. which excludes all strict
periodicities in the range 1 to 60d.
In particular the predicted times of X-ray maximum intensity with the
period of 34.1d claimed by Molteni et al.(1980) (marked by arrows in
Fig.1) do not correspond, at least in two cases, to any enhancement in
the source mean level.

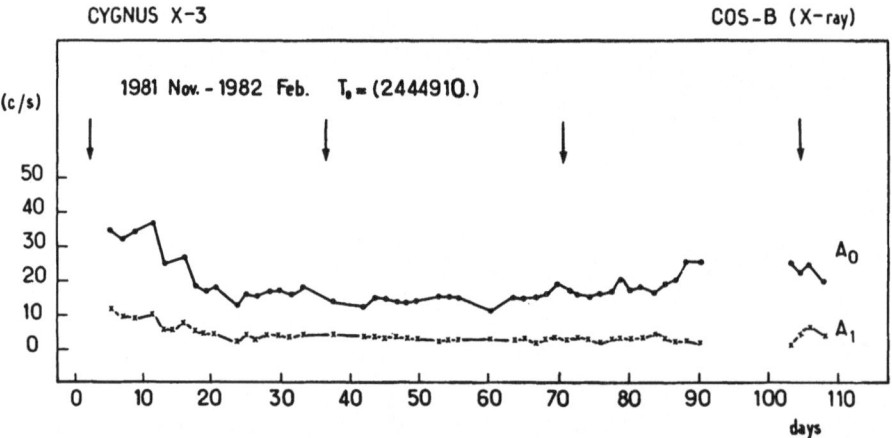

Figure 1.   The mean level A_0 and the amplitude A_1 of the 4.8h modulation
            against time. Each A_0 and A_1 point is determined from a sine
            fit to about 5 cycles of the source. Continuous coverage is
            more than 60%. Vertical units are in c/s after background
            subtraction. Statistical error bars are smaller than the
            points. There is no evidence of a 34.1d periodic intensity
            variations as proposed by Molteni et al.(1980) (the arrows
            indicate the predicted times of maximum intensity with this
            period).

2.b. X-ray minimum phase changes

The flux in each satellite orbit was first normalized by subtracting the
mean and dividing by the amplitude as derived in 2.a. A function of the
form $a-b\cos(2\pi(t-T_{min})/P)$ with P computed from the quadratic ephemeris
of van der Klis and Bonnet-Bidaud(1981), was then fitted between
JD2444912.5 and JD2445017.5. The best fitting value for T_{min} was
T_{min} = JD2444965.9688±0.00014 (one-sigma single parameter error bar).
This new determination is shown in Figure 2 and compared to the previous
quadratic ephemeris. A new determination of the period derivative using
this last point gives $\dot{P}$ = (1.09±0.08) 10^{-9} s/s with χ_{ν}^{2}=1.36 for ν =58,
while a constant period hypothesis yields χ_{ν}^{2}=4.35 at ν =59. According
to the F-test the probability for $\dot{P}$ to be zero is now less than 6 10^{-10}.
This result confirms with a higher degree of confidence the existence of
a continuous change of the 4.8h modulation period though with a rate of
change slightly less than previously quoted (Manzo et al. 1978, Lamb et
al. 1979, Elsner et al. 1980, van der Klis and Bonnet-Bidaud 1981).

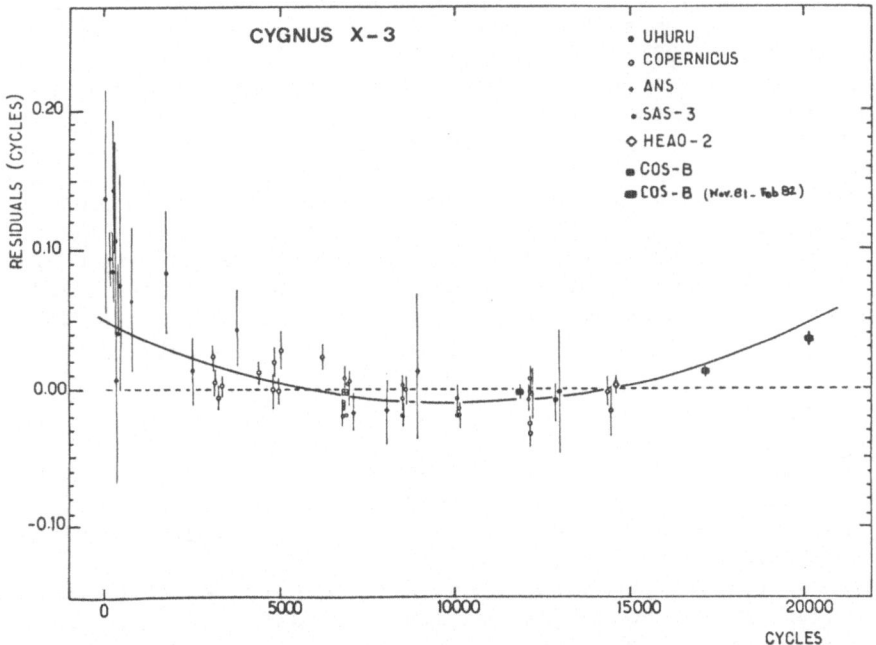

Figure 2.   The mean heliocentric time of X-ray minimum in the (Nov.81-
            Feb.82) COS-B observations compared to the nine years
            ephemeris of van der Klis and Bonnet-Bidaud (1981)
            (drawn parabola corresponds to their best value of
            $\dot{P}$ = 1.2 10^{-9} s/s).
            Introduction of the last point gives a more accurate value
            of $\dot{P}$ = (1.09±0.08) 10^{-9} s/s, with an improved level of
            confidence for the $\dot{P}$ = 0 hypothesis (formal chance for a
            constant P is less than 6 10^{-10}).

We further investigated changes in the time of X-ray minimum inside the observation. Data in each satellite orbit ($\sim$ 5 cycles of the 4.8h modulation) were folded using the above quadratic ephemeris and then cross-correlated to a template curve around minimum in order to minimize possible effects of the changes in the shape of the modulation around maximum (see Bonnet-Bidaud and van der Klis 1981). A set of 49 independent phase points were obtained which show fluctuations of $\sim \pm 2$ 10^{-3}d. A period search through those points was performed by folding them with periods in the range 10 - 30d and comparing the resulting folded curve with a flat curve using a χ^2 test. No significant peak was observed around the period of 19d quoted from previous observations (Bonnet-Bidaud and van der Klis 1981).However a simulation of a pure 19d sinusoidal signal distributed according to the data points shows that the data window will prevent to detect a modulation of the order or less than 2 10^{-3}d.

3. CONCLUSION

The source flux is not periodically modulated with periods from 1 to 60d. This probably excludes the massive binary model recently proposed by Molteni et al. (1980).
Periodic 19d changes in the time of X-ray minimum are not confirmed from the present observation possibly due to an unfavourable data coverage. The long term variation of the 4.8h modulation period is now clearly established. We compute that if the companion star is $0.4 \leqslant \dot{M} \leqslant 0.6$ $M_\odot$, the long term 4.8h period changes will be explained by a mass loss of the system of $\dot{M} \sim 5$ 10^{-7} $M_\odot \cdot yr^{-1}$. We further note that if Cygnus X-3 belongs to the class of the low-mass binaries, the observed long-term increase of the orbital period means that Roche-lobe overflow by a red-dwarf companion cannot be powering the X-ray source as the expansion of the orbit will prevent this mecanism from operating.

4. REFERENCES

Bonnet-Bidaud,J.M.,and van der Klis,M.:1981,Astron.Astrophys.101,p.299.
Elsner,R.,Ghosh,P.,Darbro,W.,Weisskopf,M.,Sutherland,P.,and Grindlay,J.:
    1980,Astrophys.J. 239,p.335.
Holt,S.,Boldt,E.,Serlemitsos,P.,and Kaluzienski,L.:1976,Nature 260,p.592.
Klis van der,M.;and Bonnet-Bidaud,J.M.:1981,Astron.Astrophys.95,L5
Lamb,R.C.,Dower,R.G.,and Fickle,R.K.:1979,Astrophys.J.229,L19.
Manzo,G.,Molteni,D.,and Robba,N.R.:1978,Astron.astrophys.70,p.317.
Molteni,D.,Rapisarda,M.,Robba,N.R.,and Scarsi,L.:1980,Astron.astrophys.
    87,p.88.

THE AM HERCULIS MAGNETIC VARIABLES

James Liebert and H. S. Stockman
Steward Observatory, University of Arizona

ABSTRACT

The observational properties of the ten known AM Her systems are reviewed. A multitude of components of continuum and line radiation from these objects are outlined. We discuss the important physical processes involved in the accretion flow and funnel shock, with emphasis on the properties of the emission line regions. Other topics include the properties of the white dwarf primaries and M dwarf secondaries, the maintenance of synchronous rotation, and the implications of "clumpings" in the known orbital periods (80-115 minutes) and magnetic field strengths (20-35 megagauss).

1. INTRODUCTION

AM Herculis objects are cataclysmic variable (CV) binary systems containing a strongly magnetic white dwarf primary. Their unique, defining observational property is strong circularly and linearly polarized radiation, at or near optical wavelengths. Since the primary star's magnetosphere apparently prevents formation of the usual CV accretion disk, material spilling over from the secondary star is funneled onto the white dwarf's surface near a magnetic pole. Most gravitational energy is released in a standing shock region in this polar accretion column, including much X-ray and polarized optical radiation. Thus the term "polars" -- first proposed by two Polish astronomers (Krzeminski and Serkowski 1977) -- has also been used to label these systems. The original polarization discoveries of AM Her and several other systems were made by Tapia -- cf. Tapia (1977).

Since AM Her was first identified as an exciting, unprecedented kind of CV, the literature on this new group has proliferated and many new systems have recently been discovered. In this review we concentrate on the observational properties of these objects, with limited attempts to point out which theoretical ideas are surviving the early tests. The previous reviews emphasizing early work on AM Her itself are those of Kruszewski (1978) and Chiappetti, Tanzi

151

D. Q. Lamb and J. Patterson (eds.), Cataclysmic Variables and Low-Mass X-Ray Binaries, 151-177.
© *1985 by D. Reidel Publishing Company.*

and Treves (1980). However, a nice discussion especially of the high
energy observational properties is included in the overall CV review of
Cordova and Mason (1982).

There are currently ten certified AM Herculis systems known, six of
these having been identified within the past two years. While only the
polarization properties may (so far) separate them uniquely from all
other types of CVs, it is appropriate to assess their other largely
common properties. The ten objects and some salient properties are
listed in Table 1. First, we note -- in column two following the
designations -- the two ways by which particular systems were first
identified for polarimetric observations, through either an X-ray (X)
source or optical (O) discovery. It was the strong emission line
spectrum -- especially He II -- which led Bond and Tifft (1974) to
identify AM Her as an interesting CV-like variable for further study;
on the other hand, the optical object was quickly identified with an
independently discovered, very strong soft X-ray source (Berg and
Duthie 1977; Hearn, Richardson and Clark 1976). Indeed, it now appears
that the AM Her systems generally show both higher relative X-ray
fluxes, and stronger optical/ultraviolet emission lines -- especially
at high excitation -- than most other CV types. They are both strong
hard and soft X-ray sources (Cordova and Mason 1982).

The objects are listed in order of decreasing period, as given in
column three. All have short binary periods, with most below the 2-3
hour CV period "gap." These are binaries of small orbital scale ($\sim$
$1R_O$), with only low mass secondary stars detected (see comments) and
comparatively low accretion rates and luminosities compared with longer
period CVs. Also given in Table 1 are a variety of photometric,
polarimetric and spectroscopic properties which will be discussed subsequently. Since the primary star is phase-locked in synchronous rotation, most observational properties are strongly modulated with binary
phase. In addition, the objects show long term (month-to-year)
variations in the accretion rate ("high" and "low" states), and largely
stochastic variations ("flickering") on short timescales of seconds to
minutes.

In Figure 1, we present a hierarchical picture of an AM Her
system, depicting the stellar components and orbital scale of the
system ($\sim$ $1R_O$), the scale of the white dwarf and broad emission line
regions ($\sim$ 10^{-1} - 10^{-2} R_O), and the bottom of the accretion funnel plus
standing shock ($\sim$ 10^{-4} - 10^{-2} R_O). The sources of continuum and line
radiation in an AM Her are numerous, complicated, and occur in all
parts of Figure 1, as we shall discuss. It is useful to present first
a descriptive overview of the radiation components (Section 2) and how
they are modulated with orbital phase. This will be followed by a physical discussion of the accretion stream/emission line region (Section
3), the accretion shock (Section 4), the stellar components and system
evolution (Section 5).

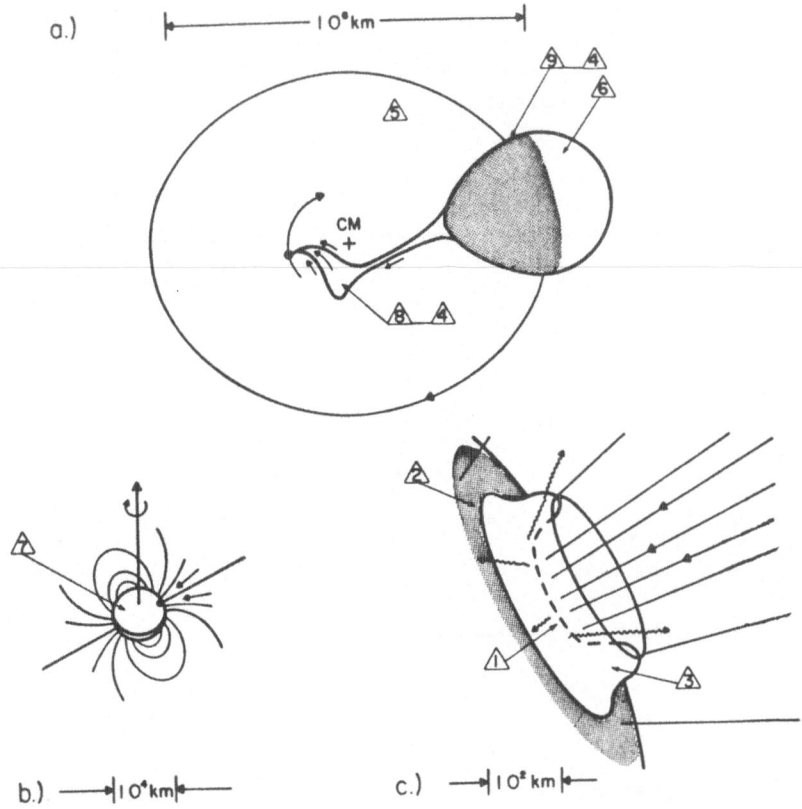

Figure 1.   Three important physical scales in an AM Her object.
Numbers indicate expected locations of continuum and line radiation
listed   below and discussed in Section 2:   (1) X-ray bremsstrahlung;
(2) soft X-ray/far UV "blackbody" component; (3) cyclotron; (4) unpolarized optical; (5) "quiescent" radio; (6) secondary star; (7) primary
star; (8) broad emission line radiation; (9) narrow emission lines.

    Some success has been achieved by assuming that the primary's
magnetic field structure is basically dipolar (Section 5).   Thus, the
discussion of the phase-dependent observational properties is facili-
tated by dividing the AM Her systems into two groups: First, there are
those which, throughout their orbital cycle, present only one magnetic
pole to our view (c.f. AN UMa, EF Eri, the one-pole geometry).   Second,
one can define a group in which both magnetic poles are alternately
viewed along our line-of-sight, though one of the poles may be
accreting all or most of the transferred matter at a given time.   The
"prototype" for the two-pole geometry is VV Pup.   Note that AM Her
itself appears to be a borderline case, as we shall discuss.

The likely components of continuous and line emission radiation in all wavelength intervals are discussed below in roughly decreasing order of luminosity. Each component is tagged with a number, and the corresponding number in Figure 1 shows from which physical region the radiation is believed to originate.

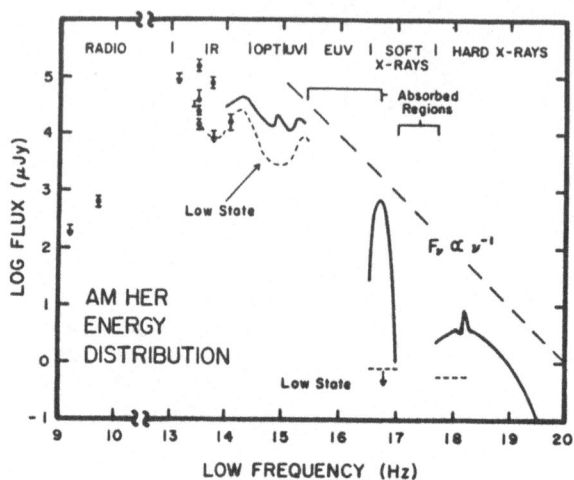

Figure 2. Energy distribution of AM Her in high and low states.

2. RADIATION COMPONENTS IN AM HER SYSTEMS

2.1 High Energy Components

The hard X-ray and gamma radiation studied in most detail for AM Her itself is believed to be due to bremsstrahlung (#1, in Figure 1) from the accretion shock region. The hard X-ray power-law spectrum (kT ~ 30 keV -- Rothschild et al. 1981) is well-documented for several of the objects. While theorists generally argue that this component should liberate the most gravitational (accretion) energy in the systems, it has been difficult to prove this from observational data.

Instead, the dominant component inferred from observations is a soft X-ray/far ultraviolet component (#2) of uncertain origin, noted by some to fit an absorbed "blackbody" kT ~ 15 - 40 ev -- Tuohy et al. 1978). The principal problem is that most of the radiation and the alleged blackbody peak should emerge at 100 - 1000A (EUV) wavelengths, a region currently unobservable. However, both the (inferred) soft X-ray and UV spectra generally turn up sharply near both EUV region boundaries. This is shown in Figure 2, an updated composite energy distribution for AM Her in the "high state."

AM Her apparently shows one accreting pole for all but 0.2 of its 3.08 hour binary period. When this pole disappears briefly behind the primary's limb, both the bremsstrahlung and soft X-ray/far UV components undergo a sharp and deep eclipse (c.f. Hearn and Richardson 1977). Thus, components (1) and (2) are localized to the accretion shock and surrounding regions. The latter may include thermalized radiation due to reprocessing of the bremsstrahlung component in a polar cap illuminated by the shock.

2.2 Optical/Infrared Radiation

The circularly and linearly-polarized radiation is dominated by a thermal cyclotron source (#3, Figure 1) which may be important at infrared, optical and ultraviolet wavelengths (see Section 4.2). Normally the optical region may cover rather high harmonics of the cyclotron fundamental frequency, as shown by the discovery of peaks equally-space in frequency for VV Puppis (Visvanathan and Wickramasinghe 1979; Stockman, Liebert and Bond 1979). This fixed the VV Pup polar field strength at ~ 32 megagauss (MG) with the fundamental wavelength near 3.34μ (Wickramasinghe and Meggitt 1982). For others, the widths of the harmonics are presumably so great that they smear together into a polarized continuum.

That this flux also comes from the accretion "shock" region becomes obvious from examining the striking phase modulation of the polarized light. In Stockman et al. (1983, see also Schmidt 1983) the circular polarization curve and other modulated properties of CW1103+254 are displayed. This presents a two-pole geometry, but only one pole -- visible for ~ 1/3 of the period -- was apparently an active accretor during these 1982 observations. The active pole showed negative circular polarization reaching -20%. When this pole passed out of view, the circular polarization dropped quickly to zero, remaining there for ~ 2/3 of the period. The photometric light curve showed a corresponding "bright" (polarized) and "faint" (unpolarized) phase, a behavior we would also expect the X-ray "light" curve to mimic.

When this accreting pole was viewed at nearly a perpendicular angle -- the column going into and out of eclipse behind the photosphere -- strong linear polarization pulses were recorded (phases 0.7 and 0.0). CW1103+254 may, however, be nearly unique among the AM Her objects in showing a kind of linear polarization "pulse" at both ingress and egress from polar eclipse. Indeed, it exhibited detectable linear polarization for all of its photometric "bright" phase. The nine other systems have usually exhibited just one well-defined pulse per period, (see, however, Bailey et al. 1978; Tapia 1982).

VV Pup has often shown a sharply-defined circular polarization curve similar to CW1103+254, with a corresponding bright phase in optical and X-ray photometry lasting for 40% of its 100-minute period (Warner and Nather 1972; Liebert et al. 1978; Patterson et al. 1983). When VV Pup was quite bright throughout the period, however, it displayed two alternating signs of circular polarization -- both poles were accreting (Liebert and Stockman 1979).

PG1550+191 (Liebert et al. 1982a) is a "textbook" example of the one-pole geometry. PG1550+191 shows one sense of (negative) circular polarization, the amount depending on the viewing angle of the pole to the line-of-sight. Due to the "beaming" effect of the cyclotron radiation, the net polarization actually decreases to zero when the view is most nearly pole-on, in nice agreement with prior theoretical predictions (c.f. Chanmugam and Dulk 1981; Meggitt and Wickramasinghe 1982). The beaming effects are greater for higher cyclotron harmonics ("blue" light) than lower harmonics ("red" light). The linear polarization pulse may be generally weak since the observed pole never gets quite perpendicular to our line-of-sight. The soft X-rays are also attenuated when a "one pole" system is viewed close to pole-on; EF Eri is a well-documented example (Bond, Chanmugam and Grauer 1979; Schneider and Young 1980; Bailey et al. 1982).

An unpolarized continuum component (Figure 1, #4) apparently contributes significantly at optical and ultraviolet wavelengths. Its origin remains ambiguous. Reprocessing of high energy radiation in the dense gas of the accretion stream and possibly at the heated stellar photospheres could contribute (see Section 3.2). The most obvious such indirect radiation at these wavelengths is the strong Balmer continuum seen in all objects during active accretion states.

2.3 Detection at Radio Wavelengths

AM Her was the first CV-like system detected at radio wavelengths (with the VLA), as reported by Chanmugam and Dulk (1982). The detection at 4.9 GHz (6 cm) was confirmed by Dulk, Bastian and Chanmugam (1983), who also observed a radio "flare" from AM Her some 20 times more intense than the quiescent (0.6 mJy) level. Attempts to detect AM Her at 1.4 and 15 GHz were unsuccessful. Likewise, only upper limits of about 0.2 mJy were obtained for 4.9 GHz radiation from VV Pup, EF Eri, PG1550+191, CW1103+254, and AN UMa. Dulk et al.'s suggested origins for the "quiescent" and "flare" radio components are discussed briefly in Section 5.2. The former is identified with loose energetic electrons on the scale of the system (Figure 1, #5).

2.4 The Low Accretion State - The Stellar Components

Since an AM Her object lacks an accretion disk reservoir, changes in the mass flow from the secondary star cause drastic, immediate consequences for the system's luminosity. AM Her, VV Pup and AN UMa have each been found for several months or more at optical brightnesses some 3 - 5 magnitudes fainter than maximum brightness. The long-term photographic history of AM Her revealed no clear pattern or significant periodicity for these changes of "state" (Hudec and Meinunger 1977; Feigelson, Dexter and Liller 1978). In particular, when the accretion flow is curtailed -- or the active accretion pole out of view -- the continuous energy components described previously may be weakened enough for two other components to appear.

At red-infrared wavelengths, the secondary star's photosphere (Figure 1, #6) has been clearly seen in AM Her (Young and Schneider 1979; Young, Schneider, Shectman 1981) and probably in VV Puppis (Liebert et al. 1978, Szkody and Capps 1980, Allen and Cherepashchuk 1982). In both cases the indicated secondary spectral type -- from both infrared photometry and optical spectrophoto-metry -- was dM4-5. A companion star was not detected in the low state spectrum of AN UMa (Liebert et al. 1982b) nor was the companion seen during the faint phase of CW1103+254 (Schmidt, Stockman and Margon 1981).

An ultraviolet-optical wavelengths, the low state of AM Herculis showed substantial radiation from the primary's photosphere (#7). The IUE spectra indicated a $T_e \sim 50,000$ photosphere or heated polar cap region (Szkody, Raymond, and Capps 1982). More strikingly, the optical spectrum revealed phase-dependent Zeeman absorption lines, not unlike those seen in single magnetic white dwarfs. The surface field strengths seems to range from ~ 13 MG to ~ 20 MG (Schmidt, Stockman and Margon 1981). Likewise, the faint phase spectropolarimetry of CW1103+254 indicated Zeeman features from a somewhat higher surface field, closer to the VV Pup value (Schmidt, Stockman, and Grandi 1983).

Similar drops in the hard and soft X-ray flux occurred during the AM Her low states of 1975 and 1980 (Fabbiano 1982; Hearn and Richardson 1977). However, while the hard X-rays scaled with the optical flux, there is a possible problem with the soft X-ray observations of 1975-1976. The problem is complicated by evidence for geometric variations in AM Her -- the disappearance of X-ray eclipse (Priedhorsky and Marshall 1982), the lack of the corresponding optical photometric minimum (Priedhorsky and Krzeminski 1978), and the reduced positive circular polarization phase (Bailey and Axon 1981; Latham, Liebert, and Steiner 1981).

Despite the exposure of both stellar components during low states, the circularly-polarized cyclotron emission from the principal accreting pole has never been found to disappear. In fact, the net circular polarizations at red-infrared wavelengths observed for AM Her, VV Pup and AN UMa during low states were comparable to or higher than their corresponding high state percentage. This behavior may be attributed to the lower optical depths in the accretion shock, resulting in higher net circular polarization from this weakened component (see Section 4.2).

On the other hand, we have previously noted that systems with two-pole geometries may exhibit essentially zero circular polarization while the "bright phase" pole is out of view. This behavior indicates that phase-lock is achieved with one magnetic pole of the primary pointed generally toward the companion. The favored pole may often collect all of the loose gas, especially during the low states.

2.5 Emission Line Components

AM Her generally exhibit large emission line-to-continuum ratios. While the strong He II λ4686 (~ Hβ) has distinguished AM Her spectra from those of other CVs, the line spectrum is a strange mix of high and low excitation lines. The characteristics are analyzed in Section 3.1. In the low states, the line emission weakens greatly -- particularly at high excitation -- and the "broad" velocity component discussed below may disappear.

The complicated emission line profiles are highly modulated with binary phase. In earlier analyses, it appeared that the profile could be divided into two components -- (1)(#8) a "broad" or "base" component, and (2)(#9) a "narrow" or "peak" component (Cowley and Crampton 1977; Greenstein et al. 1977). The origins of both have been difficult to establish, though it has usually been assumed that the narrow component comes from the vicinity of the secondary star, while the broad optically-thick component occurs in a denser region of the accretion stream. It has been customary to force sine wave fits to both components (Section 3.2). However, there is evidence from recent high resolution spectroscopy of a more complicated, multicomponent structure in the emission lines in EF Eri and VV PUP (c.f. Cowley, Crampton and Hutchings 1982, 1983). In EF Eri this structure varies from one binary period to the next (Bailey and Ward 1981; Young et al. 1982) and Hα may sometimes even go into absorption (Verbunt et al. 1980). In the following section we discuss the physical conditions and likely locations of the line emission in more detail, and discuss constraints on the secondary star masses from the radial velocity data. Coupled with our knowledge of the shock region (Section 4), we suggest a model for the accretion flow which might serve as a framework for exploiting improved spectroscopic data of these systems.

3. THE ACCRETION FLOW/EMISSION LINE REGION

3.1 Physical Conditions of the Emission Line Gas

A lower limit on the number density in the gas responsible for the broad emission lines (FWHM ~ 10^3 km/s) can be deduced from the inverted Balmer decrement (i.e. F(Hβ)/F(Hα) = 1-2 rather than 0.3). To achieve an inverted flux ratio, collisional effects between n = 2 and n = 3 levels must be important (implying $n_e > 10^{12}$ cm^{-3}) and, in general, the gas will be optically thick in Hα and probably Hβ as well (Stockman et al. 1977). The absence of semi-forbidden CIII] λ1909 in AM Her relative to the strengths of CIII λ1176 also sets a lower limit of $n_e \geq 10^{13}$ cm^{-3} (Raymond et al. 1979). In AM Her, it is possible to detect broad H_{12} and H_{13} near the Balmer series limit. For Stark broadening not to blend these lines, $n_e \leq 2 \times 10^{14}$ cm^{-3}. Unfortunately, similar quality IUE and ultraviolet spectra are not available for most of the other systems or for the narrow emission line components (F(narrow)/F(broad) ~ 0.1). In the case of the other systems, the similarity of the emission line strengths with those of AM Her suggests that this range of densities is typical: $2 \times 10^{14} \geq n_e \geq 10^{13}$.   For CW1103+254, the flux ratios of the narrow components appear similar to those of the broad components, also suggesting that high densities/optical depths are common for <u>all</u> the emission line gas. However, these generalities are not very secure and high resolution data ($\Delta\lambda \leq$ 1A) are desirable for all the brighter systems.

The temperature of the emission line gas cannot be easily measured because of density and optical depth effects.   However, on theoretical grounds, the excitation is primarily due to photoionization and thus a temperature in the range $T_e = 1 - 2 \times 10^4$ K is likely.   This agrees qualitatively with the observed Balmer continuum slope and the Planckian shape of the broad emission lines.   It is appropriate at this point to consider the most likely excitation mechanisms: X-ray heating, EUV photoionization, and hydrodynamic heating or high energy particle heating (reconnection).

<u>X-Ray Heating</u>.   Approximately half of the total system luminosity is due to bremsstrahlung radiation between 50 keV $\geq$ h$\nu \geq$ 10 keV. Relative to heating by the EUV flux, we do not expect the hard X-ray heating to be dominant in the emission line regions since these H II regions are optically thin to Compton scattering, $\tau_{comp} \sim 10^{-2}$. However, reprocessing of this radiation in the photospheric layers of the secondary star will add substantially to the total continuum luminosity of the secondary.   Using a scale height of 10^7 cm, the radiation will be reprocessed at a density of $n_e \sim (10^7 \sigma_e)^{-1} = 2 \times 10^{17}$ cm^{-3} and $\Delta L \sim$ 1% $L_X \sim 10^{31}$ erg/s $\sim$ L(dM5).

<u>EUV Heating</u>.   As a first approximation, the H II region formed by absorption of 50-150 eV photons will extend to a depth, d, given by balancing recombination to all excited levels of hydrogen with incoming photon flux:

$$n_e^2\, C(T_g)d \sim L_{EUV}\,(2.7\,kT_{EUV})^{-1}(2\pi R^2)^{-1}\ \ [kT \gg 13.6\ eV] \qquad (1)$$

$$n_e^2 d \sim 3 \times 10^{34}\, L_{33}\, R_{10}^{-2} \qquad (2)$$

where we have taken the recombination rate, $C(T_g) \sim 3 \times 10^{-13}$ appropriate for a gas temperature, $T_g = 10^4$ K; $L_{33} = L_{EUV}/10^{33}$; $R_{10} =$ gas radius/10^{10}; and $kT_{EUV} = 40$ eV.   The accretion flow/broad line emission region is characterized by $10^{13} \leq n_e \leq 2 \times 10^{14}$ and $R_{10} \sim 2$. We obtain a <u>minimum</u> ionization depth of $\overline{d_i} > 2 \times 10^5 - 8 \times 10^7$ cm. However, the actual ionization depth is probably several orders of magnitude larger.   This is due to ionization from the n = 2 level by Lyα photons.   Whenever τ(Hα) $\gg$ 1, which is suggested observationally, τ(Lyα) $> 10^5$ and trapped Lyα can photoionize the occupied n = 2 levels. The net result is that the ionization is dominated by energy loss due to the Balmer continuum rather than case B recombination.   In this case and ignoring all other sources of heating and cooling:

$$(3.4eV+kT)n_e^2\ C_2(T_g)d = L_{EUV}\ (2\pi R^2)^{-1} \tag{3}$$

$$n_e^2 d \sim 3 \times 10^{36}\ L_{33}\ R_{10}^{-2} \tag{4}$$

where C_2 is the recombination rate to the n = 2 level, C_2 (10^4 K) = 8 x 10^{-14}. Thus the upper bound to the ionization depth in the accretion flow for $n_e = 10^{13}$ is d < $d_e \sim$ 8 x 10^9 cm. We can also bound the density of the ionization zone in the atmosphere of the secondary star ($R_{10} \sim$ 10). Away from the L1 point, the atmospheric scale height is d $\sim 10^7$ cm. Thus n_e (secondary) $\sim$ 5 x 10^{12} - 5 x 10^{13} cm^{-3}, and it is quite possible that this region will also display optically thick, Planckian Balmer emission lines. Such is the case for the narrow emission line spectrum of the detached binary BE UMa (Ferguson et al. 1981, Margon, Downes, and Katz 1981).

Hydrodynamic/Reconnection Heating. The accretion flow may have considerable kinetic energy prior to its being controlled by the magnetic field, KE = 1-4 keV/nucleon. Unless the threading process is extraordinarily gentle, much of this energy will be converted into heat. One process will be the shock-heating of the infalling gas. With multiple shocks, the gas will be rapidly cooled by continuum radiation and will be almost isothermal. If only one or two principal shocks were involved, the energy would be radiated by bremsstrahlung in the energy range hν = 1-2 keV. This flux has not been observed. Another method of dissipating the kinetic energy is through reconnection and the decay of entrained magnetic field loops. If high energy particles are generated -- and they may not be in such a dense environment, -- radio and far-infrared emission may result. Quite likely, much of the energy loss will be through ohmic heating of the threaded gas and should be added to our estimate of d_i due to photoionization.

3.2 The Locations of the Line Emission Regions

As outlined in Section 2.5, the emission line profiles are actually a composite of emission spectra from several distinct regions. High resolution spectroscopy, $\Delta\lambda$ = 1-2A, and good phase coverage are required to clearly separate the multiple velocity components. Quite often the resulting picture is very complicated with many emerging and disappearing velocity systems. However, the spectra can generally be characterized by a broad ($\Delta v \sim 10^3$ km/sec) velocity component, which contains the majority of the observed flux, and a moderately strong, narrow ($\Delta v \sim 10^2$ km/sec) component whose velocity variations lag the broad component by several tenths of a phase. The prototype for such two component velocity systems is again AM Her. The spectra by Greenstein et al. (1977) show a clear separation in phase and velocity of the narrow and broad components. Additional velocity systems have been seen in EF Eri, VV Pup, and CW1103+254. These are generally in phase with the broad line system, but with somewhat higher or lower semi-amplitude, K. This produces a very asymmetric profile which complicates the detection of the out-of-phase component and any measurement of the "average" broad line semi-amplitude. In the face of these difficulties, we present in Table 2, representative semi-amplitudes and phasing for the broad and narrow components in eight AM Her systems (listed in order of decreasing period). Inclinations, i, and the phases when the accretion column is closest to "pole-on", ψ_C, are estimated from polarimetric and photometric variations. For H0139-681, we simply assume i ~ 60°. Phases for the narrow and broad components (ψ_N, ψ_B) refer to the phase of <u>maximum</u> <u>redshift</u>. Since the systems are often time variable and even the same observers change techniques for measuring K and ψ, this table serves as only a rough guide. In addition, if two or more narrow systems are observed, we have selected that system which is out of phase with the broad emission lines and we have favored those measurements, K_B, which measure the "center of light" rather than the "peak" of the broad component.

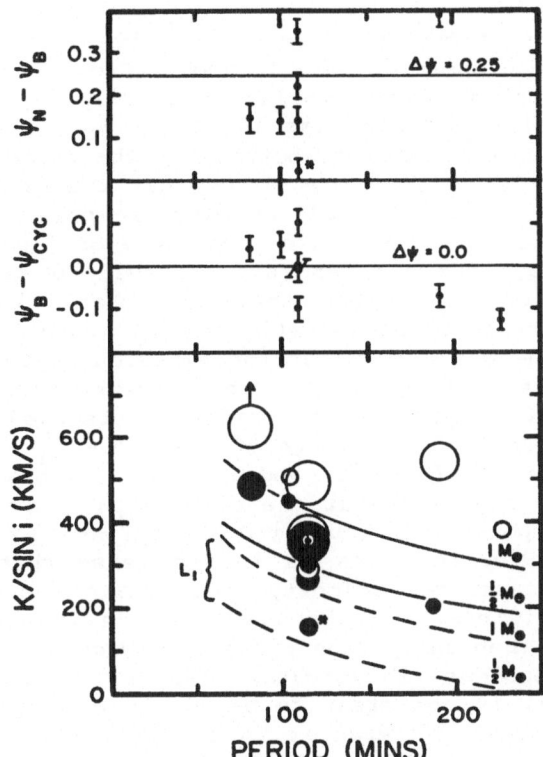

Figure 3. Radial velocities and phases for known AM Her systems as discussed in the text.

Inspection of Tables 1 and 2 reveals several interesting features. First, there is no obvious correlation between K_B and the period, P, although there may exist one between K_N and P. In the lower panel of Figure 3, we have plotted $K_B/\sin i$ (open circles) and $K_N/\sin i$ (solid circles) versus period. The solid lines represent the expected velocity of a Roche lobe-filling, ZAMS secondary with a 0.5 M_O and 1.0 M_O white dwarf primary. The dashed lines indicate the corresponding velocities for the L1 point. The run of K_N values are consistent with an origin on the illuminated portion of the secondary, approximately midway between L1 and the center of the disk. The one discrepant point, H0139-681, is marked with an asterisk. Since this component is also in phase with the broad lines, we suspect that emission from the secondary has not yet been detected. We caution that the secondary may be undermassive or and have a larger radius than the ZAMS models. This has the effect of moving the solid and dashed lines upward toward higher velocities. Thus a very wide range of white dwarf masses is still allowed, 0.4 $M_O \le M_{WD} \le$ 1.4 M_O (as is the case for other CVs, Robinson 1976).

There is a strong relationship of $\psi_B - \psi_C$ (Table 2 and the middle panel of Figure 3). The dispersion $\psi_B - \psi_C$ about zero is comparable to that allowed by the combined uncertainties (typically $\delta\psi = 0.04$) and the average of $\psi_B - \psi_C$ for eight systems is $< \psi_B - \psi_C > = 0.02$. This equality, the large observed radial velocities of the broad lines, and the phase-dependent widths of the lines argue that the broad emission lines arise in gas free-falling toward the magnetic pole of the white dwarf. Since the average line-widths are comparable or larger than the velocity semi-amplitudes, FWHM > K, and are observed to vary significantly in only fraction of the systems, they are probably due to a converging flow pattern far above the magnetic pole. We can estimate the average distance from the white dwarf of the emission region. In this picture, the average free fall velocity of the emission region is $v_{ff} \sim$ (K + $\frac{1}{2}$ FWHM)$/\sin i \sim$ (2 $GM_{WD}/R)^{1/2}$. The range of observed values is ~ 800 - 1500 km/s corresponding to a range of distances $R_g = 1-3 \times 10^{10}$ cm for reasonable white dwarf masses and periods. Table 2 and the upper panel in Figure 3 indicate that the difference in phase between the narrow and broad components is clustered about $\psi_N - \psi_B = 0.20$ ($\psi_N - \psi_B = 0.23$ without H0139-681). This is in the sense that the broad lines reach maximum recessional velocity approximately a quarter of phase before the narrow lines. If the suggested origins of these two components are correct, then on average the active magnetic pole points to within 18° of the secondary in azimuth. The dispersion about this average value, $\Delta\psi = 0.1$, is too great to be due to measurement uncertainties and indicates real differences in orbital orientations (the secondary lags the magnetic pole by ~ 60° in AM Her).

3.3 A Model for the Accretion Flow

Previously, there have been no detailed models of the accretion flow for dimensions comparable to the separation of the secondary, R_s. A popular notion was that material left the secondary star uniformly over its illuminated face. Barring extreme magnetically-induced chromospheric effects, however, it is energetically more favorable for the

escaping gas to leave near the L1 point. The resulting gas flow or
stream is denser and its path is very different from the previous uni-
form hemisphere/on-the-spot threading picture. In the remainder of
this section, we sketch this new model of the accretion stream in
terms of three separate regions, corresponding to the effects of the
white dwarf's magnetic field. These regions are bounded by the two
stars and two intermediate radii: R_P, the radius at which magnetic
pressure equals the thermal pressure of the stream; R_M, the radius at
which the magnetic pressure is equal to the ram pressure of the stream.

The Initial Acceleration Region, $R_S > R > R_P$. Before the stream
reaches the radius where the magnetic pressure equals the thermal
pressure of the gas, the stream dynamics will be similar to those in
normal CVs and is described by Lubow and Shu (1975, hereafter LS, and
references therein). As the gas leaves the secondary, it accelerates in
approximately a straight line with a ~ 20° offset from the stellar line
of centers (in the frame of the binary). The cross section remains
approximately constant, $\sigma \sim 3 \times 10^{18}$ cm^2, and the density falls with
increasing stream velocity from its initial density of $n_i \sim 2 \times 10^{15}$
L_{33}. Eventually the density will increase from a low value of n ~
$3 \times 10^{-2} n_i$ as the stream narrows. However, in the AM Her stars, the
magnetic field disrupts any steady, tubular flow near this low density
region.

The Threading Region, $R_P > R > R_M$. This intermediate region,
where the magnetic field dominates the shape and density of the flow but
not its trajectory, is physically complicated and difficult to model
accurately. Our rough calculations follow the study of accretion onto
magnetic neutron stars by John Arons and Susan Lea (Arons and Lea 1976,
1980) and earlier work and helpful suggestions by Fred and Don Lamb.
At $R_P \sim 4$-5×10^{10} cm (B ~ 200 gauss), the magnetic pressure equals the
gas thermal pressure and begins to pinch the stream. Since the magnetic pressure increases more rapidly than the gas can adjust sub-
sonically, the stream is shattered into even finer blobs until a stable
scale is reached, $\sigma_{blob} \leq 10^{16}$ cm^2. The threading of these blobs
requires several sound-crossing times with the gas almost reaching $R_M \sim$
2.0×10^{10} cm. Half-a-dozen physical processes compete during this
threading process: compression of the unthreaded blobs by the magnetic
field; Kelvin-Helmholtz shredding by the entrained surrounding gas;
ohmic diffusion; shock/reconnection heating (see Section 2.1); EUV and
X-ray heating with subsequent blackbody emission; and, as the stream
approaches R_M, tangential and radial accelerations by the stretched
field lines.

Magnetically Channeled Flow/The Line Emission Region, $R_M > R > R_{WD}$.
Near R_M, the magnetic field is sufficiently high to withstand the ram
pressure of the stream and divert the stream out of the orbital plane.
The precise location of R_M is uncertain, as it depends on the effectiveness of the threading and the degree of tangential dispersion or
"splashing" of the threaded gas. The strong radial dependence of the
magnetic pressure, $B^2 \alpha R^{-6}$, is reassuring in the face of ignorance;

and we will use $R_M \sim 1.5-2 \times 10^{10}$ cm based on the <u>undisturbed</u> densities and cross-section of the Lubow and Shu model with $L_{33} \sim 1$. Except for much higher accretion rates or clumpy accretion episodes when denser blobs are formed, $R_M > R_D \sim 1-1.2 \times 10^{10}$ cm where curvature of the stream is important and a disk would form in a less magnetic system.

The ribbon of magnetically channeled material is closer to the radiating white dwarf and has a much larger area than the material upstream. Thus, it is the probable site for the broad emission line region. Our calculations indicate areas, a range of densities, and infall velocities consistent with the observations summarized earlier in this section. Moreover, the shattering of the stream near R_P will produce a spectrum of sporadic accretion or flickering with the greatest power for timescales of $\sim 10^3$ s, reducing to negligible levels for timescales shorter than 1s due to thermal spreading in the infalling stream. The geometry of the ribbon and subsequent shock depends on the physics of the threading region and the system parameters. Generally, the shock has a small covering factor, $f \sim 10^{-4}$, and will be displaced from the "true" magnetic pole by $\sim 10°$. Multicomponent lines, accretion onto the opposite pole, and the observed variations of line strengths with orbital phase can also be addressed within the context of such a model. Thus, it would appear that the next qualitative breakthrough in our understanding of these systems lies in the detailed study of the accretion flow.

4. THE POLAR ACCRETION SHOCK

The hot ($kT \geq 20keV$) accretion shock shown in Figure 1c forms the interface between the supersonic, infalling gas and the atmosphere of the white dwarf. This shock is principally cooled by X-ray emission due to bremsstrahlung and optical emission by high cyclotron harmonics (Sections 2.1, 2.2). Because of the magnetic channeling which prevents the infalling gas from degrading the outgoing X-ray flux spectrum (Kylafis and Lamb 1978), the AM Her systems provide the best confirmation of the models of Hoshi (1973); Fabian, Pringle and Rees (1976); Katz (1977), and others for radial accretion onto white dwarfs. The importance of high harmonic, optical cyclotron emission as a coolant was recognized by Fabian <u>et al</u>. (1976) and Masters <u>et al</u>. (1977) and complete, self-consistent shock models including angle-averaged cyclotron emission have been calculated by Lamb and Masters (1979). For excellent reviews of accretion onto magnetic and non-magnetic white dwarfs, the reader should see Kylafis <u>et al</u>. (1980) and Kylafis and Lamb (1982). Elsewhere in this volume, Don Lamb (1983) addresses the current status of X-ray and reprocessed EUV emission in the AM Her and other CV systems. Thus, we comment only on two related aspects of the hot shock: the observed cyclotron energy distribution and the size and location of the X-ray/cyclotron emission regions.

4.1 The Observed Cyclotron Energy Distribution

As mentioned in Section 1, the optical continuum between 0.3 - 1.0

µm in the AM Her systems has at least two components: a flat or steeply falling polarized component ($F_\nu \propto \nu^{-\alpha}$, $\alpha > 0$) and Balmer and Paschen continuum probably originating in the broad line emission region. Because of its strong circular polarization (up to ~ 30%) and its behavior in self-eclipsing systems, the steep, polarized component is almost entirely due to high harmonic cyclotron emission and contributes 1-10% of the total system luminosity. This is not surprising since opacity calculations by Lamb and Masters (1979), Chanmugam and Dulk (1981), and Meggitt and Wickramasinghe (1982) obtain high optical depths in optical cyclotron emission for the typically assumed shock parameters, $f \sim 10^{-3}$, $L_{33} \sim 1$, $kT \sim 30$ keV. The difficulty lies in the predicted versus observed energy distribution and polarization characteristics. The basic shock model predicts a Rayleigh Jeans spectrum, $F_\nu^* \propto \nu^2$ and low polarizations up to ν^* where the shock becomes thin. The predicted ν^* for typical shocks is $\nu^* \sim 0.2$ µm. The observed peak in the optical energy distribution is $\nu_{peak} \sim 1$-2 µm. Because of the extreme dependence of the cyclotron opacity on wavelength or harmonic number, m, there are no first-order improvements to the model which can resolve the disparity. For instance, radiative transfer effects through the shock front can explain the discrete cyclotron features observed in VV Pup (Wickramasinghe and Meggitt 1982) but only exacerbate the energy distribution problem by blocking the lower harmonics (Meggitt and Wickramasinghe 1983). Similarly, directional effects can distort the energy distribution (Chanmugam 1980). But most of the observed polarization beaming effects appear more typical of an optically-thin gas. How can the basic shock model, which successfully explains the hard and soft X-ray fluxes, fail so badly at predicting the shape of the optical continuum? We suggest two areas in the standard model which should be reexamined.

The first area is the creation and maintenance of the semi-relativistic electrons, which are chiefly responsible for the high cyclotron harmonic emission. Masters (1978) considered the time required to replenish the Maxwellian tail of the electron distribution compared to the cyclotron cooling time of the shock. He found that if $kT > 16$ keV ($L/f \geq 10^{35}$) the standard shock models remained valid. However, his calculations were semiclassical and did not include a significant loss mechanism: "end-losses" through the top and bottom of the shock. Our calculations show that in bremmsstrahlung dominated shocks, > 50% of all relativistic electrons escape the shock before radiating and the problem is more serious in cyclotron-dominated shocks. The result is a truncated electron distribution which does not effect the bremsstrahlung cooling but suppresses the high harmonic cyclotron emission at the expense of heating the preshock gas and the white dwarf atmosphere. These calculation should be confirmed with relativisitic Fokker-Planck codes but are similar in spirit to the enhanced thermal conductivity calculations of King and Lasota (1980). The second suggested modification of the standard model concerns the physical structure of the hot accretion shock and, specifically, separate locations for the optical and X-ray emission.

4.2 The Geometry of the Accretion Shock

Most estimates of the height, h, and breadth ($\sim$ f) of the accretion shock are based on optical photometric and polarimetric data. The smallest optical structure h $\sim$ 0.01 R_{WD}, f $\leq$ 10^{-4}, is inferred from the rapid eclipse of VV Pup ($\Delta P/P$ = 0.02, Liebert et al. 1978) and this size is not always typical: Liebert and Stockman 1979 obtain h $\sim$ 0.1R_{WD} for data obtained at a different epoch. A similar estimate for CW1103+254 is f = 3×10^{-3} (Stockman et al. 1983). The situation for AM Her is confused by the grazing eclipse geometry, but the data suggest that the optical emission has a height of h $\sim$ 0.1 R_{WD} and is seldom totally eclipsed, while the deep X-ray eclipse ingress can be quite sharp (a factor of 5 in $\Delta P \sim 0.1P$, Touhy et al. 1978). In addition to the different eclipsing behavior of the optical and X-ray light curves in AM Her, the recent revision in the soft X-ray temperature indicates different origins for the two components (we equate soft and hard X-ray origins). For a temperature of kT_{EUV} = 50 eV, the observed flux and distance imply f $\sim$ 2×10^{-4} (Touhy et al. 1981), whereas the observed infrared peak at 1 μm requires f $\sim$ 5×10^{-3} (for a brightness temperature of 30 keV, Stockman et al. 1983). In other words, if optically thick cyclotron emission were to originate from an area of f $\sim$ 2×10^{-4}, and a temperature of kT $\sim$ 30 keV, it would meet the observed flux distribution (Figure 2) at $\lambda \sim$ 0.3 - 0.4 μm. Indeed, the predicted cyclotron emission from the small X-ray emission region can better explain the observed upturn in the UV (Lamb and Masters have $\lambda* \sim$ 0.1 μm for L/f = 10^{37} and M_{WD} = 1 M_O.

Thus, there is both direct and indirect observational evidence that the geometry of the polar accretion shock is more complicated than in the standard model. More realistic models of the infalling accretion flow, such as the one presented in Section 3.3, will force two major modifications. Firstly, the filling factor may be low, resulting in large density contrasts along a given field line. Secondly, the sides of the accretion flow will not be perfectly defined. Rather, the density may taper gradually such as $n(\rho) \alpha \rho^{-\alpha}$, where ρ is the cylindrical coordinate and $\alpha \sim$ 3. Since the shock height varies approximately inversely with the infalling density, the result of a low filling factor will be a very "choppy" shock height. Viscosity and thermal conduction, which would normally reduce the irregularities, are suppressed by the strong magnetic field. The ill-defined column sides can have a more dramatic effect. As shown in Figure 1c, the outer rim of the shock will rise to much greater heights than in the dense center. The figure indicates the shock eventually collapsing to the surface. However, the true behavior is complicated by the radiation from the central shock region which can provide significant Compton heating/cooling. Regardless, the rim of the shock will rise to considerable heights, h $\geq$ 0.1 R_{WD} before these effects become important. This rim will cool by cyclotron radiation rather than bremsstrahlung and, having a large surface area, can dominate the observed optical flux. Simple calculations using the models of Lamb and Masters (1979) indicate that a wide range of continuum slopes can be obtained corresponding to a range of filling factors and/or boundary shapes. For the simple case of spreading 1% of the total accretion over a fractional area, f ~ 0.005, the standard model yields $\lambda*$ and h remarkably close to the observed values. In addition, the tall, cooler rim can scatter and absorb the soft and hard X-ray flux from the central emission region leading to ragged and prolonged eclipse edges such as those seen in AM Her (Touhy et al. 1977) and weak or anti-correlations in the X-ray and optical flux (Szkody et al. 1980).

5. STELLAR PARAMETERS AND SYSTEM EVOLUTION

In Section 3 we discussed the observations of the emission line radial velocities and their (weak) constraints on stellar masses. The low states provide some of the best opportunities to study other pro-perties of the underlying stellar components. In the following paragraphs, we summarize some of these results, some puzzles posed by the magnetic field and orbital properties, and the implication these might hold for understanding the origin and evolution of AM Her binaries.

5.1 Masses and General Properties of the Secondary Stars

As noted in Section 1, (see also Table 1), the only detected main sequence companions in AM Her systems are of late M spectral type, and may correspond to masses of $< 0.2M_O$. This is not unusual, however, in comparison with what is known about other short period CVs. There appears to be a strong correlation between the binary period, the average accretion rate and the mass of the secondary star (Patterson 1983). Thus, the more direct question may be to ask why the AM Her binaries all have such short periods.

Of course the imprecision with which the stellar masses can be determined (see Section 2) leaves us little information on the mass, spectral type, temperature, and radius relations for the secondary stars. However, in comparison with low mass field dwarfs, the AM Her secondaries are distorted, rapidly-rotating stars, illuminated by a strong source of high energy flux. Even when the illuminating source is off, they might be expected to show very strong chromospheric-like (dMe) activity, like the RS CVn binaries, the relatively strongest stellar X-ray sources found by the Einstein Observatory (Walter and Bowyer 1981). The M star activity may be especially relevant to syste-mic properties in low accretion (nearly detached) states.

Dulk, Bastian and Chanmugam (1983) argue that the M star may be the origin of the AM Her radio "outburst" and may also supply the energetic charged particles which produce the "quiescent" radiation. They argue that the companion has a strong (solar-type) magnetic field ($\sim 10^3$ gauss) -- thus forming a kind of phase-locked Jupiter-Io system. Magnetic reconnection in the region between the stars, or within the secondary's own loops, could then drive electron-cyclotron maser

"outbursts." It will be interesting to see whether apparent differences in radio properties are tied to the intrinsic coronal/magnetic properties of the secondary stars.

5.2 The Temperature Profiles and Magnetic Geometries of the Primary Stars

The energy distributions and magnetic Zeeman absorption patterns of
AM Her primaries observed in low states potentially hold a wealth of
information on the structure of that stellar component. Szkody et
al.'s (1982) 50,000 K IUE component in the 1980 AM Her low state showed
a magnetically-broadened Lyα absorption line, and should thus be attri-
butable to some part of the primary's photosphere. But, was this the
whole star or just a heated "cap"? At optical wavelengths, the unpolarized continuum with strong Zeeman absorption looked much less hot,
leading Schmidt et al. (1981) to estimate $T_e \sim 22,000$ K. It is logical
for a larger area of the white dwarf's photosphere at a cooler T_e to
make a greater relative contribution at optical wavelengths. However,
the presence of contaminating light components, and the larger hydrogen line strengths for magnetic vs. nonmagnetic white dwarfs (at a
given T_e) leave the optical T_e estimate insecure. Schmidt et al.
(1981) and Latham et al. (1981) found evidence for magnetic absorption
components spanning the range 13-20 megagauss or greater. While Young,
Schneider and Shectman (1981) argued that the radiation was confined to
a polar cap, this would require a tangled (or purely ad hoc) polar
field geometry. More likely, the range in field strength can be taken
as evidence for a larger photospheric surface area contribution. The
phase-dependence of the Zeeman pattern did not fit predictions of
uniform-photosphere, simple dipole models. The greater contribution of
lower (~ 13 MG) field strengths at phases representing a more pole-on
view was difficult to explain. Perhaps the higher-field polar regions
were then masked by the overlying gas stream, but came into view as the
pole approached the stellar limb (the equator-on view).

There is some evidence that VV Pup lacks a readily-detectable,
"hot" polar cap in its low state (Liebert et al. 1978, 1983). Here, it
appears that most of the white dwarf is fairly cool. However, the temperature profiles and magnetic field geometries for these primaries
must be studied with more detailed, time-resolved spectropolarimetry,
especially for "two-pole" cases.

5.3 Maintenance of Rotational Synchronization

The remarkably stable modulation of the polarization and photometric properties with binary phase shows that AM Her primary stars are
phase-locked to the revolving secondary. Achievement of this synchronism
was originally attributed to a vacuum-dipole-induced, ohmic dissipation
process (Joss, Katz and Rappaport 1979). More recently the effects of
magnetohydrodynamic (MHD) torques have been argued to be more efficient
(Chanmugam and Dulk 1982; Lamb et al. 1983, and references therein).
But how rigid and how permanent is the primary's synchronism in these

systems?

While most attention has been devoted to AM Her, the best long-term photometric (photographic) evidence relevant to answering this question exists for VV Pup. Observations of the sharp photometric eclipse of the confined polar funnel extend back to Alden (1931), the early data base is nicely reviewed in Herbig (1960). There is no evidence for phase slippage exceeding 0.1 of a period over 50 years. Modern data with better time resolution may be searched for smaller departures from rigid synchronism over shorter time scales. One may argue that circular polarimetry is superior to linear polarimetry and optical photometry for this purpose, and more practical than observations at shorter wavelengths. Comparison of circular polarimetry and optical photometry suggested that the VV Pup polar ingress may show some phase "jitter" (Liebert and Stockman 1979), but it was unclear whether such small shifts couldn't be attributed to changes in the polar shock's height or geometry or even to the measurement techniques (see also Patterson et al. 1983).

The best evidence for small shifts in a magnetic pole's position comes from ongoing observations of AM Her. That some kind of geometry changes occur in the funnel radiation components is indicated by (1) circular polarization observations showing a variable length of phase for the reversal of sign (c.f. Bailey and Axon 1981), (2) changes in the linear polarization "pulses" (Bailey et al. 1978, Tapia 1983), and (3) the apparent disappearance of X-ray eclipse (Priedhorsky and Marshall 1982). Interpretation is complicated for AM Her by the extended geometry of the optical shock (see Section 4.2) and its nearness to the horizon even when it slips briefly behind the stellar limb. However, the simplest explanation for effects (1) and (3) may be that the percentage of time the main accreting pole spends behind the observer's limb may vary from 0.2 phase to zero.

The evidence for the well-studied AM Her systems points to small, month-to-year changes in either the position of the white dwarf's rotation axis or its magnetic axis. The slippage is evidently quite small (< 0.1 phase in the equatorial direction), and there is no evidence for major, long-term changes. From this we conclude that restoring forces do indeed keep the white dwarf phase-locked, and that it is likely that the geometrically-favored pole always receives most of the accreted matter. However, the observational search for long-term geometric changes in these systems is only beginning.

5.4 Clumping of Orbital Periods

All but two AM Her systems (Table 1) have orbital periods between 80 minutes and 115 minutes. The two exceptions lie just on the long side of the 2-3 hour CV orbital period "gap." Indeed, of the 25 CVs with known orbital periods within the 80-115 minute interval, eight are AM Her objects (c.f. Ritter 1982, plus updates). While such factors as X-ray selection obviously favor the discovery of AM Her systems over nonmagnetic, short-period CVs, it is still likely that the fraction of CVs in this period range with strongly magnetic primaries greatly exceeds the (~ 4%) fraction of single magnetic white dwarfs in the field.

This high fraction of magnetic CVs having short periods may not be difficult to explain. The MHD spin-orbit coupling process (c.f. Chanmugam and Dulk 1982; Lamb et al. 1983) may achieve synchronous primary rotation, efficient disposal of orbital angular momentum and a resultant shortening of the orbital period. Note that MHD torque scales with B^2/D^3 (where B is the field strength and D the separation). The period decrease stops when the time scale for angular momentum losses shortens sufficiently to become comparable with the secondary's Kelvin-Helmholtz "adjustment" time scale (Kaulkner 1971, Paczynski and Sienkiewicz 1981, Rappaport et al. 1982). Subsequent mass transfer results in adiabatic expansion of the secondary and an increase in P. Empirically, this P_{min} is 80 minutes for all CVs excepting the double helium degenerate AM CVn type. Thus, it appears that the strongly-magnetic systems may be preferentially braked to near P_{min}.

5.5 Clumping of Primary Magnetic Field Strengths

A major surprise of the recent low accretion state observations was the discovery of polar field strengths ranging from 20 to 32 megagauss in three AM Her systems (Table 1). This dispelled earlier suggestions that field strengths were ~ 10^8 gauss. Furthermore, most of the remaining seven objects appear to have similar polarized energy distributions to AM Her, VV Pup and CW1103+254, the known field cases. Indeed, the success of Chanmugam and Dulk (1981) and Meggitt and Wickramasinghe (1982) in modelling the phasing with a small range of cyclotron harmonics (n > 4-8) led these authors to the conclusion that all well-studied AM Her systems have field strengths not too different from the above range. Meggitt and Wickramasinghe (1983) find this general conclusion to be insensitive to the shock density and its detailed temperature structure.

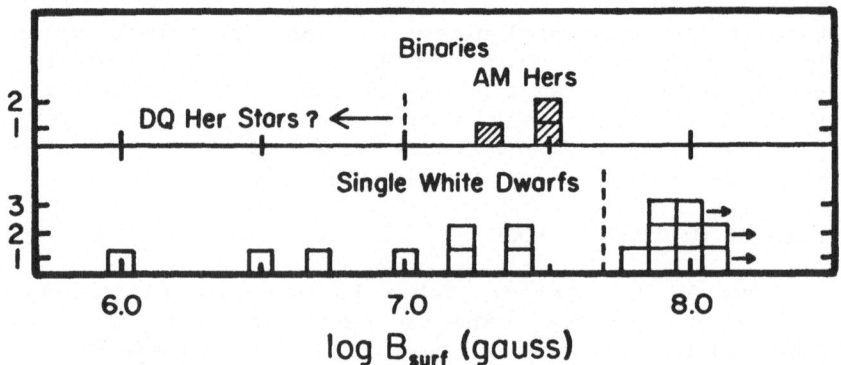

Figure 4. Histograms of magnetic field distributions for known AM Her cases and single MWDs.

In Figure 4 we show for comparison to the AM Her primary stars the distribution of magnetic field strengths for single magnetic white dwarfs (MWDs) with measured fields. (Considerable license has been used in assigning uncertain field strengths above 5×10^7 gauss, but all indicated stars lie at higher values than the dashed line.) Indeed, the recent reanalysis of MWDs by Angel, Borra and Landstreet (1981) concluded that single MWDs could exist in comparable numbers for each decade of field strength from 10^6 gauss to well above 10^8 gauss. We note in the figure that CV primaries with weaker fields (< 20 megagauss) are generally thought to be included among the DQ Her "coherent oscillators", and the so-called "intermediate polars" (Warner 1983). Actual field values are estimated for these in Lamb and Patterson (1982). It appears, however, that the known AM Her systems do not include any with fields $> 5 \times 10^7$ gauss. Is there some evolutionary reason why these are rarer, or have we not yet learned to recognize the strongest-field magnetic CVs?

Polarization surveys of likely CV candidates have now covered virtually all known, short-period cases in the Northern Hamisphere with strong X-ray or high excitation line emission (Tapia, private communication; Stockman, Liebert and Moore 1983). There seem to be no established cases showing < 1% circular polarization, though DQ Her (Kemp et al. 1971) remains a possibility. However, higher field primary stars could emit cyclotron radiation in the UV and the optical continua could be dominated by an optically-thick unpolarized component. In this case, the emission lines should be extremely strong if synchronous mass transfer is occurring. They might, in fact, look something like the VV Pup or CW1103+254 spectra in their faint phases. (So might any one-pole, 10^7 gauss cases where the active funnel remains out of our view.)

Bond and Chanmugam (1982) obtained ultraviolet energy distributions for several of the most promising, optically-unpolarized CVs, hoping to find evidence for an UV cyclotron "hump" component indicative of $\gtrsim 10^8$ gauss fields. The null results led these authors to conclude that larger field magnetic binaries probably do not exist among the well-studied CV sample.

Why, then, might the $\sim 10^8$ gauss magnetic CVs be so rare in comparison with $\sim 2.5 \times 10^7$ gauss fields? This is difficult to answer, since we do not understand the origin of the MWD fields, either as single stars or in CVs. Regardless of origin, however, the MWD fields may be subject to ohmic decay, higher order multipole components dying out first (c.f. Fontaine, Thomas and Van Horn 1973). Eichler (1982) argues that more massive and/or rapidly rotating MWDs could be less susceptible to magnetic instability, and we note that AM Her primaries could be more massive and rapidly rotating than most field WDs. Such an argument could explain the greater frequency of detectable magnetism among CV primaries (including "intermediate polars"). We might also expect that the AM Her field geometries -- like the single MWDs -- would be mostly simple and dipolar. We can only conjecture, however,

Table 1. Known AM Herculis Objects (Polars)
and a Synopsis of Properties

Names	How Found	Period (Min)	Optical Magnitude Range/Period; Low State Mag	Circular Polarization Extremes (%); Low State	Geom Type	Comments/ Reference
E 2003+225	X	222.51	~ 15 --	+7.5, 0%	1P	1
AM Her 1814+508	OX	185.65	12-14 15	-8, +3% -11, 0	2P	B=20MG M4-5; 2-6
AN UMa 1101+453	O	114.84	15-17 19	-19, 0 -35, 0	1P	7, 8
CW 1103+254	O	113.89	15-17 --	-20, 0	2P	B=32MG 9, 10
H(E) 0139-681	X	113.65	15-16 --	+15, -9	2P	11, 12
PG 1550+191	O	113.58	15-16 --	-12, 0	1P	13
E 1405-451	X	101.52	15-16 --	-30, 0	1P	14, 15
VV Pup 0812-189	O	100.44	14-16 18	-4, +10 0, +15	2P	B=32MG M4-5 16-18
E 1114+182	X	89.80	17-21 --	-35, +10:	2P?	19
EF Eri, 2A 0311-227	X	81.02	13-15 --	+16, -2	1P?	20, 21

References:

(1) Nousek et al. 1983, (2) Tapia 1977, (3) Priedhorsky et al. 1978,
(4) Schmidt et al. 1981, (5) Latham et al. 1981, (6) Rothschild et al.
1981, (7) Krzeminski and Serkowski 1977, (8) Liebert et al. 1982b, (9)
Stockman et al. 1983, (10) Schmidt et al. 1983, (11) Agrawal et al.
1981, (12) Visvanathan and Pickles 1982, (13) Liebert et al. 1982a,
(14) Mason et al. 1983, (15) Tapia 1982, (16) Liebert et al. 1978, (17)
Visvanathan and Wickramasinghe 1979, (18) Liebert and Stockman 1979,
(19) Biermann et al. 1982, (20) Bailey et al. 1982, (21) Griffiths et
al. 1979.

Table 2.

OBJECT	i(deg)	K_B	ψ_B	K_N	ψ_N	ψ_C	$\psi_B-\psi_C$	$\psi_N-\psi_B$	Reference
E2003+225	80	373+20	.37	---	---	.50*	-.13	---	1
AM Her	35	309+25	.53	116+10	.92	.60	-.07	.39	2,3
AN UMa	65	322+17	.49	271+31	.63	.50	-.01	.14	4
CW1103+254	69	465+50	.85	320+50	.20*	.86	.00	.35	5
H0139-681	60**	249+24	.15	133+15	.17	.25**	-.10	.02	6
PG1550+191	40	234+50	.60	170+20	.82	.50	.10	.22	7
VV Pup	75	488+18	.98	437+15	.12*	.93	.05	.14	8,4
EF Eri	46	450-600	.54	350+25	.69*	.50	.04	.15	9-13

* = Multiple Narrow Components; ** = Geometry uncertain.

(1) Nousek et al. 1983, (2) Greenstein et al. 1977, (3) Young and Schneider 1979, (4) Schneider and Young 1980b, (5) Stockman et al. 1983, (6) Thorstenson et al. 1983, (7) Liebert et al. 1982a, (8) Cowley et al. 1982, (9) Hutchings et al. 1982, (10) Crampton et al. 1981, (11) Young et al., (12) Schneider and Young 1980a, (13) Bailey et al. 1982.

that $\sim 10^8$ gauss cases may be rare because angular momentum losses are so efficient that the 10^8 gauss systems evolve very rapidly to the end of the CV mass transfer state.

Space limitations preclude us from thanking the many colleagues in the CV community who have provided us early preprints and valuable insight on AM Her systems. We acknowledge support from NSF grants AST 80-24324 and 82-10513. We thank Mrs. Helen Bluestein for getting this typed quickly and in camera-ready form.

REFERENCES

Alden, H. L.: 1931, A. J., 41, pp. 89.

Allen, D. A. and Cherepashchuk, A. M.: M.N.R.A.S., 201, pp. 521.

Angel, J. R. P., Borra, E. F. and Landstreet, J. D.: 1981, Ap. J., 45, pp. 457.

Agrawal, P. C., Rao, A. R., Riegler, G. R., Pickles, A. J. and Visvanathan, N.: 1981, I.A.U. Circ. No. 3649.

Arons, J. and Lea, S. M.: 1976, Ap. J., 207, pp. 914.

Arons, J. and Lea, S. M.: 1980, Ap. J., 235, pp. 1016.

Bailey, J., Jones, D. H. P., Parkes, G. E. and Mason, K. O.: 1978, M.N.R.A.S., 184, pp. 73.

Bailey, J. and Axon, D.: 1981, M.N.R.A.S., 194, pp. 187.

Bailey, J. and Ward, M.: 1981, M.N.R.A.S., 196, pp. 425.

Bailey, J., Hough, J., Axon, D., Gatley, I., Lee, T., Szkody, P., Stokes, G. and Berriman, G.: 1982, M.N.R.A.S., 199, pp. 801.

Berg, R. A. and Duthie, J. G.: 1977, Ap. J. 211, pp. 859.

Bierman, P., Kuhr, H., Liebert, J., Stockman, H., Strittmatter, P. and Tapia, S.: 1983, I.A.U. Circ. No. 3680 and work in progress.

Bond, H. E. and Tifft, W. G.: 1974, Pub.A.S.P., 86, pp. 981.

Bond, H. E., Chanmugam, G. and Grauer, A. D.: 1979, Ap. J. (Letters), 234, L113.

Bond, H. E. and Chanmugam, G. 1982, in "Advances in Ultraviolet Astronomy," NASA Conference Pub. 2238, ed. by Y. Kondo, J. Mead and R. Chapman, NASA Goddard Space Flight Center, Greenbelt, Md., pp. 530.

Chanmugam, G.: 1980, Ap. J. 241, pp. 1122.

Chanmugam, G. and Dulk, G. A.: 1981, Ap. J., 244, pp. 569.

Chanmugam, G. and Dulk, G. A.: 1982, Ap. J. (Letters), 255, L107.

Chiappetti, L., Tanzi, E. G. and Treves, A.: 1980, Space Sci. Rev., 27, pp. 3.

Cordova, F. A. and Mason, K. O.: 1982, in "Accretion Driven Stellar X-Ray Sources," eds. W. H. G. Lewin and E. P. J. van den Heuvel, Cambridge University Press, Cambridge.

Cowley, A. P. and Crampton, D.: 1977, Ap. J. (Letters), 212, L121.

Cowley, A. P., Crampton, D. and Hutchings, J. B.: 1982, Ap. J., 252, pp. 690.

Cowley, A. P., Crampton, D. and Hutchings, J. B.: 1983, Ap. J., 259, pp. 730.

Crampton, D. Hutchings, J. B. and Cowley, A. P.: 1981, Ap. J., 243, pp. 567.

Dulk, G., Bastian, T. and Chanmugam, G.: 1982, Ap. J., in press.

Eichler, D.: 1982, Ap. J., 254, pp. 683.

Fabbiano, P.: 1982, Ap. J., 262, pp. 709.

Fabian, A. C., Pringle, J. E. and Rees, M. J.: 1976, M.N.R.A.S., 173, pp. 43.

Faulkner, J.: 1971, Ap. J. (Letters), 171, L99.

Feigelson, E., Dexter, L. and Liller, W.: 1978, Ap. J., 222, pp. 263.

Ferguson, D. H., Liebert, J., Green, R. F., McGraw, J. T. and Spinrad, H.: 1981, Ap. J., 251, pp. 205.

Fontaine, G., Thomas, J. H. and Van Horn, H. M.: 1973, Ap. J., 184, pp. 911.

Greenstein, J. L., Sargent, W. L. W., Boroson, T. A. and Boksenberg, A.: 1977, Ap. J. (Letters), 218, L121.

Griffiths, R. E., Ward, M. J., Blades, J. C., Wilson, A. S., Chaisson, L. and Johnston, M. D.: 1979, Ap. J. (Letters), 232, L27.

Hearn, D., Richardson, J. and Clark, G.: 1976, Ap. J. (Letters), 210, L23.

Hearn, D. R. and Richardson, J. A.: 1977, Ap. J. (Letters), 213, L115.

Herbig, G. H.: 1960, Ap. J., 132, pp. 76.

Hoshi, R.: 1973, Progr. Theoret. Phys., 49, pp. 776.

Hudec, R. and Meinunger, L.: 1977, Mitt. Ver. Sterne Sonneberg, 7, pp. 194.

Hutchings, J. B., Cowley, A. P., Crampton, D., Fisher, W. A. and Liller, M. H.: 1982, Ap. J., 252, pp. 690.

Joss, P., Katz, J. and Rappaport, S.: 1979, Ap. J., 230, pp. 176.

Katz, J. I.: 1977, Ap. J., 215, pp. 265.

Kemp, J., Swedlund, J. and Wolstencroft, R.: 1974, Ap. J. (Letters), 193, L15.

King, A. R. and Lasota, J. P.: 1980, M.N.R.A.S., 191, pp. 721.

Krzeminski, W. and Serkowski, K.: 1977, Ap. J. (Letters), 216, L45.

Kruszewski, A. 1978, in "Nonstationary Evolution of Close Binaries," ed. A. N. Zytkow, PWN - Polish Scientific Publishers, Warszawa, pp. 55-81.

Kylafis, N. D. and Lamb, D. Q.: 1979, Ap. J. (Letters), 228, pp. 105.

Kylafis, N. D. and Lamb, D. Q.: 1980, Ap. J. (Suppl.), 48, pp. 239.

Kylafis, N. D., Lamb, D. Q., Masters, A. R. and Weast, G. J.: 1980, Proc. 9th Texas Symposium on Relativistic Astrophysics, (Ann. N.Y. Acad. Sci.), 336, pp. 520.

Lamb, D. Q.: 1982, IAU Colloq. No. 72, "Cataclysmic Variables and Related Objects," preprint.

Lamb, D. Q.: 1985, this volume.

Lamb, F. K., Lamb, D. Q., Aly, J.-J. and Cook, M. C.: 1983, submitted to Ap. J. (Letters).

Latham, D., Liebert, J. and Steiner, J.: 1981, Ap. J., 246, pp. 919.

Liebert, J., Stockman, H. S., Angel, J. R. P., Woolf, N. J. and Hege, E. K.: 1978, Ap. J., 225, pp. 201.

Liebert, J. and Stockman, H. S.: 1979, Ap. J., 229, pp. 652.

Liebert, J., Stockman, H. S., Williams, R. E., Tapia, S., Green, R. F., Rautenkranz, D. and Ferguson, D.: 1982a, Ap. J., 256, pp. 594.

Liebert, J., Tapia, S., Bond, H. E. and Grauer, A. D.: 1982b, Ap. J., 254, pp. 232.

Liebert, J., Patterson, J., Raymond, J. and Stockman, H. S.:   1983, preprint.

Lubow, S. H. and Shu, F. H.: 1975, Ap. J., 198, pp. 383.

Margon, B., Downes, R. A. and Katz, J. I.: 1981, Nature, 293, pp. 200.

Mason, K. O., Middleditch, J., Cordova, F. A., Jensen, K. A., Reichert, G., Murdin, P. G., Clark, D. and Bowyer, S.: 1983, Ap. J. 264, pp. 575.

Masters, A. R.:   1978, Ph.D. Thesis, University of Illinois.

Masters, A. R., Fabian, A. C., Pringle, J. E. and Rees, M. J.:   1977, M.N.R.A.S., 178, pp. 501.

Meggitt, S. M. A. and Wickramasinghe, D. T.:   1982, M.N.R.A.S., 198, pp. 71.

Meggitt, S. M. A. and Wickramasinghe, D. T.:   1983, preprint.

Nousek, J. A., Takalo, L. O., Schmidt, G. D., Hill, G. J., Tapia, S., Bond, H. E., Grauer, A. D. and Agrawal, P. C.:, 1983, preprint.

Paczynski, B. and Sienkiewicz, R.:   1981, Ap. J. (Letters), 248, L27.

Patterson, J.: 1983, Ap. J., in press.

Patterson, J., Fabbiano, G., Lamb, D. Q., Raymond, J., Horne, K., White, N. and Swank, J.: 1983, preprint.

Priedhorsky, W. C.: 1977, Ap. J. (Letters), 212, L117.

Priedhorsky, W. and Krzeminski, W.: 1978, Ap. J., 219, pp. 597.

Priedhorsky, W. C., Krzeminski, W., Tapia, S.: 1978, Ap. J., 225, pp. 542.

Priedhorsky, W. and Marshall, F.:   1982, Bull.A.A.S., 14, pp. 980.

Rappaport, S., Joss, P. C. and Webbink, R. F.: 1982, Ap. J., 254, pp. 616.

Raymond, J. C., Black, J. H., Davis, R. J., Dupress, A. K., Gursky, H., Hartmann, L. and Matilsky, T. A.: 1979, Ap. J. (Letters), 230, L95.

Ritter, H. 1982, "Catalogue of Cataclysmic Binaries, Low Mass X-Ray Binaries and Related Objects," MPIfPhysik u. Astrophysik Preprint, Garching, FRG.

Robinson, E. L.:   1976, Ann. Rev. of Astron. Astrophys., 14, pp. 119.

Rothschild, R. E. et al.: 1981, Ap. J., 250, pp. 723.

Schmidt, G. D., Stockman, H. S. and Margon, B.: 1981, Ap. J. (Letters), 243, L17.

Schmidt, G. D., Stockman, H. S. and Grandi, S.: 1983, Ap. J., in press.

Schmidt, G. D.: 1985, this volume.

Schneider, D. P. and Young, P. J.: 1980a, Ap. J., 238, pp. 946.

Schneider, D. P. and Young, P. J.: 1980b, Ap. J., 240, pp. 871.

Stockman, H. S., Schmidt, G. D., Angel, J. R. P., Liebert, J., Tapia, S. and Beaver, E. A.: 1977, Ap. J., 217, pp. 815.

Stockman, H. S., Liebert, J. and Bond, H. E. 1979, IAU Coll. No. 53, "White Dwarfs and Variable Degenerate Stars," ed. H. M. Van Horn and V. Weidemann, University of Rochester Press, Rochester, New York, pp. 334.

Stockman, H. S., Foltz, C. B., Schmidt, G. D. and Tapia, S.: 1983, Ap. J., in press.

Stockman, H. S., Liebert, J. and Moore, R. L.: 1982, unpublished.

Szkody, P. and Brownlee, D. E.: 1977, Ap. J. (Letters), 212, L113.

Szkody, P., Cordova, F. A., Touhy, I. R., Stockman, H. S., Angel, J. R. P. and Wisniewski, W.: 1980, Ap. J., 241, pp. 1070.

Szkody, P. and Capps, R.: 1980, A. J., 85, pp. 882.

Szkody, P., Raymond, J. C. and Capps, R. W.: 1982, Ap. J., 257, pp. 686.

Tapia, S.: 1977, Ap. J. (Letters), 212, L125.

Tapia, S.: 1982, I.A.U. Circ. No. 3685.

Tapia, S.: 1983, Bull.A.A.S., 14, pp. 980.

Thortenson, J. R., Schommer, R. A. and Charles, P. A.: 1983, submitted to PASP.

Touhy, I. R., Lamb, F. K., Garmire, G. P. and Mason, K. O.: 1978, Ap. J. (Letters), 226, L17.

Touhy, I. R., Mason, K. O., Garmire, G. P. and Lamb, F. K.: 1981, Ap. J., 245, pp. 183.

Verbunt, F., van den Heuvel, E. P. J., van der Linden, Th. J., Brand, J., van Leeuwen, F. and van Paradijs, J.: 1980, Astron. Astrophys., 86, L10.

Visvanathan, N. and Wickramasinghe, D. T.: 1979, Nature, 281, pp. 47.

Visvanathan, N. and D. T. Wickramasinghe 1979, IAU Coll. No. 53, "White Dwarfs and Variable Degenerate Stars," ed. H. M. Van Horn and V. Weidemann, University of Rochester Press, Rochester, New York, pp. 330.

Visvanathan, N. and Pickles, A.: 1982, preprint.

Walter, F. and Bowyer, S.: 1981, Ap. J., 245, pp. 671.

Warner, B. and Nather, R. E.: 1972, M.N.R.A.S., 156, pp. 305.

Warner, B.: 1985, this volume.

Wickramasinghe, D. T. and Meggitt, S. M. A.: 1982, M.N.R.A.S., 198, pp. 975.

Young, P. J. and Schneider, D. P.: 1979, Ap. J., 230, pp. 502.

Young, P. J., Schneider, D. P. and Shectman, S. A.: 1981, Ap. J., 245, pp. 1043.

Young, P. J., Schneider, D. P., Sargent, W. L. W. and Boksenberg, A.: 1982, Ap. J., 252, pp. 269.

RECENT DEVELOPMENTS IN THE THEORY OF AM HER AND DQ HER STARS

D. Q. Lamb
Harvard-Smithsonian Center for Astrophysics

ABSTRACT

We describe a number of advances in the theory of accreting magnetic white dwarfs, and discuss several unresolved issues concerning these stars.

1. INTRODUCTION

The detection in 1976 of soft X-rays (Hearn, Richardson, and Clark 1976) and polarized optical light (Tapia 1977) from AM Her signaled the long-awaited discovery of a white dwarf analogue of the accreting magnetic neutron stars. Yet the AM Her stars differ from their neutron star brethren in one important respect: their magnetic fields are strong enough to phase-lock their rotation period to that of the binary ($\simeq$ 3 hours). However, pioneering photometric studies by Patterson (1979a,b) soon discovered two fast-rotating stars, AE Aqr and V533 Her, that are fully analogous to the pulsing neutron star X-ray sources. Most recently, fast optical photometry of faint X-ray sources has revealed many such DQ Her stars with longer periods (Patterson and Price 1981; Warner, O'Donoghue, and Fairall 1981; White and Marshall 1981; Steiner *et al.* 1981; Patterson and Steiner 1982; McHardy *et al.* 1982), paralleling an earlier, similar development in observations of pulsing neutron star X-ray sources.

These observational discoveries spawned a flurry of theoretical activity that continues unabated. The AM Her and DQ Her stars have attracted theoretical interest because they provide a laboratory in which to explore the physics of hot plasmas in strong magnetic fields. We can also learn from them a great deal about the masses, internal structure, and magnetic fields of white dwarfs themselves. Potentially, these pulsing sources can provide as much information as has been obtained from the pulsing neutron star X-ray sources (Joss and Rappaport 1983).

Here we describe recent advances in the theory of these stars, and discuss several of the unresolved issues concerning them. We consider in detail the possibility that the "soft X-ray puzzle" in the AM Her stars may have disappeared, according to recently reported *Einstein* observations. We also discuss the evidence, based on these recent observations, that the region emitting the polarized optical light is separate from the X-ray emission region. Among theoretical advances we describe are extensive new calculations of the X-ray emission from accreting magnetic white dwarfs, including soft and hard X-ray pulse

179

D. Q. Lamb and J. Patterson (eds.), Cataclysmic Variables and Low-Mass X-Ray Binaries, 179–218.

profiles and time-dependent behavior; improvements in the treatment of cyclotron emission and the transfer of radiation in magnetoactive plasmas, with application to the polarized optical light from the AM Her stars; new ideas about the stream-magnetosphere interaction region that produces the forest of strong emission lines in the AM Her stars; explanation of the synchronization of the white dwarf in the AM Her stars in terms of an MHD torque; and determination of the magnetic fields in the AM Her and DQ Her stars through analysis of their spin-up behavior.

2. STATUS OF THE "SOFT X-RAY PUZZLE" IN AM HER STARS

2.1. Simple theory

AM Her stars are cataclysmic variables containing a strongly magnetic white dwarf (see, *e.g.*, Chiappetti, Tanzi, and Treves 1980; Lamb 1981, 1982; Cordova and Mason 1983; Liebert and Stockman 1985). The magnetic fields are about 2×10^7 G, sufficient to prevent the formation of a disk and to lock the rotation period of the white dwarf to that of the binary ($1.3 - 3.7$ hours). As accreting matter flows toward the white dwarf, the magnetic field funnels it onto one or both magnetic poles. There a standoff shock forms far enough above the stellar surface for the hot, post-shock matter to cool and come to rest at the stellar surface. Figure 1 shows schematically the geometry of the hot, post-shock emission region.

The resulting radiation spectrum has three components: 1) a blackbody-limited optical cyclotron component produced by the hot emission region, 2) a hard X-ray bremsstrahlung component also produced by the hot emission region, and 3) a hard UV or soft X-ray blackbody component produced by cyclotron and bremsstrahlung radiation that is absorbed by the stellar surface and re-emitted (Fabian, Pringle, and Rees 1976; Lamb and Masters 1979; King and Lasota 1979). Figure 2 shows schematically the expected spectrum.

Let f be the fractional area of the stellar surface over which accretion occurs. Figure 3 shows the predicted theoretical spectra for a 1 $M_{\odot}$ star with a magnetic field of 2×10^7 G and effective luminosities $L_f \equiv L/f = 1 \times 10^{35}$ and 1×10^{37} ergs s^{-1}, as calculated by Lamb and Masters (1979). These spectra show that the AM Her stars are expected to be strong extreme UV and soft X-ray sources, and that the relative strength and the characteristic energy of the three spectral components change with variations in the accretion rate. For example, with the decrease in accretion rate illustrated in Figure 3, the star moves from the regime where bremsstrahlung dominates the cooling of the hot, post-shock gas to the one where cyclotron emission dominates. As a result, the bremsstrahlung hard X-ray component decreases by nearly four orders of magnitude while the total accretion luminosity decreases by only two, and the blackbody component moves from the soft X-ray region into the extreme UV. These theoretical calculations indicate that AM Her stars typically lie in the bremsstrahlung-dominated regime when they are in their "high" state, and in or near this regime when they are in their "low" state, so that theoretical calculations of the X-ray emission from non-magnetic stars are also applicable (see, *e.g.*, Kylafis and Lamb 1979, 1982)

Roughly half of the cyclotron flux is emitted outward and forms the blackbody-limited optical component. Roughly half of the bremsstrahlung flux is emitted outward and forms the hard X-ray component. The other halves of the cyclotron and bremsstrahlung fluxes are emitted inward and are reflected or absorbed by the stellar surface, producing the blackbody soft X-ray component. The simple theory therefore predicts

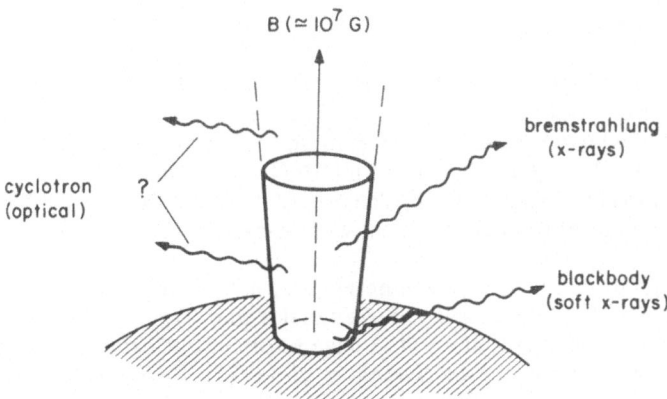

Fig. 1.–Schematic picture of the X-ray emission region in the AM Her stars, according to the simple theory. A magnetic field of $\simeq 2 \times 10^7$ G funnels the accreting matter toward the magnetic pole. Cyclotron emission and bremsstrahlung from the hot, post-shock emission region produces optical light and hard X rays. Roughly half of the cyclotron and bremsstrahlung photons are absorbed by the stellar surface and re-emitted, producing soft X rays.

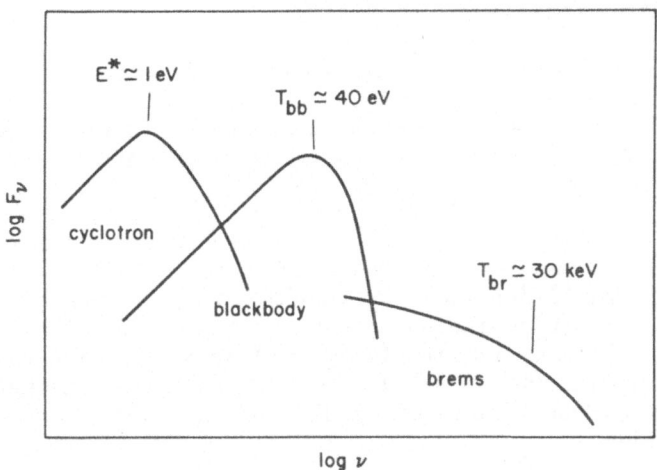

Fig. 2.–Schematic drawing of the spectrum of AM Her stars expected in the simple theory. The cyclotron component becomes optically thin at $E^* \simeq 1$ eV, while the blackbody and bremsstrahlung components peak at $T_{bb} \simeq 40$ eV and $T_{br} \simeq 30$ keV.

$$L_{bb} \simeq L_{br} + L_{cyc} . \tag{1}$$

2.2. Observational History

In 1976, Berg and Duthie (1977) suggested that AM Her was the optical counterpart of the hard X-ray source 4U1814+50. This identification was confirmed by Hearn, Richardson, and Clark (1976), who detected AM Her in soft X-rays. Soon thereafter, Szkody and Brownlee (1977) and Cowley and Crampton (1977) found that it has a binary period of 3.1 hours. More remarkably, Tapia (1977) discovered that its optical light is 10% circularly and linearly polarized, indicating that the white dwarf in the system is strongly magnetic.

Subsequent photometry and polarimetry (see, *e.g.*, Stockman *et al.* 1977; Michaelsky, Stokes, and Stokes 1977; Priedhorsky, Krzeminski, and Tapia 1978; Latham, Liebert and Steiner 1981; Bailey *et al.* 1983) show that the cyclotron optical component peaks at about 1.5 eV. Hard X-ray observations by Swank *et al.* (1977), Staubert *et al.* (1978), and most recently by Rothschild *et al.* (1981) show that the bremsstrahlung hard X-ray component has a spectral temperature $kT_{br} = 30.9 \pm 4.5$ keV. Tuohy *et al.* (1978, 1981) found the soft X-ray flux from AM Her to be adequately described by a blackbody spectrum with a temperature in the range 16 - 40 eV. Figure 4 shows the incident soft X-ray spectrum for the 1977 observation and the 90% confidence contours in the (N_H, kT)-plane for both the 1977 and the 1978 observations of Tuohy *et al.* (1981).

Figures 5 and 6 show the observed constraints on the luminosity and temperature of the soft X-ray component, and the inferred composite spectrum of AM Her. [Figures 7 and 8 show similar constraints and the inferred composite spectrum of VV Pup (Patterson *et al.* 1983.)] In Figure 6, the filled circles in the infrared and optical region are typical orbit-averaged values for the bright state from Priedhorsky *et al.* (1978). Figures 5 and 6 show the wide ranges in the temperature and the luminosity of the soft X-ray component allowed by the HEAO-1 observation.

Observations in the UV unexpectedly showed a turn-up at short wavelengths (Raymond *et al.* 1979). The most elegant explanation of the data attributed both the UV turn-up and the soft X-rays to an intense blackbody component with a temperature $kT_{bb} = 27$ eV and a luminosity $L_{bb} \simeq 3 \times 10^{34}$ erg s^{-1}, as illustrated in Figures 5 and 6. This implied

$$L_{bb} \gtrsim 50 \, (L_{cyc} + L_{br}), \tag{2}$$

which is obviously inconsistent with the approximate equality predicted by the simple theory [equation(1)]. A similar discrepancy was inferred in AN UMa (Szkody *et al.* 1981) and possibly in the other AM Her stars. This conflict has become known as the "soft X-ray puzzle". Ideas ranging from the transport of energy out of the hard X-ray emission region and into the star by electron conduction (King and Lasota 1980) or nonthermal electrons (Frank, King, and Lasota 1983) to high density filaments in the accretion flow (Kuijpers and Pringle 1982) or nuclear burning (see, *e.g.*, Fabbiano *et al.* 1981) have been put forward as possible explanations.

2.3. Implications of Iron Line Equivalent Width

Recently, Swank *et al.* (1983) have pioneered a new approach to the soft X-ray puzzle. They use modeling of the Fe K_α line to extract information about the physical properties of the X-ray emission region and the total accretion luminosity L_{acc}.

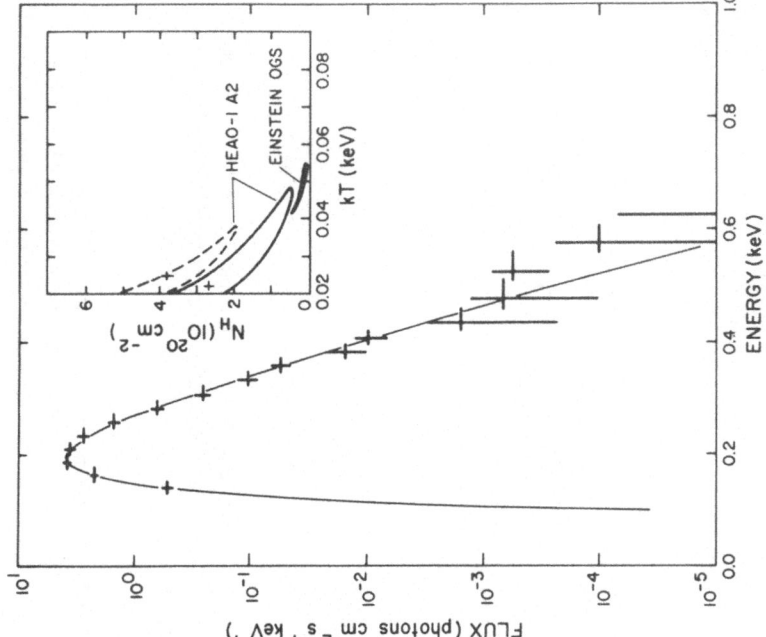

Fig. 3.–Theoretical spectra for a 1 $M_\odot$ star with a magnetic field of 2×10^7 G and effective luminosities $L/f = 1 \times 10^{35}$ and 1×10^{37} ergs s^{-1}. From Lamb and Masters (1979).

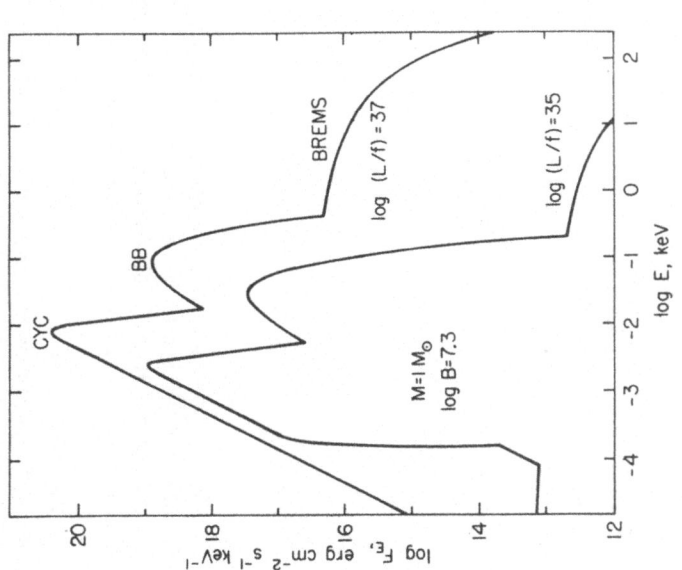

Fig. 4.–Inferred incident spectrum of AM Her for a blackbody model with $kT = 22$ eV and $N_H = 2.7 \times 10^{20}$ atoms cm^{-2} for the 1978 HEAO-1 pointed observation. The inset shows the 90% confidence contours for the HEAO-1 1977 scanning (dashed line) and 1978 pointed (solid line) observations, and the 1979 *Einstein* observation (filled ellipse). The HEAO-1 best fit values for kT and N_H are marked by crosses. From Tuohy *et al.* (1981) and Heise *et al.* (1983).

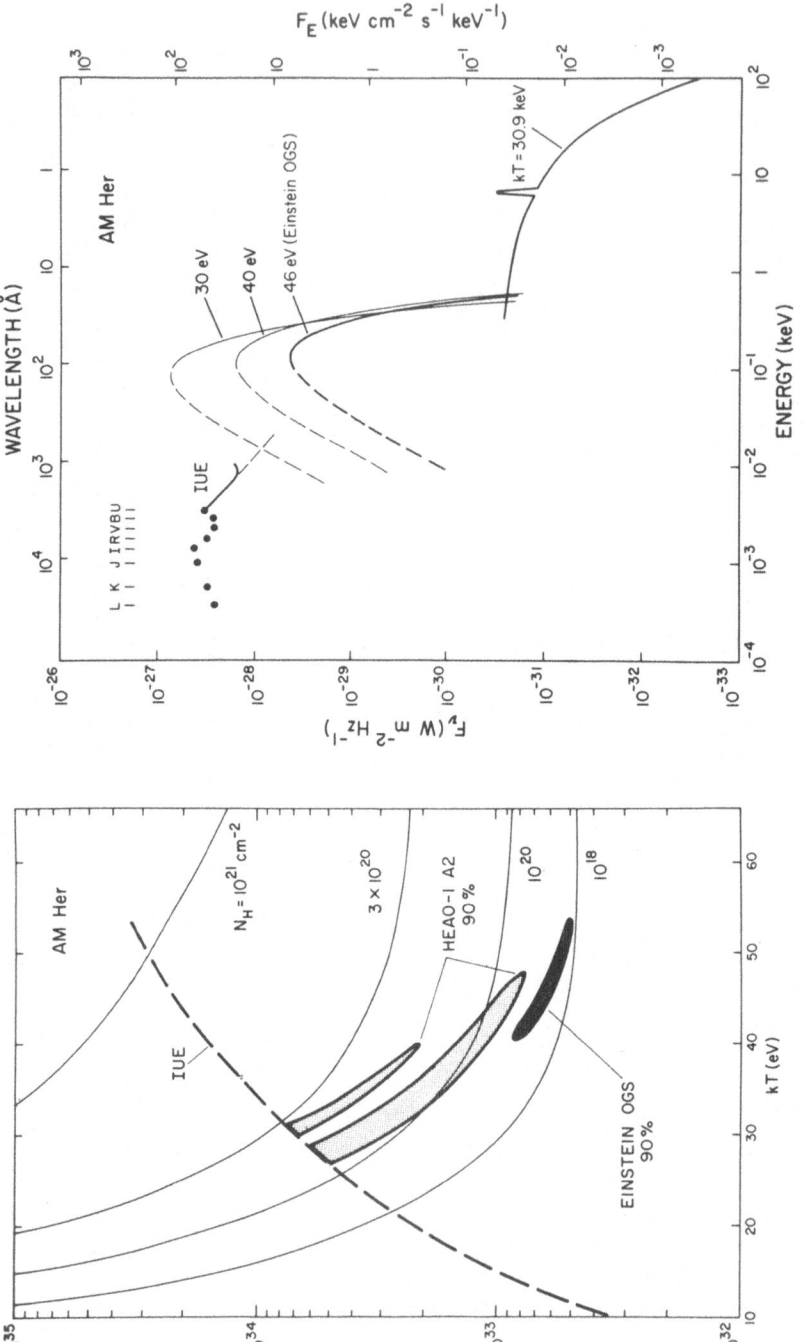

Fig. 5.—Constraints on the luminosity and temperature of the soft X-ray source in AM Her, assuming it to be a blackbody. The observed IUE spectrum constrains the source to lie below the dashed line. The 90% confidence regions for the 1977 (upper) and 1978 (lower) HEAO-1 observations are shown lightly shaded, while that for the 1979 *Einstein* observation is shown heavily shaded. Thin lines indicate the column density N_H.

Fig. 8.—Composite spectrum of AM Her. The curves labeled 30 and 40 eV correspond to the lowest and highest temperature blackbody spectra allowed by the IUE and HEAO-1 observations when taken together, while the curve labeled 46 eV (*Einstein* OGS) is the best-fit blackbody spectrum observed by *Einstein*. The filled circles in the infrared and optical region are typical orbit-averaged values for the bright state.

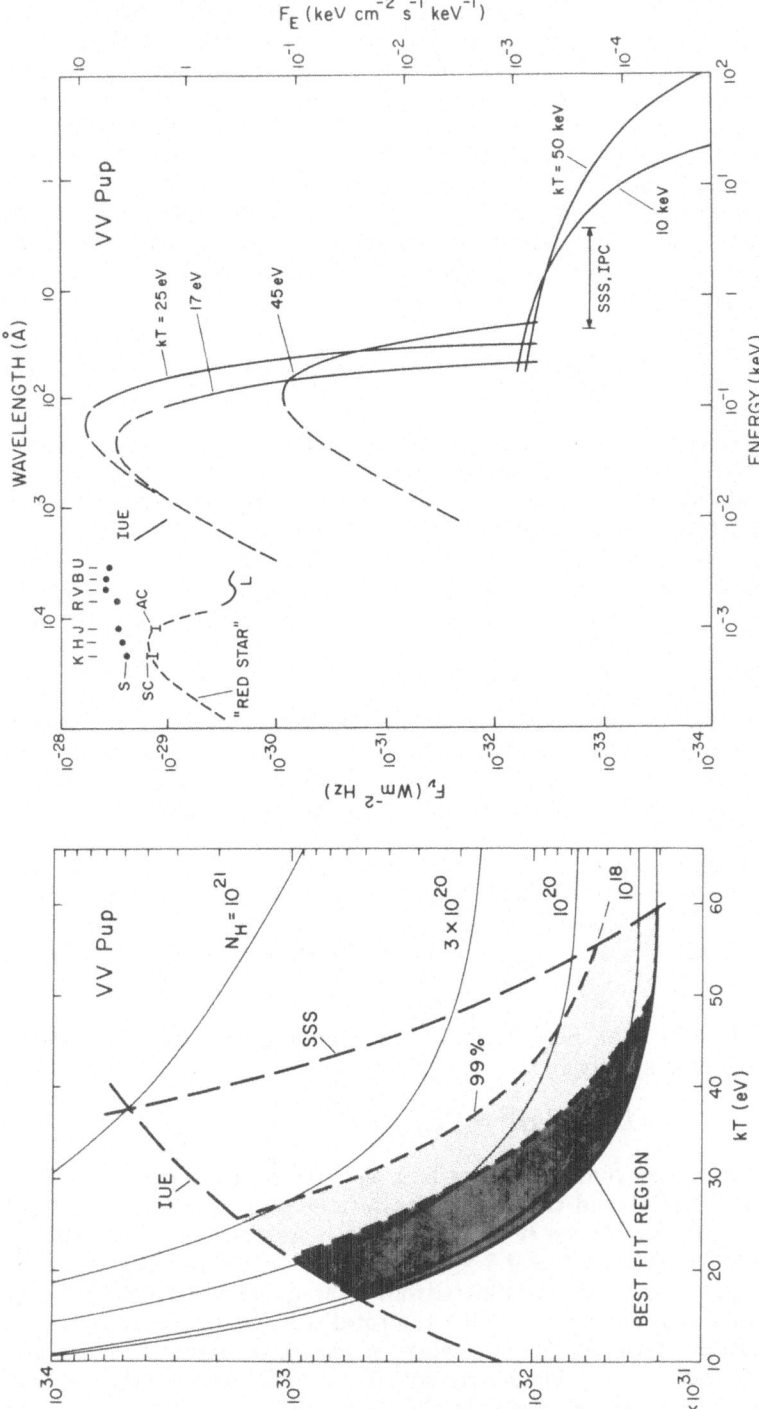

Fig. 7.—Constraints on the luminosity and temperature of the soft X-ray source in VV Pup, assuming it to be a blackbody. Nondetections by IUE and the *Einstein* Solid State Spectrometer (SSS) constrain the source to lie below the dashed lines. The allowed *Einstein* IPC spectral fits lie in the shaded region, with the heavily shaded region preferred. Thin lines indicate the corresponding column density N_H in atoms cm^{-2}. (From Patterson *et al.* 1983.)

Fig. 8.—Composite spectrum of VV Pup. The three soft X-ray spectra correspond to the three corners of the permitted region in Fig. 7. The filled circles indicate typical orbit-averaged values for the bright state, while the curved line labeled *L* represents the low state. The dashed line labelled *red star* is a cool blackbody fit to the infrared measurements in the low state. From Patterson *et al.* (1983).

As noted by Kylafis and Lamb (1982) and many others, the hard X-ray spectrum observed in AM Her is much flatter than that expected for thermal bremsstrahlung from a single-temperature emission region, let alone from an emission region with a realistic temperature profile. The equivalent width of the Fe K_α emission line in AM Her is also known to be larger than expected for purely thermal emission (Rothschild et al. 1981). Using Monte Carlo calculations, Swank et al. (1983) find that the flatness of the hard X-ray spectrum and the strength of the iron emission line can be explained by the effects of absorption and fluorescence within the accretion column. This is illustrated in Figure 9, which shows the ratio of the thermal to fluorescent components and the total equivalent width of the Fe K_α line as a function of the electron scattering optical depth across the *diameter* of the accretion column,

$$\tau_\perp = 1.6 (\frac{f}{10^{-4}})^{1/2} (\frac{L_{acc}/f}{L_E})(\frac{M}{M_\odot})^{-1/2} (\frac{R}{5 \times 10^8 cm})^{1/2}. \tag{3}$$

Whether the shape of the continuum spectrum approximates that observed, or is too flat or too steep, is also indicated. Figure 9 shows that agreement with both the equivalent width of the line and the shape of the continuum are satisfactory, if the optical depth to Thomson scattering across the diameter of the accretion column $\tau_\perp = 0.3 - 0.7$.

The implications of this result depend on the cross sectional area $A = 4\pi R^2 f$ of the emission region (and the accretion column). With $\tau_\perp$ fixed, specifying the area determines L_{acc}. The requirement

$$L_{acc} = L_{cyc} + L_{bb} + L_{br}, \tag{4}$$

then determines L_{bb}, since L_{cyc} and L_{br} are observed directly and are known. If the spot size is large, the accretion luminosity is high and L_{bb} is too large to be consistent with the simple theory. If the spot size is small, the accretion luminosity is low and L_{bb} is modest, consistent with its being reprocessed cyclotron and bremsstrahlung radiation.

What can observation tell us about the spot size? Unfortunately, estimates of its size differ by a factor of 40. If the area is determined from the shape of the hard X-ray eclipse, a value $f \simeq 2 \times 10^{-3}$ is found (Rothschild et al. 1981, Swank et al. 1983). This value yields a blackbody luminosity too large to be accommodated in the simple theory, as shown in Figure 10a.

On the other hand, if the soft X-ray component is assumed to have a blackbody spectrum, the area is determined simply by

$$L_{bb} = 4\pi R^2 f \sigma T_{bb}^4, \tag{5}$$

where L_{bb} is the inferred blackbody luminosity and R is the radius of the star. Given the temperatures found in the HEAO-1 and the recent *Einstein* OGS observations, the spot size is then small. Thus in AM Her Tuohy et al. (1981) find a spot size $f \simeq 2 \times 10^{-4}$ if $T_{bb} = 40$ eV and Heise et al. (1983) find $f \simeq 4.4 \times 10^{-5}$ for their best fit temperature $T_{bb} = 46$ eV (in these estimates the mass of the white dwarf is assumed to be $1 M_\odot$). These values yield a blackbody luminosity consistent with the total soft X-ray flux measured in the observations and consistent with the simple theory, as shown in Figure 10b. We also note that, in this case, the ratio of the blackbody soft X-ray and bremsstrahlung hard X-ray luminosities is increased by backscattering in the accretion column, as suggested by Kylafis and Lamb (1982), so that only about 40% of the initially emitted bremsstrahlung hard X-ray flux escapes to infinity (Swank et al. 1983).

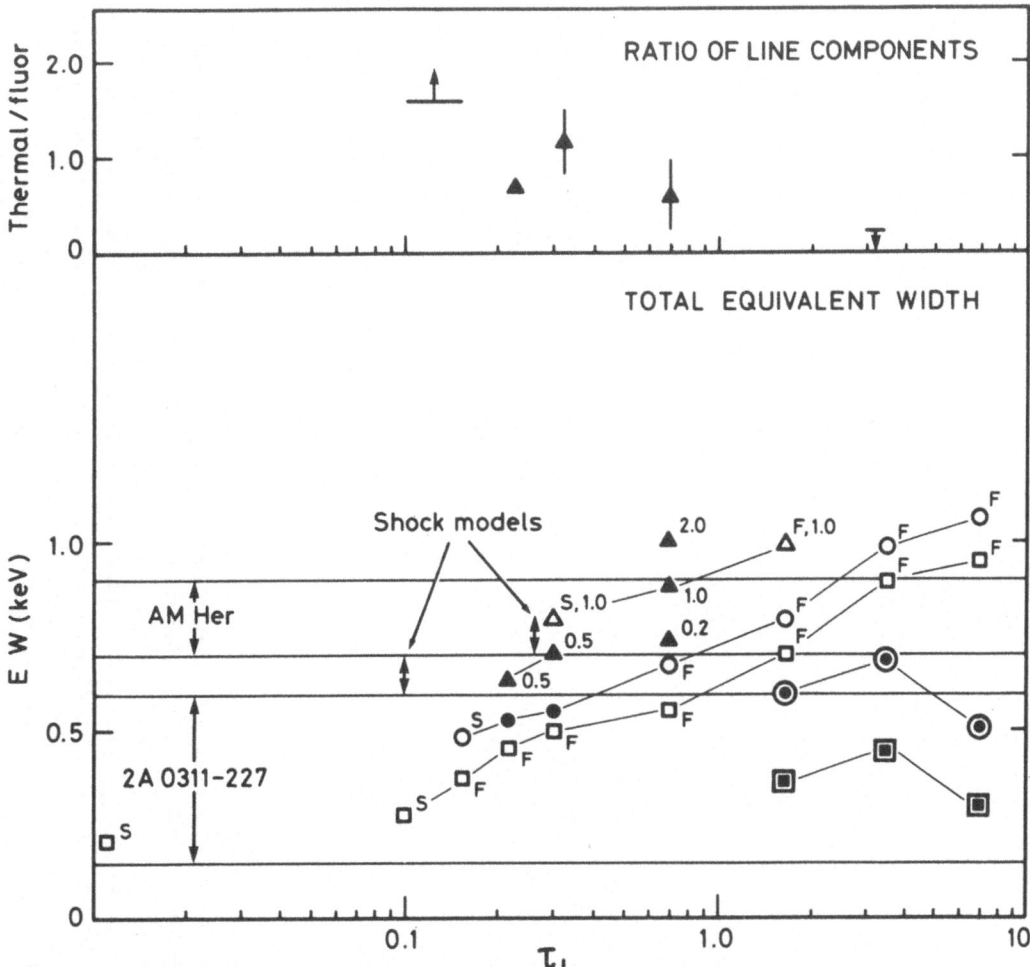

Fig. 9.–(Upper panel) Ratio of thermal to fluorescent components in AM Her. Bars denote the range of acceptable models. (Lower panel) Total equivalent width of thermal plus fluorescent Fe K_α emission as a function of the Thomson optical depth across a diameter of the accretion column just above the shock. Results are shown for a "cold" column (squares) in which only H and He are assumed ionized and for a "partially ionized" column (circles), both for a 30 keV bremsstrahlung emission spectrum. Filled symbols denote a continuum shape approximating that observed, F denotes that it is too flat, and S too steep. Filled symbols enclosed in open ones are the observed equivalent widths when 10% of the emitted X-rays are assumed to be viewed directly out the sides of a shock region of finite height. Triangles are results assuming an additional lower temperature component (5 keV) contributes to the emission spectrum, with the relative fraction indicated as a superscript. From Swank *et al.* (1983).

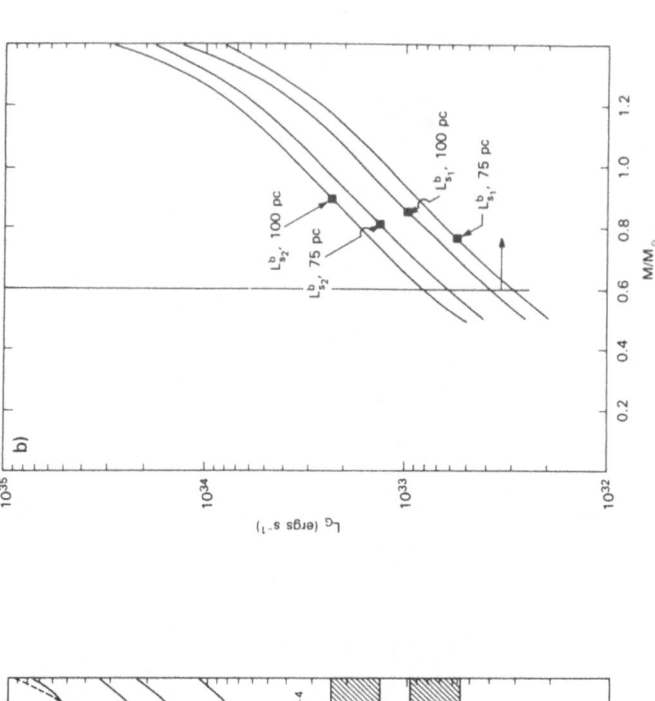

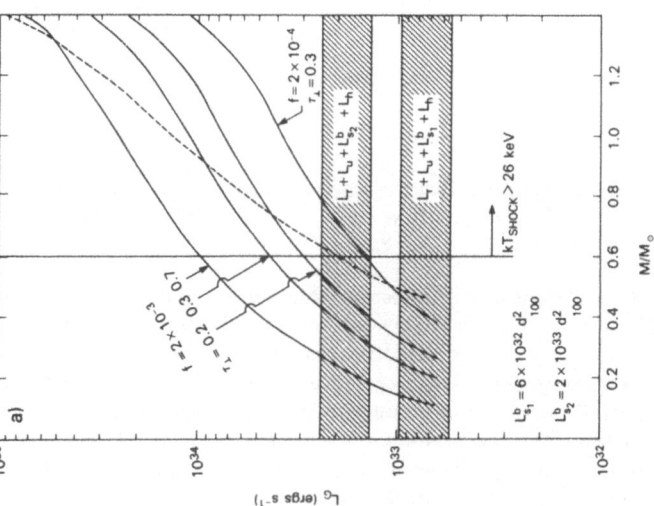

Fig. 10.—Constraints on the mass and accretion luminosity of AM Her. (a) For accretion columns with areas $f = 2 \times 10^{-3}$ and $f = 2 \times 10^{-4}$. Solutions must lie on or to the right of the vertical line labeled $kT_{shock} > 26$ keV for the temperature of the bremsstrahlung hard X-ray component to equal or exceed its observed value. The dashed curve corresponds to the accretion rate for which steady nuclear burning could occur if $f = 2 \times 10^{-3}$. The horizontal shaded areas indicate the sum of the optical (L_r), ultraviolet (L_u), soft X-ray ($L_{s1(s2)}$), and hard X-ray (L_h) luminosities for a distance between 75 and 100 pc. The lower range corresponds to $L_{s1}^b = 6 \times 10^{33} d_{100}^2$ ergs s^{-1} (the best-fit value found by the 1979 *Einstein* OGS observation), and the upper range corresponds to $L_{s2}^b = 2 \times 10^{33} d_{100}^2$ ergs s^{-1} (corresponding to a blackbody spectrum with a temperature of 40 eV, consistent with the 90% contour of the 1978 HEAO-1 observation). (b) For f determined simply from L_{bb}. Solutions of $L_{acc} \equiv L_G = L_r + L_u + L_{s1(s2)} + L_h$ are indicated by filled boxes for distances $d = 75$ and 100 pc. From Swank *et al.* (1983).

2.4. Implications of Recent Soft X-Ray Observations

Recently, the results of the *Einstein* Objective Grating Spectrometer (OGS) observation of AM Her have become available (Heise *et al.* 1983; see also Heise and Brinkman 1982). Here we closely follow the discussion of this observation given by Patterson *et al.* (1983). Figure 4 shows the 90% confidence level contour in the (N_H, kT)-plane found by Heise *et al.* (1983). As can be seen, the OGS observation favors somewhat higher temperatures and lower column densities, i.e.

$$41 \ eV < kT_{bb} < 55 \ eV$$

$$7 \times 10^{18} \ cm^{-2} < N_H < 5 \times 10^{19} \ cm^{-2},$$

than the best fits to the earlier HEAO-1 data. The soft X-ray spectrum seen in the *Einstein* OGS observation is shown in Figure 5.

Actually, these values are close to the 90% confidence contours shown by Tuohy *et al.* (1981) for the 1978 HEAO-1 pointed observation, but disagree (primarily in N_H) with the contours found in the 1977 scanning observation (Tuohy *et al.* 1978). It is not entirely clear how to interpret the discrepancy, particularly since the soft X-ray flux from AM Her is variable. But if we adopt the values of kT_{bb} and N_H found in the OGS observation, which is free from the difficult problems of spectral deconvolution, and we assume a blackbody emission model, then about half of the total radiated UV and soft X-ray flux appears in the 0.1-0.5 keV HEAO-1 bandpass. Adopting the flux observed in the 1978 HEAO-1 observation and an optically thick slab geometry, $L_{bb} \simeq (4-8) \times 10^{32} \ (d/100 \ pc)^2$ ergs s^{-1}. Assuming that the hard X rays are emitted isotropically, the simultaneous 2-150 keV measurement (Rothschild *et al.* 1981) yields $L_{br} = 4.9 \times 10^{32} (d/100 \ pc)^2$ ergs s^{-1}. If we adopt the flux from the 1975 OSO-8 observation (Swank *et al.* 1977), we obtain slightly higher (but less well determined) luminosities. In both cases, $L_{bb} \simeq L_{br}$, as predicted by the simple theory. Thus, the soft X-ray problem in AM Her itself *may* have evaporated.

Einstein Imaging Proportional Counter (IPC) observations of the AM Her stars VV Pup (Patterson *et al.* 1983) and EF Eri (2A0311-227) (Beuermann *et al.* 1984) have also recently become available. Figures 7 and 8 show the observed constraints on the luminosity and temperature of the soft X-ray component and the inferred composite spectrum of VV Pup (Patterson *et al.* 1983). It is evident that uncertainties in the values of kT_{bb} and kT_{br} do not permit a sensitive test of the simple theoretical model. We can conclude only that $L_{bb} \simeq (0.6\text{-}10) \ (L_{cyc} + L_{br})$, with a most probable value in the range 1-5.

The results in the case of EF Eri are similar. The observational constraints allow a wide range in the luminosity and temperature of the soft X-ray component and, again, they encompass the values predicted by the simple theoretical model (Beuermann *et al.* 1984). We note also that a cross-correlation analysis of the soft and hard X-ray intensities carried out on data from an extended *Einstein* observation shows a strong correlation, as predicted by the reprocessing model of the soft X-ray source (Beuermann *et al.* 1984).

In summary, it is obvious that all four original AM Her stars produce intense soft X-ray fluxes when they are in the high state. AM Her and VV Pup are the most extreme, with $F_s(0.1\text{-}0.5 \ keV) \simeq 10 F_h$ (2-6 keV) (see Table 5 of Patterson *et al.* 1983). But, if properties of the soft X-ray component in AM Her seen by the *Einstein* OGS are typical, then the observed soft X-ray flux represents most of the total emitted, while the hard X-ray data for the best-studied stars, AM Her and EF Eri, suggest that the total hard X-ray flux exceeds the 2-6 keV flux by a factor of $\simeq 7$. These arguments imply that L_{bb}/L_{br} is generally in the range 0.5-4. Also, the ultraviolet and optical fluxes are comparable to the observed soft X-ray flux, leaving open the possibility that heating by cyclotron radiation contributes significantly to the soft X-ray luminosity. *Thus, in all cases it appears possible to power the soft X-ray source by reprocessing hard X-rays and/or cyclotron radiation on the surface of the white dwarf.*

This hypothesis can be tested further by (1) more accurate determinations of T_{bb} through further OGS soft X-ray observations, (2) accurate determinations of T_{br} through more sensitive hard X-ray observations, and (3) confirmation of a strong correlation between the soft and hard X-ray intensities through simultaneous optical, soft X-ray, and hard X-ray photometry.

There remains the origin of the ν^2 ultraviolet component seen by Raymond *et al.* (1979). This question is discussed further in Section 3.5 below. However, since the observed flux in this component is relatively small, it does not constitute a major problem in energetics unless its temperature is substantially greater than the lower limit of $\simeq 10$ eV provided by the IUE data.

3. X-RAY AND OPTICAL EMISSION

3.1. X-Ray Spectra and Light Curves

The emission regions of the AM Her stars cover only a fraction $f \sim 10^{-5} - 10^{-3}$ of the stellar surface, as discussed above. The emission regions of the DQ Her stars are larger, but f is still expected to be $\ll 1$. As a result of this spherical asymmetry, at moderate and high accretion rates the optical depth across the accretion column [see equation (3)] can be much less than the optical depth vertically through it,

$$\eta_\parallel = 28(\frac{L_{acc}/f}{L_E})(\frac{M}{M_\odot})^{-1/2}(\frac{R}{5 \times 10^8 cm}). \tag{6}$$

That is,

$$\frac{\tau_\perp}{\eta_\parallel} = 0.058(\frac{f}{10^{-4}})^{1/2}. \tag{7}$$

We expect this disparity to affect both the observed spectra and the X-ray pulse profile.

Most calculations of X-ray emission from accreting white dwarfs have assumed spherical symmetry (see, *e.g,* Hoshi 1973; Aizu 1973; Katz 1976; Kylafis and Lamb 1979, 1982). Recently, however, Swank *et al.* (1983) and, especially, Imamura and Durisen (1983) have reported the results of extensive Monte Carlo calculations for the nonspherical case. Here we summarize the results of Imamura and Durisen (the results of Swank *et al.* were discussed in Section 2.3 above).

Spectra. As long as $\eta_\parallel \lesssim 1$, the accretion column has little or no effect. Thus at low effective luminosities,

$$L_{acc}/f \lesssim 5 \times 10^{36} (\frac{M}{M \odot})^{1/2} (\frac{R}{5 \times 10^8 cm})^{-1/2} ergs \ s^{-1}, \tag{8}$$

the X-ray spectra for the funneled and the spherically symmetric cases are similar. This is illustrated in the top spectra in Figure 11.

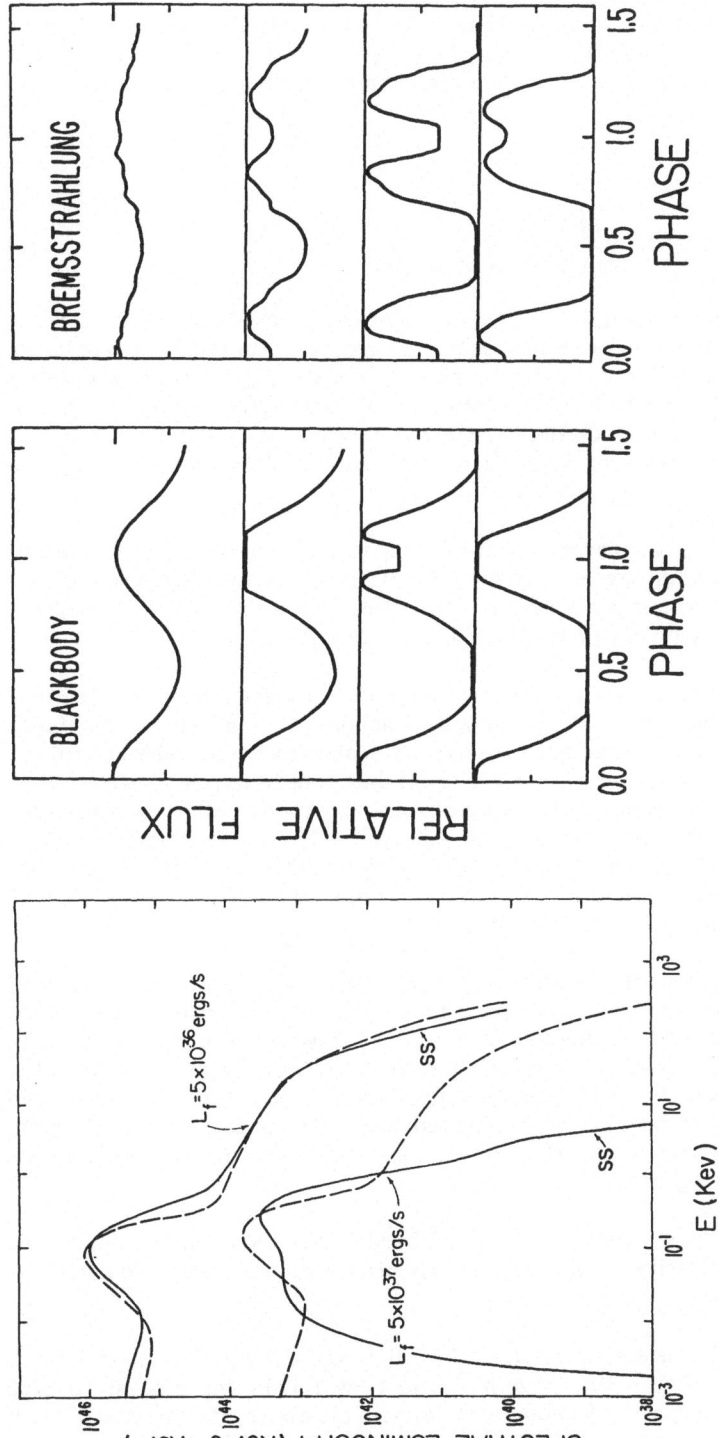

Fig. 11.—Funneled (dashed curves) and spherically symmetric (solid curves labeled SS) X-ray spectra of a $1M_\odot$ star for effective luminosities $L_f = 5 \times 10^{36}$ and 5×10^{37} ergs s^{-1}. The funneled spectra were computed for $L_{acc} = 5 \times 10^{34}$ ergs s^{-1} and $f = 10^{-2}$ and 10^{-3}, and then scaled by f^{-1}. In addition, both $L_f = 5 \times 10^{37}$ erg s^{-1} spectra are scaled by 10^{-3} for clarity of presentation. From Imamura and Durisen (1983).

Fig. 12.—Blackbody and bremsstrahlung pulse profiles of a $1M_\odot$ star for $L_{acc} = 7.5 \times 10^{33}$ ergs s^{-1} and an inclination angle $i = 60°$. Pulse profiles are shown for magnetic colatitudes $\delta = 15°$ (uppermost), $30°$, $60°$, and $90°$ (lowermost). Zero phase corresponds to the moment when the angle between the accretion column and the line of sight to the observer is smallest. From Imamura and Durisen (1983).

At higher luminosities, the spectra are no longer similar, since in the funneled case the emergent photons suffer fewer scatterings than in spherically symmetric models with the same effective accretion luminosity $L_f \equiv L_{acc}/f$. Thus the bremsstrahlung hard X-ray spectra will be harder and the ratio L_{bb}/L_{br} smaller. This is illustrated in the bottom spectra in Figure 11.

Finally, when $\tau_\perp \gtrsim 7$, i.e.

$$L_{acc}/f \gtrsim 1 \times 10^{37} (\frac{M}{M_\odot})^{1/2} (\frac{R}{5 \times 10^8 cm})^{-1/2} ergs\ s^{-1}, \qquad (9)$$

even funneled models fail to scale because of degradation of the bremsstrahlung hard X-ray component. In this regime, the effects of the nonspherical geometry on the X-ray spectrum are most easily understood as follows. Increasing the luminosity L_{acc}/f while keeping f fixed increases the ratio L_{bb}/L_{br} due to back scattering in the accretion column. Eventually, if L_{acc}/f is high enough, the bremsstrahlung hard X-ray component is degraded. Keeping L fixed while decreasing f while keeping L_{acc}/f fixed has similar effects, since it means an increase in L_{acc}/f and therefore in $\tau_\perp$ [see equation (3)].

Light curves. At low L_{acc}/f, the accretion column has no appreciable effect on the soft and hard X-rays. The bremsstrahlung hard X-rays are emitted isotropically, producing a box-like light curve. The blackbody emission is Lambertian, i.e. limb-darkened, and the soft X-rays are therefore also emitted primarily along the accretion column.

At modest L_{acc}/f, corresponding to the intermediate regime defined by equations (8) and (9), the last scattering occurs at a substantial height ($h \sim R$) above the stellar surface. As a result, the bremsstrahlung hard X-rays are scattered beyond the horizon by the accretion column and the hard X-ray light curve exhibits a peak approximately 90° to the accretion column and a dip along it. In contrast, the blackbody soft X-rays undergo little scattering in the column but are strongly self-eclipsed by the white dwarf, since they are emitted at the stellar surface. Thus the hard X-ray and soft X-ray light curves have one minimum and one maximum, but the two light curves are 180° out of phase with each other (see Figure 5 of Imamura and Durisen 1983).

At higher L_{acc}/f, the height of the emission region is small and the soft and hard X-rays are last scattered near the stellar surface. As a result, the qualitative shape of the soft and hard X-ray light curves is insensitive to L and f. However, the minimum along the accretion column is shallower in soft X-rays than in hard. This happens because the blackbody emission occurs strongly in the forward direction (i.e. it is limb darkened) and because the blackbody photons preferentially emitted near edge of the emission region, due to repeated scatterings between the accretion column and the stellar surface, and are therefore less affected by the column.

These effects lead to a wide variety of soft and hard X-ray light curves, depending not only on the effective luminosity L_f but also on the orientation of the white dwarf, as illustrated in Figure 12.

AM Her stars. The AM Her stars have $L_{acc} \simeq 1 \times 10^{33}$ ergs s^{-1} and $f \sim 10^{-4} - 10^{-3}$, so that typically $L_{acc}/f \sim 10^{36} - 10^{37}$ ergs s^{-1} and they fall in the modest to high effective luminosity regime. According to equations (8) and (9) above, the spectra of stars in the lower end of this effective luminosity range can be scaled from spherically symmetric calculations while those in the upper end cannot. However, no significant degradation of the hard X-ray spectrum occurs in either case. Throughout this effective luminosity range the height of the X-ray emission region is much smaller than the radius of the white dwarf, so that both soft and hard X-ray emission occurs preferentially along the accretion column. This is in qualitative agreement with X-ray observations of the AM Her stars.

The soft and hard X-ray light curves may be further complicated if emission occurs from both magnetic poles as apparently occurs, for example, in EF Eri (see, *e.g.*, Patterson, Williams, and Hiltner 1981; White 1981). Further, changes in the luminosity imply changes in the height of the emission region. If the effective luminosity falls as low as $\simeq 1 \times 10^{36}$ erg s^{-1}, the height of the emission region becomes comparable to the radius of the white dwarf. Then, as discussed above, the hard X-ray eclipse can actually disappear or the soft and hard X-ray minima can be 180° out of phase! It is conceivable that something like this might account for the disappearance of the soft X-ray eclipse in AM Her (Priedhorsky and Marshall 1982).

Imamura and Durisen (1983) have not included the effect of absorption of X-rays by partially ionized matter in the accretion column (see Swank *et al.* 1983). It will produce low energy absorption of the soft and hard X-ray spectrum and a deeper minimum in the soft and hard X-ray light curve when the white dwarf is viewed within an angle of $\ll 30°$ of the accretion column, according to estimates by the author based on the height of the X-ray emission region for the effective luminosities of the AM Her stars.

3.2. Time-Dependent X-Ray Behavior

Thermal instability. Langer, Chanmugam, and Shaviv (1981, 1982) have discovered that the hot, post-shock X-ray emission region of accreting white dwarfs is thermally unstable for both spherically symmetric and channeled flows when the cooling is due to thermal bremsstrahlung. The nature of the instability can be understood by considering the behavior of the accretion flow when it is perturbed by giving the shock a small outward velocity. The perturbation increases the post-shock temperature and velocity of the matter, so that it cannot cool and come to rest at the stellar surface. The accumulation of hot matter behind the shock produces an excess pressure that acts to force the shock upward. If the flow is perturbed by giving the shock a small inward velocity, this decreases the post-shock temperature and velocity of the matter, so that it requires a shorter distance to cool and come to rest. The depletion of hot matter behind the shock produces a pressure deficit that acts to collapse the shock. If the flow is to be stable, the height h of the emission region must exceed its equilibrium value whenever the shock velocity $v_s > 0$ in the rest frame of the star.

According to the numerical calculations of Langer, Chanmugam, and Shaviv (1981, 1982), given a cooling law of the form

$$\Lambda \propto n^\beta T^\alpha, \tag{10}$$

the flow is unstable for $\alpha \gtrsim 1.6$ if $\beta = 2$. Since $\alpha = 0.5$ and $\beta = 2$ for bremsstrahlung cooling, this implies that the flow is always unstable for such cooling.

Figure 13 shows the resulting oscillations in the radius of the shock, and the maximum temperature and bremsstrahlung luminosity of the post-shock gas as a function of time. The asymmetric "sawtoothed" shape of the temperature and luminosity variations is a

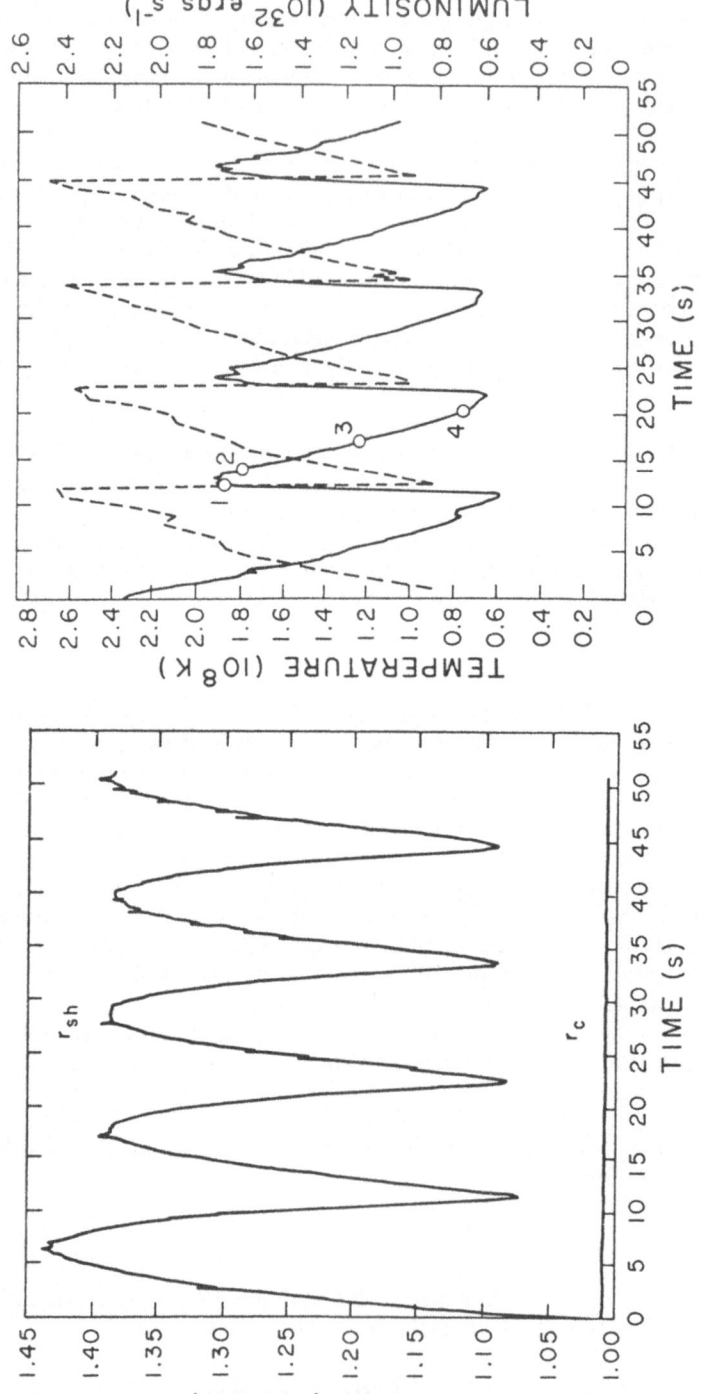

Fig. 13.—Properties of the shocked gas as a function of time for accretion onto a star with mass $0.5 M_\odot$ and radius 10^9 cm. The accretion rate is 2.5×10^{15} g s^{-1} along a dipolar magnetic flux tube with area 10^{16} cm^2 at the stellar surface. (a) The radius of the shock and the base of the cooling region. (b) The maximum temperature (solid curve) and the bremsstrahlung luminosity (dashed curve) of the post-shock gas. From Langer, Chanmugam, and Shaviv (1982).

characteristic of this instability. The period of the oscillations, 0.1 - 10 seconds, corresponds to the fundamental mode and scales as

$$P \simeq 3 t_{cool} \propto \phi_{grav} \dot{\sigma}^{-1}, \tag{11}$$

where ϕ_{grav} is the gravitational potential of the white dwarf and $\dot{\sigma}$ is the accretion rate per unit area. The calculations do not have high enough spatial or temporal resolution to see the overtones.

The hard X-ray spectrum varies a lot during an oscillation (see Langer, Chanmugam, and Shaviv 1982; Figure 5). However the time-averaged spectrum is very similar to that found in steady-state calculations (Kylafis and Lamb 1979, 1982). Both have an excess of soft photons when compared with a single temperature thermal bremsstrahlung spectrum because of the greater bremsstrahlung emissivity of the cooler matter near the stellar surface.

Langer, Chanmugam, and Shaviv (1982) have also discovered the interesting fact that there is, in the case of funneled flow, a critical effective accretion luminosity,

$$L_{acc}^{c}/f \simeq 5 \times 10^{35} (\frac{M}{M_{\odot}})^{5/2} (\frac{R}{5 \times 10^{8} cm})^{-1/2} ergs \ s^{-1}. \tag{12}$$

below which the shock moves steadily outward with time, at least until the assumptions of the model break down. This happens because the rapid increase in the area, $\propto r^{3}$, leads to a rapidly decreasing post-shock density when the shock moves more than a distance $\simeq R/2$ above the stellar surface. The resulting increase in the bremsstrahlung cooling time is so rapid that, even accounting for the increasing flow distance, the matter never has time to cool before reaching the surface.

Following the discovery of a thermal instability, Chevalier and Imamura (1982) carried out a linear analysis in planar geometry of the oscillation periods and damping rates. In contrast to Langer, Chanmugam, and Shaviv (1981, 1982), they found the flow to be unstable at the fundamental, first overtone, and second overtone for $\alpha \lesssim 0.4$, 0.8, and 0.8 if $\beta = 2$. Since, as noted above, $\alpha = 0.5$ and $\beta = 2$ for bremsstrahlung cooling, this implies that the flow is stable at the fundamental and unstable only at the overtones.

Recent nonlinear time-dependent calculations by Imamura, Wolff, and Durisen (1983) confirm these results (see Imamura *et al.* 1983), while improved calculations reported by Langer (1983) indicate a critical value for α between 0.5 and 0.6 (but definitely greater than 0.5) rather than the value 1.6 found earlier. The discrepancy between the two nonlinear numerical calculations is small and may be due to differences in numerical techniques or to the handling of the inner boundary condition. However, we think the results suggest that the linear analysis of Chevalier and Imamura (1982) is correct and that the thermal instability occurs only at the first few overtones in the case of bremsstrahlung cooling.

Stabilizing effects. Realistic calculations of the X-ray emission region of accreting white dwarfs must include Compton cooling, thermal conduction, and differing electron and ion temperatures if the mass of the white dwarf $\gtrsim 0.8 M_{\odot}$ (Imamura *et al.* 1979). All of these effects tends to suppress the thermal instability. Compton cooling tends to damp the instability because it has a weaker density dependence than ρ^{2} and a stronger temperature dependence than $T^{0.8}$. Thermal conduction acts to smooth out localized temperature variations, and its efficiency increases for smaller wavelength variations. Thus it is effective in suppressing high overtones even though it is a small effect in the X-ray emission region. Even the fundamental can be affected because conduction is enhanced during the part of the oscillation when the shock height is small. Differing electron and ion temperatures makes the electron temperature insensitive to the temperature behind the ion viscous shock and therefore to the shock velocity.

Imamura, Wolff, and Durisen (1983) have carried out a nonlinear numerical calculation including all of these effects for a $1.0 M_\odot$ white dwarf. No oscillations are found at the fundamental or at any higher overtones, confirming that these effects are indeed stabilizing and are sufficient to suppress the thermal instability. Imamura, Wolff, and Durisen suggest that the instability may still occur for white dwarfs with masses $\lesssim 1.0 M_\odot$ because Compton cooling and conduction are then unimportant. Even for low mass white dwarfs, however, the electron and ion temperatures just behind the shock differ, although they very rapidly converge thereafter. This difference may be sufficient to suppress the thermal instability for all white dwarf masses.

Imamura *et al.* (1983) have extended the linear stability analysis of Chevalier and Imamura (1982) to the case $\beta = 1$ and find critical α values of 0.1, 0.1, 0.0 for the fundamental, first overtone, and second overtone. Since $\beta \simeq 0.3$ and $\alpha \simeq 3$ for optically thick cyclotron emission, these results show that cyclotron emission strongly suppresses the thermal instability. A similar conclusion is reached by Langer (1983), who finds that the instability is damped even at the fundamental for a magnetic field strength $B = 10^7$ Gauss, $L \simeq 4 \times 10^{33}$ ergs s^{-1}, and $f = 10^{-3}$, values appropriate for the AM Her stars.

AM Her stars. In summary, recent calculations show that no thermal instability occurs for accretion onto nonmagnetic white dwarfs if the white dwarf mass is $\gtrsim 0.8 M_\odot$ and it is possible that none occurs even for lower mass white dwarfs. Other calculations indicate that no instability occurs for accretion onto magnetic white dwarfs if the magnetic field $B = 10^7$ Gauss and the effective luminosity $L_{acc}/f = 4 \times 10^{36}$ ergs s^{-1}, values appropriate for the AM Her stars. For all of these reasons, it is likely that the accretion flow is not unstable in the AM Her stars. This is consistent with the fact that, although broad bumps at a period of about 2 seconds have been seen in the *optical* power spectrum of AN UMa and E1405-451 (Middleditch 1982), no significant *X-ray* variations have been seen at periods 0.1 - 10 seconds.

3.3. Soft X-Ray Spectrum

The soft X-ray component in the AM Her and DQ Her stars is due, in the simple theory, to the reprocessing of bremsstrahlung hard X-rays and/or cyclotron optical light by the surface of the white dwarf. It has generally been taken to be a blackbody (Lamb and Masters 1979; King and Lasota 1979; Kylafis and Lamb 1979, 1982). Of course a true blackbody spectrum occurs only for a purely absorptive, constant temperature atmosphere, and all stellar atmospheres deviate from these conditions to a lesser or greater degree. It is therefore important to determine the accuracy of the blackbody approximation.

Reprocessing of bremsstrahlung hard X-rays. If the soft X-ray component arises from reprocessing of bremsstrahlung hard X-rays, most of the energy is deposited at a large absorption optical depth and merely adds to the outward flux (or luminosity) passing through the atmosphere. Heise *et al.* (1983) have pointed out that at effective temperatures as high as those of the soft X-ray component in the AM Her stars, the color temperature can be larger than the effective temperature, and this should be taken into account in fitting the observed spectra.

Generally, the color temperature is larger than the effective temperature when scattering is much more important than absorption and/or the opacity due to absorption decreases rapidly with increasing energy. The former increases the opacity (independent of energy), so that the temperature in the atmosphere must be higher in order to transport the same outward flux, while the latter tends to shift the flux to higher energies because the opacity is smaller there. Using a model atmosphere with an abundance ratio $n_{He}/n_H =$ 0.15, no metals, and a surface gravity $g = 10^8$ cm s^{-2}, Heise et $al.$ (1983) find a best fit to the $Einstein$ OGS spectrum for an effective temperature $T_e = 21$ eV. They point out that the UV flux in this model atmosphere equals that observed by Raymond et $al.$ (1979). The implied area of the emission region is $A = 8.3 \times 10^{15}$ cm^2 and the soft X-ray luminosity $L_s = 2.0 \times 10^{33}$ ergs s^{-1}. Thus the area is about sixty times larger than that implied by the blackbody fit to the same data discussed earlier, while the soft X-ray luminosity is three times larger.

However, there are two reasons to believe that the true spectrum may lie closer to that of a blackbody. First, the surface of the white dwarf is composed of matter freshly accreted from the companion star. Therefore, it is expected to consist not just of hydrogen and helium, but to have cosmic abundances. In particular, it is expected to contain metals. These metals produce absorption, mainly edges, throughout the soft X-ray band. This increases the importance of absorption relative to scattering and reduces the frequency dependence of the absorption. Both lower the flux at energies above the helium edge to values nearer those of a blackbody. Second, electron scattering and heating from above (by the hard X-rays absorbed in the outer part of the atmosphere) reduce the temperature gradient in the atmosphere (see, $e.g.$, London, McCray, and Bauer 1981). Given the presence of absorption due to metals, this also tends to bring the emergent flux closer to the blackbody value, and reduces the strength of the absorption edges. The latter is consistent with the absence of strong absorption edges in the $Einstein$ OGS spectrum (Heise et $al.$ 1983).

No model atmospheres appropriate to the effective temperatures and surface gravities encountered in the AM Her stars exist in the literature. However, we can get some idea of what to expect from the inclusion of metals, at least, by considering model atmospheres of the nuclei of planetary nebulae, which have lower temperatures but about the right surface gravity (Böhm 1969, Hummer and Mihalas 1970). Figures 14 (a) and 14 (b) show the emergent flux for model atmospheres with $T_e = 14$ eV, log g = 7.8 and $T_e = 16$ eV, log g = 7.9 from Böhm (1969). For comparison, blackbody spectra with the same temperatures are also shown. By extrapolating the flux just shortward of the helium edge to higher energies, one sees clearly that, in the absence of metals, the color temperature would be much larger than the effective temperature in both model atmospheres. However, inclusion of metallic absorption edges, particularly those due to Ne V and O VI, reduces the flux at higher energies to about the blackbody value.

Reprocessing of cyclotron optical light. If cyclotron cooling is comparable to or greater than bremsstrahlung cooling in the hot, post-shock emission region, accurate determination of the shape of the soft X-ray spectrum is more complicated still. The surface of the white dwarf is exposed to an intense optical, or possibly UV, flux with a brightness temperature equal to that of the shock-heated electrons in the X-ray emission region. This intense flux

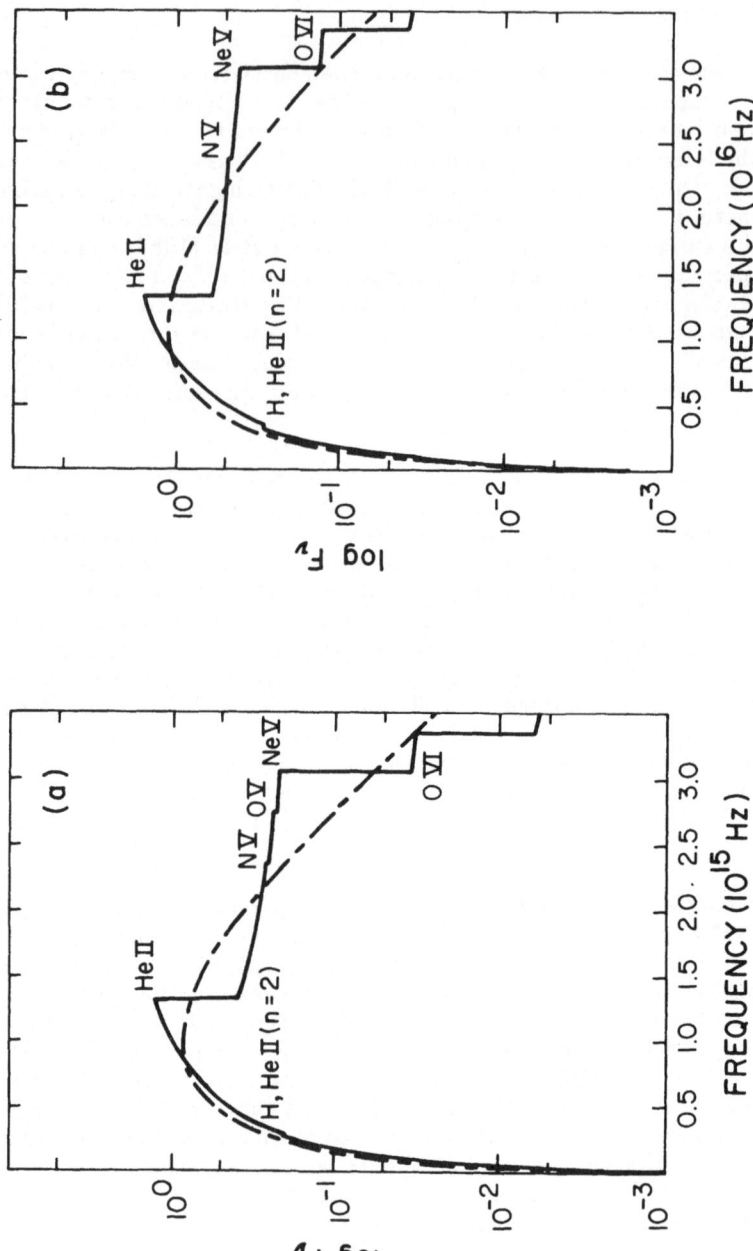

Fig. 14.—(a) Emergent flux F_ν for a model atmosphere with $T_{eff} = 1.65 \times 10^5$° K and $g = 6.5 \times 10^7$ cm s^{-1}. (b) Emergent flux F_ν for a model atmosphere with $T_{eff} = 1.8 \times 10^6$° K and $g = 8.6 \times 10^7$ cm s^{-1}. From Böhm (1969).

is absorbed at a much shallower depth in the atmosphere than before. If the absorption opacity decreases with increasing energy, as is usually the case, the flux will be absorbed at a depth which is optically thin to reradiation at soft X-ray energies. The soft X-ray component might then resemble a quasi-bremsstrahlung spectrum with emission lines. However, whether or not this occurs depends sensitively on the details of the energy deposition process, and hence on the opacity at optical and UV energies, including metallic lines. In VV Pup, cyclotron and bremsstrahlung cooling may be comparable (Patterson *et al.* 1983), so that high spectral resolution soft X-ray observations of this star may provide information about this case.

Other effects. We mention two other effects that tend to cause the color temperature to differ from the effective temperature. The first is up-scattering of soft X-ray photons in the hot, post-shock emission region and in the cold infalling matter above the shock (Kylafis and Lamb 1979, 1982). However, according to Kylafis and Lamb (1982), the change in the soft X-ray spectrum, as measured by $\delta L_{bb}/L_{bb}$, due to such up-scattering is always small. For the effective luminosities and small spot sizes appropriate to the AM Her stars, it should be unimportant.

The second effect arises from the fact that the soft X-ray spectrum is not produced by an atmosphere with a single color, or effective, temperature. Rather, it is a composite spectrum formed from a continuum of atmospheres corresponding to annuli of successively larger radius and lower temperature surrounding the emission region (see, *e.g.* Imamura and Durisen 1983). As a result, the low-energy portion of the soft X-ray spectrum is a power law that decreases toward low frequencies more slowly than does the usual Rayleigh-Jeans spectrum $\propto \nu^2$. However, calculations done by the author show that, as long as the height of the emission region is small compared to the stellar radius, the flux incident on an annulus decreases rapidly as the radius of the annulus increases. As a result, only the area immediately below the hot, post-shock emission region contributes significantly to the soft X-ray component and the lower-energy flux from annuli of larger radii is smaller than the observed UV flux. In this situation, the soft X-ray component is adequately approximated by a single-temperature atmosphere.

Clearly, detailed numerical calculations of the emergent spectrum for atmospheres heated from above both by bremsstrahlung hard X-rays and by cyclotron optical or UV light are badly needed in order to understand fully the soft X-ray spectra of the AM Her stars. In the case of bremsstrahlung heating, the H/He model atmospheres fitted to the *Einstein* OGS data by Heise *et al.* (1983) indicate that the uncertainties in the soft X-ray temperature and luminosity are less than a factor of three, but the uncertainty in the area of the emission region is much larger. In the case of cyclotron heating, the uncertainties have yet to be explored.

3.4. Polarized Optical Light from AM Her Stars

One of the features which distinguish the AM Her stars from other cataclysmic variables is the strong ($\gtrsim 10\%$) circular and linear polarization of their optical light. A striking feature of the linear polarization is its appearance as a sharp pulse, usually once but sometimes twice per rotational period. Calculations of cyclotron emission from hot plasmas show that this linear polarization behavior and the observed circular polarization can be produced by a plasma with a temperature $kT \simeq 1$ keV emitting at harmonics $m \simeq 10$ (Chanmugam and Dulk 1981) or with a temperature $kT \simeq 20$ keV emitting at harmonics $m \simeq 15$ (Meggitt and Wickramasinghe 1982).

These calculations neglected collisions. Recently Barrett and Chanmugam (1982, 1983) have extended these calculations to include Thomson scattering (treated as pure absorption) and free-free absorption using the method of Pavlov, Mitrofanov, and Shibanov (1980). The principal effect is an overall reduction in the linear and, particularly, the circular polarization. The reduction is greater for lower temperatures, higher harmonics, and smaller angles α with respect to the magnetic field.

One can understand the effects of free-free absorption and Thomson scattering on the absorption coefficient μ as follows (Barrett and Chanmugam 1983). It is well known that the cyclotron absorption coefficient decreases exponentially as a function of the angle α with respect to the magnetic field (Trubnikov 1958). It is also well known that it decreases rapidly with increasing frequency. The contribution to the absorption coefficient from collisions, on the other hand, depends only weakly on α and frequency. Therefore as α approaches 0 and/or the harmonic number m increases, collisional effects become more important and eventually dominate. At sufficiently high harmonics and/or small angles α, no harmonic structure is visible even at low temperatures. These effects are illustrated in Figure 15 (left panels).

The contributions of free-free absorption and Thomson scattering to the absorption coefficient imply, for a plasma in local thermodynamic equilibrium, a corresponding free-free emissivity and a coupling of the polarization modes due to scattering. The effects of collisions on the intensity and polarization then follow. The intensity falls off less rapidly as α approaches 0 and/or the harmonic number m increases. Also the fractional polarization is lower than in the collisionless case, and decreases as α approaches 0 and/or the harmonic m increases. Collisional effects do not change the shape of the linear polarization pulse, only its height. On the other hand, collisional effects cause the circular polarization to approach 0 or a value $< 100\%$, depending on the geometry of the emission region, rather than the previous value of 100%. This enables the theory to account for the primary minima seen in many of the AM Her stars without recourse to dilution by extraneous light. These effects are illustrated in Figure 15 (right panels).

Barrett and Chanmugam (1983; see also Meggitt and Wickramasinghe 1982) point out that the ratio,

$$Q/V = -\frac{sin^2\alpha}{2(\omega/\omega_c)cos\alpha},$$ (13)

is remarkable in that it is independent of the temperature and optical depth of the plasma, and hence can provide a valuable, simple probe of the orientation and the magnetic field of the white dwarf.

Barrett and Chanmugam (1983) have fitted the results of their calculations to observations of AM Her in its high state (Tapia 1977). They varied the magnetic field strength B and orientation of the white dwarf (defined by the inclination angle i and the magnetic colatitude δ; see Figure 17), and the temperature T and thickness l of the optical emission region to fit the visual flux I, the linear polarization Q/I, the circular polarization V/I, and the ratio Q/V. Their best fit values are $B = 2.7 \times 10^7$ Gauss, $kT = 0.2$ keV, $l = 2.6 \times 10^8$ cm, $i = 46°$, and $\delta = 55°$. Figure 16 shows the resulting theoretical curves and the observations. The fits are surprisingly good, given the simplicity of the model. The largest discrepancies occur at the onset of the primary minimum because the theoretical curves are symmetric about the minimum, whereas the observational curves are not. Interestingly, the the m_V and circular polarization curves *are* symmetric when

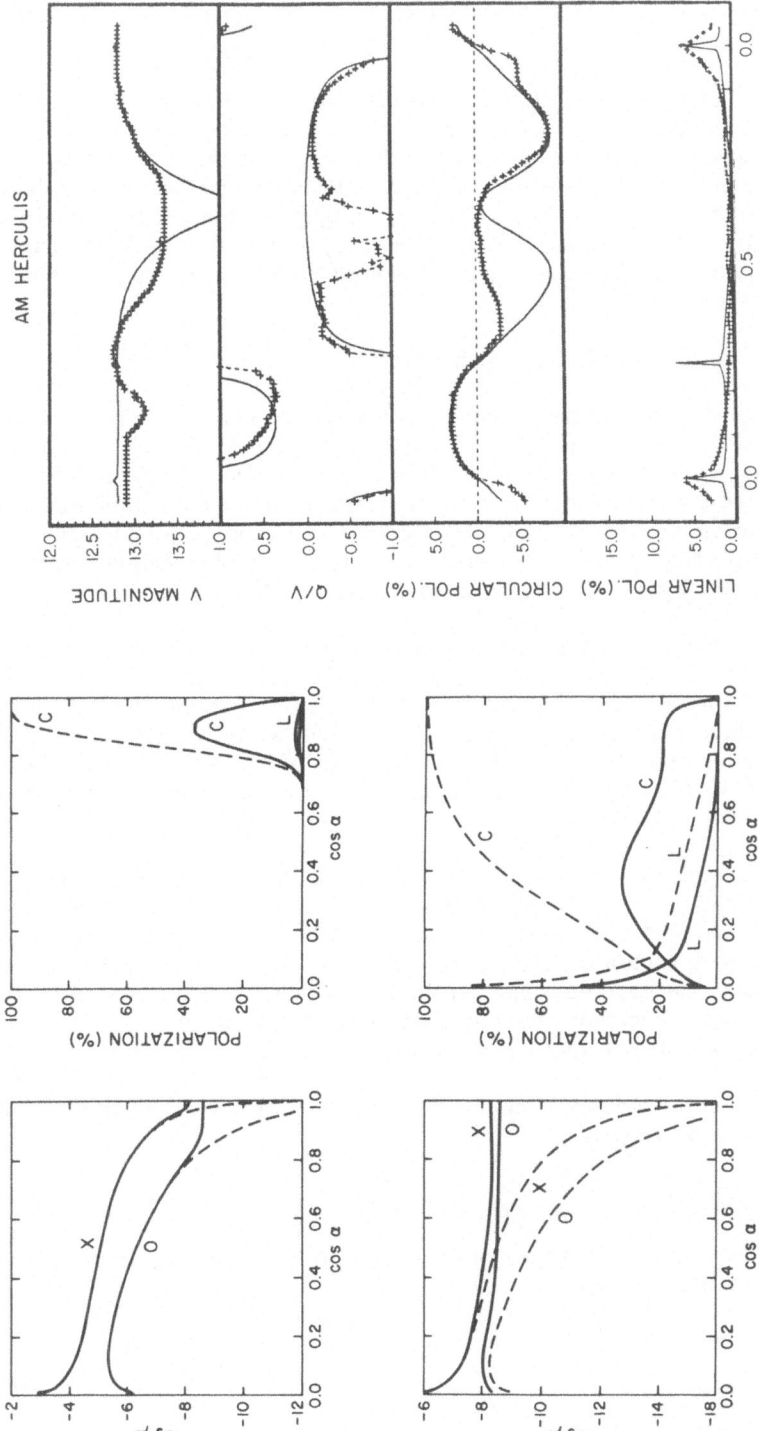

Fig. 15.—(Upper and lower left panels) Absorption coefficient versus angle for harmonic numbers 4 and 6, respectively. Ordinary and extraordinary modes are labeled O and X. Dashed lines correspond to the collisionless case. (Upper and lower right panels) Polarization versus angle for the corresponding harmonic numbers. Linear and circular polarization are denoted by L and C. The plasma has a temperature of 1 keV and a width of 10^8 cm. From Barrett and Chanmugam (1983a)

Fig. 16.—Visual magnitude, ratio Q/V, and fractional linear and circular polarization versus phase for one cycle of AM Herculis. The symbols + are observational data taken from Tapia (1977) and the solid line is the theoretical calculation. The plasma has a width $\ell = 2.6 \times 10^8$ cm, a temperature $kT = 0.2$ keV, and is oriented parallel to a magnetic field of 2.6×10^7 gauss. The inclination angle of the rotation axis is $i = 46°$ and the magnetic colatitude of the emission region is $\delta = 55°$. From Barrett and Chanmugam (1983b)

AM Her is in its *low* state (*cf.* Latham, Liebert, and Steiner 1981). Although Barrett and Chanmugam did not carry out fits to the low state data, the agreement with it would clearly be better.

The values $i = 46°$ and $\delta = 55°$ found by Barrett and Chanmugam (1983) differ somewhat from the values $i = 35°$ and $\delta = 60°$ found by Brainerd and Lamb (1983) from fits to the self-eclispe duration, as defined by the circular polarization zero crossings, and the variation of the linear polarization position angle (see Figure 18 and Secton 3.7). The difference is due largely to the fact that Barrett and Chanmugam have fitted to high state data, in which the second circular polarization zero crossing occurs at $\phi = 0.26$ (Tapia 1977), whereas Brainerd and Lamb fitted to the more symmetric low state data, in which it occurs at $\phi = 0.19$ (Latham, Liebert, and Steiner 1981).

3.5. Source of Polarized Optical Light from AM Her Stars

An important unresolved issue, alluded to above, is whether or not the region producing the optically polarized light from the AM Her stars is coincident with the hot post-shock hard X-ray emission region. It would be nice if this were true, since the simple theory described in Section 2.1 predicts cyclotron emission from the hot post-shock region, and the three principal spectral components (cyclotron, reprocessed blackbody, and bremsstrahlung) predicted by the simple theory would then correspond to the three principal observed components (polarized optical light, soft X rays, hard X rays).

However, indirect observational arguments cast doubt on this possibility. For example, the observed change in the sign of the circular polarization for a brief period at the end of the bright phase in VV Pup is naturally explained if the region emitting the circularly polarized optical light is high enough above the surface of the white dwarf ($h \simeq 0.1R$) that it briefly remains visible after the magnetic pole has passed beyond the limb of the white dwarf (Liebert *et al.* 1978), while the height of the X-ray emission region is expected to be far smaller (see Section 3.1).

What can recent observations and theory tell us about this issue? Two principal arguments can be brought to bear on it (see also Liebert and Stockman 1983): (1) Does the area required to produce the observed (optically thick) polarized optical flux agree with the area of the X-ray emission region inferred from soft X-ray observations? (2) Can the energy E^* (or, equivalently, the harmonic number m^*) at which the cyclotron emission from the X-ray emission region becomes optically thin be made to fall in the optical (at $\simeq 1.5$ eV; see Section 2.1)? Let us consider each argument in turn.

Areas of the cyclotron and blackbody emitting regions. The area required to emit an optically thick flux f_ν at frequency ν is,

$$A_{cyc} = 4.6 \times 10^{15} \left(\frac{f_\nu}{10^{-28} Wm^{-2} Hz^{-1}}\right)\left(\frac{kT}{10 keV}\right)^{-1}\left(\frac{\nu}{eV}\right)^{-2} cm^2, \qquad (14)$$

where here and below we assume a distance of 100 pc. The ratio of the areas of the cyclotron and blackbody emitting regions is then

$$\alpha \equiv \frac{A_{cyc}}{A_{bb}} = \frac{f_\nu^{cyc}}{f_\nu^{bb}} \frac{T}{T_{bb}}$$
$$= 29 \left(\frac{f_\nu^{cyc}}{10^{-28} Wm^{-2} Hz^{-1}}\right)\left(\frac{kT_e}{10 keV}\right)^{-1}\left(\frac{\nu}{eV}\right)^{-2}\left(\frac{L_{bb}}{10^{33} ergs\ s^{-1}}\right)^{-1}\left(\frac{T_{bb}}{50 eV}\right)^4, \qquad (15)$$

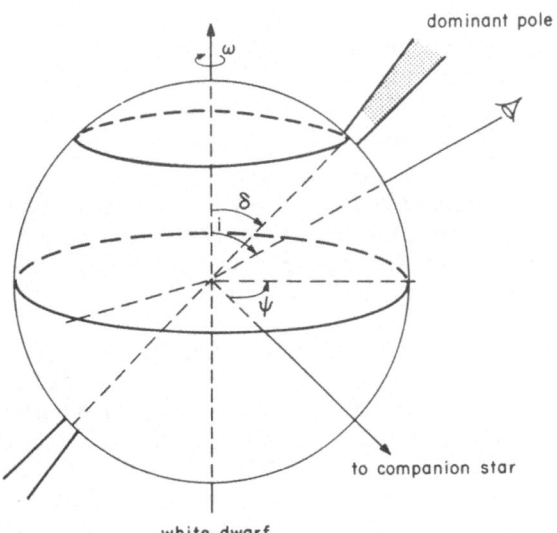

Fig. 17.–Definitions of the inclination angle i, the magnetic colatitude δ, and the magnetic longitude ψ. The rotation axis of the white dwarf and that of the binary are assumed to be parallel. From Brainerd and Lamb (1983).

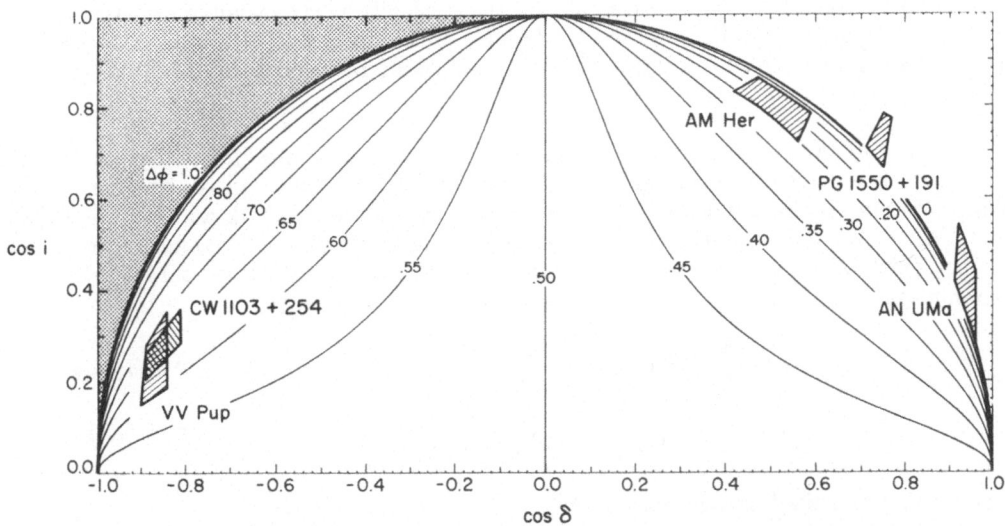

Fig. 18.–Distribution of AM Her stars in the (cos δ, cos i)-plane. The shaded boxes show the regions of allowed values of i and δ. The curves show the locations of stars having eclipse lengths $\Delta\phi = 0-1$, in intervals of 0.05; note that throughout much of the plane, the eclipse duration is near 0.5. The shaded region at the upper left corresponds to stars in which the dominant pole is never visible; the analogous region at the upper right corresponds to stars in which it never approaches the limb of the star. Note that all 5 AM Her stars lie on or near the semicircle $cos^2\delta + cos^2i = 1$. From Brainerd and Lamb (1983).

where in the second line we have used equation (5) for the blackbody component.

Applying these expressions first to VV Pup, we find that for a flux $f_\nu^{cyc} \simeq 1.5 \times 10^{-29}$ W m^{-2} Hz^{-1} at 0.7 eV (see Figure 9) and $T = 30$ keV, the cyclotron emitting area $A_{cyc} \simeq 1.6 \times 10^{15}$ cm^2 or $f \simeq 5 \times 10^{-4}$, where here and below we assume $R = 5 \times 10^8$ cm. The results of Patterson et $al.$ (1983) and equation (15) show that $\alpha \lesssim 1$ only if $T_{bb} \lesssim 30$ eV, while $T_{bb} \gtrsim 35$ eV is needed if the prediction $L_{bb} \simeq L_{br} + L_{cyc}$ of the simple theory is to be satisfied and the spectrum of the soft X-ray component is approximately a blackbody (see Section 3.3). Although uncertainties in the values of T and T_{bb} do not permit a sensitive test, the results suggest that the cyclotron and X-ray emission regions cannot be coincident without violating the prediction of the simple theory.

Applying the above expressions now to AM Her, we find that for a flux $f_\nu^{cyc} \simeq 4.0 \times 10^{-28}$ W m^{-2} Hz^{-1} at 1.5 eV (see Figure 7) and $T = 30$ keV, the cyclotron emitting area $A_{cyc} \simeq 2.7 \times 10^{15}$ cm^2 or $f \simeq 8.6 \times 10^{-4}$. This compares with the values $A_{bb} \simeq 8.3 \times 10^{15}$ cm^2 or $f \simeq 2.6 \times 10^{-3}$, assuming a hydrogen/helium atmosphere with no metals, and $A_{bb} \simeq 1.4 \times 10^{14}$ cm^2 or $f \simeq 4.4 \times 10^{-5}$, assuming a blackbody spectrum, found in the recent $Einstein$ OGS observation (Heise et $al.$ 1983). Thus $\alpha = 0.3 - 20$, but $\alpha = 1$ is definitely excluded if the prediction $L_{bb} \simeq L_{br} + L_{cyc}$ of the simple theory holds and the spectrum of the soft X-ray component is approximately a blackbody. On the other hand, for $f_\nu^{UV} \simeq 1.6 \times 10^{-28}$ W m^{-2} Hz^{-1} at $\gtrsim 10$ eV appropriate to the turn-up in the UV seen by Raymond et $al.$ (1979) (see Figure 7) and $T = 30$ keV, the cyclotron emitting area $A_{cyc} \simeq 3.7 \times 10^{13}$ cm^2 or $f \simeq 1.1 \times 10^{-5}$. Then $\alpha = 0.25 - 5 \times 10^{-3}$ compared with the two different soft X-ray emitting areas inferred by Heise et $al.$ (1983). Thus $\alpha \sim 1$ if the prediction of the simple theory holds and the spectrum of the soft X-ray component is approximately a blackbody; otherwise $\alpha \ll 1$. These results are very suggestive that the turn-up in the UV might, after all, be the optically thick cyclotron component from the X-ray emission region predicted by Lamb and Masters (1979) and that the polarized optical light has a separate origin. Similar conclusions have been reached by Liebert and Stockman (1983).

$Energy$ of the $cyclotron$ $emission$ $peak.$ Let us now consider the energy at which the cyclotron flux from the X-ray emission region peaks (or, equivalently, becomes optically thin). We can estimate it using the approximate expression of Dulk and Marsh (1982; see also Lada et $al.$ 1979) for the corresponding harmonic number m^*, which represents a fit to the cyclotron emissivity (see, $e.g.$, Masters 1978, Lamb and Masters 1979). Neglecting the moderate angular dependence, the peak occurs at an energy,

$$
\begin{aligned}
E^* \equiv m^* \hbar \omega_c &\simeq 1.5 (\frac{\Lambda}{10^7})^{0.05} (\frac{T}{3keV})^{0.5} (\frac{B}{2 \times 10^7 G}) \, eV, \qquad 1keV \lesssim T \lesssim 10keV \\
&\simeq 6.0 (\frac{\Lambda}{10^7})^{0.1} (\frac{T}{30keV})^{0.7} (\frac{B}{2 \times 10^7 G}) \, eV, \qquad 10keV \lesssim T \lesssim 100keV,
\end{aligned}
\tag{16}
$$

where the dimensionless size parameter,

$$
\Lambda \equiv \frac{\omega_p^2}{\omega_c} \frac{l}{c} = 3 \times 10^7 (\frac{n}{10^{16}cm^{-3}})(\frac{l}{10^7 cm})(\frac{B}{2 \times 10^7 G})^{-1}.
\tag{17}
$$

Equation (16) shows that the peak occurs in the UV for values of the temperature and magnetic field appropriate to the known AM Her stars, and confirms the results of Lamb and Masters (1979). Equations(16) and (17) also show that E^* is terribly insensitive to the column density, and therefore to the density and the size of the emission region. Thus, even if the area of the X-ray emitting region is large enough to produce the observed cyclotron optical flux (*i.e.* the prediction $L_{bb} \simeq L_{br} + L_{cyc}$ of the simple theory does not hold and/or the spectrum of the soft X-ray component does not resemble a blackbody), it is still nearly impossible to make the cyclotron emission peak (*i.e.* become optically thin) at $\simeq 1.5$ eV, as required by observation. This fact strongly argues that the polarized optical light has a separate origin.

Liebert and Stockman (1985) suggest that the Maxwellian tail of the electron velocity distribution might be depleted due to "end losses" through the top and the botton of the X-ray emission region. This would suppress high harmonic cyclotron emission (radiation at harmonic m^* is produced primarily by electrons with energy $E \gtrsim m^* kT$), and might make it peak in the optical rather than the UV. We can estimate the electron-electron mean free path using the relationship $l_{e-e}(E) \simeq l_{e-e} \times (E/kT)^2 \simeq l_{i-i} \times (E/kT)^2$, where l_{i-i} is the ion-ion mean free path (Spitzer 1962). Using the expression $d \simeq v_{ff} t_{br}/4$ (see, *e.g.*, Kylafis and Lamb 1979), we find

$$\frac{l_{e-e}}{d} \simeq \frac{l_{i-i}(\frac{E}{kT})^2}{\frac{1}{4} v_{ff} t_{br}} \simeq \frac{t_{i-i}}{t_{br}} (\frac{E}{kT})^2$$
$$\simeq 1.0 \times 10^{-2} (\frac{ln\Lambda}{10}) (\frac{T}{10 keV})^{1/2} (\frac{M}{M_\odot})^{-3/2} (\frac{R}{5 \times 10^8 cm})^{3/2}, \tag{18}$$

where we have used equations (A6) and (A17) of Kylafis and Lamb (1982) to evaluate t_{br} and t_{i-i}. Equation (18) shows that $l_{e-e} \gtrsim d$ only when $E \gtrsim 10$ kT so that we don't expect dramatic losses for the electrons that produce cyclotron emission at $m^* \simeq 7$. We conclude that "end losses" are not likely the answer.

Then where might the polarized optical light originate? Liebert and Stockman (1985) also suggest that it might come from an extended low-density rim surrounding the X-ray emission region. But equation (16) shows that the column density through the rim would have to be 10^6 times smaller than through the X-ray emission region. The cyclotron luminosity produced in such a rim then falls far short of that observed. Given the insensitivity of E^* to the column density and the fact that the magnetic fields of three AM Her stars have been measured to be $\simeq 2 \times 10^7$ Gauss (Liebert and Stockman 1983), the solution must be that the polarized optical light comes from a lower temperature region $[kT \lesssim 3$ keV; see equation (16)]. Chanmugam and Dulk (1981) recognized this fact and suggested that it comes from hot electrons in the shock precursor produced by electron conduction. However, the precursor is always smaller, usually much smaller, than the hard X-ray emission region, and the electron temperature in the precursor is less than that in the hard X-ray emission region only in a region that is much smaller still. This makes it difficult to produce the observed optical flux, even if the area of the X-ray emission region is much larger than in the simple theory with an approximately blackbody soft X-ray spectrum.

Instead, the polarized optical light may be produced in that portion of the accretion column heated by Compton scattering and extending up to heights $\sim R$ [*cf.* Kylafis and Lamb 1982, equation (3)]. This region has both the large extent ($h \sim R$) needed to produce the observed optical flux and the relatively low temperature ($T \sim 0.1 - 1$ keV) needed to make the polarized light peak in the optical.

3.6. Broad Emission Line Region in the AM Her Stars

One of the singular features of the AM Her stars is the forest of broad hydrogen emission lines in the optical and UV. These lines are thought to be produced in the stream-magnetosphere interaction region, whose properties also determine the accretion torque and the location and geometry of the optical and X-ray emission region (since the latter are fixed by the footpoints on the stellar surface of the field lines on which plasma enters the magnetosphere).

Liebert and Stockman (1985) have made an important contribution by working out for the first time a detailed picture of the mass transfer process in the AM Her stars. According to their picture, the process begins in the usual way, via a gas stream from the companion star. The stream penetrates the magnetic field of the white dwarf until magnetic stresses halt it, where heating and turbulence create a large, hot interaction region. Since the gas lacks sufficient angular momentum to form a disk, it cools *in situ*, enters the magnetosphere of the white dwarf, and plunges toward one or both magnetic poles. As the plasma stream flows along the converging magnetic field lines, it narrows until at the white dwarf it covers only a fraction $f \sim 10^{-3} - 10^{-4}$ of the stellar surface. This picture is shown in Figure 19.

We can estimate the minimum distance r_i^{min} of the interaction region from the white dwarf by equating the ram pressure of the gas stream and the pressure due to the magnetic field of the white dwarf, *i.e.*

$$\frac{B^2(r_i^{min})}{8\pi} \simeq \frac{1}{2}\rho_s v_s^2. \tag{19}$$

where $B(r_i^{min})$ is the magnetic field of the white dwarf at r_i^{min}, and ρ_s and v_s are the density and velocity of the stream. This gives

$$r_i^{min} \simeq 9 \times 10^9 \left(\frac{n}{10^{13} cm^{-3}}\right)^{-1} \left(\frac{B}{2 \times 10^7 Gauss}\right)^{2/5} \left(\frac{M}{M_\odot}\right)^{-1/5} \left(\frac{R}{5 \times 10^8 cm}\right)^{6/5} cm, \tag{20}$$

where n is the number density in the stream (Lubow and Shu 1975) and we have taken $v_s = v_{ff}(r)$. The accretion column is stable only up to the height above the stellar surface at which the magnetic field ceases to dominate the kinetic energy of infall of the gas in the column. We can therefore estimate the maximum distance r_i^{max} of the interaction region from the white dwarf by equating the energy density in the magnetic field with that of the accreting matter. This gives (Tuohy *et al.* 1981),

$$r_i^{max} \simeq 1.3 \times 10^{10} \left(\frac{f}{10^{-4}}\right)^{2/5} \left(\frac{L}{10^{33} ergs \, s^{-1}}\right)^{-2/5}$$
$$\left(\frac{B}{2 \times 10^7 Gauss}\right)^{4/5} \left(\frac{M}{M_\odot}\right)^{1/5} \left(\frac{R}{5 \times 10^8 cm}\right)^{8/5} cm. \tag{21}$$

These estimates bracket the distance of the interaction region from the white dwarf.

At the location of the interaction region, the gas stream from the companion star produces an indentation or cavity in the magnetosphere of the white dwarf. The gas in the cavity is heated by a standoff shock and/or by turbulence. In principle, its temperature could be as high as the freefall temperature,

$$T_{ff}(r_i) \sim \frac{R}{r_i}T_{ff}(R) \sim 1\left(\frac{M}{M_\odot}\right)\left(\frac{R}{5 \times 10^8 cm}\right)^{-1} keV. \tag{22}$$

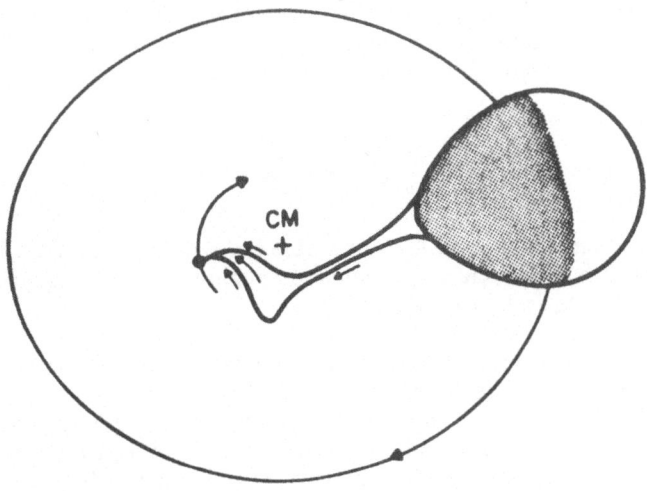

Fig. 19.–Schematic picture of the mass transfer process in the AM Her stars, showing the gas stream from the companion star, the interaction region (where the broad emission lines are formed), and the accretion column onto the white dwarf. From Liebert and Stockman (1983).

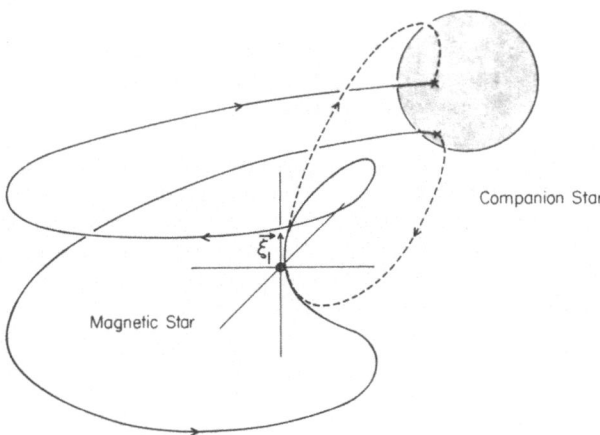

Fig. 20.–Perspective view of the binary system illustrating how a given flux tube is twisted as the magnetic star rotates with respect to its companion. Dashed curve: initial shape; solid curve: shape at a later time. The resulting Maxwell stress on the magnetic star produces a torque that opposes the rotation. For simplicity, slippage of field lines through the plasma has been neglected in this figure. In reality, the twist of the flux tube will not increase indefinitely but will be limited by large scale magnetohydrodynamic instabilities. From Lamb *et al.* (1983).

However, there appears to be no observational evidence for temperatures of the order of the free fall temperature. The gas is therefore probably somewhat optically thick. The hot plasma may cool by continuum or line emission, depending upon conditions in the region (Liebert and Stockman 1985). The gravitational potential energy of the infalling matter supplies a luminosity

$$L_i \sim \frac{GM}{r_i} \dot{M} \sim \frac{R}{r_i} L_{acc} \sim 5 \times 10^{-2} (\frac{r_i}{20R})^{-1} L_{acc}, \qquad (23)$$

where $\dot{M}$ is the mass accretion rate. A comparable amount is produced by photoabsorption of the extreme UV flux from the X-ray emission region of the white dwarf (Liebert and Stockman 1985). Thus L_i is $\sim 10^{32}$ ergs s^{-1}.

Since the pressure exerted by the hot plasma in the magnetospheric cavity is comparable to that of the magnetic field confining it, the hot plasma can spread somewhat. The degree to which it does is a function of the depth of the cavity, and the cooling and entry rates of the hot plasma into the magnetosphere. Judging by the size of the interaction region inferred by Liebert and Stockman (1985) from the broad emission lines, the spreading may be substantial.

Cyclotron emission is unlikely to be important in cooling the hot plasma in the magnetospheric cavity, since the plasma there is not fully threaded by the magnetic field of the white dwarf. However, cyclotron cooling of the accreting matter once it has entered the magnetosphere can easily produce detectable radio emission at a frequency,

$$\nu_c(r_i) \simeq (\frac{r_i}{R})^{-3} B(r_i) \simeq 2.1 \, (\frac{r_i}{30R})^{-3} (\frac{B}{2 \times 10^7 Gauss}) \, GHz. \qquad (24)$$

Whether such emission could be related to the radio emission from AM Her (Chanmugam and Dulk 1982) is problematical, since the plasma frequency $\nu_p \simeq 2.8(n/10^{13} \, cm^{-3})^{1/2}$ GHz and the radio emission may be unable to escape.

The location and the geometry of the polarized optical and X-ray emission regions are determined, as mentioned above, by the footpoints on the stellar surface of the field lines threading the interaction region. According to the picture of this region that we have outlined above, the optical and X-ray emission regions do *not* lie exactly at the magnetic pole(s), since the interaction region lies at a finite distance from the white dwarf. The amount by which they are offset is determined by

$$sin\theta_b = (\frac{R}{r_i})^{1/2} \, , \qquad (24)$$

so that $\theta_b \simeq 13°$ or $11°$ for $r_i = 20$ or $30R$. The location of the optical and X-ray emission regions thus depends on the orientation and the accretion luminosity of the white dwarf, since changes in the accretion rate alter the value of r_i. Large changes in the luminosity could produce changes in the phases of the linear polarization pulse(s) and the circular polarization zero crossings as large as $\simeq 0.1$; this effect almost certainly accounts for the differences in the length of the self-eclipse and in the phases of the circular polarization zero crossings between the high and low states of AM Her (recall Section 3.4), a system whose self-eclipse appears to be nearly grazing. The fact that the field lines threading the optical and X-ray emission regions also are not exactly vertical may account for the frequent absence of one of the two expected linear polarization pulses, since if the region emitting the polarized optical light lies not too far above the stellar surface one of the pulses will be beamed into the surface.

According to the picture of the interaction region described above, the accretion column will not be circular in cross-section. Rather, it is like a garden trowel stuck into the ground at an odd angle, *i.e.* a curved sheet intersecting the stellar surface at an oblique angle. This could resolve the conflict discussed in Section 2.3, since if the emission region has such a shape, it is long in one direction (which might account for the shape and the duration of the hard X-ray eclipse) yet its area is relatively small (as required for a blackbody luminosity L_{bb} consistent with the simple theory).

Such an emission region may also be capable of accounting for other puzzling and apparently contradictory aspects of the optical and X-ray emission from the AM Her stars, such as the severe asymmetry of the optical light curve in VV Pup and the lack of concordance between the length of self-eclipse determined from the optical light curve and from the circular polarization zero crossings (see Patterson *et al.* 1983, Figure 8).

3.7. White Dwarf Orientation in the AM Her Stars

The orientation of the white dwarf in the AM Her stars is needed in order to compare theoretical calculations and observations of the intensity and angular dependence of the polarized optical light and the soft and hard X-rays from these stars. Knowledge of the distribution of orientations can also discriminate among theoretical models of the magnetic coupling between the white dwarf and the companion star (see below). Assuming that the rotation axes of the degenerate dwarf and of the binary are parallel, the orientation of the degenerate dwarf is determined by three angles: the inclination angle i of the binary system, the magnetic colatitude δ (the angle between the magnetic moment $\boldsymbol{\mu}$ and the rotation axis $\hat{\boldsymbol{\omega}}$ of the degenerate dwarf), and the magnetic longitude ψ relative to the companion star. These angles are defined in Figure 17.

The radial velocity curve of the narrow (or sharp) emission lines give information about the magnetic longitude ψ, if these emission lines originate on the face of the companion star (*cf.* Liebert and Stockman 1985). Fits to the duration of self-eclipse, as determined by the circular polarization zero crossings and the optical and X-ray light curves, and the variation of the linear polarization position angle with phase can be used to determine i and δ (*cf.* Stockman 1977, Meggitt and Wickramasinghe 1982, Brainerd and Lamb 1983)

Brainerd and Lamb (1983) have determined i and δ values for five of the ten known AM Her stars. Figure 18 shows their results plotted on the (cos i, cos δ)-plane. Also shown are curves of fixed eclipse length $\Delta\phi$. The shaded region at the upper left corresponds to stars in which the dominant pole is never visible, while the analogous region at the upper right corresponds to stars in which the dominant pole never approaches the limb of the star; these possibilities have been noted previously by Bond, Chanmugam, and Grauer (1979). The AM Her stars should be uniformly distributed in this plane if the white dwarf has no preferred orientation. Visual inspection suggests, and the Kolmogorov-Smirnov statistical test employed by Brainerd and Lamb confirms, that the stars are significantly concentrated on or near the semicircle $cos^2\delta + cos^2 i = 1$.

Brainerd and Lamb suggest that the non-uniform distribution of the known AM Her stars is a result of the way in which they have been identified observationally, and the angular properties of their X-ray and polarized optical emission. Essentially, their argument is as follows. Both X rays and polarized optical light are emitted in fan beams from the magnetic pole(s) and are thus most intense when the magnetic pole is on or near the limb of the star. The stars that lie along the semicircle $cos^2\delta + cos^2i = 1$ are just the ones in which the dominant and secondary magnetic poles spend an inordinate fraction of time near the limb, so that strong X-ray emission and strongly polarized light are almost always visible. These stars are therefore most easily discovered and confirmed to be members of the AM Her class. However, one might argue that the stars in the central region of the (cos δ, cos i)-plane ought still to be detected. They have eclipse durations $\Delta\phi \simeq 0.5$, so that their optical and X-ray light curves should be striking. And even if most of the time their polarization is much less than the $\gtrsim 10\%$ typical of the known AM Her stars, their polarization should be detectable since current observations are sensitive to polarizations $\gtrsim 0.1\%$.

An alternative possibility is that one or more of the assumptions made in analyzing the observations may not be valid. Brainerd and Lamb (1983) have pointed out that among the most important of these are (1) the magnetic field is a centered dipole and (2) the emission region is a point lying directly above the magnetic pole. Clearly, any off-center displacement of the stellar magnetic moment produces poles of different strengths and/or poles that do not lie 180° from one another. It is also clear that if the optical emission region does not lie exactly at the magnetic pole and/or if it lies significantly above the stellar surface, the magnetic field in the emission region will not be oriented exactly along the direction of the magnetic moment. And if the optical emission region extends over a large area, the width of the linear polarization pulse will be greater length, while the duration of any self-eclipse and the percentage of circular polarization will be reduced. Thus if either or both assumptions are invalid, the central region of the (cos δ, cos i)-plane, where $\Delta\phi \simeq 0.5$, would be underpopulated. Whether or not this is sufficient to account for the observed nonuniform distribution is unclear, but certainly both assumptions must be carefully examined. For example as discussed above, the location and the geometry of the stream-magnetosphere interaction region imply that the X-ray and the optical emission regions will not lie exactly at the magnetic pole.

4. SYNCHRONISM, SPIN-UP, AND MAGNETIC FIELDS

4.1. Synchronism in AM Her Stars

Synchronous rotation of the magnetic white dwarf with the binary period is one of the singular features of the AM Her stars. Such synchronism is surprising since viscous coupling between the white dwarf and the companion star is expected to be negligible, and the accretion spin-up torque is large while the moment of inertia of the white dwarf is small.

Two separate questions then arise: how does the white dwarf become synchronized and, once synchronism is achieved, how is it maintained? A widespread conjecture is that both are somehow the result of magnetic coupling between the white dwarf and the companion star. However, the nature of the magnetic coupling in the two cases need not be the same.

Achieving synchronism. In their pioneering study, Joss, Katz, and Rappaport (1979) proposed that synchronism is achieved by a vacuum dipole-dipole (VDD) interaction between the magnetic white dwarf and the companion star. According to this model, the companion star experiences a time-varying magnetic field due to asynchronous rotation of the magnetic white dwarf. The resulting ohmic dissipation in the surface of the companion star, which is a good conductor, produces a torque,

$$N_{VDD} \simeq \frac{15}{8} \delta R_2^2 (\mu_1/D)^2$$
$$\simeq 8 \times 10^{28} \left(\frac{\xi}{5 \times 10^{-4} s^{-1}} \right) \left(\frac{\mu_1}{2 \times 10^{34} Gauss\ cm^3} \right) \left(\frac{R_2}{3 \times 10^{10} cm} \right) \left(\frac{D}{10^{11} cm} \right) dyne\,cm.$$

$$(26)$$

Here ξ is the synodic frequency of the white dwarf, μ_1 is its magnetic moment, R_2 is the radius of the companion star, and D is the binary separation. The magnitude of this torque is much less than that of the accretion torque,

$$N_{acc} \simeq \dot{M} v_{orb} D \simeq 1 \times 10^{34} \left(\frac{M/M_{red}}{1.67} \right) \left(\frac{L}{10^{33} ergs\ s^{-1}} \right) \left(\frac{M}{M_\odot} \right)^{-1/3} \left(\frac{R}{5 \times 10^8 cm} \right) dyne\ cm,$$

$$(27)$$

expected for an accretion luminosity (Patterson 1983) of approximately 10^{33} ergs s^{-1}. Therefore, as noted by Joss, Katz, and Rappaport (1979), synchronization can occur by this process only in the absence of accretion onto the white dwarf. Even then, the VDD interaction would require more than

$$t_{VDD} \simeq 1 \times 10^{10} \left(\frac{\xi}{5 \times 10^{-4} s^{-1}} \right) \left(\frac{\mu_1}{2 \times 10^{34} Gauss\ cm^3} \right)^{-2}$$
$$\left(\frac{I_1}{10^{50} g\ cm^2} \right) \left(\frac{R_2}{3 \times 10^{10} cm} \right)^{-2} \left(\frac{D}{10^{11} cm} \right)^6 yr,$$

$$(28)$$

where I_1 is the moment of inertia of the white dwarf, to change the present spin periods of the white dwarf in the AM Her systems by even a factor of two. This is almost certainly much longer than the evolutionary time available for synchronization. The VDD torque is therefore too weak to explain how synchronism is achieved in the AM Her systems. Its weakness was not as apparent in 1979 because at that time the magnetic field strength in the AM Her stars was widely believed to be a factor of 5 larger than the value of approximately 2×10^7 Gauss that has since been measured (Schmidt, Stockman, and Margon 1981; Latham, Liebert, and Steiner 1981).

Recently, Campbell (1983) has reexamined the VDD interaction. He correctly points out (see Lamb, Aly, Cook, and Lamb 1983) that since the companion star is largely or completely convective, and the turbulent diffusivity due to convection is as much as five orders of magnitude larger than the Ohmic diffusivity, the time-varying magnetic field due to asynchronous rotation of the magnetic white dwarf penetrates much further into the companion star than thought by Joss, Katz, and Rappaport (1979). However, he then assumes that the *energy dissipation* within the companion star can also be estimated simply by replacing the ordinary ohmic diffusivity by a phenomenological turbulent diffusivity (which is, as noted above, orders of magnitude larger than the ohmic one), *i.e.* that the energy dissipated by turbulent diffusion is the same as the energy that would be dissipated by ohmic diffusion at the same rate. But turbulent diffusion, like the reconnection process that forms a part of it, normally proceeds with little or no dissipation of the associated electrical currents. Since the VDD torque is due to dissipation of electrical currents within the companion star, Campbell's estimate of the VDD torque is orders of magnitude larger than the actual VDD torque and the estimate of Joss, Katz, and Rappaport. He therefore wrongly concludes that the VDD torque is able to achieve synchronism in the presence of accretion and in a time scale short compared with the evolutionary time available.

Lamb, Aly, Cook, and Lamb (1983; see also Lamb and Lamb 1979) have proposed an alternative possibility, that synchronism is achieved by a magnetohydrodynamic (MHD) torque. They point out that, as a result of the turbulent diffusivity due to convection in the low-mass companion star, the component of the magnetic field of the white dwarf parallel to its spin axis threads the companion star and that, for any plausible plasma density (*i.e.* $n \gtrsim 10^{-2}$ cm^{-3}), the volume between the two stars is electrodynamically not a vacuum. Then asynchronous rotation of the magnetic white dwarf stresses the magnetic field lines and drives large field-aligned currents between the two stars, producing a torque on the white dwarf. This situation is illustrated in Figure 20.

The MHD torque is given by

$$
N_{MHD} \simeq \alpha \gamma D R_2^2 (\frac{\mu_1}{D^3})^2
$$

$$
\simeq 8 \times 10^{34} (\frac{\alpha}{1})(\frac{\gamma}{1})(\frac{\mu_1}{2 \times 10^{34} Gauss\ cm^3})(\frac{R_2}{3 \times 10^{10}\ cm})^2(\frac{D}{10^{11}\ cm})^{-5} dyne\ cm. \quad (29)
$$

Thus the MHD torque is larger than the accretion torque [see equation (103)] and can therefore synchronize the white dwarf even in the presence of accretion. It can do so on a time scale

$$
t_{MHD} \simeq 2 \times 10^5 \gamma^{-1} (\frac{\xi}{5 \times 10^{-4} rads^{-1}})(\frac{\mu_1}{2 \times 10^{34} Gauss\ cm^3})^{-2}
$$
$$
(\frac{I_1}{10^{50} g\ cm^2})(\frac{R_2}{3 \times 10^{10}\ cm})^{-2}(\frac{D}{10^{11}\ cm})^5\ yr, \quad (30)
$$

which is much less than the evolutionary time available.

The MHD torque proposed by Lamb, Aly, Cook, and Lamb (1983) is analogous to the unipolar inductor torque proposed by Goldreich and Lynden-Bell (1967) for the Jupiter-Io system, but differs from it in important respects. For example, in the Jupiter-Io system, the major dissipation and slippage of field lines occurs in the Jovian atmosphere, where the conductivity is relatively low. Chanmugam and Dulk (1982) similarly suggest that in the AM Her stars, the major dissipation occurs in the atmosphere of the white dwarf. However, the atmosphere of the white dwarf is, unlike the Jovian atmosphere, highly conducting if evaluated at the proper depth. As a result, most of the dissipation occurs in the volume between the white dwarf and the companion star. Therefore arge-scale magnetohydrodynamic instabilities, not the conductivity of the white dwarf atmosphere, likely limit the pitch angle of the stressed field lines connecting the two stars, and hence the strength of the MHD torque.

Maintaining synchronism. Joss, Katz, and Rappaport (1979) have proposed that synchronous rotation is maintained in the AM Her stars by the magnetostatic (MS) torque resulting from the interaction of the induced magnetic field of the companion star with the magnetic moment of the white dwarf. Such a torque would have a strength

$$
N_{MS} \simeq \frac{3}{2} R_2^3 (\frac{\mu_1}{D^3})^2 \simeq 6 \times 10^{33} (\frac{R_1}{3 \times 10^{10}\ cm})^3 (\frac{\mu_1}{2 \times 10^{34} Gauss cm^3})^2 (\frac{D}{10^{11}\ cm})^{-6} dynecm,
$$
$$
\quad (31)
$$

which is comparable to the accretion torque [equation (27)].

However, Campbell (1983) has correctly pointed out that the decay time for the induced magnetic field of the companion star due to Ohmic resistive diffusion is not $\sim 10^{11}$ yr as stated by Joss, Katz, and Rappaport (1979), but is rather

$$t_{Ohm} \simeq \frac{R_2^2}{\pi^2 \eta_{Ohm}} \simeq 5 \times 10^8 (\frac{R_2}{3 \times 10^{10} cm})^2 (\frac{\eta_{Ohm}}{10^7 cm^2 s^{-1}})^{-1} \ years, \qquad (32)$$

where η_{Ohm} is the Ohmic diffusivity, which is comparable to the evolutionary time of the system. Campbell (1983) further points out that if the companion star is fully convective, as expected, the induced magnetic field decays in a time

$$t_{tur} \sim 2 \ (\frac{R_2}{3 \times 10^{10} cm})^2 (\frac{\eta_{tur}}{10^{12} cm^2 s^{-1}}) \ years, \qquad (33)$$

which is far too short to be of importance.

Lamb, Aly, Cook, and Lamb (1983) have proposed that the MHD torque might not only bring about synchronism but also maintain it. It is certainly strong enough to do so.

A third possibility, which has received passing mention [see, *e.g.*, Joss, Katz, and Rappaport (1979), Campbell (1983)], is that synchronism is maintained by a magnetostatic torque arising from the interaction between an intrinsic magnetic field of the companion star and the dipole moment of the white dwarf. We note that an intrinsic field of only $\sim$ 100 gauss would produce a torque stronger than the MHD torque and therefore also able to dominate the accretion torque.

Determination of the magnetic longitude of the white dwarfs in the AM Her systems, similar to the determinations of magnetic latitude and inclination angle carried out by Brainerd and Lamb (1983) and others and discussed above, can discriminate among these various possibilities. For example, the induced MS torque predicts that the magnetic longitude will be $\pi/2$ or $3\pi/2$, the intrinsic MS torque predicts 0 or π, while the MHD torque, in its present form, predicts that no value of magnetic longitude will be preferred.

4.2. Spin-Up and Magnetic Fields in DQ Her Stars

Liebert and Stockman (1985) have emphasized that the AM Her stars all have magnetic field strengths in a narrow range about 2×10^7 Gauss, while isolated white dwarfs have magnetic fields that range from $10^6 - 2 \times 10^8$ Gauss. This could be just an observational selection effect, since the cyclotron emission peak is proportional to the magnetic field [see equation (16)] and a change in the magnetic field by a factor of two either way would be sufficient to move the peak out of the optical window. Yet it is worrying that no AM Her stars with polarizations of $\sim 1\%$ have been found, despite the fact that a search has been made of most of the strong emission line cataclysmic variables (Liebert and Stockman 1985).

It also seems possible that DQ Her stars change into AM Her stars, and *vice versa*. For example, if the mass transfer rate in a DQ Her star temporarily decreases or stops, the white dwarf could become phase-locked; conversely, if the mass transfer rate in an AM Her star temporarily increases, the accretion torque could overwhelm the magnetic torque and the white dwarf break free. This may have recently happened in the case of EX Hya, although a resonant interaction might also be occuring between the rotation period of the white dwarf and the binary period in this system. As another example, the DQ Her stars have larger mass transfer rates (Patterson 1983) and longer binary periods (Robinson 1982). However, their mass transfer rates and periods are decreasing as they evolve (Pacsynski and Sienkiewicz 1981; Rappaport, Joss, and Webbink 1982). Both effects increase the strength of any type of magnetic torque relative to the accretion torque, suggesting that DQ Her stars may eventually become phase-locked and thus evolve into AM Her stars.

In order to clarify the relationship between the AM Her stars and the DQ Her stars, it is important to try to estimate the magnetic fields of the white dwarfs in the DQ Her stars. Lamb and Patterson (1982) have done so by analyzing their spin-up behavior using the magnetohydrodynamic theory of disk accretion onto magnetic stars developed by Ghosh and Lamb (1978; 1979a,b). According to this theory (see also Pringle and Rees 1972; Lamb, Pethick, and Pines 1973; Davidson and Ostriker 1973), the spin-up rate is given by

$$
\begin{aligned}
-\dot{P} = 4 \times 10^{-11} n(\omega_s)[(\frac{P}{10^3 \, sec})(\frac{L}{10^{34} \, ergs \, s^{-1}})^{3/7}]^2 \\
(\frac{B}{10^6 \, Gauss})^{2/7}(\frac{M}{M_\odot})^{-3/7}(\frac{R}{5 \times 10^8 \, cm})^{12/7}(\frac{I}{10^{50} \, gm \, cm^2})^{-1} \, s \, s^{-1},
\end{aligned}
\tag{34}
$$

where the dimensionless torque function $n(\omega_s)$ and the fastness parameter ω_s are given in Ghosh and Lamb [1979b; equation (10)].

The results are shown in Figure 21, which shows the theoretical spin-up rate for white dwarfs and neutron stars and the observed upper limits or measured values of the spin-up rate for the DQ Her stars and the nine pulsing neutron star X-ray sources for which the spin-up rate has been measured. The observed upper limits or measured values of the spin-up rate for the DQ Her stars are in good agreement with the theoretical curves for white dwarf, and therefore supports the theory.

Lamb and Patterson (1982) also determined the range of magnetic fields allowed by the spin-up and pulsing behavior of the DQ Her stars. They found that the magnetic fields of the DQ Her stars are typically an order of magnitude smaller than those of the AM Her stars. This is illustrated in Figure 22, which shows a histogram of the magnetic fields of the AM Her and the DQ Her stars. They also found that there is an apparent correlation between the magnetic field strength and the binary period, such that shorter period, and hence older, systems have stronger magnetic fields.

This research was supported in part by NASA grant NAGW 246.

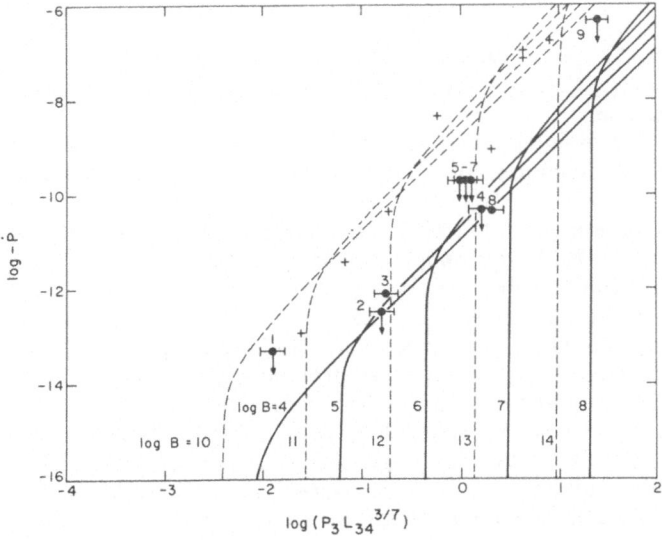

Fig. 21.–Curves of the spin-up rate for white dwarfs (solid lines) and for neutron stars (dashed lines). Also plotted are the observed upper limits or measured values of $\dot{P}$ for the nine known DQ Her stars and the nine pulsing neutron star sources for which $\dot{P}$'s have been measured. From Lamb and Patterson (1983).

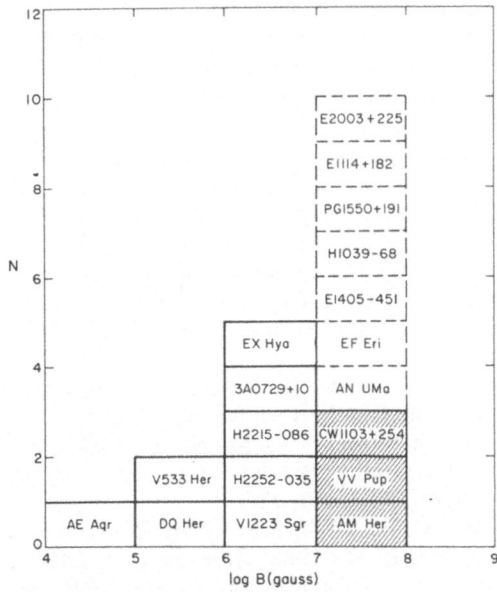

Fig. 22.–Histogram of the stellar magnetic field strengths of the known magnetic cataclysmic variables. The shaded boxes are AM Her stars for which the magnetic field has been directly measured; the dashed boxes are the remaining AM Her stars (assumed to have similar fields). The unshaded boxes are DQ Her stars; their fields have been inferred from their spin-up and pulsing behavior. From Lamb and Patteson (1983).

REFERENCES

Aizu, K.: 1973, *Prog. Theor. Phys.* **49**, 1184.

Bailey, J., Hough, J. H., Gilmozzi, R., and Axon, D. J.: 1983, to be submitted to *Mon. Not. R. Astron. Soc.*.

Barrett, P. E., and Chanmugam, G.: 1982, in *Cataclysmic Variables and Related Objects, Proceedings IAU Colloquium No. 72*, ed. M. Livio and G. Shaviv (Reidel, Dordrecht), p. 223.

Barrett, P. E., and Chanmugam, G.: 1973, submitted to *Astrophys. J.*.

Berg, R., and Duthie, J.: 1977, *Astrophys. J.* **211**, p. 859.

Beuermann, K. *et al.*: 1983, in preparation.

Böhm, K. H.: 1969, *Astron. Astrophys.* **1**, 180.

Bond, H. E., Chanmugam, G., and Grauer, A. D.: 1979, *Astrophys. J. (Letters)* **234**, L114.

Brainerd, J. J., and Lamb, D. Q.: 1983, these proceedings.

Campbell, C. G.: 1983, submitted to *Mon. Not. R. Astron. Soc.*.

Chanmugam, G., and Dulk, G. A.: 1981, *Astrophys. J.* **244**, 569.

Chanmugam, G., and Dulk, G. A.: 1982, *Astrophys. J. (Letters)* **255**, L107.

Chanmugam, G., and Dulk, G. A.: 1982, in *Cataclysmic Variables and Related Objects, Proceedings IAU Colloquium No. 72*, ed. M. Livio and G. Shaviv (Reidel, Dordrecht), p. 223.

Chevalier, R. A., and Imamura, J. N.: 1982, *Astrophys. J.* **261**, 543.

Chiappetti, L., Tanzi, E. G., and Treves, A.: 1980, *Space Sci. Rev.* **27**, p. 3.

Cordova, F. A., and Mason, K. O.: 1983, in *Accretion Driven Stellar X-Ray Sources*, ed. W. H. G. Lewin and E. P. J. van den Heuvel (Cambridge, England: Cambridge University Press),

Cowley, A., and Crampton, D.: 1977, *Astrophys. J. (Letters)*, p. L121.

Davidson, K., and Ostriker, J. P.: 1973, Astrophys. J. **179**, 585.

Dulk, G. A., and Marsh, K. A.: 1982, *Astrophys. J.* **259**, 350.

Fabbiano, G., Hartmann, L., Raymond, J., Steiner, J., Branduardi-Raymont, G., and Matilsky, T.: 1981, *Astrophys. J.* **243**, p. 911.

Fabian, A. C., Pringle, J. E., and Rees, M. J.: 1976, *Mon. Not. R. Astron. Soc.* **173**, p. 43.

Frank, J., King, A. R., and Lasota, J. P.: 1983, *Mon. Not. R. Astron. Soc.*, **202**, p. 183.

Ghosh, P., and Lamb, F. K.: 1978, *Astrophys. J. (Letters)* **223**, L83.

Ghosh, P., and Lamb, F. K.: 1979a, *Astrophys. J.* **232**, 259.

Ghosh, P., and Lamb, F. K.: 1979b, *Astrophys. J.* **234**, 296.

Goldreich, P., and Lynden-Bell, D.: 1969, *Astrophys. J.* **156**, 59.

Hearn, D. R., Richardson, J. A., and Clark, G. W.: 1976, *Astrophys. J. (Letters)* **210**, p. L23.

Heise, J., and Brinkman, A. C.: 1982, in *Galactic X-Ray Sources*, ed. P. W. Sanford, P. Laskarides, and J. Salton (Chichester: Wiley), p. 393.

Heise, J., Kruszewski, A., Chlebowski, T., Mewe, R., Kahn, S., and Seward, F. D.: 1983, to be submitted to *Astr. and Astrophys.*.

Hoshi, R.: 1973, *Prog. Theor. Phys.* **49**, 776.

Hummer, D. G., and Mihalas, D.: 1970, *Mon. Not. R. Astron. Soc.* **147**, 339.

Imamura, J. N., and Durisen, R. H.: 1983, *Astrophys. J.* **268**, 291.

Imamura, J. N., Chevalier, R. A., Durisen, R. H., and Wolff, M. T.: 1983, preprint.

Imamura, J. N., Chevalier, R. A., Durisen, R. H., and Wolff, M. T.: 1983, these proceedings.

Imamura, J. N., Wolff, M. T., and Durisne, R. H.: 1983, submitted to Astrophys. J..

Joss, P. C., and Rappaport, S. A.: 1983, in *Accretion Driven Stellar X-Ray Sources*, ed. W. H. G. Lewin and E. P. J. van den Heuvel (Cambridge, England: Cambridge University Press).

Joss, P. C., Katz, J. I., and Rappaport, S. A.: 1979, *Astrophys. J.* **230**, 176.
King, A. R., and Lasota, J. P.: 1979, *Mon. Not. R. Astron. Soc.* **188**, p. 653.
Katz, J. I.: 1977, *Astrophys. J.* **215**, 265.
King, A. R., and Lasota, J. P.: 1980, *Mon. Not. R. Astron. Soc.* **191**, p. 721.
Kuijpers, J., and Pringle, J. E.: 1982, *Astr. and Astrophys.*, **114**, p. L4.
Kylafis, N. D., and Lamb, D. Q.: 1979, *Astrophys. J. (Letters)* **228**, p. L105.
Kylafis, N. D., and Lamb, D. Q.: 1982, *Astrophys. J. (Supp.)* **48**, p. 239.
Lamb, D. Q.: 1974, *Astrophys. J. (Letters)* **192**, L129.
Lamb, D. Q.: 1981, in *X-Ray Astronomy in the 1980's*, ed. S. Holt (NASA TM-83848), p. 37.
Lamb, D. Q.: 1982, in *Cataclysmic Variables and Related Objects, Proceedings IAU Colloquium No. 72*, ed. M. Livio and G. Shaviv (Reidel, Dordrecht), p. 299.
Lamb, D. Q., and Lamb, F. K.: 1979, *Bull. AAS* **11**, 463.
Lamb, D. Q., and Masters, A. R.: 1979, *Astrophys. J. (Letters)* **234**, p. L117.
Lamb, D. Q., and Patterson, J.: 1982, in *Cataclysmic Variables and Related Objects, Proceedings IAU Colloquium No. 72*, ed. M. Livio and G. Shaviv (Reidel, Dordrecht), p. 229.
Lamb, F. K., Lamb, D. Q., Aly, J.-J., and Cook, M. C.: 1983, submitted to *Astrophys. J.*.
Lamb, F. K., Pethick, C. J., and Pines, D.: 1973, *Astrophys. J.* **184**, 271.
Langer, S. H.: 1985, this volume.
Langer, S. H., Chanmugam, G., and Shaviv, G.: 1981, *Astrophys. J. (Letters)* **245**, p. L23.
Langer, S. H., Chanmugam, G., and Shaviv, G.: 1982, *Astrophys. J.* **258**, 289.
Latham, D., Liebert, J., and Steiner, J.: 1981, *Astrophys. J.* **246**, p. 919.
Liebert, J., and Stockman, H. S.: 1979, *Astrophys. J.* **229**, p. 652.
Liebert, J., and Stockman, H. S.: 1985, this volume.
Liebert, J., Stockman, H. S., Angel, J. R. P., Woolf, N. J., and Hege, E. K.: 1978, *Astrophys. J.* **225**, p. 201.
London, R., McCray, R., and Auer, L. H.: 1981, *Astrophys. J.* **243**, 970.
Lubow, S. H., and Shu, F. H.: 1975, *Astrophys. J.*, **198**, 383.
Masters, A. R.: 1978, Ph.D. Thesis, University of Illinois, unpublished.
Meggitt, S. M. A., and Wickramasinghe, D. T.: 1982, *Mon. Not. R.Astron. Soc.* **198**, 71.
McHardy, I. M., Pye, J. P., Fairall, A. P., Warner, B., Allen, S., Cropper, M., and Ward, M. J.: 1982, *IAU Circ.* No. 3687.
Michaelsky, J. J., Stokes, G. M., and Stokes, R. A.: 1977, *Astrophys. J. (Letters)*, p. L35.
Middleditch, J.: 1982, *Astrophys. J. (Letters)* **257**, L71.
Paczynski, B., and Sienkiewicz, R.: 1981, *Astrophys. J. (Letters)*, **248**, L27.
Patterson, J.: 1979a, *Astrophys. J. (Letters)* **233**, L13.
Patterson, J.: 1979a, *Astrophys. J.* **234**, 978.
Patterson, J.: 1983, submitted to *Astrophys. J.*.
Patterson, J., Beuermann, K., Lamb, D. Q., Fabbiano, G., Raymond, J. C., Swank, J., and White, N. E.: 1983, submitted to *Astrophys. J.*.
Patterson, J., Branch, D., Chincarini, G., and Robinson, E. L.: 1980,
Patterson, J., and Price, C. M.: 1981, *Astrophys. J. (Letters)* **243**, L83.
Patterson, J., and Steiner, J. E.: 1983, *Astrophys. J. (Letters)*, in press.
Pavlov, G. G., Mitrofanov, I. G., and Shibanov, Yu. A.: 1980, *Astrophys. Space Sci.* **73**, 63.
Priedhorsky, W. C., Krzeminski, W., and Tapia, S.: 1978, *Astrophys. J.* **225**, p. 542.
Priedhorsky, W. C., and Marshall, F. E.: 1982, *Bull. Am. Astron. Soc.* **14**, 980.
Priedhorsky, W. C., Matthews, K., Neugebauer, G., Werner, M., and Krzeminski, W.: 1978, *Astrophys. J.* **226**, 397.
Pringle, J. E., and Rees, M. J.: 1972, *Astron. Astrophys.* **21**, 1.

Rappaport, S., Joss, P. C., and Webbink, R.: 1982, *Astrophys. J.* **254**, 616.

Raymond, J. C., Black, J. H., Davis, R. J., Dupree, A. K., Gursky, H., Hartmann, L., and Matilsky, T. A.: 1979, *Astrophys. J. (Letters)* **230**, p. L95.

Robinson, E. L.: 1982, in *Cataclysmic Variables and Related Objects, Proceedings IAU Colloquium No. 72*, ed. M. Livio and G. Shaviv (Reidel, Dordrecht), p. 1.

Rothschild, R. E., *et al.*: 1981, Astrophys. J. **250**, 723.

Schmidt, G. D., Stockman, H. S., and Margon, B.: 1981, *Astrophys. J. (Letters)* **243**, L17.

Spitzer, L.: 1962, *Physics of Fully Ionized Gases*, (2nd ed.; New York: Interscience).

Staubert, R., Kendziorra, E., Pietsch, W., Reppin, C., Trümper, J., and Voges W.: 1978, *Astrophys. J. (Letters)* **225**, p. L113.

Steiner, J. E., Schwartz, D. A., Jablonski, F. J., Busko, I. C., Watson, M. G., Pye, J. P., and McHardy, I. M.: 1981, *Astrophys. J. (Letters)* **249**, L21.

Stockman, H. S.: 1977, *Astrophys. J. (Letters)* **218**, L57.

Stockman, H. S., Schmidt, G. D., Angel, J. R. P., Liebert, J., Tapia, S., and Beaver, E. A.: 1977, *Astrophys. J.* **217**, p. 815.

Swank, J. H., Fabian, A. C., and Ross, R. R.: 1983, submitted to *Astrophys. J.*

Swank, J. H., Lampton, M., Boldt, E. A., Holt, S. S., and Serlemitsos, P. J.: 1977, *Astrophys. J. Letters* **216**, p. L71.

Szkody, P., and Brownlee, D. E.: 1977, *Astrophys. J. (Letters)* **212**, p. L113.

Szkody, P., Schmidt, E., Crosa, L., and Schommer, R.: 1981, *Astrophys. J.* **246**, 233.

Tapia, S.: 1977, *Astrophys. J. (Letters)* **212**, p. L125.

Tuohy, I. R., Lamb, F. K., Garmire, G. P., and Mason, K. O.: 1978, *Astrophys. J. (Letters)* **226**, p. L17.

Tuohy, I. R., Mason, K. O., Garmire, G. P., and Lamb, F. K.: 1981, *Astrophys. J.* **245**, p. 183.

Wada, T., Shimizu, A., Suzuki, M., Kato, M., and Hoshi, R.: 1980, *Progr. Theor. Phys.* **64**, 1986.

Warner, B., O'Donoghue, D., and Fairall, A. P.: 1981, *Mon. Not. R. Astron. Soc.* **196**, 705.

White, N. E., and Marshall, F. E.: 1981, *Astrophys. J. (Letters)* **249**, L25.

CW1103+254:   THE AM HER OBJECT THAT HAS EVERYTHING

Gary D. Schmidt
Steward Observatory, University of Arizona

1.   INTRODUCTION

   With a dozen AM Her systems now known, each new object becomes less important in its own right, but each holds promise for presenting unusual variations or new perspectives on the phenomenon. The recently discovered binary CW1103+254 is particularly simple in its observed behavior, yet is proving to be a great aid in refining our understanding of accretion in these CVs with magnetic primaries.

2.   OBSERVATIONS

   The optical light curve of CW1103+254 (Figure 1) is distinguished by its extreme variations. The bright phase lasts 1/3 of the 104 min orbital cycle and is unique in showing both strong linear and circular polarization. During the remainder of the cycle, the object is two magnitudes fainter and unpolarized. These remarkable characteristics can be understood in the context of our current knowledge of AM Her variables as resulting from a perspective which offers an unusually clear view of each important constituent at one time or another through the cycle. Using the analytic relations of Stockman (1977) and Meggitt and Wickramasinghe (1982), we have derived the system geometry of CW1103+254 from the very well-determined variations in polarization and brightness. The result, depicted in Figure 2, is characterized by the orbital inclination $i = 69°$, the latitude of the accreting magnetic pole $\Delta = -56°$, and the phase angle Φ, which we define to be zero with the occurrence of a linear polarization pulse at the end of the bright phase. The strong and smoothly varying polarization arises in a corotating cyclotron funnel which transits our hemisphere of the primary during the bright phase. As our line of sight is never aligned with the B field axis to better than 77°, linear polarization is present whenever the funnel is in view. For the remaining two-thirds of the orbit, the intense funnel is hidden behind the white dwarf limb and we are able to view the stellar continuum diluted only by gaseous emission in the system.

D. Q. Lamb and J. Patterson (eds.), Cataclysmic Variables and Low-Mass X-Ray Binaries, 219–223.
© *1985 by D. Reidel Publishing Company.*

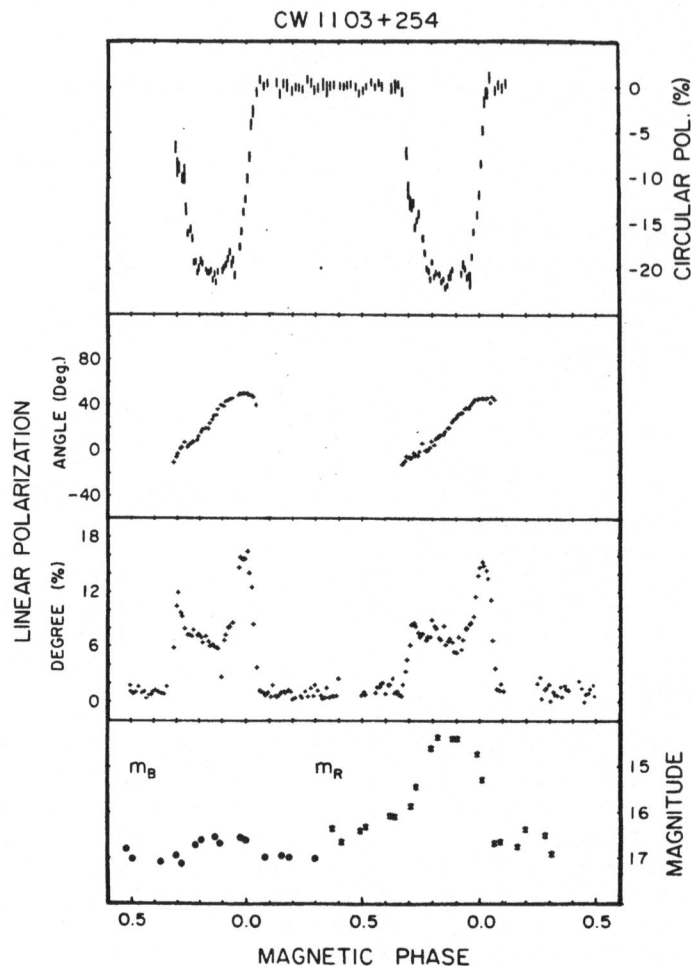

Figure 1. Linear and circular polarization and blue and red light curves of CW1103+254.

Optical spectropolarimetry obtained during this latter period (Figure 3) reveals Zeeman-split photospheric features due to H in a field 19 ± 2 MG. The inferred polar value, 30 ± 5 MG, is similar to that deduced for VV Pup (Stockman, Liebert, and Bond 1979; Visvanathan and Wickramasinghe 1979) and only somewhat above the ~ 20 MG measured with the same technique for AM Her (Schmidt, Stockman, and Margon 1981).

3. EMISSION IN AN ACCRETION FUNNEL

Our unusually accurate knowledge of the field strength and viewing geometry permit other characteristics of CW1103+254 to be analyzed in

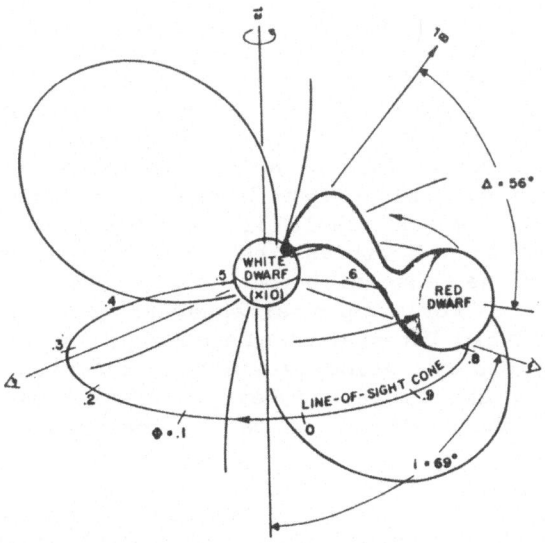

Figure 2.   System geometry.

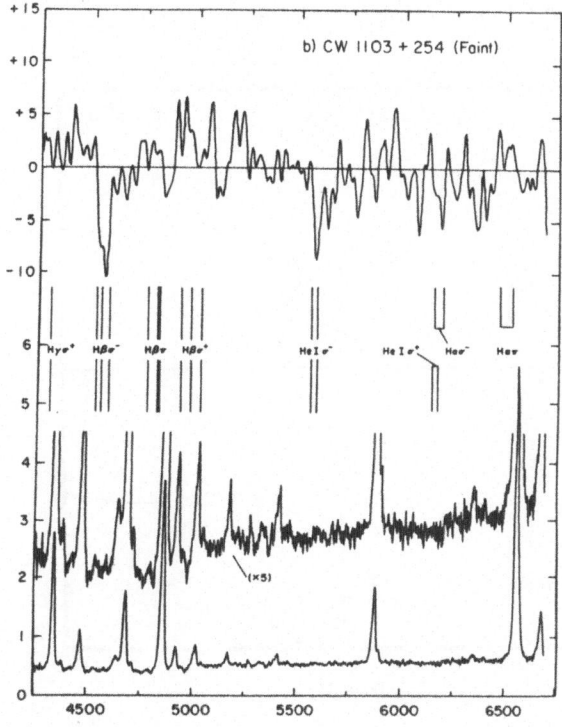

Figure 3.   Circular polarization (in percent) and spectral flux of the faint phase.

detail. In particular, the dramatic emission-line variations of this system present a challenge (Figure 4). Several mechanisms have been proposed for broadening the lines in AM Her variables, but we feel that variations as extreme as nearly 800 km s^{-1} in radial velocity and 1000 km s^{-1} in line width can only be due to supersonic infall in the accretion funnel. To simulate this behavior, we approximate the funnel as a cone of infalling streamlines which fan out into a full opening angle 2β around an axis which itself is a radius vector of stellar latitude Δ'. As the funnel orbits the primary, variations in the aspect of the streamline ensemble are solely responsible for the emission-line behavior. For simple arrangements of the streamlines within the funnel, analytic integration yields expressions for the projected infall velocity and rms velocity width as functions of the total velocity of infall and the system geometry. When added to the projected orbital motion, these variations can be compared with the observations.

A cone whose opening is uniformly pierced by streamlines cannot mimic the large and highly variable line widths while maintaining only modest radial velocity variations. However, a model which places the streamlines only on the surface of a cone produces rather good agreement with the observations. The particular results depicted in Figure 4 represent such a hollow-cone model, with total infall velocity v = 1000 km s^{-1}, β = 50°, Δ' = -35°, and the inclination i = 69° deter-

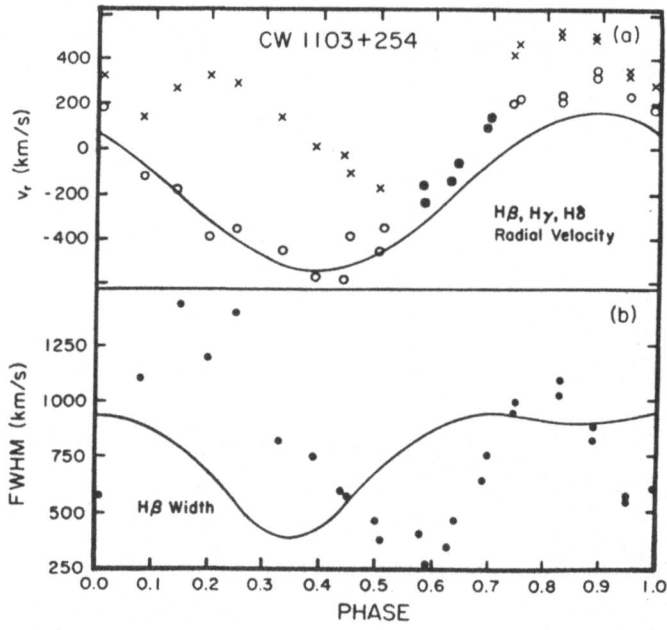

Figure 4. Broad emission line behavior (circles) and model results (solid lines).

mined above. (The streamlines originate on the secondary, yet must climb out of the equatorial plane to impact the primary at a latitude = -56° on the magnetic pole. The angle $\Delta' = -35°$ reflects the effective latitude of broad Balmer emission.) As the funnel orbits the primary, the radial velocity executes a sinusoidal variation with net blueshift -132 km s^{-1} caused by our viewpoint below the primary equator. Line widths undergo a more complicated behavior, with maxima when the cone is seen from the side, and minima when it is best aligned with our view.

The fact that our simple model can simultaneously reproduce the basics of both curves suggests that our premise of velocity broadening in an orbiting accretion funnel is correct. However, the predicted secondary minimum is too shallow; furthermore, the observed width variations lag the calculations by anywhere between 0.1 and 0.2 phase. Such phase shifts can occur in curved streams; specifically a lag of the line width relative to the radial velocity would arise in a system where the source (the secondary) lags in its orbit the accreting magnetic pole. Independent evidence of such phase shifts exist in the data of CW1103+254.

While our simple calculations suggest the basic picture is correct, more detailed models should be attempted to refine our understanding of mass flow in these systems. In particular, realistic trajectories and probable departures from the cone symmetry should be included, and an attempt be made at evaluating the perturbed magnetic field geometry at these large distances from the primary. Spectroscopic study of CW1103+254 at higher resolution would provide the ideal data base for comparison with more sophisticated models.

This work was done in conjunction with H. S. Stockman, C. B. Foltz, S. Tapia, and S. A. Grandi. Support was provided by the NSF through grant AST 82-10513. The figures presented here will appear in the Astrophysical Journal.

REFERENCES

Meggitt, S. M. A. and Wickramasinghe, D. T. 1982, M.N.R.A.S., 198, 71.
Schmidt, G. D., Stockman, H. S. and Margon, B. 1981, Ap. J. (Letters), 243, L157.
Stockman, H. S. 1977, Ap. J. (Letters), 218, L57.
Stockman, H. S., Liebert, J. and Bond, H. E. 1979, IAU Colloquium, 53, "White Dwarfs and Variable Degenerate Stars," ed. H. M. Van Horn and V. Weidemann (Rochester: University of Rochester), p. 334.
Visvanathan, N. and Wickramasinghe, D. T.1979, IAU Colloquium, 53, "White Dwarfs and Variable Degenerate Stars," ed. H. M. Van Horn and V. Weidemann (Rochester: University of Rochester), p. 330.

GENERAL DISCUSSION ON AM HERCULIS STARS

TAPIA: Does it make any sense to continue to model AM Herculis sources without considering the magnetic field of the M dwarf? In considering synchronization, the magnetic field of the M dwarf should be very important.

D. LAMB: I would rather defer that question until after Fred's talk on phase locking. Certainly if we concentrate on the kinds of things I was talking about, phenomena that are occurring very close to the surface of the white dwarf, the field of the white dwarf will entirely dominate and the M dwarf's field should not be an important factor.

CHANMUGAM: I have a brief comment on this. Tim Bastien will describe the radio outburst of AM Herculis, and it is quite possible that the outburst occurred on the M dwarf. And it could have, at least in the emitting region, a field of a thousand gauss.

VERBUNT: I have the impression that the original idea of synchronization in these sources was from when we thought that one of the eclipses was from the companion. But I think that almost all the electromagnetic radiation received comes from the white dwarf, so I would like to know what direct information we have on the period of the companion.

STOCKMAN: Well, there is one piece of direct information, which is the work that Young and Schneider have done on the absorption lines in AM Herculis. They were able to phase them with the correct period and semi-amplitude. Indirectly we know that the synchronization of the orbit with the rotation of the white dwarf can't be far off; that is to say, a part in 10^8, because VV Puppis, which has been looked at for 40 years, has never been observed to flip states: that is, to show a bright state for 60% of its period and a dark state for 40% of the period. That's what you'd expect if the poles flipped and you started accreting down the other pole.

SCHMIDT: There are two more points. Biermann's star and John Nousek's new star both show eclipses, presumably by the secondaries.

D. LAMB: Yes, but how long have these eclipses been monitored?

SCHMIDT: The limits on the synchronization in those systems are not great, but they are not wildly out of synchronism.

TAPIA: Why haven't we observed the cyclotron harmonics seen in VV Puppis in any other systems? Why is this so rare?

CHANMUGAM: Maybe the observers aren't looking hard enough!

STOCKMAN: For one thing, it's rare in VV Pup itself. VV Puppis has one of the highest fields. It's also a system in which we view the pole almost perpendicularly, and that's the only view where you have a prayer

D. Q. Lamb and J. Patterson (eds.), Cataclysmic Variables and Low-Mass X-Ray Binaries, 225–229.
© *1985 by D. Reidel Publishing Company.*

of seeing these features, because they get blurred out by relativistic doppler effects at any other angle. That was pointed out by Visvanathan and Wickramasinghe.

TAPIA: We probably have another prayer, because according to Mason and Cordova, E1405-45 has a light curve that indicates that the accretion column is pointing towards the Earth for part of the period. And it looks pretty good in terms of the polarization.

STOCKMAN: As does CW1103+254. But it's a combination of fields, low temperature, and high inclination to the pole that gives you that kind of thing.

FABBIANO: Don Lamb has given us the warm feeling that finally we understand the AM Herculis systems, and everything is hunky-dory. But there are some facts which are still unexplained. The time variability of the hard and soft components is, to say the least, questionable. In a paper that just came out in Ap. J. I wrote about the low state of AM Her, comparing the new Einstein observation with the old SAS observation. It shows that the soft component goes up and down in an absolutely erratic way, which shouldn't happen if we have reprocessing. Also, the data that Priedhorsky found seem to support that view. There is something really strange going on in that system--maybe there is a third component. The fact that the energy balance works, and at the same time we have a system in which these strange things go on, makes me feel that the fact that the energy balance works is just accidental.

D. LAMB: I consider the very good correlation between the soft and hard components in 2A0311-226 to be a very strong piece of evidence that reprocessing is the right idea.

FABBIANO: For that system.

D. LAMB: And I would generalize that quite easily to the other systems. I feel that the conflict you talk about between the new data and the old SAS data requires many steps in the chain of reasoning, and I don't consider it to be such a direct measurement.

FABBIANO: Also, there is no direct correlation between the hard and soft components in AM Herculis, as far as I know.

D. LAMB: It's never been looked at, and I highly recommend it.

[Ed: Faint voice from the back.]

FABBIANO: They're anti-correlated, Jean says.

D. LAMB: They're anti-correlated!?

SWANK: I missed the earlier talk, so I don't know what time scales you're talking about.

D. LAMB: Seconds.

SWANK: On time scales of minutes it's significant, and they're anti-correlated.

D. LAMB: That's just as good! But I'm not talking about minutes. I want seconds.

PATTERSON: Let me say that there is a continuum of states between Pepi's view and Don's view, and I'm sort of in the middle. I want to emphasize the importance of carrying out this kind of covariability analysis on very large collections of data. The data on 2A0311-226 that showed such nice covariability, high cross-correlation at zero lag, required 40,000 X-ray photons with essentially no background. When I first looked at the X-ray data, I said, "Boy, Lamb and Masters are really in trouble. The high energy peaks are not in phase with the low energy peaks." It wasn't until all the data were examined by Klaus Beuermann that this beautiful correlation at about 70% cross-correlation coefficient came out. I hope that Bill's data, many photons but with more background, is adequate. I hope that people don't come to firm conclusions based on rather small segments of data.

LIEBERT: Can I get a word in edgewise? I think that AM Her, despite the fact that it's the best studied of these systems, is a special case. Prototypes usually are. In this special case, the pole that we study most of the time comes perilously close to the limb, and usually goes past the limb. But there is evidence from the polarization change in 1979, and from Bill Priedorsky's very interesting X-ray data from November 1976, that that pole may sometimes not slip past the limb. There may be a migration in magnetic latitude, a little bit of nodding motion. Otherwise, we need to invoke two poles. The second pole has to almost exactly balance and account for the lack of an eclipse.

TAPIA: I would like to say that about six years of polarization data that I have collected indicates that a dipole is not a good representation of the magnetic field of AM Herculis. Looking to long-term variations of these objects, I think we are going to be forced to mature and to consider more complicated magnetic fields, perhaps a quadrupole or at least magnetic moment off-center.

LANGER: My intuition would be that the theory that you use to calculate the emission and to infer the magnetic field is not good enough to <u>tell</u> you whether it's a dipole or not, particularly if there is nonuniform flow down the accretion column or other potential complications. Personally, I don't think the field is exactly dipolar--it would be surprising if it were--but the theory is not ready to tell you.

PACZYNSKI: When AM Herculis was discovered six years ago, it was claimed by observers that circular polarization vanished below 4000 Å. Is it true still, and does theory explain that?

<u>A</u>: It does still vanish.

PACZYNSKI: Does theory understand that?

<u>A</u> (four-part harmony): No.

STOCKMAN: It does the same in several other objects: CS 1103+254. . .

D. LAMB: That was the last issue I referred to in my talk. You expect that at the point where the cyclotron emission goes optically thin, it will fall off very rapidly. For the conditions that you infer in the emission region, you would not expect that fall off to come until something like 7 or 10 eV, not 1 or 2 eV, as is observed. How that is explained is not clear. Either it's along the lines of what Pete was saying, that the electron distribution is not Maxwellian in its high energy tail, which radiates these high harmonics, or something else.

PACZYNSKI: Is there some simple-minded explanation for why you all believe that the cyclotron emission is optically thick?

D. LAMB: If you use theoretical cyclotron calculations, you find that the emission will indeed be optically thick. The opacity falls rapidly with increasing frequency but is insensitive to the density and magnetic field strength. You have to change the parameters by many orders of magnitude to change the harmonic where it becomes optically thin by a factor of 5 or 10. Secondly, observationally. . .

STOCKMAN: Observationally, I'd say that it looks thin. If it looks thick anywhere, it looks thick at about a micron.

PACZYNSKI: That was my impression.

STOCKMAN: That's what we're trying to discuss here, that it <u>doesn't really fit</u>.

D. LAMB: Theoretically, it is always thick below the peak and thin about it, so I would interpret what Pete is saying as that the transition occurs in the infrared, not in the optical. So in the optical you are seeing the point where it starts falling, and it is thin there.

PACZYNSKI: But if it is optically thick, it should radiate as a black body at 40 keV. Is it that bright?

STOCKMAN: Yes. If you calculate the size of the spot on the star by looking at, say, 2 microns, and, say, that the brightness temperature we see at that distance is 40 kilovolts, you get the right size--as inferred from other things. So essentially it is optically thick at a few microns.

PRIEDHORSKY: Do we even really know the orbital phasing of AM Her yet? The last I heard there were two conflicting analyses of the phase of the secondary motion of the system, differing by about 180°. There is the high state Young and Schneider measurement and a low state measurement from Cowley and others.

LIEBERT: But there are low state measurements by three other groups that disagree with Cowley's interpretation at that time, and agree with the high state interpretation.

D. LAMB: The data have never changed. It was just a disagreement in interpretation.

LIEBERT: There was also some difficulty in identifying absorption features in the secondary star in the low state.

KATZ: Does anyone believe that we can estimate the white dwarf mass from the hardness of the hard X-ray emission?

D. LAMB: I think that you can find a lower bound to it, namely, that the temperature of the hard X-ray component cannot exceed the free-fall temperature.

KATZ: But why is it a lower bound and not an actual value?

D. LAMB: Only for AM Herculis do we have a reasonable measurement of the spectral shape, and in that system the spectral shape does not agree with an optically thin bremsstrahlung. There must be other effects going on. In the other systems we have spectra, but they are not of sufficient quality to tell whether they fit a thin bremsstrahlung spectrum or not.

SCHMIDT: There are ways, though, of estimating the white dwarf mass from the dynamics of the system, and the inclination comes out of the polarization.

LIEBERT: And they're not very accurate.

GRINDLAY: What are the numbers that come out?

SCHMIDT: There are two cases that I know about. In the case of CW 1103+254 the estimate is 0.4 to 0.8 $M_{\odot}$, and Jim must have the estimate for PG 1550+191.

LIEBERT: Yes, but the problem is that the narrow line component cannot be assigned to the orbital motion of the secondary star. We have to allow more than just the offset due to the distortion due to Roche geometry.

SCHMIDT: The primary confusion comes from identifying which line component comes from the secondary.

AM HERCULIS: AN OUTBURST AT 4.9 GHz

T.S. Bastian and George A. Dulk
Department of Astro-Geophysics
University of Colorado, Boulder, U.S.A.

G. Chanmugam
Department of Physics and Astronomy
Louisiana State University, Baton Rouge, U.S.A.

ABSTRACT

We report the results of radio observations of AM Her with the Very Large Array (VLA). The quiescent emission first discovered by Chanmugam and Dulk (1982) at 4.9 GHz is confirmed and upper limits to the flux density at 1.5 GHz and 15 GHz obtained. We also report the discovery of a remarkable outburst at 4.9 GHz which was essentially 100% circularly polarized. The outburst is probably due to an electron-cyclotron maser which operates near the red-dwarf companion in a region where the magnetic field is ≈1000 gauss.

1. OBSERVATIONS

The VLA* was used on 7-9 July 1982 to search for radio emission from AM Her type binaries. For AM Her (α_{1950} = $18^h14^m58\overset{s}{.}63$, δ_{1950} = $49°50'55\overset{''}{.}06$) a mean flux density of 0.55±0.05 mJy was detected, confirming the 4.9 GHz emission discovered by Chanmugam and Dulk (1982). There is no evidence for periodic variation of the radio emission in time. The quiescent radiation is evidently unpolarized. Observations of AM Her at 1.4 GHz and 15 GHz yielded upper limits (3σ) to the flux density of 0.24 mJy and 1.11 mJy, respectively.

The most remarkable aspect of the 4.9 GHz observations of AM Her was the discovery of an outburst lasting roughly 10 min (Figure 1). The finest time resolution of 10 s revealed a peak flux density of 9.7±2.3 mJy at 09:05:09 UT, 8 July 1982. The radiation was 100% right hand circularly polarized.

---

*The National Radio Astronomy Observatory Very Large Array is operated by Associated Universities, Inc., under contract with the National Science Foundation.

D. Q. Lamb and J. Patterson (eds.), Cataclysmic Variables and Low-Mass X-Ray Binaries, 231–235.
© 1985 by D. Reidel Publishing Company.

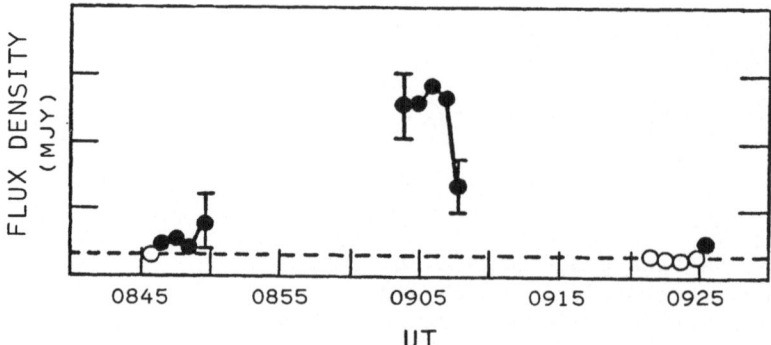

Figure 1. Five minute averages of the 4.9 GHz emission from AM Her during the outburst. The two gaps represent those times during which 15 GHz observations were being made. At no time was 15 GHz emission detected. The points plotted represent RH circularly polarized radiation, where the open points represent 1σ upper limits. The dashed line indicates the LH circularly polarized noise level.

2. INTERPRETATION

The quiescent radiation from AM Her is adequately described in terms of gyro-synchrotron radiation resulting from energetic electrons (≈ 500 keV) spiraling around the white dwarf's magnetospheric field lines at radii $\approx 7-9 \times 10^{10}$ cm and emitting at harmonics $s = f/f_B = 30-50$, where $f = 4.9$ GHz and $f_B = eB/2\pi mc$ with B the magnetic field strength. A more detailed description of the quiescent emission and upper limits for the fluxes from AN UMa, VV Pup, EF Eri, CW 1103+254, and PG 1550+191 may be found in Dulk, Bastian, and Chanmugam (1983).

Of more immediate interest here are possible source locations and mechanisms for the outburst observed at 4.9 GHz. The brightness temperature of a radio source at frequency f is defined by the equation

$$S_\nu = 2kT_B \frac{f^2}{c^2} \Omega$$

where Ω is the solid angle subtended by the source and S_ν is the observed flux density. Assuming AM Her to be 100 pc distant and the emitting region to have a projected area of πq^2, we find that

$$T_B = 4 \times 10^{12} q_{10}^{-2} \text{ K}$$

where q_{10} is in units of 10^{10} cm and we have used the peak flux value of 9.7 mJy. Even with the generous assumption of $q \approx 10^{11}$ cm, it follows that $T \gtrsim 10^{10}$ K. This, coupled with the extremely high degree of circular polarization observed during the outburst, implies a coherent emission mechanism.

Plasma emission (Melrose 1980) and electron-cyclotron maser action (Melrose and Dulk 1982) have each been proposed to explain stellar and solar radio bursts of high brightness. In this particular instance preference is given to the electron-cyclotron maser operating on or near the red dwarf companion. Emission near the white dwarf is discounted because the magnetosphere is optically thick (Chanmugam and Dulk 1982, Dulk, Bastian, and Chanmugam 1983). Emission from the accretion column could only occur in a region of sufficiently low number density that $f = 4.9$ GHz $> f_p$ where f_p is the plasma frequency. This requirement is fulfilled near the origin of the accretion column, that is, in the vicinity of the red dwarf (Dulk, Bastian, and Chanmugam 1983). Plasma emission from the red dwarf, while not entirely ruled out, is unlikely since free-free absorption in the red dwarf atmosphere would strongly attenuate the radiation. Hence we propose electron-cyclotron maser action at the second harmonic of the gyrofrequency in the red dwarf corona as the mechanism responsible for the observed outburst. Under stellar and solar conditions, radiation at the fundamental frequency cannot escape (Melrose and Dulk 1982). The mechanism hence requires that $f_B > f_p$, a condition that appears to be fulfilled, for example, in the low solar corona (Dulk and Melrose 1983). The magnetic field strength required with $s = 2$, $f = 4.9$ GHz is moderately strong ~1000 gauss.

We envisage two possible causes for the outburst:
  i) A flare occurred in a magnetic active region, but with much greater intensity than is typical of a flare star. In this case, the binary nature of the system is irrelevant.
  ii) Interactions inducing magnetic reconnection between the field lines of the two stars in a volume where the field strengths are comparable could provide the necessary free energy to drive the maser. In this case the binary nature of the system would play a crucial role.

3. CONCLUSIONS

The most intriguing implication of the observed 4.9 GHz outburst is the presence of a magnetic field, and possibly a corona (e.g., Linsky 1981; Golub 1983), on the red dwarf companion. Its existence provides possibilities for a wide variety of energetic phenomena taking place both on the red dwarf and in the volume where magnetic interactions take place between the two stars. Field line reconnection provides a mechanism for charged particle acceleration in the magnetosphere. An evolving red dwarf magnetic field almost surely plays a role in modulating the mass tranfer rate to the white dwarf, particularly if the red dwarf undergoes any sort of magnetic cycle analogous to that of the Sun. Finally, such modulation may play a part in the observed high and low states of the AM Her system.

This work was supported in part by NASA under grants NSG-7287 and NAGW-91 to the University of Colorado, and by NSF under grant AST 8025250 to Louisiana State University.

REFERENCES

Chanmugam, G., and Dulk, G.A.: 1982, Astrophys. J. (Letters), 255,
    L107.
Dulk, G.A., Bastian, T.S., and Chanmugam, G.: 1983, submitted to
    Astrophys. J.
Dulk, G.A., and Melrose, D.B.: 1983, Proc. 4th CESRA Workshop on Solar
    Noise Storm, A. Benz (Ed.), in press.
Golub, L.: 1983, IAU Colloq. No. 71, Activity in Red Dwarf Stars,
    M. Rodono and P. Byrne (Eds.), D. Reidel Publ. Co., Dordrecht,
    Holland, p. 83.
Linsky, J.L.: 1981, X-ray Astronomy in the 1980's, S.S. Holt (Ed.),
    NASA Tech. Memo 83848, 13.
Melrose, D.B.: 1980, Space Sci. Rev., 20, p. 3.
Melrose, D.B., and Dulk, G.A.: 1982, Astrophys. J., 259, 844.

DISCUSSION

Q: How long did the flare stay high?

BASTIAN: Well, it's hard to say. We have a five minute scan and it was
at a fairly high level, though variable. Electron cyclotron masing
would predict a great deal of variation on very short timescales, and,
in fact, it has been observed on millisecond timescales on the Sun. Of
course we didn't have that sort of time resolution.

TAPIA: Is there any objection in the audience for the 1000 gauss of
the M dwarf? I'm not going to go back to observe until I know what is
the magnetic field of the M dwarf!

CHANMUGAM: The field is in a local region. The Sun, for example, has
fields of 1000 gauss in a sunspot.

BASTIAN: Presumably it's in the form of flux tubes.

CHANMUGAM: If it is on the M dwarf. Another possibility is that it's
in between the white dwarf and the M dwarf and that there is some sort
of reconnection going on.

BASTIAN: I didn't really have time to go into specific mechanisms.

BODE: What phase was that?

BASTIAN: The peak was at phase 0.4.

CHANMUGAM: Phase 0.4 corresponds, coincidentally I think, to the
second linear polarization pulse now seen by Tapia. But we believe that
the linear polarization comes from near the white dwarf, while the radio
flaring must come from fairly far away, near the surface of the red
dwarf. So these phenomena are probably unrelated.

SHAFTER: Have you looked at AE Aquarii?

BASTIAN: No.

SHAFTER: There is a model for that system by Chincarini and Walker. They claim that the red star underfills its Roche lobe and that mass transfer in that system is caused by flaring on the red star. This might be a good candidate for observing.

CHANMUGAM: I should mention that Cordova, Mason, and Hjellming have made radio observations of six cataclysmic variables not known to have magnetic fields, and they've got upper limits. TT Ari was one of them.

BASTIAN: I'd like to add that I think the thing killing us on the observations of these other five stars is the fact that they have very short periods--which means their separation is much less and therefore the source region is much smaller.

SHAVIV: Is it correct that these M dwarfs in the binary systems are very different in their magnetic activity from the single M dwarf stars, which are known to be very active?

CHANMUGAM: We don't know. Could be.

JENSEN: One thing is that they're rotating faster, and there is a correlation between coronal activity and rotation. So naively, you'd expect even more activity.

SYNCHRONIZATION OF MAGNETIC WHITE DWARFS IN CLOSE BINARY SYSTEMS

F. K. Lamb[1,2], J.-J. Aly[1], M. C. Cook[1], and D. Q. Lamb[3]

[1]Department of Physics, University of Illinois at Urbana-Champaign, Urbana, IL 61801
[2]Department of Astronomy, University of Illinois at Urbana-Champaign, Urbana, IL 61801
[3]Harvard-Smithsonian Center for Astrophysics, 60 Garden Street, Cambridge, MA 02138

ABSTRACT

Asynchronous rotation of strongly magnetic white dwarfs in close binary systems drives substantial field-aligned electrical currents between the magnetic star and its companion. The resulting magneto-hydrodynamic torque is able to account for the heretofore unexplained synchronous rotation of the strongly magnetic degenerate dwarf component in systems like AM Her, VV Pup, AN UMa, and EF Eri. The electric fields produced by even a small asynchronism are large enough to accelerate electrons to high energies, and may lead to radio emission. The total energy dissipation rate in systems with white dwarf spin periods as short as 1^m may reach 10^{33} ergs s^{-1}. Total luminosities of this order may be a characteristic feature of such systems.

1. INTRODUCTION

A key feature of systems like AM Her, VV Pup, AN UMa, and EF Eri is that the strongly magnetic white dwarf component is observed to be spinning synchronously with the orbital motion (see Liebert and Stockman 1985 and references therein). Synchronous rotation of these white dwarfs is surprising, for (1) evolution of these systems leads naturally to asynchronism, (2) the viscous coupling between the white dwarf and its companion is expected to be negligible, (3) the accretion torque tending to spin up the white dwarf is typically strong, while (4) the moment of inertia of the white dwarf is small.

A common conjecture is that synchronism in these systems is somehow the result of magnetic coupling. The first detailed calculation of magnetic coupling in magnetic white dwarf systems was the pioneering study of Joss, Katz, and Rappaport (1979, hereafter JKR), who assumed a vacuum dipole-dipole (VDD) interaction between the white dwarf and its companion and carefully worked out the resulting torques. However, this

237

D. Q. Lamb and J. Patterson (eds.), Cataclysmic Variables and Low-Mass X-Ray Binaries, 237–245.
© 1985 by D. Reidel Publishing Company.

model does not appear capable of explaining the observations, for two reasons. First, for any plausible plasma density (i.e., n $\gtrsim 10^{-2}$ cm^{-3}) the volume between the two stars is electrodynamically not a vacuum. As a result, asynchronous rotation of the white dwarf drives large electrical currents between the two stars, qualitatively changing the nature of their interaction. Second, even if the VDD interaction were relevant, it would probably be too weak to explain the data. In the absence of accretion, the VDD torque would require $> 10^{10}$ yr to change the spin period of the white dwarf in systems like AM Her and VV Pup even by a factor of two. This is almost certainly much longer than the evolutionary time available for synchronization. (The extreme weakness of the VDD synchronization torque was not as apparent in 1979 because at that time the magnetic field strength in AM Her stars was believed to be a factor ~5 larger than the value ~2×10^{7} G that has since been measured (Schmidt, Stockman, and Margon 1981; Latham, Liebert, and Steiner (1981)). The VDD synchronization torque would also be too weak to explain the synchronism observed in magnetic A star systems with orbital periods $\lesssim 7^{d}$ (Abt et al. 1968, Wolf 1973), a fact noted by JKR. This suggests that some other mechanism is operating in binaries containing magnetic A-type stars, a mechanism that may also be effective in magnetic white dwarf systems.

An alternative possibility is that synchronism is due to the magnetohydrodynamic (MHD) torque resulting from currents flowing between the two stars. A preliminary estimate of the likely size of the MHD torque showed it to be ~10^{4} times stronger than the VDD synchronization torque (Lamb and Lamb 1979). We have now carried out a detailed study of MHD spin-orbit coupling in binary stellar systems. Here we summarize the results for AM Her-type systems. The full details of our analysis and a more complete discussion of the implications for magnetic stars in binary systems generally will be published elsewhere.

2. THE MODEL

The quantity of magnetic flux linking the magnetic star to its companion depends on the outcome of several physical processes acting over the entire evolutionary history of the system. These processes include flux linkage at the time the two stars are formed, diffusion of flux into the companion star due to Coulomb collisions and turbulence, pumping of flux into the companion by convection, and reconnection (if the companion has its own intrinsic magnetic field). In order to simplify the description that follows, we assume that there is initially no flux linking the two stars. The resulting torque is therefore a lower bound on the torque at any later time. We assume that the spin axis of the magnetic star is parallel to the angular momentum of the system. We further assume that the white dwarf has an axisymmetric magnetic field aligned with its spin axis, and that the companion star has no intrinsic magnetic field. Some of the effects of relaxing the latter assumptions are discussed briefly at the end of this section.

The physics of the MHD coupling is as follows. The spinning motion of the white dwarf creates a $\underline{v}\times\underline{B}$ electric field within it, which produces a potential difference between points on its surface that are at different magnetic latitudes. To the extent that the electric potential is constant along a given field line, field lines threading the plasma outside the star impress a potential drop across it. If the white dwarf is spinning asynchronously, the potential drop between the field lines threading the companion star drives cross-field electrical currents inside the companion. The circuit is closed primarily by field-aligned currents flowing between the two stars, and by cross-field currents inside the white dwarf. The resulting $j \times B$ forces within the two stars produce torques that alter their angular velocities and that of the system. Some of these phenomena are similar to those that are thought to occur in the Jupiter-Io system (see Scarf et al. 1982 and references therein).

An alternative but equivalent and perhaps more easily visualized description is that the field-aligned currents produce a toroidal component of magnetic field between the two stars, as seen from the white dwarf (see Fig. 1). The resulting magnetic stresses, when integrated over the surfaces of the two stars, give the torques on them and the system.

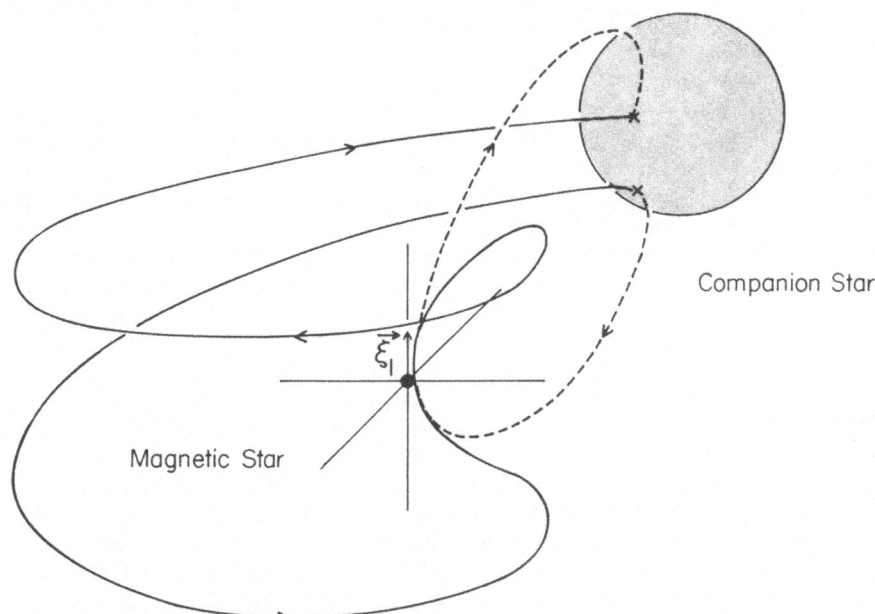

Fig. 1 Perspective view of the binary system illustrating how a given flux tube is twisted as the white dwarf rotates with respect to its companion. Dashed curve: initial shape; solid curve: shape at a later time. The resulting Maxwell stress on the white dwarf produces a torque that opposes the rotation. For simplicity, slippage of field lines through the plasma has been neglected in this figure. In reality, the twist of the flux tube will not increase indefinitely but will be limited by the processes described in the text. (To appear in The Astrophysical Journal)

We have estimated the MHD torque on the white dwarf by integrating the magnetic stresses over a surface enclosing it. The magnetic field configuration that we have assumed is motivated by detailed axisymmetric calculations. The result is

$$N_{MHD} \approx \alpha \gamma D R_2^2 \ (\mu_1/D^3)^2 , \tag{1}$$

where α is the fractional area of the companion star threaded by magnetic flux, γ is the pitch of the magnetic field linking the two stars, D is the binary separation, R_2 is the radius of the companion star, and μ_1 is the magnetic moment of the white dwarf.

The pitch of the magnetic field linking the two stars is limited by dissipation of the field-aligned currents that maintain it, by magnetic flux reconnection, and by large-scale MHD instability of the field configuration. Our calculations indicate that dissipation within the magnetosphere by current-driven plasma instabilities there is less important in limiting the pitch than other effects if the particle density there exceeds $10-50$ cm^{-3}. We assume that this is the case and that the pitch is limited by large-scale MHD instabilities. Stability analyses of sheared-field configurations similar to those that arise in this system (see Low 1982; Aly 1983) indicate that MHD instabilities limit γ to a value ≈ 1.

The fractional area α of the companion star threaded by flux depends principally on the extent of convection in the companion star. We have computed the evolution of α with time for companion stars in the mass range $0.1-2.9$ $M_\odot$ using eleven ZAMS stellar models constructed by Webbink (private communication). This complex physical process was approximated by a scalar diffusion equation. Where present, turbulent transport by convective motions was modeled by an effective magnetic diffusivity $\eta_m = 0.15 \ u_t \ell_t$. Here u_t and ℓ_t are the convective velocity and the mixing length. The models with masses less than 0.3 $M_\odot$ are fully convective. In these models α grows rapidly due to flux pumping and turbulent diffusion, reaching ≈ 1 in a time much shorter than 10^5 yr. The models with masses in the range 0.43 to 1.0 $M_\odot$ have substantial surface convection zones; here α initially grows to $\sim 0.4-0.8$ in much less that 10^5 yr and then grows more slowly as flux continues to diffuse inward by ordinary resistive diffusion. The 1.5, 2.0, and 2.9 $M_\odot$ models are radiative outside a small convective core. In these models α grows slowly by resistive diffusion, reaching ~ 0.2 after $\sim 10^8$ yr (the flux does not reach the convective core until much later). The results for the stars that are not fully convective are shown in Figure 2. A more detailed discussion of the threading of the companion star will be given elsewhere.

The time to reach synchronism is plotted on the right-hand vertical axis of Figure 2, in units of the characteristic synchronization time

$$\tau_s \equiv \frac{I_1 \xi_o}{N_{MHD}}$$

$$= 1.76 \times 10^5 \ \gamma^{-1} \left(\frac{\xi_o}{5 \times 10^{-4} \, \text{rad s}^{-1}} \right) \left(\frac{\mu_1}{2 \times 10^{34} \, \text{G cm}^3} \right)^{-2}$$

$$\times \left(\frac{I_1}{10^{50} \, \text{g cm}^2} \right) \left(\frac{R_2}{3 \times 10^{10} \, \text{cm}} \right)^{-2} \left(\frac{D}{10^{11} \, \text{cm}} \right)^5 \text{yr.} \qquad (2)$$

Here I_1 is the inertial moment of the white dwarf and $\xi_o = \Omega_1(0) - \Omega_b(0)$ is its initial synodic frequency, where $\Omega_1(0)$ and $\Omega_b(0)$ are the initial angular frequencies of the white dwarf and the binary system. Equation (2) assumes that the synchronization torque is constant, an assumption that is approximately correct for the time scales of interest. The second expression on the right is scaled in units appropriate to systems like AM Her.

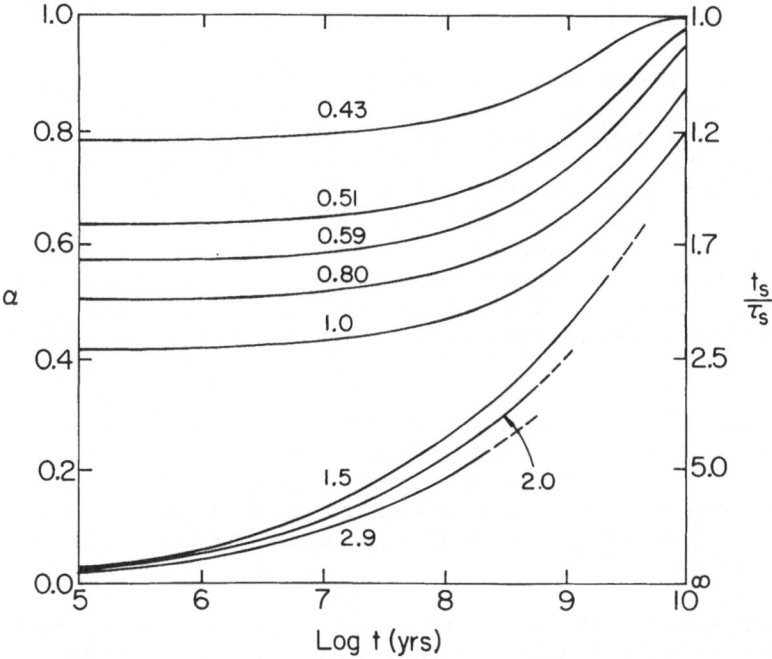

Fig. 2   Increase in the strength of the MHD spin-orbit coupling with time for eight models of companion stars, assuming no flux threads the companion star initially. The vertical scale on the left shows the fractional area α of the companion threaded by magnetic flux, while that on the right shows the synchronization time t_s in units of the characteristic synchronization time τ_s, assuming that the torque is constant. The curves are labeled with the mass of the star in solar masses. Those for the three most massive stars end at 10^9 yr because the ZAMS models used in the calculations are inaccurate representations of these stars at such late times. (To appear in The Astrophysical Journal)

We have also considered the size of the MHD torque on white dwarfs with magnetic field configurations more general than the aligned rotator described above. We find that the torque produced by the aligned component $\mu_\parallel$ of the magnetic moment is comparable to the expression for N_{MHD} given above, with μ replaced by $\mu_\parallel$. In contrast, the torque produced by the perpendicular component $\mu_\perp$ is much less, except for synodic periods $\gtrsim 10^3$ yr, because α for this component is limited by the fact that the magnetic field associated with $\mu_\perp$ varies with the synodic frequency. If the companion star also has an intrinsic magnetic field, flux linkage between the two stars will occur via reconnection in the volume between the two stars. The magnetic pitch will be limited by precisely the same physical processes discussed above, and we expect the resulting torque to be comparable to N_{MHD}. In this case α will depend on the strength and orientation of the companion star's magnetic field as well as the evolutionary history of the system.

3. APPLICATIONS TO MAGNETIC STARS IN BINARY SYSTEMS

In the absence of accretion, equation (1) shows that the MHD torque can synchronize the white dwarf in AM Her-type systems in as little as 10^5 yr. Even white dwarfs with initial spin periods as short as $\sim 1^m$ can be synchronized in $\sim 10^8$ yr in close systems. For white dwarfs with periods as short as this, the resulting energy dissipation rate $N_{MHD}\xi$ is $\sim 10^{33}$ ergs s^{-1} for $\sim 10^3$ yr. If they exist, such systems would be an interesting new class of relatively bright but short-lived low mass binaries. We expect most of the energy to be dissipated in regions of low plasma density between the two stars. Thus, luminosities of this order from low-density plasma may be a characteristic feature of such systems.

The electric fields produced by even a small asynchronism are large enough to accelerate electrons to high energies, and may lead to radio emission. In AM Her a temporary synodic frequency as small as 10^{-4} of the orbital frequency is large enough to account for the luminosity of the radio emission observed by Chanmugam and Dulk (1982), assuming an efficiency $\sim 10^{-2}$.

The MHD synchronization torque can dominate the accretion torque for typical system parameters. Thus, for a system like AM Her, with $M_1 = 0.39\ M_\odot$, $R_1 = 1.1\times 10^9$ cm, $B_1 \approx 2\times 10^7$ G, $M_2 = 0.26\ M_\odot$, and $D = 6.4\times 10^{10}$ cm (Young, Schneider, and Shectman 1981), one has $\mu_1 \approx 10^{34}$ G cm^3 and $I_1 \approx 2.7\times 10^{50}$ g cm^2. The resulting MHD synchronization torque is $\sim 8\times 10^{34}$ dyne cm, whereas the accretion torque is only $\approx 10^{34}$ dyne cm for an accretion luminosity (Patterson 1983) of $\approx 10^{33}$ ergs s^{-1}.

The AM Her stars appear to have very similar magnetic fields. For a given field strength, the strength of the MHD torque increases rapidly as the binary separation decreases and the companion star becomes more fully convective. This dependence offers a natural explanation for the fact that most systems with orbital periods $>4^h$, which are wider and contain companions with radiative cores (see Rappaport, Joss and Webbink 1982) are asynchronous (see Lamb and Patterson 1983 and references therein), whereas most systems with orbital periods $\lesssim 4^{h}$, which are closer and contain fully convective companions, are synchronized.

This same torque is strong enough to account for the observed synchronous rotation of magnetic A-type stars in binary systems with periods as long as 7^{d}, as we have discussed elsewhere (Lamb et al. 1983). In neutron star systems, the MHD synchronization torque is too small to counter even a weak accretion torque, in agreement with the observed asynchronism of these systems.

ACKNOWLEDGEMENTS

This research was supported in part by NSF grant PHY 80-25605 (at Illinois), NASA grants NSG 7653 (at Illinois) and NAGW 246 (at CFA), and by CNRS and NSF under a US-France exchange grant.

REFERENCES

Abt, H. A., Conti, P. S., Deutsch, A. J., and Wallerstein, G.: 1968, Ap. J., 153, 177.
Aly, J.-J.: 1983, in preparation.
Chanmugam, G., and Dulk, G. A.: 1982, Ap. J., 255, L107.
Iben, I.: 1967, Ann. Rev. Astr. Ap. 5, 571.
Joss, P. C., Katz, J. I., and Rappaport, S. A.: 1979, Ap. J., 230, 176.
Lamb, D. Q., and Lamb, F. K.: 1979, B.A.A.S., 11, 463.
Lamb, D. O., and Patterson, J.: 1983, in Cataclysmic Variables and Related Objects, IAU Colloq. No. 72, ed. M. Livio and G. Shaviv (Reidel, Dordrecht), p. 229.
Lamb, F. K., Aly, J.-J., Cook, M. C., and Lamb, D. Q.: 1983, submitted to Ap. J. (Letters).
Latham, D. W., Liebert, J., Steiner, J.: 1981, Ap. J., 246, 919.
Liebert, J., and Stockman, H. S.: 1985, this volume.
Low, B. C.: 1982, Rev. Geophys. Space Phys., 20, 145.
Patterson, J.: 1983, Ap. J., in press.
Rappaport, S. A., Joss, P. C., and Webbink, R.: 1982, Ap. J., 254, 616.
Scarf, F. L., Coroniti, F. V., Kennel, C. F., and Gurnett, D. A.: 1982, Vistas in Astronomy, 25, 263.
Schmidt, G. D., Stockman, H. S., Margon, B.: 1981, Ap. J., 243, L157.
Wolf, S. C.: 1973, Ap. J., 186, 951.
Young, P., Schneider, D. P., and Shectman, S. A.: 1981, Ap. J., 245, 1043

DISCUSSION

McCLINTOCK: Is this pretty much independent of the density of the material between the stars, assuming it's above some critical level? How
does it scale with the density?

F. LAMB: At very low densities, the kind of effects that tend to limit
the magnetic pitch are current-driven plasma instabilities. If you get
to too low a density you're limited by that effect. If you go up to
higher densities, the densities are high enough to support the currents
that are needed to wrap up the field, but then you get into magnetohydrodynamic instabilities in which the magnetosphere blows up and the
pitch goes to zero again. It's very similar to what's been found in
arches in the solar chromosphere, and apparently there it's related to
the flare phenomenon. So there are two basically different physical
effects which limit the pitch. In this calculation, we've assumed the
desnity was high enough so that we could ignore the dissipation due to
plasma instabilities, but we've given estimates of the density at which
that should come in.

McCLINTOCK: What is the mean density between the stars then?

F. LAMB: Well, in order to ignore the current-driven instabilities you
need a density of the order of a few per cubic centimeter.

TAPIA: What is the magnetic moment of the M dwarf?

F. LAMB: This calculation assumes no magnetic field on the M dwarf. If
there is a field on the M dwarf, the expected scenario is actually very
similar. In this picture what happens is the following: As the dwarf
moves through the magnetosphere, magnetic flux is being connected in
the forward hemisphere of the motion and disconnected in the backward
hemisphere due to these magnetohydrodynamic instabilities. So, if you
replace that companion with one with a significant intrinsic magnetic
field, the intrinsic field will then connect to the degenerate dwarf's
magnetic field in a similar way. The domain in which the reconnection
occurs moves away from the companion star as the strength of its field
is increased, but the basic physics is very similar. The current
system and all the rest of it remains very similar.

WADE: In a synchronized system, is there going to be a preferred
orientation of the magnetic axis of the white dwarf with respect to the
secondary star?

F. LAMB: Almost surely.

WADE: Is that supported by the observations? I thought we heard
earlier that those angles tend to be random.

F. LAMB: I've focussed here on the synchronization problem. You're
addressing the nature of the synchronous state. Here the problem is

that there are several effects which are quite competitive and which tend to go in opposite directions. I'll name them. One is the magnetostatic interaction that Joss, Katz, and Rappaport considered, and that tends to make the magnetic moment take a direction perpendicular to the line of centers. Another effect is the tidal interaction with the bulge in the degenerate dwarf produced by its magnetic field, which tends to cause the same orientation. On the other hand, the MHD torque, which is comparable in size with the other two, tends to make the degenerate dwarf line up with its field pointing at the companion star. So it's a complicated problem to decide in any given case which of these effects is going to predominate.

ORIENTATIONS OF AM HER STARS FROM THEIR POLARIZATION PROPERTIES: THE CASE OF THE MISSING AM HER STARS

J. J. Brainerd and D. Q. Lamb
Harvard-Smithsonian Center for Astrophysics

ABSTRACT

We use observation of the polarization properties of the AM Her stars to determine their orientations. We find that the AM Her stars are randomly distributed with respect to their inclination angle i but not with respect to their magnetic colatitude δ. The stars are concentrated along the semicircle $\cos^2\delta + \cos^2 i = 1$, where δ is the angle between the magnetic moment and the spin axis of the degenerate dwarf. This result implies that the discovery of AM Her stars is strongly affected by observational selection effects. We suggest that these effects are a result of the way in which the AM Her stars have been identified observationally, and of the angular properties of high harmonic cyclotron emission. We estimate that there are about three times as many AM Her stars as have been found so far.

1. INTRODUCTION

AM Her stars are binary systems containing a magnetic degenerate dwarf rotating synchronously about a companion M dwarf. The optical light is strongly circularly polarized. Soft and hard X-rays are observed, and the optical and X-ray light curves have the same period. The optical and X-ray light curves of some systems show a minimum accompanied by a change in the sign of the circular polarization. This is interpreted as due to eclipse of the dominant emission region by the degenerate dwarf itself. Often a single linear polarization pulse is observed at the onset or end of the eclipse. All of these features can be explained by accretion onto predominantly one pole of a magnetic degenerate dwarf having a field strength of 2-3 x 10^7 G (for reviews, see Chiappetti, Tanzi, and Treves 1980; Lamb 1981, 1982; and Liebert and Stockman 1983).

Assuming that the rotation axes of the degenerate dwarf and of the binary are parallel, the orientation of the degenerate dwarf is determined by three angles: the inclination angle i of the binary system, the magnetic colatitude δ (the angle between the magnetic moment μ and the rotation axis $\hat{\omega}$ of the degenerate dwarf), and the magnetic longitude ψ relative to the companion star. These angles are shown in Figure 1.

The angles i and δ are needed in order to compare theoretical calculations and observations of the intensity and angular dependence of the polarized optical light from these systems. In this paper we use observation of the polarization properties of the AM

247

D. Q. Lamb and J. Patterson (eds.), Cataclysmic Variables and Low-Mass X-Ray Binaries, 247–256.

Her stars to determine them. We find that the AM Her stars are randomly distributed with respect to their inclination angle i but are concentrated along the semicircle $\cos^2 \delta + \cos^2 i = 1$, implying that their discovery is strongly affected by observational selection effects. We suggest that their non-uniform distribution is a result of the way in which they have been identified observationally, and of the angular properties of high harmonic cyclotron emission. We estimate that there are about three times as many AM Her stars as have been found so far.

2. CALCULATIONS

Two pieces of information are needed to determine the angles i and δ. One is the behavior of the linear polarization position angle with phase. In systems in which the emission region is self-eclipsed, the eclipse duration provides a second contraint. We calculate it from the times at which the circular polarization crosses zero. These crossings give a value for the eclipse duration that is consistent with the behavior of the linear polarization, but is generally not the same as the length of the minimum in the optical flux. In systems in which the emission region is not self-eclipsed, another constraint must be sought. All of the latter observed so far exhibit a linear polarization pulse (this need not be so; see below), which means that the sum of i and δ must equal nearly 90° at this phase (Chanmugam and Dulk 1981, Meggitt and Wickramasinghe 1982, Brainerd and Lamb 1983). This requirement provides the needed second constraint.

The polarized optical light comes from an emission region near, but above, one or both magnetic poles of the degenerate dwarf (cf. Liebert and Stockman 1983). We assume the following in our calculations: (1) We approximate the magnetic field by a centered dipole. We therefore neglect any off-center displacement of the stellar magnetic moment, which produces poles of different strength and/or ones that do not lie 180° from one another, or any complex field geometry in the emission region. (2) We approximate the emission region by a point lying directly above one or both magnetic poles. We therefore neglect the fact that the emission region has a finite size and possibly an irregular shape, that it may not be located exactly at the magnetic pole, and that, if it lies significantly above the stellar surface, curvature of the field lines means that those in the emission region will not be oriented exactly along the direction of the magnetic moment. Note that, with these assumptions, our results are unaffected by any contribution of polarized light from the secondary pole.

The variation of the linear polarization position angle α with phase can then be calculated as follows. Consider a coordinate system in which $\hat{z}$ is parallel to the spin axis $\hat{\omega}$, and $\hat{x}$ and $\hat{y}$ lie in the equatorial plane of the degenerate dwarf. Then the direction of the magnetic moment μ is

$$\hat{\mu} = \hat{x} \sin \delta \cos(\omega t + \psi) + \hat{y} \sin \delta \sin(\omega t + \psi) + \hat{z} \cos \delta, \tag{1}$$

and $\phi \equiv (\omega t + \psi)/2\pi$ is the phase. We assume that the rotation axes of the degenerate dwarf and of the binary system are parallel and rotate the coordinate system in the $\hat{x}$-$\hat{z}$ plane by an angle i, so that $\hat{z}'$ lies along the line of sight. The projection of the direction of the magnetic moment on the plane of the sky is

$$\hat{\mu}_{sky} = \hat{x}'[\sin \delta \cos i \cos(\omega t + \psi) - \sin i \cos \delta] + \hat{y}' \sin \delta \sin(\omega t + \psi). \tag{2}$$

For sufficiently low temperatures ($kT \lesssim 20$ keV) in the cyclotron emission region, the linear polarization of the emitted light is parallel to the magnetic field (Brainerd and Lamb 1983).

Then the linear polarization position angle α is given by

$$\tan \alpha \equiv \hat{\mu}_{y'}/\hat{\mu}_{x'} = \pm \sin \delta \sin \omega t / [\sin \delta \cos i \cos \omega t - \sin i \cos \delta], \qquad (3)$$

where we set $\psi = 0$ so that α is zero at $t = 0$. The positive sign applies when $\cot i > \cot \delta$, and the negative sign when $\cot i < \cot \delta$.

Figure 2 shows α as a function of phase for $i = 45°$ and for ten different values of δ. For $\delta > i$, α varies by $180°$ at most, and usually by much less. For $\delta < i$, α changes by a full $360°$ in one period. The slope of these curves gives the rate of change of the position angle $\dot{\alpha}$. The behavior of the slope depends upon i: if i is small, the slope varies little with phase, but if i is large, the slope changes a great deal. For most inclination angles, the change is large. Since the linear polarization pulse can occur at any phase, using the value of $\dot{\alpha}$ at the linear polarization pulse to determine i and δ (cf. Meggitt and Wickramasinghe 1982) can produce inaccurate results. Instead, the variation of α itself over the widest possible phase interval should be used whenever possible.

3. RESULTS

We fit our theoretical curves for the linear polarization position angle α to the observations as follows. We define zero phase to be the time when we most nearly look down the field line passing through the optical emission region, and rescale the observations so that $\alpha = 0$ at this time. We then find values of δ, as a function of i, that are consistent with the observed variation of α with phase. The observed eclipse duration or, if the system does not eclipse, the constraint $i + \delta \simeq 90°$ gives a second relation between δ and i. The intersection of the two curves determines i and δ.

The uncertainties in the values of i and δ we derive are clearly dominated by systematic errors, such as those that may arise from approximating the magnetic field as a centered dipole, from approximating the dominant emission region by a point directly above the magnetic pole, from determining the eclipse length by the circular polarization zero crossings, and from variations in the observed behavior of α from one observational run to another. We have therefore attempted to estimate the uncertainties by allowing a liberal error for the eclipse length (generally ± 0.05 in phase) or the nearness of $i + \delta$ to $90°$ (generally $10°$), and by accepting fits (by eye) to the variation of the linear polarization position angle that are fairly extreme. The two constraints then lead to two allowed bands in the (δ, i)-plane, whose intersection is the region of allowed values of i and δ.

AM Her and AN UMa illustrate the application of these procedures to a self-eclipsing system and a non-self-eclipsing system, respectively. Figure 3 shows theoretical curves fitted by eye to observations (Tapia 1981) of the polarization position angle α for AM Her. The solid curve shows the best fit and the two dashed curves show the most extreme fits we have accepted, with the eclipse length $\Delta \phi$ fixed at 0.15 of the period. This eclipse length agrees with that derived from the circular polarization zero crossings when AM Her is in its low state (Latham, Liebert, and Steiner 1981; see also Stockman and Sargent 1979), and is consistent with the duration of the X-ray eclipse (Tuohy et $al.$ 1978, 1981). Figure 5 shows the overall constraints on i and δ. The curves labeled $\Delta \phi = 0.10, 0.15$, and 0.20 show the constraints from the eclipse duration, while the unlabeled curves show the constraints imposed by the observations of α. The shaded region shows the allowed values of i and δ.

Figure 4 shows theoretical curves fitted by eye to observations (Tapia 1981) of the

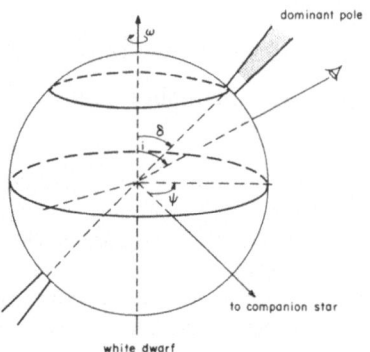

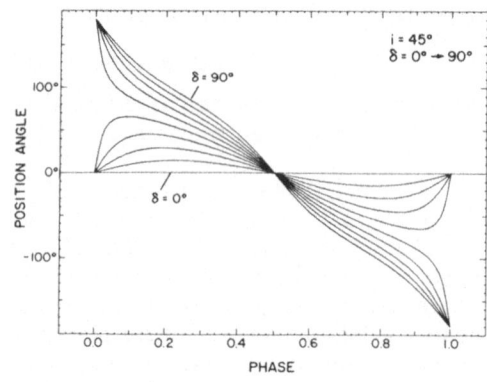

Fig. 1.–Definitions of the inclination angle i, the magnetic colatitude δ, and the magnetic longitude ψ. We assume that the rotation axes of the degenerate dwarf and that of the binary are parallel.

Fig. 2.–Variation of the linear polarization position angle with phase for $i = 45°$ and $\delta = 0° - 90°$ in intervals of $10°$. The behavior for $\delta = 90° - 180°$ is given by the same curves with the substitutions $\phi \rightarrow \phi + 0.5$ and $\delta \rightarrow 180° - \delta$.

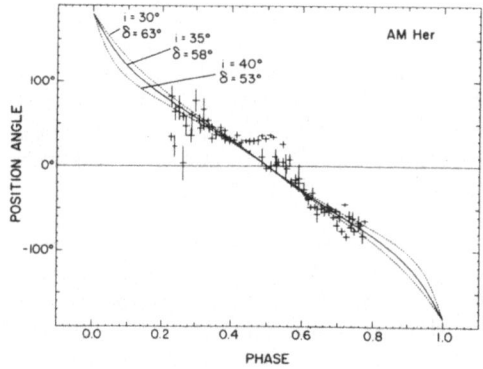

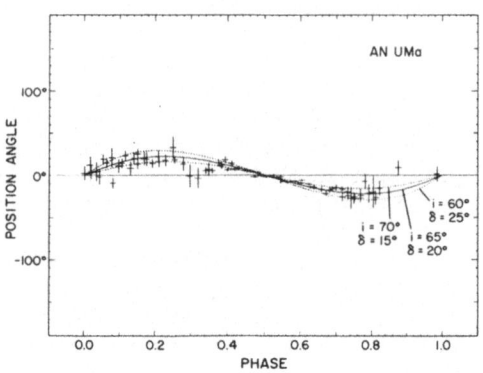

Fig. 3.–Comparison of the fitted theoretical curves and the observational data for the change in the linear polarization position angle with phase for AM Her. The eclipse length $\Delta\phi$ is fixed at 0.15 in all three fits.

Fig. 4.–Same as Figure 3, but for AN UMa. The sum $i+\delta$ is fixed at $85°$ in all three fits.

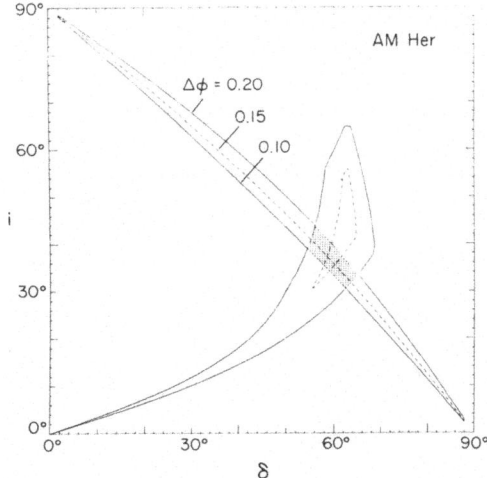

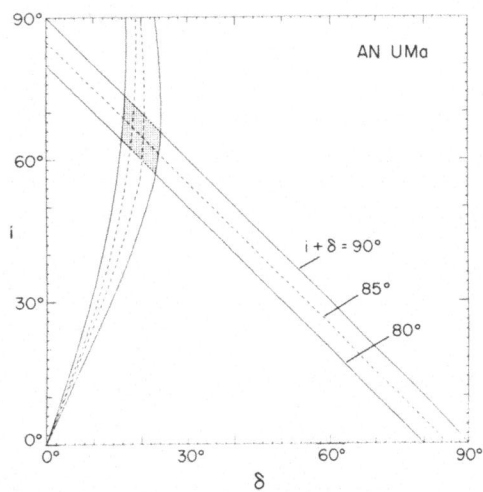

Fig. 5.–Constraints on i and δ for AM Her. The curves labeled $\Delta\phi = 0.10$, 0.15, and 0.20 show the region allowed by the eclipse length 0.15 ± 0.05, while the unlabeled curves show the region allowed by the observed change in the linear polarization position angle with phase.

Fig. 6.–Constraints on i and δ for AN UMa. The curves labeled $i + \delta = 80°, 85°$, and 90° show the region consistent with observation of a linear polarization pulse, while the unlabeled curves show the region allowed by the observed change in the linear polarization position angle with phase.

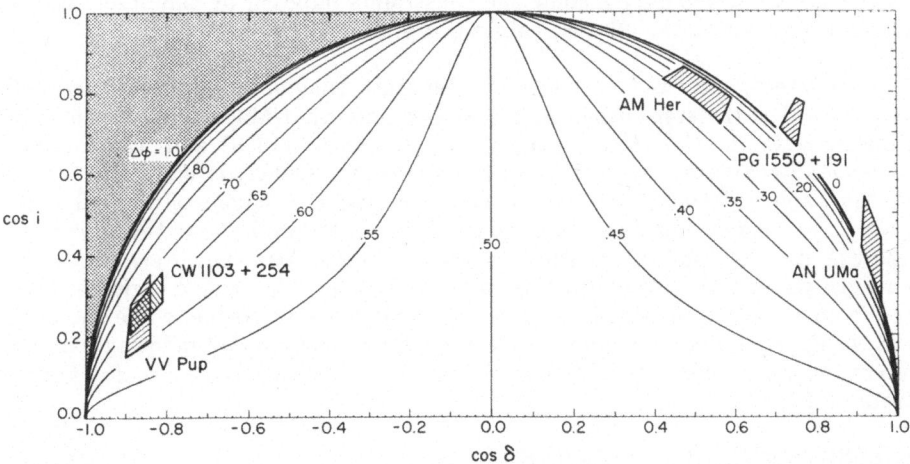

Fig. 7.–Distribution of AM Her stars in the $(\cos\delta, \cos i)$-plane. The shaded boxes show the regions of allowed values of i and δ. The curves show the locations of stars having eclipse lengths $\Delta\phi = 0-1$, in intervals of 0.05; note that throughout much of the plane, the eclipse duration is near 0.5. The shaded region at the upper left corresponds to stars in which the dominant pole is never visible; the analogous region at the upper right corresponds to stars in which the dominant pole never approaches the limb of the star. Note that all 5 AM Her stars lie on or near the semicircle $\cos^2\delta + \cos^2 i = 1$.

TABLE 1.   ORIENTATIONS OF AM HER STARS

Star	$\Delta\phi$	i	δ
AM Her	0.15 ± 0.05	$35° \pm 5°$	$58° \pm 5°$
VV Pup	0.64 ± 0.10	$75° \pm 5°$	$148° \pm 5°$
AN UMa	---	$65° \pm 5°$	$20° \pm 5°$
PG 1550+191	---	$45° \pm 5°$	$40° \pm 5°$
CW 1103+254	0.64 ± 0.05	$75° \pm 5°$	$148° \pm 5°$

linear polarization position angle α for AN UMa. AN UMa does not undergo a self-eclipse and there are therefore no circular polarization zero crossings (cf. Krzeminski and Serkowski 1977). But AN UMa does exhibit a linear polarization pulse (Tapia 1981). Such a pulse requires $i + \delta \geq 80°$, and probably $\geq 85°$. The solid curve in Figure 4 shows the best fit and the two dashed curves show the most extreme fits we have accepted, with $i + \delta$ fixed at 85°. Figure 6 shows the overall constraints on i and δ. The curves labeled $i + \delta = 80°$, 85°, and 90° show the constraints from observation of the linear polarization pulse, while the unlabeled curves show the constraints imposed by the observations of α. The shaded region shows the allowed values of i and δ.

We have determined i and δ for five AM Her stars using observations of their polarization properties. For the stars other than AM Her and AN UMa, we have primarily used the following data: for VV Pup, Liebert and Stockman (1979) and Tapia (1981); for CW1103+254, Stockman et al. (1983); for PG 1550+191, Liebert et al. (1983). The results are listed in Table 1. The ±5° errors quoted encompass the regions of allowed values of i and δ, which are smaller but have a complicated shape. Figure 7 shows the allowed regions for these five stars are plotted on the (cos i,cos δ)-plane. Also shown are curves of fixed eclipse length $\Delta\phi$, ranging from 0 to 1, in intervals of 0.05. The shaded region at the upper left corresponds to stars in which the dominant pole is never visible, while the analogous region at the upper right corresponds to stars in which the dominant pole never approaches the limb of the star; these possibilities have been noted previously by Bond, Chanmugam, and Grauer (1979).

4. DISCUSSION AND CONCLUSIONS

The AM Her stars should be uniformly distributed in the (cos δ, cos i)-plane, if the degenerate dwarf itself has no preferred orientation and there are no observational selection effects. Visual inspection of Figure 7 suggests that the AM Her stars are uniformly distributed in cos i but not in cos δ: all of the stars have $|\cos\delta| \gtrsim 0.4$. We have tested these hypotheses using the Kolmogorov-Smirnov test (see e.g., von Mises 1964). Ranked by cos i, the cumulative, normalized distribution of the 5 stars in Table 1 has a maximum deviation of 0.259 from a uniform distribution. The probability of such a deviation is 0.33 for 5 objects drawn randomly from a uniform distribution. However, ranked according to $\cos\delta$, the 5 stars have a maximum deviation of 0.566. Such a deviation has a probability of only 0.024. These tests confirm that the spin axes of the degenerate dwarfs are randomly oriented with respect to our line of sight, as far as present observations can tell, but that their magnetic moments tend to be aligned with their spin axes. This lack of orthogonal rotators may partly reflect the fact that the MHD torque thought to be responsible for phase-locking in the AM Her systems is proportional to the component of μ parallel to the spin axis of the degenerate dwarf, and hence is proportional to $\cos^2\delta$ (Lamb et $al.$ 1983). Other things being equal, aligned rotators will tend to become synchronized while orthogonal rotators will not. The former may tend to become AM Her stars and the latter, if they have been spun up by accretion, will tend to continue rotating rapidly and may look like DQ Her stars. However, the effect is not especially strong, and can be easily compensated for by a stronger magnetic field and/or a shorter binary period (i.e. a smaller separation).

Closer inspection of Figure 7 shows that all 5 AM Her stars have eclipse lengths near zero or one (see also Table 1), although the duration is near 0.5 throughout much of the ($\cos\delta$, $\cos i$)-plane. This suggests that the lack of AM Her stars with $|\cos\delta| \lesssim 0.4$ reflects only the tendency of the stars to lie on or near the semicircle $\cos^2\delta + \cos^2 i = 1$. To test this hypothesis, we define the independent variables r and θ, where r is the magnitude of the radius vector from ($\cos\delta$, $\cos i$) = (0,0) to a star and θ is the angle this vector makes with the $\cos\delta$-axis, omit the shaded region of Figure 7, and again apply the Kolmogorov-Smirnov test. Ranked by r, the 5 stars have a maximum deviation of 0.719 from a uniform distribution. The probability of such a deviation is only 0.0019. Ranked by θ, the maximum deviation is 0.319; the probability of such a deviation is 0.23. These tests confirm that the AM Her stars are indeed significantly concentrated on the semicircle $\cos^2\delta + \cos^2 i = 1$, but are not significantly clumped along it. (We note that, even if the region $|\cos\delta| \lesssim 0.4$ is excluded, the probability of having such a concentration in r is only 0.018; this result is further evidence that the concentration is not just due to a lack of orthogonal rotators among AM Her stars.) Such a non-uniform distribution implies that the discovery of AM Her stars is strongly affected by observational selection effects and has important implications for the search for more such stars.

We suggest that the non-uniform distribution of the known AM Her stars is a result of the way in which they have been identified observationally, and the angular properties of their X-ray and polarized optical emission. Consider first their identification. To date, AM Her candidates have been culled from lists of known cataclysmic variables or from lists of faint X-ray sources. Spectroscopic observations have then been used to screen these candidates for stars having strong emission lines of H, He I, and especially He II. Such an emission-line star is admitted to the AM Her class if follow-up polarimetric observations detect strong circular and/or linear polarization; otherwise it is not.

Consider now the properties of X-ray emission from a small polar cap, and high-harmonic cyclotron emission from above the cap. Both soft and hard X-rays are deflected sideways by scattering in the accretion column above the cap, and are emitted in a wide fan beam (Imamura and Durisen 1983). The effect is more pronounced for the hard X-rays than for the soft, since the escaping soft X-rays tend to originate more from the rim of the polar cap. Nevertheless, both soft and hard X-rays are more visible when the emission region is near the limb. High-harmonic cyclotron optical emission is strongly beamed perpendicular to the magnetic field (cf. Trubnikov 1958, Stockman 1977, Chanmugam and Dulk 1981, Meggitt and Wickramasinghe 1982). As a result, the polarized light from AM Her stars is emitted in a fan beam and strong circular polarization is visible only when the emission region is near the limb. Linearly polarized light is emitted in an extremely narrow fan beam, and a linear polarization pulse is visible only when the emission region is essentially on the limb.

With these facts in mind, let us consider again the various regions in Figure 7. The shaded region at the upper left corresponds to stars in which the dominant pole is never visible; X-ray emission as well as circularly and linearly polarized light optical light from the dominant pole is not visible. These stars can be identified as AM Her stars only if emission from the secondary pole is detectable. The analogous region at the upper right corresponds to stars in which the dominant pole is always visible, but never approaches the limb of the star. X-ray emission may well be weak, and the optical light is expected to exhibit little circular polarization and no linear polarization pulse. Most of the area within the semicircle $cos^2 \delta + cos^2 i = 1$ corresponds to stars in which the emission region is self-eclipsed for approximately half the time. During this time, X-rays and polarized optical light from the dominant pole cannot be seen. During much of the remaining time, the dominant pole is visible but far away from the limb. Again, X-ray emission may well be weak and the optical light is expected to exhibit only weak circular polarization and no linear polarization pulse. Only for perhaps 20 min, as the dominant pole crosses the limb of the star at eclipse ingress and egress, is strong circular polarization, and a narrow linear polarization pulse, expected. Thus AM Her stars in these three regions of Figure 7 are expected to exhibit strong optical polarization only for brief periods, if at all. Attempts to verify that they are AM Her stars through several short polarimetric observations could well fail.

Consider now the stars that lie along the semicircle $cos^2 \delta + cos^2 i = 1$. These stars are ones in which the dominant magnetic pole has an eclipse duration near zero or one and spends an inordinate fraction of time near the limb, so that strongly polarized light is almost always visible! Because the dominant and secondary magnetic poles are related by $(i, \delta) \rightarrow (i, 180° - \delta)$, the secondary pole also lies on the semicircle, has an eclipse duration near zero or one, and spends an inordinate fraction of time near the limb. Strongly polarized light is therefore also expected from the secondary pole, if it becomes active and contributes to the optical light. Such stars are easily confirmed to be members of the AM Her class, and indeed they make up the known sample of AM Her stars.

How many AM Her stars have been missed? The known AM Her stars lie roughly in a strip of width 0.15 about the semicircle $cos^2 \delta + cos^2 i = 1$ in Figure 7. Omitting both the upper left and right corners, this strip covers a fraction 0.36 of the (cos δ, cos i)-plane. We conclude that there are three times as many AM Her stars among cataclysmic variables as have been discovered so far. (If the nonuniform distribution of known AM Her stars is due to the combined effect of some process that excludes stars from the region $|cos \delta| \lesssim 0.4$ and of observational selection effects elsewhere, the total number of AM Her stars is only about double the number discovered so far; however, the fact that the known stars are not significantly clumped <u>along</u> the semicircle $cos^2 \delta + cos^2 i = 1$ is evidence against this possibility.) Unlike the polarized optical light, the strong He II line emission characteristic of AM Her stars is not narrowly beamed and is not eclipsed by the degenerate dwarf, since it originates in an extended region fairly far away from the degenerate dwarf. We therefore expect that the missing AM Her stars will show strong (equivalent width $\gtrsim$ 20 Å) He II line emission. This signature could· be very useful, since the number of known stars with strong He II line emission is only a few dozen (Patterson, private communication). Extensive polarimetric coverage may be needed to catch the missing AM Her stars during the brief periods of strong polarization at eclipse ingress and egress. We therefore encourage further polarimetric observations of binaries showing strong He II emission, preferably covering a considerable fraction of their orbits. Good candidates would appear to include E1013-477, V603 Aql, V425 Cas, V2051 Oph, V Sge, WZ Sge, and VW Vul.

What are the X-ray and optical characteristics expected for the 20 or so missing AM Her stars? We expect most of them to be faint X-ray sources, although the expected wide fan beam of the X-ray emission may make them more difficult to detect. (We note that the missing AM Her stars can be accomodated among the unidentified faint soft X-ray sources in the HEAO-1 catalogue, but just barely.) As already noted, we also expect them to exhibit strong emission lines, especially of He II. Some other characteristics can be deduced from our earlier discussion of the different regions in Figure 7. A few will be very difficult to identify, since only emission from the secondary pole will be visible. A comparable number will be weak X-ray sources and will show only weak circular polarization and no linear polarization pulse. The vast majority will have X-ray and optical eclipses lasting approximately half their period, and will emit strong circularly polarized light together with a very narrow linear polarization pulse during a brief period at eclipse ingress and egress.

Each AM Her star discovered so far has added in some important way to our understanding of these remarkable systems. But our analysis shows that, in one sense, all of these systems are the same: we view all of them from nearly the same direction. Discovery of the missing AM Her stars would not only add to their number, but would allow us to observe new aspects of their X-ray and, especially, polarization properties.

We are grateful to Santiago Tapia for providing linear polarization data on AM Her, AN UMa, and VV Pup in advance of publication. We thank Joe Patterson for lengthy conversations that helped us to refine our ideas. One of us (DQL) acknowledges support from the Fluid Research Fund of the Smithsonian Institution. This research was supported in part by NASA grants NAG-8310 and NAGW-246.

REFERENCES

Bond, H. E., Chanmugam, G., and Grauer, A. D.: 1979, *Ap. J. (Letters)* **234**, p. L114.
Brainerd, J. J., and Lamb, D. Q.: 1983, in preparation.
Chiappetti, L., Tanzi, E. G., and Treves, A.: 1980, *Space Sci. Rev.* **27**, p. 3.
Chanmugam, G., and Dulk, G.: 1981, *Ap. J.* **244**, p. 569.
Imamura, J. N., and Durisen, R. H.: 1983, *Ap. J.*, in press.
Krzeminski, W., and Serkowski, K.: 1977, *Ap. J. (Letters)* **216**, p. L45.
Lamb, D.Q.: 1981, in *X-Ray Astronomy in the 1980's*, ed. S. Holt (NASA TM-83848), p. 37.
Lamb, D.Q.: 1982, in *Cataclysmic Variables and Related Objects, Proceedings IAU Colloquium No. 72*, ed. M. Livio and G. Shaviv (Reidel, Dordrecht), p. 299.
Lamb, F. K., Aly, J.-J., Cook, M. C., and Lamb, D. Q. 1983, submitted to *Ap. J. (Letters)*.
Latham, D., Liebert, J., and Steiner, J.: 1981, *Ap. J.* **246**, p. 919.
Liebert, J., and Stockman, H. S.: 1979, *Ap. J.* **229**, p. 652.
Liebert, J., and Stockman, H.S.: 1985, this volume.

Liebert, J., Stockman, H. S., Angel, J. R. P., Woolf, N. J., and Hege, E. K.: 1978, *Ap. J.* **225**, p. 201.

Liebert, J., Stockman, H. S., Williams, R. E., Tapia, S., Green, R. F., Rautenkranz, D., Ferguson, D. H., and Szkody, P.: 1983, *Ap. J.* in press.

Meggitt, S. M. A., and Wickramasinghe, D. T.: 1982, *M.N.R.A.S.* **198**, p. 71.

von Mises, R.: 1964, *Mathematical Theory of Probability and Statistics* (Academic Press, New York), pp. 490-492.

Stockman, H. S.: 1977, *Ap. J. (Letters)* **218**, p. L57.

Stockman, H. S., Foltz, C. B., Schmidt, G. D. and Tapia, S.: 1983, *Ap. J.*, in press.

Stockman, H. S., and Sargent, T. A.: 1979, *Ap. J.* **227**, p. 197.

Tapia, S.: 1981, private communication.

Trubnikov, B. A.: 1958, Ph.D. thesis, Moscow Institute of Engineering and Physics, AEC-tr-4073 (Washington: Office of Technical Services).

Tuohy, I. R., Lamb, F. K., Garmire, G. P., and Mason, K. O.: 1978, *Ap. J. (Letters)* **226**, p. L17.

Tuohy, I. R., Mason, K. O., Garmire, G. P., and Lamb, F. K.: 1981, *Ap. J.* **245**, p. 183.

TIME DEPENDENT ACCRETION FLOWS IN THE AM Her SYSTEMS

Steven H. Langer
Departments of Physics and Astronomy, University of Illinois

We consider time dependent accretion onto white dwarfs with strong
magnetic fields. For certain conditions this flow is unstable and
leads to large periodic variations in the hard X-ray luminosity. If
the magnetic field is large enough, electron cyclotron radiation will
stabilize the flow. We also discuss the probable observational appearance of the oscillations found in our calculations.

I. INTRODUCTION

        The cataclysmic variables contain a white dwarf accreting matter
lost by its binary companion. An AM Her system occurs when the white
dwarf has a magnetic field so strong (B ~ 10^7 gauss) that the accretion flow is forced to follow the magnetic field lines to the polar
cap of the white dwarf. The AM Her systems do not contain the accretion disk that is seen in the other cataclysmic variable systems. Instead, almost all of the accretion energy is released near the surface
of the white dwarf at frequencies ranging from the optical to hard X-
rays. We investigate how the kinetic energy of the infalling matter is
thermalized, and then radiated, as the accretion flow is brought to a
halt at the surface of the white dwarf. In two earlier papers (Langer,
Chanmugam, and Shaviv 1981, hereafter Paper I, and 1982, hereafter Paper II) we reported numerical solutions to the time-dependent hydrodynamic equations for the accretion flow. The most striking result we
obtained was the discovery of a thermal instability in the accretion
flow that led to periodic oscillations in the hard X-ray luminosity of
the source. I will report on some of our recent results, and discuss
the expected observational appearance of these oscillations.

        Our results are more easily understood if we first summarize some
of the basic characteristics of the flow. The infalling matter is in
free-fall and is highly supersonic as it approaches the white dwarf.
At the surface, the velocity goes to zero and the temperature drops to
the photospheric temperature of the white dwarf. This behavior can occur only if there is a strong standoff shock in the accretion column.
The shock converts most of the kinetic energy of the infalling matter

D. Q. Lamb and J. Patterson (eds.), Cataclysmic Variables and Low-Mass X-Ray Binaries, 257–260.
© 1985 by D. Reidel Publishing Company.

into heat. The height of this shock is such that the hot postshock gas
cools to the surface temperature in the time required to flow from the
shock to the surface. An important property of the gas is thus the
radiative cooling time just below the shock. The cooling time is of
order one second, and is proportional to the square root of the post-
shock temperature divided by the density. The postshock velocity is
roughly 10^8 cm/sec, and the resulting shock height is of order 10^8 cm.
This discussion is based on the assumption of a static flow. If the
shock moves upward, the postshock temperature increases, which results
in a longer cooling time. When the shock moves downward, both the
postshock temperature and the cooling time are less than for the static solution. If the shock is moving downward as it crosses the equilibrium position, the cooling time is less than the flow time to the
surface, and the shock continues to move downward. If the shock is
moving upward when it crosses the equilibrium height, the gas does not
have time to cool before it reaches the surface, and the shock continues to move upward. The question of whether these motions damp out,
or attain some large and steady amplitude, is one of the points addressed by our models.

II. RESULTS

In papers I and II the only radiative cooling included was free-
free radiation. In the hot (T ~ 10^8 K), ionized postshock gas, two additional cooling processes that may be important are Compton cooling,
which occurs when photons gain energy in scattering off hot electrons,
and electron cyclotron emission, which is important when the magnetic
field is large. We have now included cyclotron emission in an approximate fashion (see the discussion in Paper II), and find that it can
strongly damp the oscillations if the magnetic field is large enough.
Figure 1 compares the dependence of the shock height on time for two
models that are identical, except for the magnetic field. The model
with no magnetic field exhibits undamped oscillations, as reported
previously, but in the model with B=10^7 G the oscillations are strongly damped. The cyclotron losses depend on the first power of the density, while the free-free losses depend on the square of the density.
As a consequence, cyclotron emission is most important for low accretion rates, where the density is low, and for high magnetic fields. We
will summarize the exact conditions under which cyclotron emission can
stabilize the flow in a future paper.

Another way to study the stability of the flow is to use a radiative cooling law which is proportional to the square of the density
and to an arbitrary power of the temperature, α. Chevalier and Imamura
(1982) have carried out a linear stability analysis of the steady-
state flow solution. They find that for $\alpha > 0.4$ the flow is stable to
perturbations in the fundamental mode (the mode with a frequency close
to that found in our calculations). There are also modes with higher
frequencies that are unstable for $\alpha < 0.8$. In Paper I we reported that
the oscillations damped out for $\alpha > 1.6$. However, numerical considerations led us to cut off the radiative cooling at temperatures below

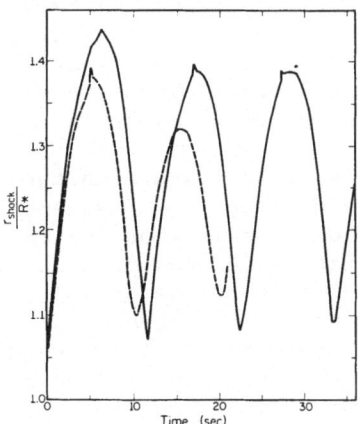

Fig. 1.   The shock radius is plotted as a function of time for a model
in which $M_* = 0.5\ M_\odot$, $R_* = 10^9$ cm, $\dot{M} = 2.5 \times 10^{15}$ g s^{-1}, and
the polar car area is 10^{16} cm^2. The solid curve is for $B = 0$,
the dashed line for $B = 10^7$ G.

2×10^7 K, resulting in the rapid accumulation of a layer of matter at
2×10^7 K sitting on top of the white dwarf. The presence of this layer
prevented a clean determination of the amplitude of the oscillation,
and thus made it difficult to determine when damping occurred. We have
now repeated this calculation with the revised model presented in Paper II, which did not accumulate a layer of hot gas, and find that for
$\alpha = 0.6$ the oscillations clearly tend to damp out, while for $\alpha = 0.5$ there
is no evidence for damping after 5 oscillation periods. This suggests
that the α for which stability occurs is somewhere between these two
values. Imamura, Wolf, and Durisen (1983) have also solved the hydrodynamic equations numerically, and find that the oscillations in the
fundamental mode tend to damp out for α somewhere between 0.33 and
0.5. They find that oscillations at the higher eigenfrequencies persist until α exceeds, roughly 0.8 (our models did not have fine enough
spatial resolution or enough time steps to see oscillations in these
higher modes). The disagreement between the two numerical models is
small enough that it may be the result of the slightly different physical approximations made in the two calculations. Further study of the
stability of the flow and the amplitude of the oscillation in the different modes is certainly justified.

A recent observational result obtained by Middleditch (1982)
found a broad bump in the power spectrum of E1405−451 and AN UMa at a
period of roughly 1.7 seconds. Similar features were not seen in
AM Her, and in AN UMa in its low state. It is tempting to identify
this pulsation with the thermal instability present in our models, but
we considered only the hard X-ray emission. A periodic variation in
the optical luminosity could arise when hard X-rays are absorbed in

cooler regions of the accretion column above the shock and re-radiated as optical radiation. The cyclotron emission below the shock might also contribute to the optical radiation, while still being too weak to stabilize the flow. Future models should directly calculate the optical radiation, but it seems plausible that there would be some variability in the optical at the same period as in the hard X-rays.

The observations find a broad feature in the power spectrum, instead of a peak at a single well defined frequency. This is not surprising for two reasons. The first is that the magnetic field of the white dwarf is so strong that matter can move only along field lines, and any interaction between blobs of gas on adjacent field lines is strongly suppressed. Consequently, the accretion column is likely to be composed of several independently oscillating flux tubes. The second point is that the matter lost by the binary companion is unlikely to become attached to the magnetic field lines of the white dwarf in such a fashion as to have a constant accretion rate per unit area when it reaches the surface of the white dwarf. The period of the oscillation depends upon the mass flux, so there will be a spread of frequencies in the various flux tubes filling the accretion column. The observed radiation will consist of the sum of the emission from several different flux tubes oscillating with a range of frequencies and random relative phases. An additional effect will arise in connection with the flickering on time scales of roughly a hundred seconds that is present in the AM Her systems. We interpret this flickering as a variation in the accretion rate which in turn will disrupt the oscillations previously occurring in the various flux tubes, and re-excite flux tubes where the previous oscillations had died out. The resulting light curve will contain a range of frequencies, instead of being strictly periodic.

In summary, our numerical models of the accretion flow onto a white dwarf with a strong magnetic field show strong oscillations in the hard X-ray luminosity under certain circumstances. If the magnetic field is much greater than 10^7 G, the oscillations tend to damp out. The oscillations in the hard X-rays can plausibly lead to similar oscillations in the optical luminosity, although at a lower amplitude. These oscillations may correspond to the feature in the power spectrum of the optical light of two of the AM Her systems reported by Middleditch. Further study of the accretion flow in the AM Her systems will lead to models that can be directly compared to observations in an effort to determine some of the parameters of the system.

REFERENCES

Chevalier, R. A., and Imamura, J. N. 1982, Ap. J. 261, 543.
Imamura, J. N., Wolff, M. T., and Durisen, R. H. 1983, preprint.
Langer, S. H., Chanmugam, G., and Shaviv, G. 1981, Ap. J. (Letters) 245, L23.
Langer, S. H., Chanmugam, G., and Shaviv, G. 1982, Ap. J. 258, 289.
Middleditch, J. 1982, Ap. J. (Letters) 257, L71.

STABILITY OF RADIATIVE SHOCK WAVES

J.N. Imamura, R.A. Chevalier
Department of Astronomy, University of Virginia

R.H. Durisen, M.T. Wolff
Department of Astronomy, Indiana University

1. INTRODUCTION

Radially accreting degenerate dwarfs are thought to produce hard X-rays in radiative shocks formed when the accreting plasma strikes their surfaces. Recent nonlinear time-dependent numerical calculations of funneled accretion onto magnetic degenerate dwarfs found that such shocks are unstable under certain circumstances (Langer, Chanmugam, and Shaviv 1982a, hereafter LCS). LCS found that radiative shocks with power law cooling functions proportional to $\rho^2 T^\alpha$ oscillate in shock thickness x_s and luminosity L_s with a period of $3t_{cool}$ whenever $\alpha < 1.6$. Here ρ is the mass density, T is the temperature, and t_{cool} is the postshock plasma cooling time scale. Because bremsstrahlung emission ($\alpha = 0.5$) may be the dominant loss process for many accreting degenerate dwarfs, LCS suggested that pulsed X-ray emission might be a common property of these sources. In this talk, we review our study of this instability. We first present results of our own linear and nonlinear analyses. We then compare our results to those of LCS and discuss the observational implications of our work. Lastly, we summarize our major conclusions.

2. RESULTS

We have performed linear stability analyses of planar radiative shocks using cooling functions of the form $\rho^\beta T^\alpha$ for both $\beta = 1$ and 2 (for $\beta = 2$, see Chevalier and Imamura 1982, hereafter CI). We find that the shocks can oscillate in several modes, which we call, in order of increasing oscillation frequency, the Fundamental (F), First Overtone (1O), Second Overtone (2O), and so on. For $\beta = 2$, the shocks are unstable to linear perturbations in the F, 1O, and 2O for $\alpha < 0.4$, 0.8, and 0.8, respectively, while for $\beta = 1$, the respective limits are 0.1, 0.1, and 0.0. These results suggest that, for fixed β, increasing α tends to stabilize the shocks, and that, for fixed α, increasing β tends to destabilize the shocks. For bremsstrahlung dominated shocks, i.e., $\beta = 2$ and $\alpha = 0.5$, the F is stable while the 1O and 2O are unstable.

D. Q. Lamb and J. Patterson (eds.), Cataclysmic Variables and Low-Mass X-Ray Binaries, 261–265.
© *1985 by D. Reidel Publishing Company.*

The F, 10, and 20 have oscillation periods of $3t_{cool}$, $1.1t_{cool}$, and $0.6t_{cool}$, respectively. In the linear regime, stable modes are exponentially damped, while unstable modes are overstable and grow exponentially.

We also performed nonlinear, time-dependent numerical calculations of spherical radiative shocks produced by accretion onto a 1 $M_{\odot}$ non-magnetic degenerate dwarf (Imamura, Wolff, and Durisen, 1983, hereafter IWD). We used the following sets of assumptions. <u>Case 1.</u> We considered power law cooling functions proportional to $\rho^2 T^{\alpha}$, equal ion and electron temperatures and no electron thermal conduction. <u>Case 2.</u> We considered additional physics relevant to accretion onto dwarfs of mass $M_{\star} \gtrsim 1\ M_{\odot}$, such as Compton cooling, relativistic and quadrupole corrections to the bremsstrahlung cooling rate, unequal ion and electron temperatures and nonzero electron thermal conduction. For both cases, we considered shocks where the shock thickness (x_s) was much less than the stellar radius ($R_{\star}$). Such shocks are effectively planar shocks as in CI.

Our Case 1 nonlinear results are in quantitative agreement with our linear analysis. They yield the same critical α values for stability and have oscillation frequencies within ± 10% of the linear values for the F and 10 (see Figure 1 for plots of $x_s(t)$ and $L_s(t)$ for $\alpha = 1/3$, 1/2, and 1). The stable modes damp in an exponential manner with damping times which agree to within a factor of two with the linear analysis. Our Case 2 results show that for accretion onto dwarfs of $M_{\star} \gtrsim 1\ M_{\odot}$, all oscillation modes are stable.

3.   COMPARISON WITH PREVIOUS WORK

The IWD results quantitatively agree with the linear analysis of CI, but disagree with the nonlinear analysis of LCS. The IWD calculations exhibit overtone oscillations, and give instability of the F only for $\alpha < 0.4$, while LCS find instability for $\alpha < 1.6$ and detect the F only. It is possible that LCS do not see oscillations in the overtones because the overtones saturate at amplitudes negligible compared to the very large amplitudes LCS find in the F. Explaining the discrepancy between the behaviors of the F is more difficult, however. A crucial difference between the IWD and LCS calculations is the existence of nonlinear pressure waves which are generated near the dwarf's surface in the IWD calculation. Such waves do not form in the LCS calculations. IWD believe that these waves control the nonlinear behavior of both damped and undamped modes of oscillation. Because the waves are generated near the dwarf, the handling of the inner boundary may play a crucial role in their formation. Neither the LCS nor IWD calculations correctly treats the physics in this region. LCS represent the dwarf by an isothermal, hydrostatic atmosphere while IWD use a rigid wall. Thus the LCS calculations have a slightly "mushy" inner boundary while the IWD calculations do not. The linear analysis of CI also uses a

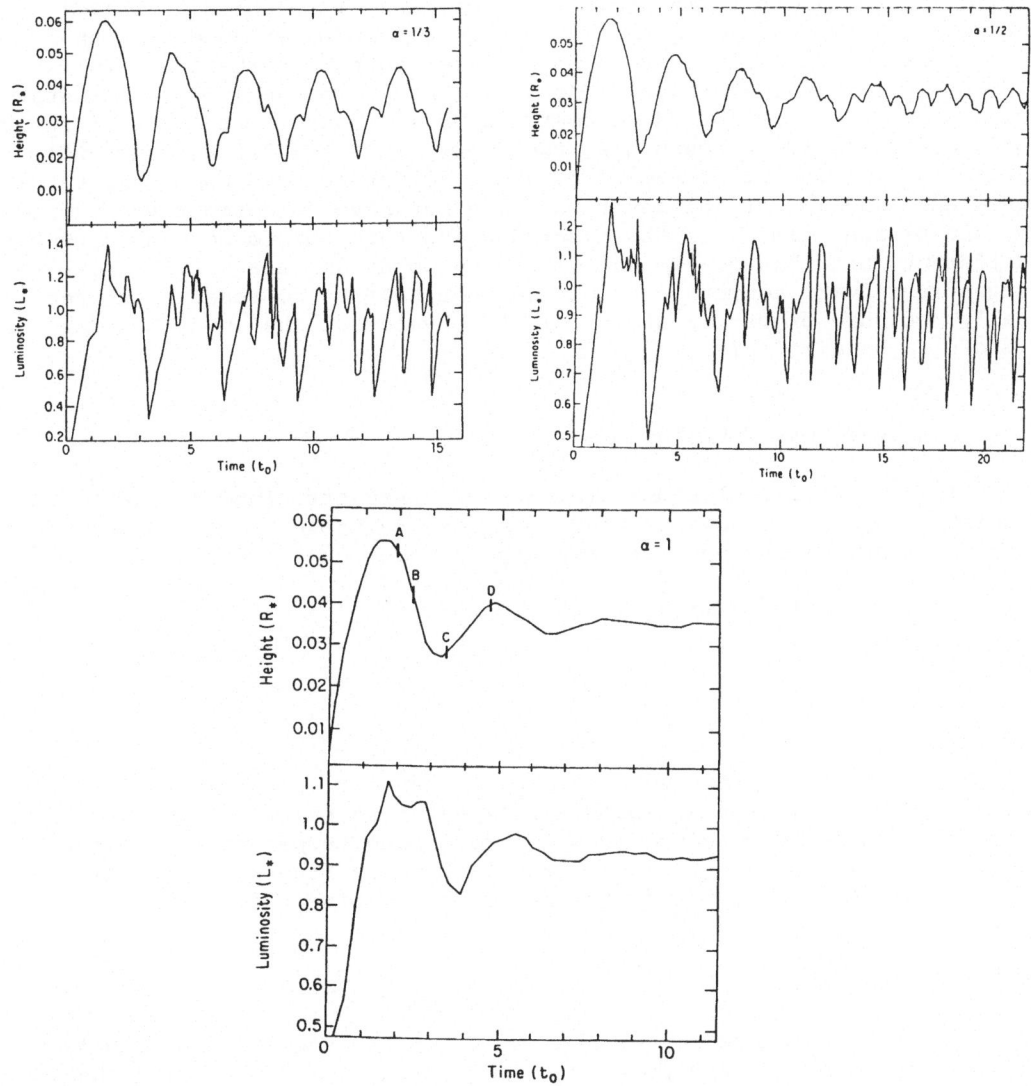

Figure 1.   The shock height and luminosity evolutions of β = 2 and
α = 1/3, 1/2 and 1 shocks.   The shock heights are normalized in units
of the stellar radius, the luminosities are normalized in units of
$GM_{\star}\dot{M}/R_{\star}$ and the times are normalized in units of $2\pi x_{s}/v_{in}$.   The times
$2\pi x_{s}/v_{in}$ are 0.89, 0.86, and 0.98 t_{cool}, for α = 1/3, 1/2, and 1, respectively.   In the α = 1/3 evolution, an oscillation in the F is
readily apparent as well as a lower amplitude oscillation in the 1O.   In
the α = 1/2 evolution, the F damps, giving way to the 1O at large times.
In the α = 1 evolution there are no hints of the overtones.   The evolution is dominated by the damping F.

rigid inner boundary. In addition, there are differences in the
numerical techniques of each group. LCS use an Eulerian grid with a
"numerical filter" to suppress short wavelength perturbations, while
IWD use a Lagrangian grid with no filtering. Because IWD use a Lagrangian grid, they are forced to continually add zones to the top of their
grid and to coalesce zones near the dwarf. LCS and IWD are currently
investigating questions raised by these differences in numerical techniques and handling of the inner boundary but have no definitive answers
at the present time. If the discrepancies turn out to be caused by the
different inner boundary conditions, the physical behavior for real
degenerate dwarf accretion shocks may not be easy to resolve theoretically, because proper treatment of the physics at the inner boundary is
likely to be difficult.

4. OBSERVATIONAL IMPLICATIONS

If the IWD calculations are correct, bremsstrahlung dominated
radiative shocks for dwarfs of $M_* \lesssim M_\odot$ are unstable only to overtone
oscillations. The lowest order unstable modes have periods of

$$t_{10} = 0.013 (\dot{M}_E/\dot{M}) (M_*/M_\odot) \text{ s}$$

and

$$t_{20} = 0.6 \, t_{10}.$$

Shocks for $M_* \gtrsim 1 \, M_\odot$ are stable to perturbations in all modes. Shocks
dominated by cyclotron emission are expected to be stable because of
the weak density dependence ($\beta < 1$) and strong temperature dependence
($\alpha > 2$) of cyclotron emission. Calculations by Langer, Chanmugam, and
Shaviv (1982b) for cyclotron dominated shocks give the same result.

5. CONCLUSIONS

We have performed linear and nonlinear analyses of the stability
of radiative shocks with power law cooling functions. We find that our
linear and nonlinear results are in quantitative agreement and that our
nonlinear results disagree with the nonlinear calculations of LCS. At
the present time, we have no definitive explanation for the disagreement
between the nonlinear calculations, but we suspect that it may be due to
differences in numerical techniques or to the handling of the inner
boundary condition. Both calculations find instabilities in bremsstrahlung dominated shock structures for $M_* \lesssim 1 \, M_\odot$, but differ in the
oscillation modes which they find to be unstable. We find that for
$M_* \gtrsim 1 \, M_\odot$ additional physics omitted by LCS stabilizes all modes. Both
groups agree that cyclotron dominated shocks should be stable.

REFERENCES

Chevalier, R.A. and Imamura, J.N. 1982, Ap. J., 261, p. 543.
Imamura, J.N., Wolff, M.T., and Durisen, R.H. 1983, Ap. J., submitted.
Langer, S., Chanmugam, G., and Shaviv, G. 1982a, Ap. J., 258, p. 289.
_____. 1982b, published in
    IAU Coll. 72: "Cataclysmic Variables and Related Objects," eds.
    M. Livio and G. Shaviv (Reidel).

GENERAL DISCUSSION ON SHOCK WAVE INSTABILITIES

JENSEN: In the Middleditch observations the pulsed fraction of the rapid oscillations is quite low, only a few percent. It may be that other stars just have a lower pulsed fraction. Also, the pulsed fraction of the X-rays in E1405-45 (not simultaneous with the optical) gives an upper limit of 45%.

D. LAMB: I had a remark and also a question. The remark is that the Middleditch observations are in the optical, so some caution is in order, because what is being described here are oscillations in the X-ray luminosity. My question is: If you have this picture of so many separate flux tubes, how do we know that there are going to be any oscillations at all? It seems that you could have a very large number of flux tubes, and with random phases, the whole phenomenon will wash out.

LANGER: Indeed, you have to get some sort of a calculation of the area of one of those individual flux tubes, and thereby get an idea of how many you have. You're quite right, the optical radiation that's correlated with this oscillation must come from reprocessing of those X-rays someplace in the system, or reprocessing of cyclotron emission from the hot post-shock gas, which would also vary but is presumably at a lower level because bremsstrahlung is dominating the cooling.

KROLIK: How does the post-shock pressure compare to the magnetic field pressure? Can you really maintain those independent flux tubes?

LANGER: The estimates indicate that it's not a particular difficulty if the fields are $\geq 10^7$ gauss, as expected. If the field is much smaller, as perhaps would be the case in some of the DQ Her stars, where otherwise you would expect this oscillation still to occur, you might find that you are running out of magnetic field pressure. We saw something like a magnetic field of 10^5 gauss suggested earlier today. It's not clear that's adequate.

D. LAMB: Just to clarify a point. If you insist on field-aligned flow, you cannot get into a situation where the gas pressure is comparable to the magnetic pressure. Such a flow is hydromagnetically unstable and could not arise in the first place.

PRIEDHORSKY: It would seem profitable for somebody to go back to the HEAO-1 pointings--soft X-ray data with 80 millisecond time resolution and several hundred counts a second--and look specifically for the sort of phenomena that are being theorized here.

IMAMURA: We tried it with the A-1 data.

PRIEDHORSKY: Does that have adequate count rates in hard X-rays?

IMAMURA: I think so. It's marginal, though. We haven't looked at it very closely.

267

D. Q. Lamb and J. Patterson (eds.), Cataclysmic Variables and Low-Mass X-Ray Binaries, 267–268.
© *1985 by D. Reidel Publishing Company.*

KATZ: Didn't I read a preprint by one of your groups that included conduction in these calculations and found it strongly stabilizing?

[Ed: Langer and Imamura look at each other blankly.   Deafening laughter.]

IMAMURA: We've never run a case where conduction was very important in the emission regions.   We could look at that.

CHANMUGAM: King and Lasota wrote some things.

LANGER: King and Lasota have a model where they manage to haul almost all of the accretion energy downward by conduction.

KATZ: I know that, and I don't believe it.   But I could swear, because it was this preprint that dissuaded me from setting a student on the problem, that someone did a stability analysis like yours including conduction, and I thought it was you.

IMAMURA: I've heard of that, but I never saw it.   Two or three years ago?

KATZ: No, very recently; I got it maybe two months ago.

JENSEN: In the Middleditch observation of E1405-45, the pulsed fraction is pretty stable; it didn't seem to change around the orbit.   Every section of data he analyzed seemed to be pretty much the same, so any model that requires some kind of orbital inclination effect might be in trouble.

INTERMEDIATE POLARS

Brian Warner

Department of Astronomy, University of Cape Town, Rondebosch, 7700. South Africa.

INTRODUCTION

The identification (Charles *et al.*, 1979) of the hard X-ray source 2A0526-328 with a previously unknown star of the cataclysmic type, having a spectrum almost identical to those of AM Her and EF Eri but lacking detectable circular polarisation, marked the start of an exciting new development in studies of cataclysmic binary systems. Motch (1981) found two photometric periods in TV Col (2A0526-328): 5^h11^m5 and 4^d042. The latter turned out to be a beat period with the spectroscopic period of 5^h29^m2 (Hutchings *et al.*, 1981), and suggested, by analogy with models developed for DQ Her, that TV Col contains an asynchronously rotating degenerate star from which a beam of high energy radiation emerges, illuminating gas distributed within the binary system.

This model was strongly supported when the next such star to be identified, AO Psc (H2252-035), was found to possess an orbital period of 3^h35^m5, an X-ray period of 13.4 min and the related beat period of 14.3 min (White and Marshall 1980, 1981; Patterson and Price 1980; Warner *et al.*, 1981).

In the past two years three more similar systems have been discovered as new identifications of hard X-ray sources (H2215-086, 4U1849-31 and 3A0729+103) and a further system, EX Hya, long known as a hard X-ray source and well-studied optically as a cataclysmic variable, has been recognised as belonging to the same general class.

The structural similarity of these systems to those of the rapidly rotating (periods 28-71 secs) DQ Her stars (comprising WZ Sge, DQ Her, V533 Her and AE Aqr: Patterson 1979) suggests that they may be grouped in the same class (Patterson and Steiner 1983). Alternatively, the longer period objects, intermediate in

269

D. Q. Lamb and J. Patterson (eds.), Cataclysmic Variables and Low-Mass X-Ray Binaries, 269–279.
© *1985 by D. Reidel Publishing Company.*

asynchronism between the phase-locked AM Her stars (also known as Polars: Kruszewski 1978) and the rapidly rotating DQ Her stars, may also be described as Interlopers (Patterson and Price 1981) or Intermediate Polars (Warner 1982). The distinction between Intermediate Polars and DQ Her stars may fade if objects are found whose periods lie between them. For the present review we will retain the term Intermediate Polars, noting that, like the Polars, their special characters have in general been identified as a result of their hard X-ray emission and they have rotation periods and optical modulations an order of magnitude larger than those of the DQ Her stars. If there is a continuum of sources filling the gap between the two groups, we must be prepared to seek objects having low amplitude optical modulations with periods in the range 2-5 mins. Many photometric surveys, and probably all spectroscopic surveys, have been insensitive to the detection of such objects so it is possible that some of the already known cataclysmic variables are members of this group.

THE INTERMEDIATE POLARS

The six systems currently recognised as Intermediate Polars are listed in Table I.
We will briefly review the present state of knowledge for each of these objects.

TABLE I

Variable Star	X-ray	$<m_v>$	Orbital Period	Other Period(s)
TV Col	2A0526-328	13.5	$5^h29^m.2$	$5^h11^m.5$ ($4^d.0$)
	H2215-086	13.5	4 01.5	20.9 ($19^m.2$)
AO Psc	H2252-035	13.3	3 35.5	14.3 (13.4)
V1223 Sgr	4U1849-31	13.2	3 21.9	13.2 (14.2)
	3A0729+103	14.5	3 14.2	15.2
EX Hya	2A1251-29	13.0	1 38.3	67.0

TV Col

The optical spectrum closely resembles those of nova remnants or Polars (Hutchings *et al.*, 1981, Watts *et al.*, 1982). Large variations in line strength and profile occur, not obviously synchronised with the spectroscopic period. Total line widths of up to 3000 km s^{-1} occur, suggesting at least transient formation of an accretion disc close to the surface of the degenerate component. On the other hand, the rapid variations in line profile are similar to those seen in the Polars (Schneider and Young 1980a,b) and imply extensive and unstable gas streaming.
In the far ultraviolet, IUE observations (Coe and Wickramasinghe 1981, Mouchet *et al.*, 1981) show emission lines, which are

more similar to those of SS Cyg in outburst (Heap *et al.*, 1978),
with P-Cyg structures in the CIV and SiIV lines, than those of
the Polars.

Coe and Wickramasinghe (1981) deduced $E_{B-V} \lesssim 0.04$ from absence of the 2200Å absorption feature, but Mouchet (1982) does
detect a small amount of interstellar absorption, amounting to
$E_{B-V} \simeq 0.06$. At the galactic latitude of TV Col this implies a
distance 700 pc and an absolute magnitude $M_V \sim 4.3$. This is
similar to the luminosity of dwarf novae during outburst (Warner
1976) which, together with the IUE spectra, is consistent with a
well-established accretion disc.

The continuum in the far ultraviolet is well represented by
a standard accretion disc model with T = 200 000K (Mouchet 1982).
This, however, leads to an unreasonably large distance and luminosity for TV Col. The two-component model of Mouchet *et al.*
(1981), comprising a 9000 K blackbody and an optically thick ac-
retion disc, fits both the ultraviolet continuum and UBV measurements satisfactorily and reduces the estimated distance.

The X-ray spectrum in the range 2-10 keV is interpreted as
thermal bremsstrahlung with a temperature of $\sim$20 keV (Watts *et al.*,
1982). The absence of soft X-rays, or of a Flux $\propto \nu$ component at
the shortward end of the IUE measurements, shows that no strong
FUV component is present of the type detected in AM Her and 2A0311-
227 (Fabbiano *et al.*, 1981).

In the model of TV Col proposed by Hutchings *et al.*, (1981)
the spectroscopic period is the orbital period, the short photometric period arises from rotation of the magnetic degenerate
star and the 4^d00 period is caused by the beam from the primary
sweeping across the secondary. Watts *et al.* (1982), however, associate the 4-day period with orbital motion and assign the spectroscopic period to the primary's rotation. On several grounds this
model appears less probable: (i) it is necessary for the degenerate star to rotate retrogradely, (ii) the large orbital period
requires an evolved giant to fill the secondary's Roche Lobe: by
analogy with other cataclysmic variables the spectrum of such a
secondary would be expected to dominate in the red region, whereas
no secondary has yet been detected spectroscopically or through
infrared photometry.

Adopting the model by Hutchings *et al.* (1981) and their ephe-
merides for the three periodicities, we find that at the minimum
of the 4^d00 modulation (i.e. when the beam from the primary is
pointing directly away from the secondary) maxima of the short
(5^h11^m5) photometric period occur when the secondary is at superior conjunction. Interpretation of this is left for the later
section on a general model for the Intermediate Polars.

In summary, there is strong evidence for a well developed
disc in TV Col, with added gas streaming activity as shown in the
variations of line profile. There is no clear evidence for a hot
spot on the rim of the disc, but this may be a result of the low
inclination of the system. The origin of the flickering in hard

X-rays (Watts *et al.*, 1982) and in the optical region (Warner 1980) may be in the disc rather than a hot spot.

AO Psc

This is the most intensively studied of the Intermediate Polars; we will review here only the more important observations that are relevant to the structure of the system.

Time-resolved spectroscopy (Warner *et al.*, 1981, Wickramasinghe *et al.*, 1982, Cordova *et al.*, 1983) shows complex emission line profile variations, with double structure at some phases and occasional broad Hβ absorption. Wickramasinghe *et al.* find $K_1 \sim$ 60 km s^{-1} and an S-wave distortion with $K \sim$ 350 km s^{-1}. The combination of these accounts for the value $K_1 \sim$ 140 km s^{-1} found by Patterson and Price (1981). Total emission line widths of up to 3000 km s^{-1} are observed. All of these results point to an extensive accretion disc in AO Psc.

The flux distribution from 1200 to 2200Å is well fitted by a two-component spectrum comprising α = 2.74 power law and T = 13400K black body (Mouchet 1982, Cordova *et al.*, 1983). 2200Å interstellar absorption implies E_{B-V} = 0.10, giving a distance $\sim$400 pc and $M_V \sim$ 5.0.

The X-ray flux distribution is well represented by thermal bremsstrahlung with T > 20keV (White and Marshall 1981). There is no evidence for a strong soft X-ray or FUV component.

In optical photometry the dominant variation is an 859 sec periodic oscillation, accompanied by an 805 sec oscillation of small and varying amplitude and a 3^h35^m5 orbital modulation (Warner *et al.*, 1981, Patterson and Price 1981, Motch and Pakull 1981, Wickramasinghe *et al.*, 1982). It is clear that the 859 sec periodic optical component results from reprocessing of the 805 sec X-ray beam (White and Marshall 1980, Patterson and Garcia 1980), emitted by the rotating primary, from a structure at rest in the rotating frame of the binary system. Patterson and Price (1981) suggested that the reprocessing site is the atmosphere of the secondary star, but Hassall *et al.* (1981) rejected this on theoretical grounds and proposed the inflated hot spot region instead. Wickramasinghe *et al.* (1982), however, show that the objections of Hassall *et al.*, concerning the low amplitude of the orbitally modulated component, are overcome when the luminosity of the accretion disc and its shadowing effects on the secondary are taken into account. Motch and Pakull (1981) find (i) that the 805 and 859 sec pulsations are in phase coherence near minimum of the orbitally modulated light curve, (ii) the spectral gradient of the orbitally modulated component and the 859 sec component are very similar, whereas the 805 sec component is much bluer, and (iii) the flux distributions of the first two sources agree well with that calculated for an X-ray illuminated atmosphere.

The 805 sec X-ray and optical pulses are coincident in phase; the latter has a spectral distribution consistent with the Rayleigh-Jeans tail of a hot blackbody (Motch and Pakull 1981).

A possible source is the X-ray heated surface of the primary immediately below the accretion column.

From these observations we conclude that the 805 sec component
probably arises directly from the surface of the primary, whereas
the 859 sec and orbitally modulated components are from reprocessed radiation. Whether the reprocessing site is the atmosphere of
the secondary or the region of the hot spot is not decided: the
phase coherence at minimum of the orbital modulation merely requires that the same site gives rise to both reprocessed components.
The coincidence of minimum light with inferior conjunction of the
line emitting region (Patterson and Price 1981) does not necessarily imply that the secondary is at superior conjunction at this
time: if a significant amount of line emission arises in the vicinity of the hot spot this can distort the radial velocity curve
in the direction of causing the apparent moment of superior conjunction to be actually that of the hot spot.

V1223 Sgr

Spectroscopically, V1223 Sgr is similar to AO Psc (Steiner
et al., 1981) and total line widths of $\sim$4000 km s^{-1} are seen. The
ultraviolet and visible flux distribution is well fitted by a disc
spectrum (Mouchet 1982, Bonnet-Bidaud *et al.*, 1982) and reddening
of $E_{B-V} \sim 0.15$ is present. This amount of reddening requires a
distance $\sim$600 pc and implies $M_V \sim 3.7$.

Optical pulsations with a period of 794 secs were discovered
by Steiner *et al.* (1981). Preliminary results from observations
made in 1981 gave an orbital period of $3^h22^m.8$ (Warner 1982). A
complete discussion of the observations made in 1981-82 is now
available (Warner and Cropper, 1983) with the following conclusions:
(a) The 794 sec oscillation is coherent with no significant
     change in period over the one year baseline.
(b) The orbital period, derived from modulation of the brightness
     of the system, is $3^h21^m.9$.
(c) Occasional low amplitude coherent oscillations are seen at
     a period of 850 secs. This is the orbital side band of the
     794 sec period.
The 850 and 794 sec oscillations are in phase coherence at
the maximum of the orbitally modulated light curve (in contrast
to AO Psc described above).

3A0729+103

This star is the most recent to have been identified as an
Intermediate Polar (McHardy *et al.*, 1982). Detailed analysis is
under way. The 1982 optical photometry shows the dominant presence of a 913.5 sec periodicity and a $3^h14^m.2$ orbital modulation,
but no orbital sideband has yet been detected.

H2215-086

Spectroscopic and photometric observations of H2215-086
(Patterson and Steiner 1981, 1983; Shafter and Targon 1982)

_how that this star is very similar to AO Psc, V1223 Sgr and
3A0729+103. Pakull (Patterson and Steiner 1983) reports a low
amplitude orbital sideband at 19.2 min to the dominant optical
periodicity at 20.9 min. H2215-086 is therefore similar to AO Psc.

EX Hya

EX Hya possesses both soft X-rays (Cordova and Riegler 1979)
and hard X-rays (Watson *et al.* 1978). From spectra and eclipses
the orbital period is known to be 98.3 mins, but the dominant optical modulation is at 67.0 mins (Vogt *et al.* 1980). Both the
soft X-rays and emission line intensities are modulated with the
67.0 min period and both show a weak modulation at 33.5 mins (Kru-
zewski *et al.* 1982, Gilliland 1982) implying the visibility of two
actively accreting magnetic poles. The higher energy X-rays are
not significantly modulated (Swank 1981).

Unlike the other Intermediate Polars, observations of EX Hya
extend over a long enough baseline to show that the primary's rotation period is decreasing on a timescale of 2.8×10^6 yrs (Gilliland 1982) which is satisfactorily accounted for by the accretion
torque acting on the primary if it is a white dwarf (Warner 1982,
Lamb and Patterson 1982).

The optical, infrared and X-ray 67 min modulations are all
in phase (Kruszewski *et al.*, 1982). The existence of a hot spot
and extensive accretion disc is established from the presence of
an S-wave and total line widths of 7000 km sec^{-1} (Cowley *et al.*,
1981). Eclipses were thought to be the hot spot, itself periodically illuminated by a beam from the primary (Warner and McGraw
1981), but the X-ray periodicity requires that the (probably grazing) eclipses are actually of the central disc region (Warner 1982).

AN ILLUMINATING MODEL

The striking contrast in behaviour between AO Psc and V1223
Sgr, in which, in the former, the orbital sideband is the dominant optical periodicity and this is in phase coherence with the
beam from the primary when the secondary (or the hot spot) is at
inferior conjunction, whereas in the latter the relative amplitudes of primary and sideband are interchanged and phase coherence
occurs at superior conjunction of the secondary, suggests the
following model.

We will adopt, for illustration, a model in which the reprocessing site is the atmosphere of the secondary. The X-radiation
that sweeps across the secondary could be restricted in altitude
either by a beaming effect or because of the location of the X-ray
emitting spot on the primary. Thus in AO Psc (Figure 1a) the beam
is so located that little of it falls on the disc but the secondary
is well-illuminated. Alternatively, if the radiation is not beamed, then the spot must be situated at such a high latitude that
it is invisible from the disc but just visible from the secondary;
then the variation of illumination of the secondary is a consequence

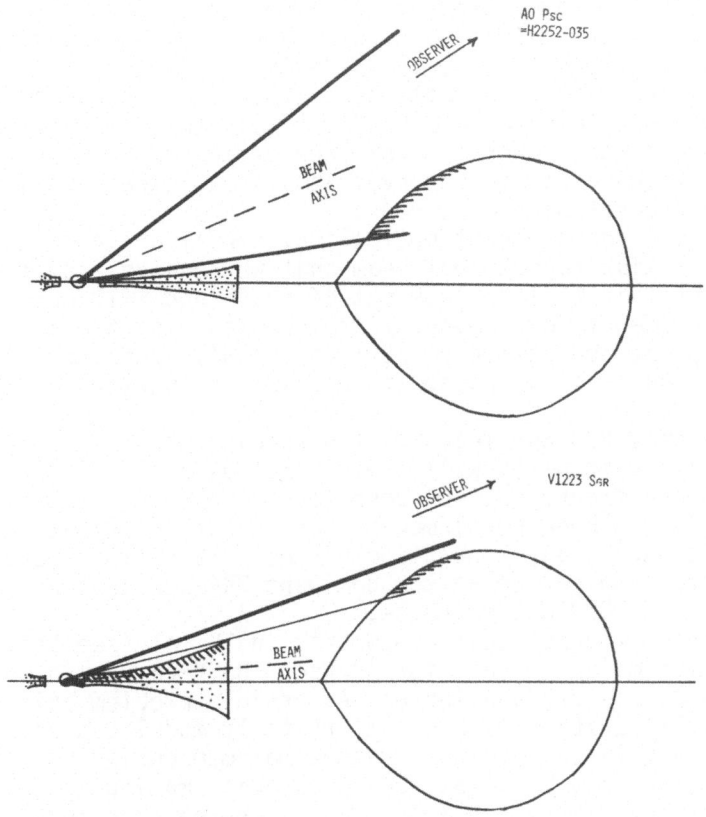

only of the varying aspect ·of the spot as seen at the secondary.
This geometry is so improbable (and is required again for H2215-
086) that it is preferable to describe the model in terms of a
beam.

The two short period oscillations that we see in AO Psc arise
from a direct beam from the surface of the white dwarf and reprocessed radiation from the secondary. These two will be in phase
coherence when the secondary is at inferior conjunction. As the
amplitude of the 859 sec reprocessed component is not significantly
modulated around the orbit (see Fig. 1 of Warner *et al.*, 1981) we
evidently view the system at quite low inclination. The different
spectral gradients observed by Motch and Pakull for the 805 and
859 sec components result from the radically different conditions
in the two emitting sites.

In V1223 Sgr, on the other hand, we suppose that the beam
axis is inclined $\sim 90^\circ$ to the rotation axis of the primary (Fig.
2b). The inner face and edge of the disc are then strongly

illuminated but only a small (and highly variable if there are changes in disc thickness) amount of illumination reaches the secondary. The disc is axially symmetric as seen from the primary, but there can be a strong front-back asymmetry as seen by us if (i) the inclination is high so that the back of the concave disc projects greater area than the near side or (ii) the dominant reprocessing occurs at the inner (Alvén) face of the disc, where the part that is closer to the observer is always (except at i = 0°) obscured by the nearside of the disc.

For V1223 Sgr, the dominant optical pulse will arise from reprocessing in the disc (any direct beam that we receive from the surface of the primary will be 180° out of phase and will merely reduce the amplitude of the component reprocessed from the disc). The components processed respectively from the disc and the secondary will obviously be in phase at superior conjunction of the latter.

This model of V1223 Sgr leads to two predictions. Firstly, all three periodic components will have a shallow spectral gradient (in contrast to the steep gradient seen for the 805 sec component in AO Psc). This has been confirmed by van Paradijs (private communication). Secondly, any modulation of the X-rays from V1223 Sgr will have the same period as the dominant 794 sec optical component, but will be 180° out of phase.

The phasing of the periodic components in TV Col (see above) requires a model similar to that for AO Psc. No X-ray period or spectral gradients are yet available. A possible problem arises from Motch's (1981) statement that the orbitally modulated brightness (detected only in the U band) reaches maximum during spectroscopic phases (with zero phase at radial velocity maximum as observed by Hutchings *et al.*, 1981) 0.47-0.67, whereas the model requires maximum at phase 0.75.

In H2215-086, the appropriate model is again similar to that for AO Psc. However, it should be noted that if the light curve shown in the upper part of Fig. 3 of Patterson and Steiner (1983) is typical, then there is almost no constant component of illumination of the secondary. Also, the strong orbital modulation of the amplitude of the 21 min oscillation (resulting from changing aspect of the surface of the secondary) implies a higher inclination of the orbital axis than is the case for AO Psc.

The EX Hya model is also similar to that for AO Psc. In EX Hya, however, there is no certain detection of any orbital modulation which would result from constant illumination of the secondary. Furthermore, if the 67 min rotating beam succeeds in illuminating the secondary, an orbital sideband at 3^h31^m would result. This is not obvious in the light curve published by Warner and McGraw (1981) but should be looked for in a Fourier analysis of the extensive data described by Sterken *et al.* (1983).

DISCUSSION

The beamed radiation model is successful in relating the otherwise unconnected observations of phase coherence, relative amplitude of orbital sideband and spectral gradient. For illustrative reasons we have used a fairly narrow beam, but it is now time to examine the reality or necessity of such a confined stream of radiation.

It is evident from the X-ray pulses in AO Psc (White and Marshall 1981) and EX Hya (Kruszewski *et al.*, 1982) and from the optical pulse shapes in TV Col (Motch 1981), AO Psc (Warner *et al.*, 1981), H2215-086 (Patterson and Steiner 1983), and V1223 Sgr and 3A0729+103 (Warner 1982), that the beam is not strongly confined in the azimuthal (primary rotation) direction. With the exception of the 2-5keV X-rays in AO Psc (White and Marshall 1981) there is no evidence for eclipse of the radiating source (as seen by us or by the reprocessing site) by disappearance behind the primary as it rotates. This argues that for all of these systems either the source has great extent in longitude around the primary, and there is no restriction in longitude of the radiation from this source, or the beam is from a small source but the reprocessing site is extended in longitude (e.g. a combination of the secondary, the hot spot and/or a ridge extending around the edge of the disc near the hot spot), or a combination of both. Note, however, that the times of phase coherence set limitations on the degree of asymmetry that the reprocessing site may have. These limitations may be such that an extended source is the only solution. The information contained in the optical and X-ray pulse shapes has still to be extracted; the opportunity afforded by the Intermediate Polars, in contrast to the phase-locked Polars, of determining the cross-section of the emitted beam has still to be taken advantage of.

In its simplest form, the model as illustrated in Figure 1 requires the beam to be restricted in altitude (perpendicular to the orbital plane). If there were no restriction, and the magnetic axes were randomly inclined to the rotation axes, then we would expect the disc in general to be well-illuminated whereas in fact we find only one such case (V1223 Sgr) but several in which the disc reprocessing is not readily detected. Some relaxation of the requirement of beaming is obtained if we allow what we have termed the direct beam from the primary to be in fact the net effect of the direct beam and reprocessing from the disc. That is, the dominant pulse comes from the primary, but there is a reprocessed pulse 180° out of phase but with lower amplitude. This is the situation in the model developed by Petterson (1980) for UX UMa. Variations in the structure of the disc would lead to a varying amplitude of the reprocessed component which could account, for example, for the varying amplitude of the 805 sec optical modulation in AO Psc. If such is the case, then changes of 180° in phase of the 805 sec modulation could arise according to which component dominates. The spectral gradient of this modulation would also vary according to the relative fractions of direct and reprocessed radiation.

ACKNOWLEDGEMENTS

The writer acknowledges the assistance of M.S. Cropper in reduction of the V1223 Sgr observations and is grateful to Drs. D.Q. Lamb and J. Patterson for fruitful discussions.

REFERENCES

Bonnet-Bidaud, J.M., Mouchet, M. and Motch, C.: 1982, *Astr. Astrophys. 112*, 355.
Charles, P.A., Thorstensen, J., Bowyer, S. and Middleditch, J.: 1979, *Ap. J. Letts. 231*, L131.
Coe, M.J. and Wickramasinghe, D.T.: 1981, *Nature, 290*, 119.
Cordova, F.A. and Riegler, G.R.: 1979, *M.N.R.A.S. 188*, 103.
Cordova, F.A., Fenimore, E.E., Middleditch, J. and Mason, K.D.: 1983. Preprint.
Cowley, A.P., Hutchings, J.B. and Crampton, D.: 1981, *Ap. J. 246*, 489.
Fabbiano, G., Hartmann, L., Raymond, J., Steiner, J., Branduardi-Raymont, G. and Matilsky, T.: 1981, *Ap. J., 243*, 911.
Gilliland, R.L.: 1982, *Ap. J., 258*, 576.
Hassall, B.J.M., Pringle, J.E., Ward, M.J., Whelan, J.A.J., Mayo, S.K., Echevarria, J., Jones, D.H.P., Wallis, R.E., Allen, D.A. and Hyland, A.R.: 1981, *M.N.R.A.S. 197*, 275.
Heap, S.R. *et al.*: 1978, *Nature, 275*, 385.
Hutchings, J.B., Crampton, D., Cowley, A.P., Thorstensen, J. and Charles, P.A.: 1981, *Ap. J. 249*, 680.
Kruszewski, A., Mewe, R., Heise, J., Chlebowski, T., van Dijk, W. and Bakker, R.: 1982, *Binary and Multiple Stars as Traces of Stellar Evolution*, ed. Z. Kopal and J. Rahe, Reidel, p. 457.
Lamb, D.Q. and Patterson, J.: 1983, *I.A.U. Colloq. No. 72*. In Livio and Shaviv (eds.), p. 229.
McHardy, I.M., Pye, J.P., Fairall, A.P., Warner, B., Allen, S., Cropper, M.S. and Ward, M.: 1982, *I.A.U. Circ. No. 3687*.
Motch, C.: 1981, *Astr. Astrophys. 100*, 277.
Motch, C. and Pakull, M.W.: 1981, *Astr. Astrophys. 101*, L9.
Mouchet, M.: 1983, *I.A.U. Colloq. No. 72*. p. 173.
Mouchet, M., Bonnet-Bidaud, J.M., Ilovaisky, S.A. and Chevalier, C.: 1981, *Astron. Astrophys. 102*, 31.
Patterson, J.: 1979, *Ap. J. 234*, 978.
Patterson, J. and Garcia, N.: 1980, *I.A.U. Circ. No. 3514*.
Patterson, J. and Price, C.: 1980, *I.A.U. Circ. No. 3511*.
Patterson, J. and Price, C.: 1981, *Ap. J. Letts. 243*, L83.
Patterson, J. and Steiner, J.E.: 1983, *Ap. J. Letts. 264*, L61.
Patterson, J. and Steiner, J.E.: 1981, *B.A.A.S. 13*, 817.

Petterson, J.A.: 1980, *Ap. J. 241*, 247.
Schneider, D.P. and Young, P.: 1980a, *Ap. J. 238*, 946.
Schneider, D.P. and Young, P.: 1980b, *Ap. J. 240*, 871.
Shafter, A.W. and Targon, D.M.: 1982, *A.J. 87*, 655.
Steiner, J.E., Schwartz, D.A., Jablonski, F.J., Busko, I.C.,
     Watson, M.G., Pye, J.P. and McHardy, I.M.: 1981, *Ap. J.
     Letts. 249*, L21.
Sterken, C., Vogt, N., Freeth, R.F., Kennedy, H.D., Marino, B.F.,
     Page, A.A. and Walker. W.S.G.: 1983, *I.A.U. Colloq. No. 72.*
     p. 51.
Swank, J.: 1981, Reported at Santa Cruz Workshop on Cataclysmic
     Variable Stars.
Vogt, N., Krzeminski, W. and Sterken, C.: 1980, *Astr. Astrophys.
     85*, 106.
Warner, B.: 1976, *I.A.U. Symp. No. 73*, p. 85.
Warner, B.: 1980, *M.N.R.A.S. 190*, 69P.
Warner, B.: 1983, *I.A.U. Colloq. No. 72.* p. 155.
Warner, B. and McGraw, J.T.: 1981, *M.N.R.A.S. 196*, 59P.
Warner, B., O'Donoghue, D. and Fairall, A.P.: 1981, *M.N.R.A.S.
     196*, 705.
Warner, B. and Cropper, M.S.: 1983. In preparation.
Watson, M.G., Sherrington, M.R. and Jameson, R.F.: 1978, *M.N.R.A.S.
     184*, 79P.
Watts, D.J., Greenhill, J.G., Hill, P.W. and Thomas, R.M.: 1982,
     M.N.R.A.S. 200, 1039.
White, N.E. and Marshall, F.E.: 1980, *I.A.U. Circ. No. 3514.*
White, N.E. and Marshall, F.E.: 1981, *Ap. J. Letts. 249*, L25.
Wickramasinghe, D.T., Stobie, R.S. and Bessell, M.S.: 1982,
     M.N.R.A.S. 200, 605.

GENERAL DISCUSSION ON DQ HER STARS

JENSEN: Anticipating what I was going to say tomorrow about TT Arietis, there is some controversy as to whether or not it is really a polar. It has a very low ratio of X-ray to visual flux, orders of magnitude lower than the other intermediate polars; and if it is an intermediate polar, somebody's going to have to explain that.

WARNER: Yes. Of course, TT Arietis has a very different spectrum. It has a very well-established absorption spectrum of an optically thick disk, so it isn't typical of the other ones.

JENSEN: Also, it's much fainter now, which means the field will be dominating the system. We should be looking to see whether it's a magnetic field dominated system now, before it gets brighter.

KATZ: More of a comment than a question. I'm impressed at how many of these objects seem to be close to synchronism, as opposed to, say, DQ Her, which is very far from it. Maybe we will be able to observe the processes by which synchronism is achieved, or broken.

PATTERSON: I would like to sound a note of skepticism about the distinction between the long period and short period systems. I can say for myself that when I was having fun finding these things at McDonald Observatory, I was highly biased towards the short period systems. I now work at observatories with second-rate photometric capabilities, and I'm now highly biased toward long periods, where I can accumulate enough signal-to-noise to actually see the thing pulse by pulse in the light curve. If you want to play statistics with saying that there is some fundamental distinction, you ought, at the very least, to remove every star that I have found from that list. Then play statistics with the remainder.

PACZYNSKI: When they are in the low state, you don't see any reflection effect on the secondary, or do you?

WARNER: Individual polars have not been studied long enough. None of them has gone into a low state in the last couple of years since they were found from X-rays. Looking at the archival material, at least three of them have been through minima, but, of course, there are no observations to say anything about reflection effects.

PACZYNSKI: If for some reason they were burning hydrogen at the surface, this couldn't switch on and off on such a short timescale. So if you were switching off the accretion rate, the nuclear burning should be still going on, and the reflection effect should still be there. So if these are those objects I was asking about, they should show reflection effects even at minimum light.

WARNER: We don't know at the moment. They haven't gone down.

D. Q. Lamb and J. Patterson (eds.), Cataclysmic Variables and Low-Mass X-Ray Binaries, 281–282.
© *1985 by D. Reidel Publishing Company.*

THORSTENSEN: Phil Charles and I have some old data which languished for years on an illegible tape, which we finally read. For about a two-hour string, H2252-035 showed no oscillations at all in the visible. So they do change. I don't know what its photometric state was then.

WADE: Are these oscillations accurately sinusoidal, or if not, what can you say about the profile of the beam? And can anybody tell me why there should be beats?

WARNER: They're pretty accurately sinusoidal. I can't see any second harmonic in the power spectrum, but that's not very sensitive because of noise in the power spectrum.

STOCKMAN: For anyone interested, we looked at TT Ari when it was in a low state. The polarization was low, 0.2% or less.

JENSEN: In the infrared?

STOCKMAN: Optical.

A COMPARISON OF H2252-035/AO PSC AND 4U1849-31/V1223 SGR

J. van Paradijs[1], S. van Amerongen[1], M. de Kool[1],
M. Pakull[2] and H. van der Woerd[1]

[1]Astronomical Institute "Anton Pannekoek"
University of Amsterdam

[2]Institute for Astronomy and Astrophysics
Technical University of Berlin

ABSTRACT

From five-colour photometry of the triple-periodic systems H2252-035 and V1223 Sgr, which have very similar orbital and pulsation periods, we find a significant difference in the wavelength dependence of their short period pulsations.Our results indicate that in H2252-035 this pulse originates in a hot region, probably on the white dwarf (Motch and Pakull, 1981), whereas in V1223 Sgr both pulses are the result of reprocessing of X-rays (in the far side of the accretion disk, and in the companion star, respectively), as suggested by Warner (1985).

INTRODUCTION

H2252-035/AO Psc and 4U1849-31/V1223 Sgr are DQ Her type systems, i.e. close-binaries, consisting of a low-mass star undergoing mass loss through Roche-lobe overflow and a white dwarf, whose magnetic field is sufficiently strong to disturb the inner parts of the accretion disk and channel the inflowing matter onto its magnetic pole(s). (For a review of DQ Her systems see Warner, this volume). Unlike in the AM Her systems the white dwarf does not co-rotate with the orbital motion.
X-ray pulsations (P$\sim$805 sec) were discovered in H2252-035 by White and Marshall (1981). Optical variability with periods of $\sim$859 sec and $\sim$3.6 hours was discovered by Patterson and Price (1981). The frequency difference between the 805-sec period and the 859-sec period equals the orbital frequency, which indicates that the 859-sec pulsation is the result of reprocessing of X-rays in matter which co-rotates with the orbital motion. Possible sites for the reprocessing are the companion star (Patterson and Price, 1981) or a bulge in the outer parts of the disk (Hassall et al. 1981). The 805-sec period was subsequently detected in optical brightness variations by Warner et al. (1981).
The source V1223 Sgr is remarkably similar to H2252-035. Optical brightness modulations with a period of $\sim$795 sec were discovered by Steiner et al. (1981). An orbital period of $\sim$3.3 hours and a beat period of

D. Q. Lamb and J. Patterson (eds.), Cataclysmic Variables and Low-Mass X-Ray Binaries, 283–286.
© *1985 by D. Reidel Publishing Company.*

850 sec have been reported by Warner (1985).
In this paper we report on photometric observations of these two sources
and make a comparison of their properties. More detailed discussions of
our results will be presented elsewhere.

OBSERVATIONS

H2252-035 was observed with the Walraven photometer at the 90-cm
Dutch telescope at ESO during 12 nights between July 30 and September 5,
1981, and 3 nights in October, 1982. Using the same instrument we observed
V1223 Sgr on 9 nights between April 16 and May 29, 1982. The Walraven
photometer provides simultaneous brightness measurements in five passbands
between 3250 and 5500 Å. All measurements were made through a 16"
diaphragm, and integration times of 16 seconds were used. Two comparison
stars and sky background were measured at intervals of 30 minutes. The
raw data were reduced to sky-corrected ratios of brightness of the
variable star to one of the comparison stars. A total of ~4500 such data
points (five colours each) were obtained on both sources.

PERIOD DETERMINATION

We made period searches by folding the data at a series of period
values and minimizing the variance with respect to the average light curve.
By combining the epochs of maximum light in the three light curves of
H2252-035, as obtained from the 1981 and 1982 data, and values given by
Motch and Pakull (1981) and Patterson and Price (1981) we find the
following periods: P_{805}= 805.2060 ± 0.0007 s, P_{859}= 858.6862 ± 0.0012 s,

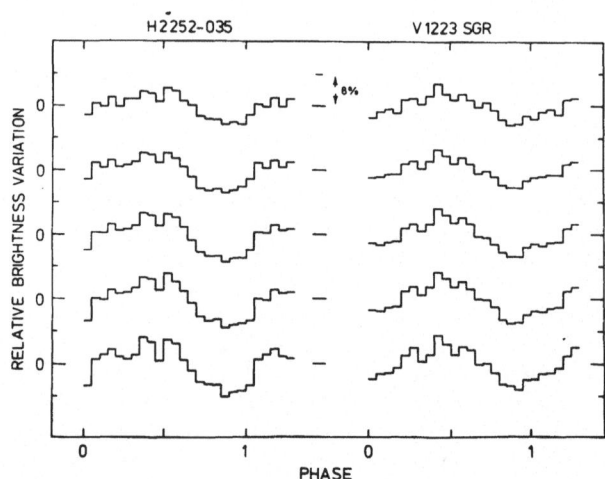

Fig. 1a   *Percentage orbital brightness variation (relative to average
brightness) for H2252-035 and V1223 Sgr. The light curves in
the V, B, L, U and W channels are shown from top to bottom.
The zero point of the phases has been chosen arbitrarily.*

and P_{orb} = 12927.80 ±0.17 s. The ephemeris for the 859-sec pulsation is
consistent with the epoch of maximum light that can be inferred from the
information given by Warner et al. (1981). The ephemeris for the 805-sec
pulsation agrees with Patterson's (pr. comm.) result that the optical
and X-ray pulsations at 805 sec are in phase. From the deviations of the
observed times of maximum light with respect to the predicted values a
limit $\dot{P}$ < 1.0E-10 can be placed on the variation of the rotation period
of the compact object in H2252-035. It is therefore unlikely that it is
a neutron star, a possibility suggested by White and Marshall (1981).

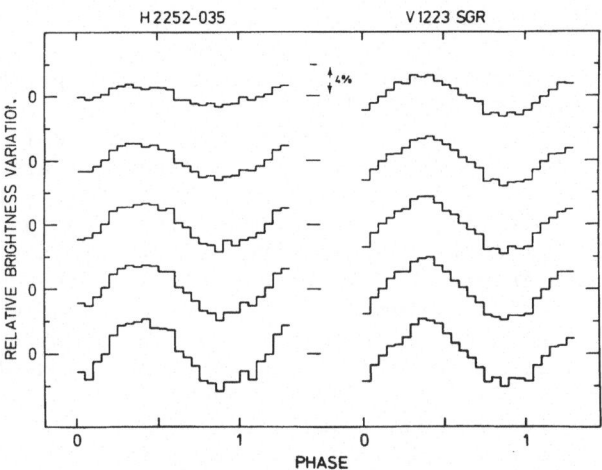

*Fig. 1b   Same as Fig. 1a, for the short-period pulsations of H2252-035
and V1223 Sgr.*

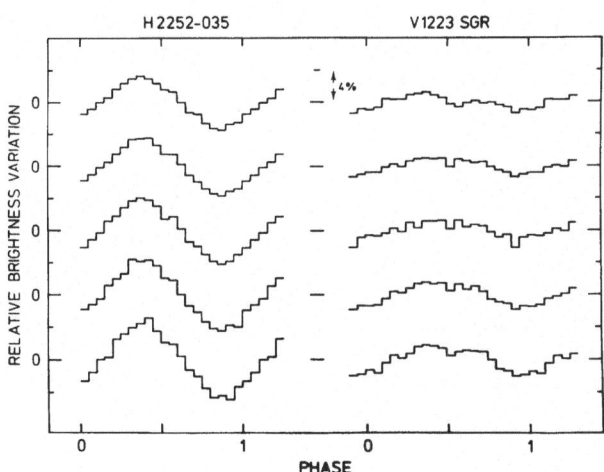

*Fig. 1c   Same as Fig. 1a, for the long-period pulsations of H2252-035
and V1223 Sgr.*

Our data on V1223 Sgr do not allow an unambiguous determination of the
orbital period, due to uncertainty in the cycle count per day. The data
of Warner (1983) resolve this ambiguity. Using his information we derive
for V1223 Sgr periods equal to 794.381 ±0.008 s, 850.150 ±0.009 s and
12113.28 ±1.73 s. Using the ephemeris of Steiner et al. (1981) we can
refine the first period to 794.3802 ±0.0013 s.

LIGHT CURVES

The average light curves of H2252-035 and V1223 Sgr are shown in
Fig. 1a-c. The amplitudes of all light curves increase toward shorter
wavelengths. In H2252-035 the amplitude of the long-period pulse is
larger than that of the short-period pulse. In V1223 Sgr the opposite is
the case. The two pulsations of H2252-035 are in phase at orbital minimum
light, confirming a result first found by Motch and Pakull (1981). This
is expected if the short pulse comes from near the surface of the white
dwarf and the long pulse is due to reprocessing of X-rays in either a
companion star or a bulge at the edge of the disk. The phasing of the two
pulsations in V1223 Sgr shows the opposite behaviour; they are in phase
near orbital maximum. As proposed by Warner (1983) this can be understood
if the short pulse in this system does not come from the white dwarf, but
is due to reprocessing in the disk regions on the far side of the white
dwarf. Support for this interpretation comes from the wavelength dependence
of the pulsation amplitudes in H2252-035 and V1223 Sgr. The short-period
pulse in H2252-035 (steady signal subtracted) has a spectral shape which
is well described by a Rayleigh-Jeans spectrum, indicating that it
originates in a hot region. It is attractive to identify this region with
a relatively small area on the surface of the white dwarf. Such a region
has been inferred from the blackbody component in soft X-rays observed
from e.g. AM Her (Rothschild et al. 1981). The wavelength variation of
the amplitudes of the long pulse in H2252-035 and both pulses in V1223 Sgr
interpreted as temperature variations of a blackbody, indicate average
temperatures near 18,000 K for H2252-035 and < 20,000 K for V1223 Sgr.
This is consistent with the values inferred from the spectral shape of
the average steady emission components.

REFERENCES

Hassall, B.J.M., Pringle, J.E. et al. 1981, M.N.R.A.S. 197,275.
Motch, C. and Pakull, M.W. 1981, Astron. Aph. 101, L9.
Patterson, J. and Price, C.M. 1981, Ap. J. (Letters) 243, L83.
Rothschild, R.E., Gruber, D.E. et al. 1981, Ap. J. 250, 723.
Steiner, J.E., Schwartz, D.A. et al. 1981, Ap. J. (Letters) 249, L21.
Warner, B., O Donoghue, D. and Fairall, A.D. 1981, M.N.R.A.S. 196, 705.
Warner, B. 1985, this volume.
White, N.E. and Marshall, F.E. 1981, Ap. J. (Letters) 249, L25.

ULTRA-VIOLET RADIATION FROM DWARF NOVA DISCS

B. J. M. Hassall
Institute of Astronomy, Cambridge, England

ABSTRACT

The discs in dwarf novae are the major source of ultra-violet light. The UV observations are reviewed and the continuum is compared with several steady disc models, at quiescence and outburst. The results are discussed in terms of the mass accretion rates, and the time-dependent behaviour of the continuum during the rise is presented.

INTRODUCTION

Dwarf novae are interacting binaries in which the orbital period can range between approximately 1.5 and 15 hours. Although in the upper end of this range, the secondary star can often make a significant contribution to the infra-red or even the optical flux, it is of negligible importance in the UV. Instead the accretion disc about the white dwarf primary dominates in the UV part of the spectrum. This is especially true at outburst when the disc is bright, although other sources, such as the hot spot, secondary star, boundary layer, emission line region do contribute to the overall radiation. The ultra-violet wavelength range has nonetheless been important in the study of dwarf nova discs.

During the review, we will consider briefly the following aspects:-
(1) The ultra-violet observations available.
     (These are mainly from IUE, together with some ANS data.)
(2) Comparison with steady disc models.
(3) The evolution of the spectral continuum during outburst.
(4) Ultra-violet line features in quiescence and outburst.

1. OBSERVATIONS

Table 1 contains 30 dwarf novae which have been observed with IUE, and whose spectra have been published, or were taken by the Cambridge group.

D. Q. Lamb and J. Patterson (eds.), Cataclysmic Variables and Low-Mass X-Ray Binaries, 287–297.
© *1985 by D. Reidel Publishing Company.*

DWARF NOVA	E(B-V) +Ref.	DWARF NOVA	E(B-V) +Ref.	References
AR AND	<0.03,(3)	EX HYA	~0,<0.04 (5)	1) Szkody (1982)
RX AND	~0(1,3),0.1(2)	VW HYI	<0.02(5),0(10)	2) Szkody (1981)
SS AUR	0.03,(4)	WX HYI	<0.08(U),0(10)	3) Klare et al. (1982)
Z CAM	~0(2),0.2(3)	X LEO	0(U)	4) Wargau et al. (1982)
BV CEN	~0,<0.1(5)	AY LYR	0(1)	5) Bath et al. (1980)
V436CEN	~0(U)	TU MEN	0.08(3)	6) Rayne & Whelan (1981)
WW CET	0.06(3)	CN ORI	0(U)	7) Fabbiano et al. (1981)
Z CHA	~0(6)	RU PEG	0(11),0.05(U)	8) Heap et al. (1978)
YZ CNC	~0(2)	KT PER	0.15(U)	9) C. La Dous (1982)
SY CNC	0(2)	TZ PER	0.24(U),0.3(3)	10) Hassall et al. (1982)
EM CYG	0(2)	UZ SER	0.32(U),0.1(12)	11) Krautter et al. (1981)
SS CYG	~0(7,9),<0.08(8)	WZ SGE	0,<0.02(13),0(14)	12) Echevarria et al.(1981)
AB DRA	0.1(2,4)	EK TRA	0(10)	13) Friedjung (1980)
U GEM	~0(7,9)	SU UMA	0(2)	14) Holm et al. (1979)
AH HER	0.03(U),0(2)	TW VIR	0(15)	15) Cordova & Mason (1982)
				(U) Unpublished

Table 1. Dwarf Novae   Observed & Interstellar Reddening.

It is not therefore a complete list of observed dwarf novae, but only
of spectra published to date (Jan. 1983). For completeness, I have
included the reddening determined by the authors from the 2200Å dust
feature. The technique frequently used relies upon increasing the value
of E(B-V), until the dip near 2200Å has disappeared, forcing the local
continuum to be a power law. The reddening correction is of interest
since it affects the continuum distribution used for comparison with
model disc spectra. However, in practice, we find that the value is
small (i.e. close to the errors in the technique) except in the cases
of TZ Per, UZ Ser and KT Per with values of E(B-V) of 0.24, 0.32 and
0.15 respectively.

The 18 dwarf novae that have been observed in quiescence are given
in Table 2. The continuum has been characterised by the spectral index,
alpha, given by $f_\lambda \propto \lambda^{-\alpha}$. Where given in the original reference, the
quoted value is tabulated; otherwise it has been estimated from the
published spectrum. Usually the continuum is relatively flat at
quiescence, with α lying between -2 and -1.3. In some objects the
spectrum is steeper for wavelengths shortward of 1500Å, and in 2 cases,
U Gem and SS Cyg the effect is very marked.

The UV lines seen in quiescent dwarf novae are mainly resonance
lines of high ionisation species, such as CIV(1550), SiIV(1400) and
NV(1240), usually but not always in emission. Frequently MgII and CIV
still appear weakly in emission when other lines are in absorption. An
asterisk in Table 2 denotes that the optical spectrum has strong
emission lines according to Warner (1976) or Oke & Wade (1982). There
is no obvious correlation between the strength of the observed optical

DWARF NOVA	α $(f_\lambda \propto \lambda^{+\alpha})$	ULTRAVIOLET SPECTROSCOPIC BEHAVIOUR
Emission only		
YZ CNC	-2	Many emission lines
* SS CYG	-1.3[1]	Strong emission lines (CIV phase dependent)
* AB DRA	-1.4	Emission at CIV,SiIV,NV,MgII (near quiescent)
* EX HYA	-2.2	Many strong emission lines
* WX HYI	-1.75	Strong emission lines
* RU PEG	-1.7,-1.65	Strong emission lines
* SU UMA	-2.0,or less	Many emission lines
Emission weak, some abs		
BV CEN	-0.3	Emission only at MgII,CIV
* V436CEN	-	MgII emission. No strong SWP features[2]
EM CYG	-1.8	Emission MgII,CIV,SiIV; Absorption 1300,NV
* AH HER	-1.6,-2	Weak emission CIV,MgII
VW HYI	-1.75	Weak emission CIV,MgII; w. absorption 1300
EK TrA	-1.5	Weak emission CIV,MgII; w. absorption elsewhere
No emission, some abs		
* Z CHA	-1.7	Weak absorption, possible CIV emission
* U GEM	-1.3[1]	Weak absorption 1300,CII,CIV,HeII
* X LEO	-	Weak absorption near quiescence [2]
AY LYR	-	No strong features [2]
KT PER	-2.0	No strong features

[1]   much steeper at short λ
[2]   weak IUE exposure
 *   strong optical emission   (Warner 1976, Oke & Wade 1982)

Table 2. Quiescent Dwarf Novae.

lines and those in the UV, indicating that they probably originate in regions of different temperature regimes.

At outburst, 25 objects have been observed, (Table 3). In general, once maximum has been reached, the continuum is steeper than at quiescence, with the spectral index α lying between -2.5 and -1.5. The superoutburst spectra of the SU UMa subclass, and the standstill spectra of the Z Cam class are not significantly different from the normal outbursts. The line features are seen in absorption, or are absent altogether. The only exceptions to this are MgII(2800) observed only in emission even at maximum, and CIV(1550), SiIV(1400) and NV(1240) which occasionally have P Cygni characteristics. The line behaviour of a dwarf nova, as it goes into outburst is typified by WX Hydri (Figure 5, Hassall et al., 1983), the emission lines of quiescence being replaced by strong absorption lines at maximum. Table 4 summarises the occurrence of P Cygni features in dwarf nova outbursts. There seems to be a tendency for them to appear during standstills and superoutbursts.

DN	α	Spectroscopic Behaviour
RX AND	−1	Emission during early rise
	−	Strong abs. near peak } P Cyg CIV
	−2.3	Absorption in decline, weakening
AR AND	−2.4	Strong absorption
SS AUR	−2.1	−
WW CET	−1.4	Absorption (A)III, MgII emission CIV P Cyg
YZ CNC	−2.3	P Cyg Features
SY CNC	−2.3	Absorption at SiIV, CIV only
EM CYG	−2.3	Weak MgII emission, abs. elsewhere
SS CYG	−2.3	Strong absorption CIV P Cyg(?)
AH HER	−2.5 SWP / −2 LWR	Strong absorption P Cyg at CIV, Rise and decline
	−1.6	Weak abs. CIV emission Later decline
VW HYI	−1.75	Rise, like Q, CIV MgII emission
	+0.6	Late Rise weak absorption features
	−2.3 SWP / −2.0 LWR	Peak Strong absorption
	−2.3	Late decline. No features except MgII emission
X LEO	−2.3	Weak absorption
	−2.1	Decreasing during decline
AY LYR	−2.3	Broad weak absorption
CN ORI	−	Medium strong absorption, diminishing in strength in decline
RU PEG	−2.0	Strong absorption except MgII emission, CIV P Cyg
TZ PER	−2.1	Broad weak absorption CIV P Cyg
UZ SER	−2.9 SWP / −2.1 LWR	Broad absorption, CIV sharp, perhaps blue shifted
	−2.4	Broad weak absorption later decline
SU UMA	−2	Medium absorption
TW VIR	−2.3	Absorption blue shifted on all lines CIV P Cyg

SUPEROUTBURST

WX HYI	−1.5	Strong absorption P Cyg at SiIV, CIV
	−2.0	V. strong abs. at peak. P Cyg at CIV
AY LYR	~ −2.3	Broad weak absorption
TU MEN	−2.4	Broad weak absorption, possible, P Cyg at CIV
EK TRA	−1.75	Absorption. P Cyg at CIV (strong emission comp.)

STANDSTILL

RX AND	−1.7	Strong absorption (w emission at MgII) P Cyg at CIV
Z 'CAM	~ −2.3	Absorption MgII emission, no P Cyg
	−1.9	Medium abs. P Cyg at CIV
TZ PER	−2.0	Medium absorption P Cyg at CIV

Table 3. Outburst Dwarf Novae.

DWARF NOVA	OUTBURST TYPE	SUBCLASS	INC i	PERIOD hours	P CYG
RX AND	O	Z	29	5.08	√
WW CET	O	Z	m	3.83	√
YZ CNC	O	S	m	2.22	√
SS CYG	O	U	38	6.63	√
AH HER	O	Z	?	6.19	√
WX HYI	SUPER	S	~40	1.80	√
TU MEN	SUPER	S	?	2.82	?
RU PEG	O	U	44	8.90	√
TZ PER	O	Z	?	?	√
UZ SER	O	U	m	?	??
EK TRA	SUPER	S	~40	1.55	√
TW VIR	O	U	m	?	√
RX AND	STAND	Z	29	5.08	√
Z CAM	STAND	Z	59	6.96	√(?)
TZ PER	STAND	Z	?	?	√
AR AND	O	U	?	?	X
SY CNC	O	Z	?	7.7	X
EM CYG	O	Z	63	6.98	X
VW HYI	O	S	60	1.78	X
X LEO	O	U	m	?	X
AY LYR	O	S	?	1.81	X
CN ORI	O	Z	?	?	X
WZ SGE	O?	S?	81	1.36	X
SU UMA	O	S	m	~1.9	X

KEY   O   ordinary   m   medium
        U   U GEM   ?   uncertain
        S   SU UMA   √   present
        Z   Z CAM   X   absent

Table 4.   Occurrence of P Cyg Feature.

Otherwise, there is no obvious correlation with subclass, inclination or orbital period.

2. COMPARISON WITH STEADY DISC MODELS

One of the first papers to consider optical and UV results was Bath et al. (1980), in which a $f_\lambda \propto \lambda^{-2.33}$ (i.e. $\nu^{1/3}$) spectrum was found for VW Hydri during decline from an ordinary outburst. The authors concluded that the maximum temperature in the blackbody disc was in excess of 80,000K with $\dot{M} > 2.3 \times 10^{18}$ g s^{-1} and the ratio of inner to outer radii $R_{out}/R_{in} > 90$. Such a disc is too large to fit within the primary star's Roche lobe, the orbital separation in this case being only of the order of 5.2×10^{10} cm. In general, a power law distribution can only be achieved if a large range of temperatures is present in the black-body disc. The equation for the temperature distribution (e.g. equation 4.1 of Bath et al., 1980) shows that this is only possible if R_{out}/R_{in} is large. Therefore in practice, the relatively small Roche lobe size, especially for ultra-short period systems, precludes the possibility of a power law spectrum in this model.

Williams (1980) and Tylenda (1981) have each considered the free-free radiation from the optically thin outer disc, which exists at low mass accretion rates and should be taken into account for dwarf novae at quiescence. However, the mass accretion rates necessary to produce significant deviation in the UV, of Tylenda's model (his Figure 2) from the black body case is $\dot{M} \sim 10^{15}$ g s^{-1}, lower than expected for dwarf novae even at minimum.

Several papers treat discs by stellar atmosphere methods (Kiplinger, 1979; Mayo et al. 1980). Herter et al. (1979), for example, consider just H and He opacities, truncating their disc at an outer temperature of about 10,000K. The resultant model disc spectra are very steep in the IUE region, typically with α of about −3.0, and the Balmer jump is large compared with that in the observed spectra. The discrepancy with the observed optical continuum is largely due to the neglect of the cooler outer regions of the disc in the model. Mayo et al., on the other hand, include temperatures down to 5000K, and discuss the effect of the outer radius on the calculated UBV disc colours. However, they point out that their own UV fluxes may be wrong by up to a factor of 2, (due to not having included scattering and certain high temperature opacities), and therefore the models are unsuitable for comparison with the IUE data.

Cordova & Mason (1982) compared the spectrum of TW Vir with fluxes of a single temperature atmosphere modelled by Kurucz (1979). They concluded that an atmosphere at approximately 30,000K was the best match, but that it could not adequately fit the entire spectrum. A similar approach was adopted by Wu & Panek (1982) for their ANS data. Their model discs consisted of annuli at the effective temperature calculated from the standard steady state distribution, replacing the black body spectrum with the corresponding main sequence stellar atmosphere spectrum.

The phase-dependent observations of VW Hydri and U Gem suggested that
the hot spot may contribute 0.2 mag in the UV and that the white dwarf
flux is important at short wavelengths, consequently providing poor tests
of disc models

Wade (in preparation) discusses the use of Kurucz's model atmospheres
in building discs. Angle-averaged fluxes exist for temperatures between
6000 and 50,000K. At higher temperatures, for wavelengths greater than
about 1000A, the black body approximation is adequate. The flux distribution is dominated by the Lyman absorption edge so that the peak always
lies near 1000A. The effect of increasing the maximum temperature in
the disc is to steepen the UV continuum, and to decrease the Balmer
jump. The atmosphere treatment takes some account of the vertical
structure in the disc, ignored by the single zone black body model, and
is capable of giving a power law type spectrum without invoking a large
range of temperatures within the disc or a large outer radius.

I will now take some observations of EK TrA in superoutburst and
nearly quiescent as examples of fitting atmosphere disc models. By
considering the UV range only, the observed spectral index α gives the
following maximum temperatures in the disc:-
Outburst $\alpha = -1.75$ T_{max} ~ 20,000K
Quiescent $\alpha = -1.5$ T_{max} ~ 17,000K
These models fit the data well in the UV but give poor agreement in the
optical and near the Balmer jump. Improvements can perhaps be achieved
by taking the correct inclination for the disc, (rather than using
angle-averaged fluxes) and by adding some recombination radiation from
the optically thin outer regions to produce the Balmer jump in emission,
as opposed to absorption, as seen in atmosphere models. The effect of
the optically thin component in the model spectra is to flatten out the
UV continuum for any one T_{max}. The maximum temeperature in the disc
for EK TrA at superoutburst could then be 30,000K. Complete compensation
for the Balmer jump in this way, however, predicts an excessively strong
emission line flux.

As we see, there is quite a contrast between these results and those
obtained assuming a black body disc, where the characteristic temperatures
are determined from the turnovers in the spectrum at long and short
wavelength. In atmosphere models, the temperature is derived from the
gradient, tending to give low maximum disc temperatures and associated
low mass transfer rates. Conversely, the lack of short wavelength
cut-off would lead us to assume a much higher disc temperature (over
40,000K) in the black body case. Since $\dot{M}$ is a strong function of
temperature (e.g. Equation 4.2, Bath et al., 1980), the method of
modelling the continuum can affect drastically our conclusions about the
mass transfer rate. The effect is summarised below for a $0.5 M_{\odot}$ white
dwarf with radius 10^9 cm.

$T_{max}/1000K$	40	30	20	15	10
$\dot{M}/10^{16}$ g s^{-1}	26	8	2	0.5	0.1
$\dot{M}/10^{-10} M_{\odot}$ yr^{-1}	41	13	3	0.8	0.2

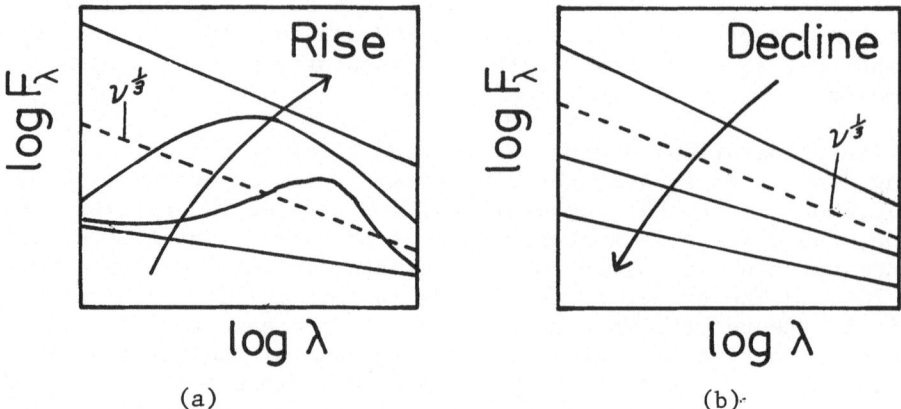

Figure 1.   Schematic behaviour of continuum during rise (a) and
             decline (b).

        The Kurucz discs require T_{max} of just 20,000K for EK TrA as
compared with over 40,000K for the black body case when no turnover is
observed, and a corresponding reduction in $\dot{M}$ by a factor of order 10.
For comparison for VW Hyi, Bath et al. found $\dot{M} > 2 \times 10^{18}$ g s^{-1} which
is about 100 times the amount suggested above. Perhaps in the past we
have been over-estimating the mass transfer rate.

3.   CONTINUUM BEHAVIOUR DURING OUTBURST

The development of the continuum has now been observed for a number of
dwarf nova outbursts, including VW Hydri, CN Ori, X Leo, AH Her and
RX And.   Figure 2 in Hassall et al. (1983) illustrates the behaviour in
the early rise to normal maximum of VW Hydri. Here the optical flux,
approximately one day into outburst had increased by 2 magnitudes, with
a peak in the flux at 4300Å, whereas the UV flux in the SWP region was
unchanged from its quiescent level. From the observations of several
normal outbursts we can build up a schematic picture (Figure 1) of the
time dependent behaviour. During the rise, seen twice for VW Hyi, the
flux increases first at long wavelengths and the peak moves through to
the UV in about a day, reaching a steep distribution approximating to a
power law near maximum. A gradual decline occurs at all wavelengths, as
observed in several sequences of spectra in different objects. The
power law becomes flatter, sinking through $\lambda^{-2.3}$ on the way back to
quiescence.

       From this we can conclude that the outburst starts in the cooler
parts of the disc. If the steady disc model is applicable, the
temperature distribution indicates that the outburst begins in the outer
regions. However, no such constraints on the position of the triggering
exist if the disc is not steady.

Can the theories for the outburst mechanism produce this type of spectral development? The contenders fall into two main categories: increased mass transfer through L_1 and increased accretion onto the white dwarf due to collapse of the disc. Bath & Pringle (1981 and 1982) deal respectively with bursts of mass transfer and a viscous instability in the disc. Alternatively, the disc may never be stable due to the existence of a thermal instability (Faulkner, 1983, AAS Meeting, Boston).

Having presented the observational material, I think this is a suitable point to stop, to allow the theorists to elaborate upon the different mechanisms and continue the discussion.

REFERENCES

Bath, G.T. & Pringle, J.E.: 1981, Mon.Not.R.astr.Soc. 194, p.967.
Bath, G.T. & Pringle, J.E.: 1982, Mon.Not.R.astr.Soc. 199, p.267.
Bath, G.T., Pringle, J.E. & Whelan, J.A.J.: 1980, Mon.Not.R.astr.Soc. 190, p.185.
Cordova, F.A. & Mason, K.O.: 1982, Astrophys. J. 260, p.716.
Echevarria, J., Jones, D.H.P., Wallis, R.E., Mayo, S.K., Hassall, B.J.M., Pringle, J.E. & Whelan, J.A.J.: 1981, Mon.Not.R.astr.Soc. 197, p.565.
Fabbiano, G., Hartmann, L., Raymond, J., Steiner, J., Branduardi-Raymont, G. & Matilsky, T.: 1981, Astrophys. J. 243, p.911.
Friedjung, M., Rocca-Volmerange, B. & Debeve, G.: 1980, Proc. Second IUE Conf., Tubingen p.85.
Hassall, B.J.M., Pringle, J.E., Wade, R.A. & Whelan, J.A.J.: 1982, Proc. Third IUE Conf., Madrid p.179.
Hassall, B.J.M., Pringle, J.E., Wade, R.A., Whelan, J.A.J., Schwarzenberg-Czerny, A. & Hill, P.W.: 1983, Mon.Not.R.astr.Soc. (In press).
Heap, S. et al.: 1978, Nature 275, p.385.
Herter, T., Lacasse, M.G., Wesemael, F. & Winget, D.E.: 1979, Astrophys. J. Supp. 39, p.513.
Holm, A.V.: 1979, Proc. IAU Symp. 88.
Kiplinger, A.L.: 1979, Astrophys. J. 234, p.997.
Klare, G., Krautter, J., Wolf, B., Stahl, O., Vogt, N., Wargau, W. & Rahe, J.: 1982, Astr. Astrophys. 113, p.77.
Krautter, J., Klare, G., Wolf, B., Duerbeck, H.W., Rahe, J., Vogt, N. & Wargau, W.: 1981, Astr. Astrophys. 102, p.337.
Kurucz, R.L.: 1979, Astrophys. J. Supp. 40, p.1.
La Dous, C.: 1982, Diplomarbeit, Munich.
Mayo, S.K., Wickramasinghe, D.T. & Whelan, J.A.J.: 1980, Mon.Not.R.astr. Soc. 193, p.793.
Oke, J.B. & Wade, R.A.: 1982, Astron. J. 87, p.670.
Rayne, M.W. & Whelan, J.A.J.: 1981, Mon.Not.R.astr.Soc. 196, p.73.
Szkody, P.: 1981, Astrophys. J. 247, p.577.
Szkody, P.: 1982, Astrophys. J. 261, p.200.
Tylenda, R.: 1981, Acta Astron. 31, p.127.
Wade, R.A.: 1983, In preparation.

Wargau, W., Drechsel, H. & Rahe, J.: 1982, Proc. Third IUE Conf.,
    Madrid p.215.
Warner, B.: 1976, Proc. IAU Symp. 73, p.85.
Williams, R.E.: 1980, Astrophys. J. 235, p.939.
Wu, C.-C. & Panek, R.J.: 1982, Astrophys. J. 262, p.244.

DISCUSSION

SZKODY: At what wavelength did the flux reach a peak on the rise to
eruption?

HASSALL: In one case it was 5000 $\overset{\circ}{A}$, but of course it moves. Presumably it will be reddest when you catch it the earliest.

STARRFIELD: Do you see any comparable behavior in the optical emission
lines through an outburst?

HASSALL: In VW Hyi on the rise, the optical emission lines disappeared,
i.e., they went into absorption at a time when the UV emission lines
were still strong and the UV continuum distribution hadn't changed.

VERBUNT: I would like to ask John Faulkner a question about his model.
If the instability in his model can occur anywhere in the disk, doesn't
that contradict the fact that we always see the optical rise before the
ultraviolet?

FAULKNER: There just isn't a one-to-one correspondence between temperature and position in the disk, for non-steady disks. If you equate
"cool" with "outer part of the disk", you're being seduced by the
implications of the steady-state model.

McCLINTOCK: By how much was the peak of the 1000-2000 $\overset{\circ}{A}$ light curve
delayed from the peak of the optical light curve?

HASSALL: Approximately one day.

SMAK: A question: what is it that we see at the far UV, near 1000 $\overset{\circ}{A}$?

[Ed: A great many disjointed remarks follow from every corner of the
room. No coherent answer emerges.]

LANGER: You said that the reason you thought we were overestimating
the accretion rate was that we attempt to get the spectrum to go as a
power law right on through the UV. Would we run into the same difficulty
with the simple theory if the accretion rate were higher, say about what
we've been associating with the classical novae, or would it already be
a power-law through the UV, in which case it wouldn't make any difference?

HASSALL: You're referring to the blackbody models?

LANGER: Well, yes. That's the simplest thing you could do to try to come up with a disk spectrum. Would you infer the correct accretion rate, or would you still have this difference by a factor of 10 or more?

HASSALL: I think the same problem would apply.

ON THE NATURE OF DWARF NOVAE

J. Smak
Joint Institute for Laboratory Astrophysics
National Bureau of Standards and University
of Colorado, Boulder, CO 80309

ABSTRACT

Observational data on mass-transfer rates and radii of disks
indicate that the outer parts of disks in novae and nova-
like binaries are sufficiently hot for stationary accretion;
those in dwarf novae are too cool to avoid an accretion instability, while these in Z Cam systems are the borderline
cases. The mass ratios of novae and nova-like binaries with
main-sequence secondaries appear - at a given orbital period
- to be systematically larger than those of dwarf novae,
implying that higher mass ratios are responsible for higher
mass-transfer rates.

This paper appeared in The Astrophysical Journal, September
1, 1983.

D. Q. Lamb and J. Patterson (eds.), Cataclysmic Variables and Low-Mass X-Ray Binaries, 299.

DWARF NOVAE ERUPTIONS

V. J. Mantle
Oxford University

It is now well established that the eruptions characteristic of dwarf novae result from changes in the luminosity of the accretion disc centred around the white dwarf component of these systems. These luminosity variations are most successfully accounted for by variations in the accretion flow through the disc but the mechanism responsible for this accretion modulation remains unsettled. Two possibilities are :-
1) Instability within the companion star resulting in a variable mass input rate to the disc.
2) Instability within the disc resulting in a steady input of matter from the companion being accreted at a variable rate.
I shall show how observed features of dwarf novae eruptions, in particular the eruption decay time, spectral evolution and stream impact bright-spot behaviour are accounted for by the bursting mass transfer model and leave the disc instability model to be described elsewhere in this volume.

It is normally assumed that the quiescent bright-spot results from shock-heating of the stream/disc gas at the point of impact of the stream with the disc. In practice the mass transfer stream will penetrate the disc on a 'quasi-elliptical' orbit. How deeply the stream penetrates before all the stream material has been sheared off by shocks driven by disc material in circular orbits is unknown. However there is evidence (e.g. Stover 1981) that extensive penetration by the stream occurs in some systems. An approach to the mass input problem which includes mass and angular momentum sources implicitly in the conservation equations for disc structure is described below.

If the mass flux fed from the stream into the disc at radius R and time t is $\partial(\dot{m}(R,t))/\partial R$ then, in the thin disc approximation, conservation of mass gives

$$\frac{\partial \Sigma}{\partial t} = -\frac{1}{R}\frac{\partial}{\partial R}\left(R\Sigma v_R\right) + \frac{1}{2\pi R}\frac{\partial \dot{m}}{\partial R}$$

301

D. Q. Lamb and J. Patterson (eds.), Cataclysmic Variables and Low-Mass X-Ray Binaries, 301–305.
© 1985 by D. Reidel Publishing Company.

where Σ is the surface density and U_R the radial velocity. Conservation of angular momentum implies,

$$\frac{\partial}{\partial t}\left(\Sigma R^2 \Omega\right) = -\frac{1}{R}\frac{\partial}{\partial R}\left(R^3 \Sigma\left(U_R \Omega - \nu\frac{\partial\Omega}{\partial R}\right)\right) + \frac{R_K^2 \Omega_K}{2\pi R}\frac{\partial\dot{m}}{\partial R}$$

where ν is the effective kinematic viscosity. Combining (1) and (2), eliminating U_R, and assuming Keplerian angular velocities, leads to an evolution equation for the surface density as a function of time of the form,

$$\frac{\partial\Sigma}{\partial t} = \frac{3}{R}\frac{\partial}{\partial R}\left(R^{1/2}\frac{\partial}{\partial R}\left(\nu\Sigma R^{1/2}\right)\right)$$

$$+ \frac{1}{2\pi R}\frac{\partial\dot{m}}{\partial R}$$

$$+ \frac{1}{\pi R}\frac{\partial}{\partial R}\left(R\left(1 - \frac{R_K^{1/2}}{R^{1/2}}\right)\frac{\partial\dot{m}}{\partial R}\right)$$

The first term is the basic diffusion term describing matter and angular momentum redistribution due to viscous stress, and is the fundamental expression controlling disc evolution (see e.g. Pringle 1981). The second term is the stream source term from equation (1). The third term describes the influence on the evolution of Σ of the angular momentum of the mass transfer stream. The first, diffusion, term drives material in (and out) into a disc, whilst the third term decribes the tendency of new input material with angular momentum appropriate to an orbit at R_K, to squeeze the disc towards an annulus at radius R_K.

To solve equation (3) the rate at which matter is stripped off the stream and redirected into circular orbits; $\partial\dot{m}/\partial R$, must be specified. In order to make progress and examine the effect on disc evolution of deep disc penetration the following parametrization has been adopted. We assume that the stream is stripped at a rate which is a fraction $\beta\sin\theta$ times the circular mass flux in the disc where β is a numerical constant and θ is the angle between the stream and the relative motion of material in the disc i.e.

$$\frac{\partial\dot{m}}{\partial R} = \beta\sin\theta\,\Sigma\left(\frac{GM_1}{R}\right)^{1/2}$$

We have assumed the stream follows a parabolic trajectory defined by the specific angular momentum appropriate to a circular orbit at R_K. This gives

$$\sin\theta = \left\{\left(1 - \frac{R_K}{2R}\right)\Big/\left(3 - 2\left(R_K/R\right)^{1/2}\right)\right\}^{1/2}$$

The effects of different values of β have been explored with numerical models. Low values of β allow deep stream penetration whereas high values of β ($\beta = 1$) approach instantaneous mixing (Bath, Edwards & Mantle, 1983).

1) Eruption Decay Time.

Bailey (1975) first pointed out that the eruption decay-time increased as the binary period of cataclysmic variables increased. This relation has recently been confirmed by Mattei and Klavetter, and their results are shown in Fig 1. The eruption decay time is determined by the size of the viscosity, and the decay time provides a sensitive test of its quantitative value. Theoretical models, shown for comparison, fit the observed data for a value of α =1.5 (Mantle & Bath, 1983). The observed spread can be accounted for by white dwarf masses in the range 0.5 $M_{\odot}$ to 1.4 $M_{\odot}$. The allowed variation of α is small. With 1 $M_{\odot}$ white dwarfs the observational points all lie within the range 0.5 < α <3.0. Values of α < 0.5 are incompatible with observed decay timescales. In particular values of α = 0.01 lead to decay times approaching a year. The physical explanation of the relation discovered by Bailey is simply the increased time taken for material to diffuse through, and drain out, of larger discs in longer period systems.

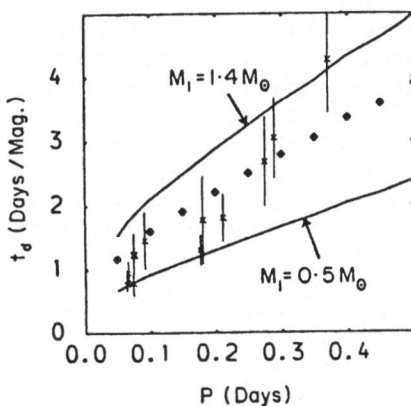

Fig 1. Decay time t_d, versus binary period P, for computed models using a value of α = 1.5 (solid diamonds) and the distribution of observed systems (open squares). The computed points were determined using a white dwarf mass of 1 $M_{\odot}$, and the solid lines show the effect of varying the white dwarf mass. Variations in α within the range 1.0 < α < 3.0 would produce a similar range of scatter (reprinted from Mantle & Bath, 1983)

2) Spectral Evolution.

The continuum spectral evolution of VW Hydri throughout an outburst cycle in the spectral range 1225 Å to 7500 Å is described elsewhere in this volume by Hassall. The most distinctive feature is that the continuum increases above the quiescent distribution primarily at optical wavelenths with the short wavelength UV spectrum remaining unaffected until about half a day into the optical outburst. Fig 2 shows the spectral evolution of our model using disc parameters approximately equal to those of VW Hydri and using a value of α = 1.5. A mass transfer burst lasting 10^5 sec at 10^{19} gs^{-1} into a steady state quiescent disc with an accretion rate of 10^{16} gs^{-1} generates the eruption. The characteristic 'spectral reddening' is evident and results as a direct consequence of enhanced dissipation in the cool outer disc regions as the mass transfer burst mixes into the outer regions and then diffuses inward through viscous transport.

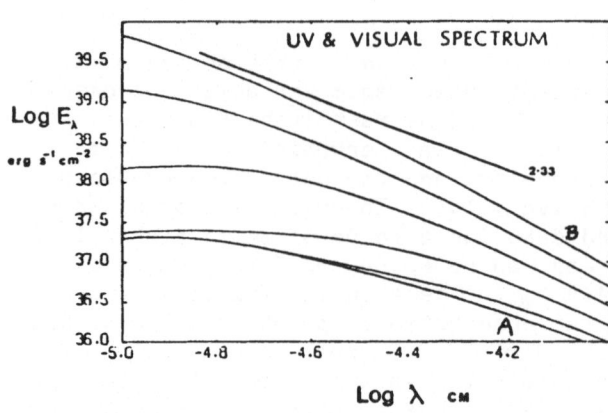

Fig 2. Spectral evolution during the rise in the mass transfer rate. Curve labelled A corresponds to the steady state condition preceeding the burst and B to the state when the maximum mass transfer rate is achieved. The intermediate states are separated by ΔT = 2 x 10 sec. The straight line shows an $E_\lambda \propto \lambda^{-2.33}$ power law.

3) Bright Spot Behaviour.

It is well established that during the rise the hump luminosity does not change in proportion to the overall increase in luminosity from the disc. Indeed it is often claimed that this indicates no change of the mass flux in the stream, and that the outbursts must therefore be produced by disc instabilities. However this conclusion is based on the naive assumption that all the available stream kinetic energy is released within a distance of order the disc height from the disc edge. If penetration occurs this is not the case. In Fig 3 the light curve of VW Hyi in outburst (Warner, 1975) is shown together with the hump luminosity. The delay of the rise in the hump behind the overall outburst is evident.

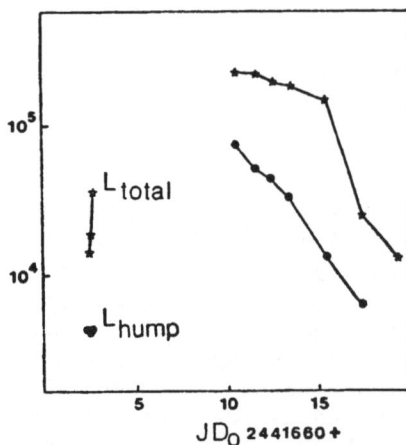

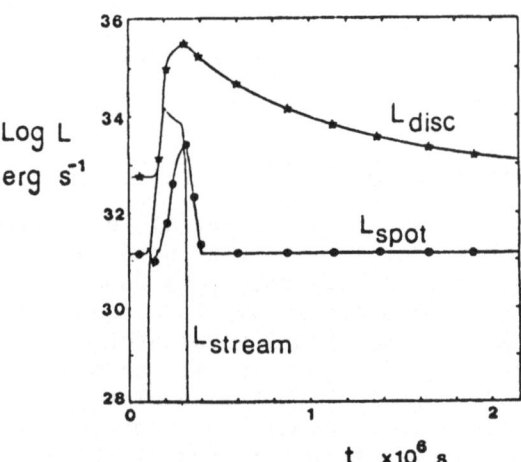

Fig 3. Observed total flux variations and hump luminosity through outburst in VW Hydri (to appear in Bath, Edwards & Mantle, 1983).

Fig 4. Changes in disc, spot and stream luminosity in an α = 1.5 β = 0.3 disc (to appear in Bath, Edwards & Mantle, 1983).

In Fig 4 the changes in the disc, spot and stream luminosity are shown for a β = 0.3 disc. The spot luminosity L_{SPOT} is computed as that fraction of the stream-disc impact energy liberated within a region of thickness H of the disc edge (i.e. that proportion of the impact flux which could produce an asymmetric radiation pattern with the properties of the bright spot). It is evident that the spot luminosity does not rise simultaneously with the outburst rise, but is delayed. This is simply due to the fact that until the surface density at the outer disc edge has grown by viscous transport of material deposited interior to the edge, the stream continues to penetrate and the spot flux is unaffected by the outburst.

We conclude that there is no observational support for the view that mass transfer variations cannot be driving the outbursts. With $\alpha \sim$ 1.0 and $\beta \sim$ 0.5 all the major observed features presently known to. be associated with the disc component of dwarf novae are reproduced.

References.

Bailey, J., 1975. J.Br.astr.Ass., <u>86</u>, 30.
Bath, G.T., Edwards, A.C. & Mantle, V.J., Mon.Not.R.astr.Soc.
                                                    In press (1983).
Hassall, B., Hill, P., Czerny, A., Pringle, J.E., Wade, R.,
                    Whelan, J.A.J., Mon.Not.R.astr.Soc. In press (1983).
Mantle, V.J. & Bath, G.T., 1983. Mon.Not.R.astr.Soc., <u>202</u>, 151.
Mattei, J.A. & Kalvetter, J.K., In press (1983).
Pringle, J.E., 1981. A.Rev.Astr.Astrophys., <u>19</u>, 137.
Stover, J.R., 1981. Astrophys.J., <u>249</u>, 673.
Warner, B., 1975. Mon.Not.R.astr.Soc., <u>170</u>, 219.

ACCRETION INSTABILITY MODELS FOR DWARF NOVAE AND X-RAY TRANSIENTS

John K. Cannizzo, J. Craig Wheeler
University of Texas at Austin

Pranab Ghosh
Tata Institute

Steady state and time dependent calculations are presented of a model for accretion instability in which matter accumulates in a cold torus and then undergoes a thermal instability causing the matter to heat and flow down onto the central star. This model is compared to the observations of dwarf novae and certain examples of both hard and soft X-ray transients.

We have proposed a model of accretion instability leading to dwarf novae and related phenomena in which mass is considered to be transferred from a companion star at a constant rate. The mass is stored in a cold torus from which the accretion timescale is longer than the transfer timescale. The density builds up until a critical point is reached at which the rate of viscous heating exceeds the cooling and a thermal instability ensues. The resulting hot torus spreads out into a disk and onto the central compact star causing the outburst. The accretion is then envisaged to rebuild the torus.

Steady state, geometrically thin, alpha model accretion disks are constructed by explicit vertical integration using realistic opacities to identify points of thermal instability. The behavior of the post-instability hot torus is calculated with a vertically averaged time dependent code again using realistic opacities. This work has been introduced in Cannizzo, Ghosh and Wheeler (1982) and Cannizzo, Wheeler and Ghosh (1982). Many of the basic principles are discussed in Meyer and Meyer-Hofmeister (1981, 1982), Smak (1982a,b,c), and Bath and Pringle (1981, 1982). The work is similar in spirit to that discussed in this workshop by Faulkner, although there are important differences in application.

The loci of steady state models in the effective temperature, surface density plane are calculated for different values of radius and alpha. Portions of the optically thick steady state locus for which the effective temperature varies as a negative power of the surface density are thermally unstable and hence not accessible to

307

D. Q. Lamb and J. Patterson (eds.), Cataclysmic Variables and Low-Mass X-Ray Binaries, 307–313.
© 1985 by D. Reidel Publishing Company.

evolving disks (Bath and Pringle 1982). There is general agreement that partial ionization induces thermal instability in models with effective temperature of about 5000 – 7000 K. For steady state solutions computed at a typical dwarf nova outer disk radius of 3 x 10^{10} cm, however, the rate of accretion at these effective temperatures, > 10^{17} g/s, exceeds the deduced transfer rate by about an order of magnitude. For r = 10^{10} cm the rate of accretion associated with this feature is smaller, but the amount of material stored is also smaller. We have argued that this partial ionization instability can only operate at small radii to give frequent, low luminosity bursts. Otherwise, the accretion rate prior to the outburst is greater than the transfer rate, and steady accretion at the transfer rate should occur.

We have chosen to invoke a change in alpha at the point where convection first breaks out. This also serves to induce a thermal instability. At these low temperatures, about 2000 K, the accretion rate is less than 10^{16} g/s and storage of an appreciable amount of matter at reasonable radii can occur before instability . We find values of alpha of order 0.01 in the cold state and alpha of order several tenths in the hot state reproduce the observations of a variety of astrophysical objects.

The steady state models have been used to estimate timescales and delineate regions of steady and unsteady accretion, as shown in Figure 1. For cataclysmic variables with orbital periods of several hours we predict that transfer at a rate in excess of 10^{17} g/s should result in steady accretion since the accretion timescale after the instability is less than the transfer rate and steady accretion at the transfer rate should be driven. At very low transfer rates, the accretion rate in the cold material should exceed the transfer rate, so that the matter should leak steadily in. The upper boundary, in particular, seems to divide the steadily accreting novae and nova-like variables from the dwarf novae. The classic soft X-ray transient Aquila X-1 has a larger orbital period and longer repetition time than dwarf novae. We estimate that it falls well within our instability region. A plausible case can also be made for Cen X-4. There are some reasons to be encouraged to pursue a similar argument for the hard X-ray transients such as 4U0115+63, although the existence of Be star companions complicate these systems considerably. Her X-1 falls within our instability region for nominal parameters, but does not display the type of unstable behavior we are discussing. It would be predicted to be stable if the disk began at slightly larger radii, or the accretion rate were a bit less.

For the time dependent models we construct initial models of the hot torus and then calculate its evolution, including the spectrum and the total luminosity. The material spreads out into a configuration resembling a steady state disk, but only approximately. The accretion rate, for instance, is not independent of radius, but decreases linearly with increasing radius. Although we do not trust the

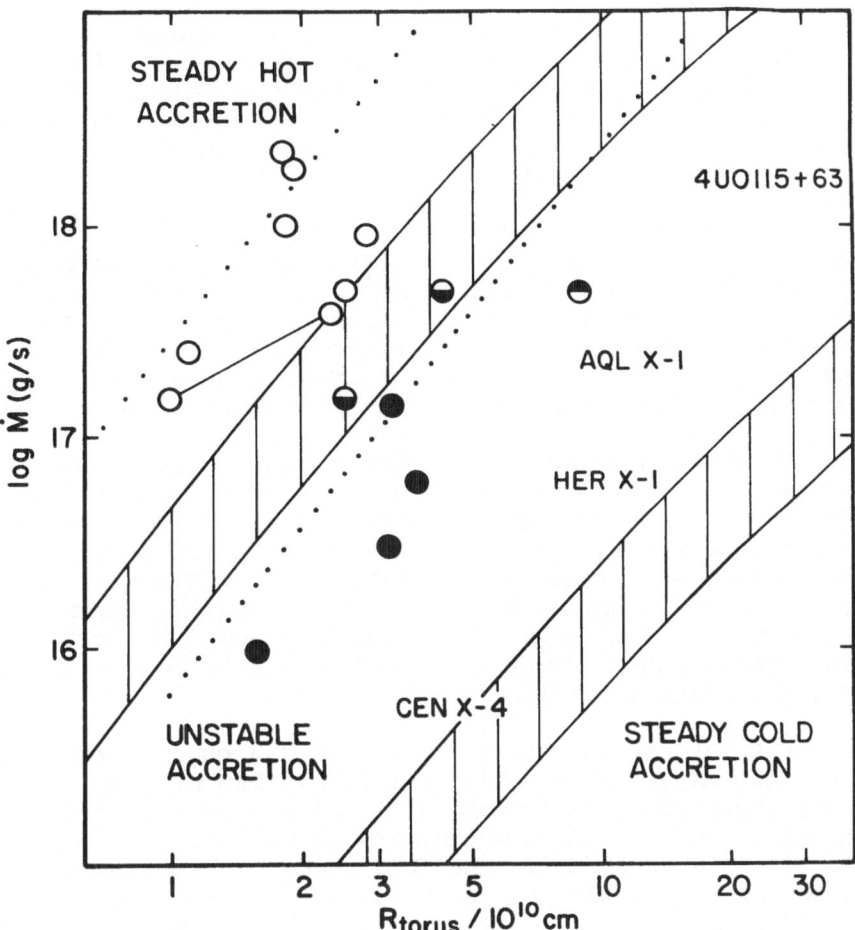

Figure 1. The stability of accretion flow is examined in the plane characterized by the steady-state mass transfer rate and the distance of the quiescent torus from the central star. An instability strip is defined such that the time between bursts is less than the spreading time in the accumulating torus (lower bound) and longer than the burst duration (upper bound). The dotted boundaries correspond to the opacity peak instability. The convective onset instability is indicated by the broad dashed bands, the widths of which represent theoretical uncertainties. Inferred parameters for a number of cataclysmic variables are taken from Smak (1982b), suppressing the large uncertainty associated with individual points. Open circles are steadily accreting novae, closed circles, dwarf novae. Partially filled circles are objects with both steady and transient behavior. Cen X-4, Aql X-1, and 4U0115+63 are X-ray transients that lie in the convective onset instability strip. Her X-1 does not show periodic bursts but also lies in the strip for canonical parameters. The opacity peak strip lies at higher steady-state transfer rates that do not conform as well to observations.

particular results because, among other things, we have not yet
included radial heat transport in the models, the code does iterate to
find material in the cold state after the density drops and the hot
state is no longer accessible in thermal equilibrium. We have not yet
explicitly calculated a complete cycle including the rebuilding phase
and repetitive outbursts.

Smak (1982b) estimates the transfer rate in SS Cyg to be 1.5×10^{17} g/s, and the outer edge of the disk in outburst to be $2-5 \times 10^{10}$
cm. The mean time between bursts is about 31 days and the visual band
e-folding time after outburst is about 4 days. Using the steady state
models and analytic estimates of timescales we estimate that alpha in
the cold state should be about 0.004 and in the hot state 0.5. We
estimate the luminosity after one e-fold to be about 10^{35} erg/s.

As shown in Figure 2 the time dependent models give a reasonable
representation for the decline rate of an optical outburst of SS Cyg
with the torus at 3×10^{10} cm and alpha of 0.7. This agrees with Bath
and Pringle (1981). We find a peak bolometric luminosity of about 2×10^{35} erg/s and, assuming a distance of 100 pc, a maximum visual
magnitude of slightly in excess of +8, both in good agreement with
observations. Kiplinger (1979) finds a peak luminosity of 7.5×10^{35}
ergs/s for SS Cyg. The models have some difficulty accounting for the
range in outburst profiles.

Aql X-1 has an orbital period of 1.3 days, a repetition time of
about 400 days, a luminosity e-folding time of about 45 days, and an
X-ray luminosity of about 10^4 solar luminosities. With these
parameters we estimate the value of alpha in the hot state to be 0.32.
The estimate of alpha in the cold state is sensitive to the accretion
efficiency, distance and torus radius to the powers 2, 4, and 6,
respectively. It could plausibly be of order 0.01. The transfer rate
should be about 2×10^{17} g/s. A time dependent model with the torus
at 10^{11} cm and a distance of 4 kpc gives X-ray and optical light
curves which resemble those of Aql X-1. If anything the model gives
too much power, but the similarity between the predicted and observed
decline of the X-ray light curves is striking. Since reddening and X-
ray reprocessing have been ignored and a maximum distance has been
assumed, the surfeit of optical light may be an embarrassment to this
model, although an inclination effect would temper this.

Cen X-4 has an orbital period of 8.2 hours, a repetition time of
about 10 years, a luminosity e-fold of about 90 days, and a
characteristic X-ray luminosity of about 10^4 solar luminosities (1969
outburst). For these parameters we estimate values of alpha of ~ 0.01
and .05, and a transfer rate of about 10^{16} g/s, putting Cen X-4 on the
lower edge of our instability strip. The 1979 outburst of Cen X-4
apparently had a shorter decay time, and a peak luminosity a factor of
ten smaller than the 1969 outburst. Such changes, are difficult to
reconcile with the model in its current incarnation.

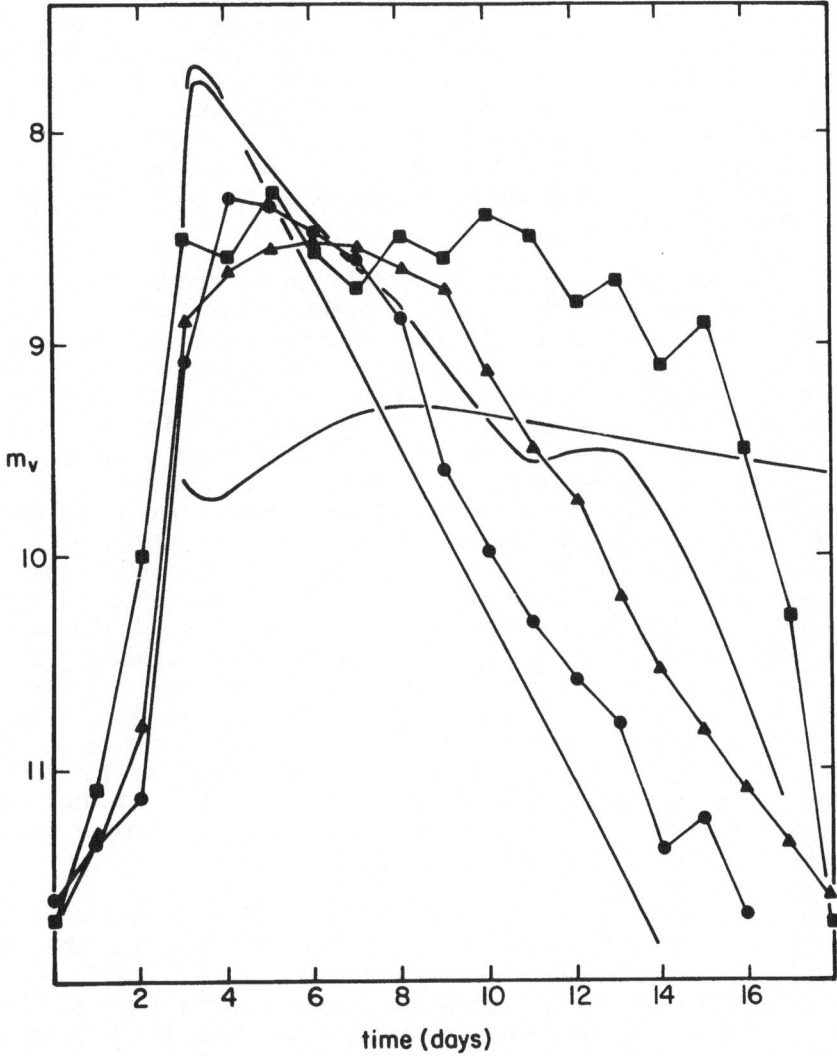

Figure 2. The apparent visual magnitudes of three outbursts of SS Cygni (Journal of the AAVSO (vol. 5, no. 1, 1976, p. 54; bursts #4 (squares) and #5 (circles) on 1st row, burst #4 (triangles) on 2nd row) are compared with time-dependent theoretical models of outburst based on the convective onset mechanism. The two theoretical curves which peak at $m_v = 7.8$ are for $\alpha = 0.7$ and 1.0, the latter being the more rapid. The curve with a peak at $m_v = 9.5$ has $\alpha = 0.1$. The parameters for the initial torus are $r_{torus} = 3.2 \times 10^{10}$ cm, $\delta r = 6.4 \times 10^9$ cm, and $\Sigma_{max} = 590$ g cm^{-2} - the critical surface density for $\alpha_{cold} = 0.01$. A distance of 100 pc is assumed. The luminosity associated with the theoretical curves is of order 10^{35} ergs/s - in accord with the observed power and a factor of order 10 higher than that typical of the opacity peak mechanism.

With a 7.8 hour period (McClintock et al. 1983), a repetition time of 60 years, a luminosity e-fold of 30^d and mean luminosity of 3 x 10^4 L_O, A0620−00 would have a transfer rate of about 2 x 10^{15} g/s and values of α ~ .01 and .08. This places it very near the low boundary of the instability strip. A repetition time less than 60 years due to missed outbursts, or an accretion efficiency less than 0.1 would yield parameters very similar to those deduced for the 1969 outburst of Cen X-4.

An accretion instability model can be reconciled at some level of approximation with the optical outburst and repetition time of the hard X-ray transient 4U0115+63. Starting with a torus at 3 x 10^{11} cm, about half the Roche radius of the neutron star, gives an optical outburst from the disk alone about two and a half magnitudes less than that observed for 4U0115+63. The peak output is very sensitive to the initial radius, so that is not a fundamental problem. The disk is very much brighter than for ordinary cataclysmic variables. More troubling is the fact that although there is a delay between the optical and X-ray outbursts, it is only a few days, not the two months which is observed (Kriss et al. 1983). The model also predicts an accretion rate which exceeds the Eddington limit 100 km from the neutron star surface. Whether this is a fatal limitation of the model or an important physical development is unclear. This model can rather naturally account for the repetition time and the luminosity of the optical outburst. The best alternate model is one in which the Be star flares and belches and the neutron star collides with the excretion disk (Rappaport and van den Heuvel 1982). In this picture the repetition time and the magnitude of the optical flare must be chosen ad hoc, with the latter considerably larger than the flares routinely observed in Be stars.

With standard orbital and model parameters and a transfer rate of 10^{-9} solar masses per year we predict that Her X-1 should be unstable with a repetition time of about 4 years and a burst time of about 40 days. Although Her X-1 displays various sorts of transient phenomena, this behavior is not seen. Perhaps the simplest way to reconcile the model with the observations would be to invoke a transfer rate smaller by about a factor of two and a disk outer radius a bit larger than half the Roche radius which we estimated to be 10^{11} cm. Then the outer parts of the disk should be cool, but still viscous enough to accrete steadily at the transfer rate. Whether the flow would be steady through the partially ionized regions and all the way down to the magnetosphere remains to be investigated.

In summary, we find that a model in which matter piles up in a cold, 2000 K, torus and undergoes a thermal instability when convection breaks out reproduces many of the basic properties of dwarf novae and X-ray transients. The task remains to show that a complete repetitive cycle can be produced in this manner and that the model is in accord with all relevant observations.

We thank Ed Nather, Rob Robinson, Lynn Cominsky and Saul Rappaport for discussions of the observational constraints on this model. This work was supported in part by NASA and by the R. A. Welch Foundation.

References

Bath, G. T. and Pringle, J. E.: 1981, Monthly Notices Roy. Astron. Soc., 194, 967.

Bath, G. T. and Pringle, J. E.: 1982, Monthly Notices Roy. Astron. Soc., 199, 267.

Cannizzo, J. K., Ghosh, P., and Wheeler, J. C.: 1982, Astrophys. J., 260, L83.

Cannizzo, J. K., Wheeler, J. C., and Ghosh, P.: 1982, in J. P. Cox and C. J. Hansen (eds.), Pulsations in Classical and Cataclysmic Stars, University of Colorado, Boulder, 13.

Kiplinger, A. L.: 1979, Astrophys. J., 234, 997.

Kriss, G. A., Cominsky, L. R., Remillard, R. A., Williams, G., and Thorstensen, J. R.: 1983, Astrophys. J., 266, 806.

McClintock, J. E., Petro, L. D. Remillard, R. A. and Ricker, G. R.: 1983, Astrophys. J., 266, L27.

Meyer, F. and Meyer-Hofmeister, E.: 1981, Astron. Astrophys., 104, L10.

Meyer, F. and Meyer-Hofmeister, E.: 1982, Astron. Astrophys., 106, 34.

Rappaport, S. and van den Heuvel, E. P. J.: 1982, in M. Jaschek and and H.-G. Groth (eds.), Be Stars, Reidel, Dordrecht, 327.

Smak, J.: 1982a, Acta Astron., in press.

Smak, J.: 1982b, Acta Astron., in press.

Smak, J.: 1982c, in preparation.

DWARF NOVAE - A HOT DAM INSTABILITY

John Faulkner and D.N.C. Lin
Lick Observatory, Board of Studies in Astronomy & Astrophysics
University of California, Santa Cruz

and

J.C.B. Papaloizou
Queen Mary College, London

ABSTRACT

When mass input rates are low enough, accretion disks cease to be fully ionized. The strong and oppositely directed temperature dependences of opacity (e.g. in hydrogen ionization regions) drive a <u>thermal</u> instability which modulates mass transfer throughout the disk.

The conventional thin-disk treatment breaks down in cool regions; instead, the surface boundary condition dominates the structure of these largely radiative zones. Convection, when it occurs, is inefficient and essentially irrelevant. We modify the conventional treatment to produce useful approximations to local vertical energy losses.

Our local cooling modification is then incorporated into global disk computations. Such disks are indeed thermally unstable. Large areas can be cool and optically thin between outbursts. Non-steady mass transfer within the disk allows the instability to be triggered at different radial positions. True hysteresis occurs, a new outburst depending on an unrelaxed memory of the previous outburst. Outbursts of alternating character, as in SS Cyg, are found quite naturally.

1. INTRODUCTION

This paper summarizes work described in two lengthy papers submitted for publication elsewhere (Faulkner, Lin and Papaloizou, 1983; Papaloizou, Faulkner and Lin, 1983; hereafter FLP and PFL respectively).

Our work on this problem owes a large debt to a talk by Jim Pringle (1981) at the previous CV Workshop (UC Santa Cruz 1981 Summer Workshop in Astronomy and Astrophysics: "Cataclysmic Variables and Related Systems"). In particular, Pringle described an analysis with Geoffrey

315

D. Q. Lamb and J. Patterson (eds.), Cataclysmic Variables and Low-Mass X-Ray Binaries, 315–322.
© *1985 by D. Reidel Publishing Company.*

Bath (later published in Bath and Pringle, 1982; hereafter BP) which
pointed out the necessity, for repeated disk outbursts, of a hysteresis-
like "steady-state relationship" between $\mu (\equiv \nu\Sigma)$ and Σ. Here Σ is the
integrated surface-density and ν is the characteristic local viscosity
(i.e., μ is the integrated local viscosity). However, although we found
this analysis most compelling, no physical example was proposed by BP.

It occurred to us that significantly different cooling behaviors,
possible for material which is partially or fully ionized, might provide
the physical realization of the BP idea. This possibility, familiar
enough to classical stellar structure theorists, had not been fully ex-
ploited by disk enthusiasts. While other investigators have now written
on the same subject, we differ in one or more major ways with them.
Those differences will emerge below.

2. THE LOCAL THERMAL-INSTABILITY TRIGGERING MECHANISM

2.1 A basic, but misunderstood, point

A basic point, not always appreciated, is that the (μ, Σ) "relationship" (e.g. as illustrated by Fig. 1c of BP; see also our Fig. 1) is not
a functional relationship, but is set by equating heating and cooling.
If thermal time-scales are very much shorter than viscous time-scales,
the "hookback" (region of negative slope) in the generalized S-shape will
be physically inaccessible. Corresponding "functional dependences"
(e.g. of $\mu(\Sigma)$) will be illusory and unattainable in practice. Bath and
Pringle expressed this very clearly. However, Meyer and Meyer-Hofmeister
(1981) failed to notice that their own viscous instability was constrained
to lie on an inaccessible relationship, and that even if it could, it
would only result in accumulation at one or other turning point and not
in the BP cycle which they illustrated. Smak (1982) on the other hand,
while independently producing the same viscous instability analysis,
clearly recognized that thermal instability might yet vitiate it.

To complete the discussion in general terms, the (μ, Σ) plot suppresses a possible dependence on an additional degree of freedom (related
to vertical dispersion at a given Σ, through the scale height to temperature, and thus ultimately to mass transfer rate). The latter may then
play the role of a control parameter in general catastrophe theory.

2.2 The importance of hydrogen ionization and opacity behavior

Ionization is strongly temperature dependent; so, too, are related
properties such as opacity and therefore cooling efficiency. If a modest
change in scale height for a given Σ permits access to very different
cooling behavior, the stage is set for a possible thermal instability.
Such is the case with hydrogen ionization. As stellar structure investigators are aware, the most dramatic opacity changes occur for
$T \gtrsim 10^4$ K. A tolerably accurate approximation (FLP) gives

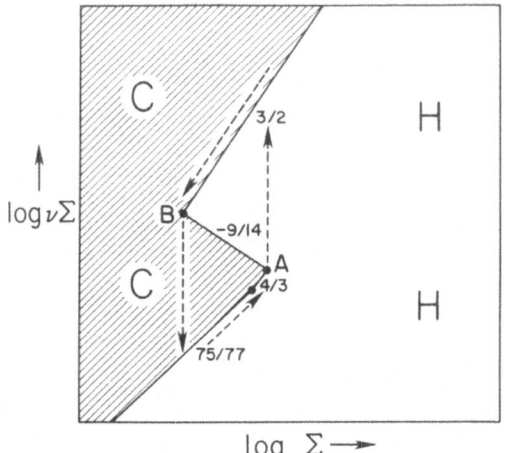

Figure 1. Analytical $(\nu\Sigma, \Sigma)$ "relationship" obtained with the modified, local, thin-disk treatment, with α-model viscosity (α constant). Heating (H) or cooling (C) dominates in regions marked. Logarithmic slopes for four major equilibrium regimes are indicated.

$$\kappa_1 \sim 10^{-36} \rho^{1/3} T^{10} \quad \text{or} \quad \kappa_2 \sim 1.5 \times 10^{20} \rho T^{-2.5} \qquad (1)$$

below and above a fairly sharp ridge line (given approximately by $T/1.2 \times 10^4 K = (10^8 \rho)^{.053}$) representing mid-ionization.

In effect, this sharp opacity ridge puts a very strong kink across the heating versus cooling diagram, causing the hookback from point B to point A in Figure 1. The analytical results in Figure 1 were obtained using the opacities given in equation (1) and our modified thin-disk treatment (see below). These results are in good agreement with full vertical, radiative integrations (Figure 2a) using Cox-Stewart tabulated opacities (with modifications for molecules, if needed, according to Alexander, 1975, as updated by Bodenheimer et al., 1980).

2.3 A modification of canonical thin-disk treatments

A subtle point has been missed in previous canonical thin-disk treatments. It has generally been taken as axiomatic, following Shakura and Sunyaev (1973), that the effective temperature T_e is related to central plane conditions for radiative disks by an equation of the form:

$$T_e^4 \sim 4T_c^4/3 \, \tau_c \qquad (2)$$

where T_c is the mid-plane temperature and τ_c is an appropriate estimate of mid-plane optical depth. However, this ceases to be correct when T_e, so calculated, becomes somewhat less than 10^4 K. This may be appreciated by examining the physical content of equation (2). It essentially states that the flux out of the disk's surface (LHS) equals the flux passing

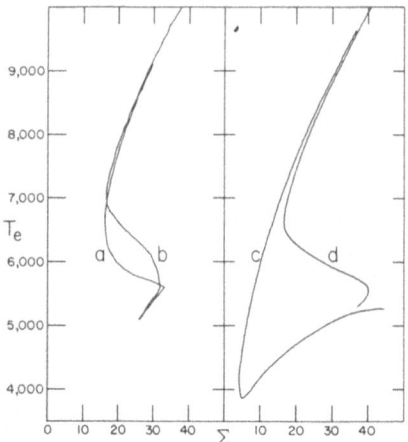

Figure 2. (T_e, Σ) "relationships" for $\alpha = 1$ at the outer edge of a disk with $M_{WD} = 1\ M_\odot$, $R = 3 \times 10^{10}$ cm. Left-hand panel, full vertical integrations (a: radiative transport only; b: mixing-length convection allowed) Right-hand panel, approximate one-zone treatments (c: classical thin-disk method using solely central-plane conditions; d: our full modified thin-disk approach). All curves employ Cox-Stewart etc. opacities.

between mid-plane and that surface (RHS). The latter involves the difference of an integrand evaluated at T_e and T_c. When κ is a decreasing function of T, the mid-plane expression indeed dominates. This has been the conventional situation.

For cool enough disks, however, κ is a strongly increasing function of T near the surface. The flux integral (RHS of equation (2)) is then dominated by the hitherto unappreciated and unexpressed surface term, until the whole disk approaches optically thin conditions. The full behavior is explored in FLP. The important point is that the hookback is determined by the condition

$$\tau_e \sim 1 \qquad\qquad\qquad (3)$$

This is identical to the surface boundary condition of classical stellar structure, which is so important in understanding the structure of late-type stars!

The condition (3) turns out to be <u>crucial</u> in promoting continued outbursts in dwarf nova accretion disks. With it, the hookback has a negative slope, preventing any steady states for the associated range of $3\pi\ \nu\Sigma$ ($\stackrel{\sim}{=} \dot{m}$ in a steady state). The canonical thin-disk treatment (e.g. Figure 2c) produces a hookback of positive slope. As a consequence, this implies at least one, and sometimes two, steady states for all values of $\dot{m}$. This is, incidentally, the major problem with Hōshi's (1979) work, which is otherwise a scientific precursor to these ideas.

We remark that convection plays a minor role in all this. We expected that this would be the case for two reasons. First, in any mixing-length type of treatment in which mixing-length $\sim$ scale height, convection is unable to work up a full head of steam. Thin disks are $\sim$ one scale height thick, and there is no net transport across the mid-plane. Thus the efficiency of convective transport is reduced because of the natural dimensional limitations on it. Second, densities are such that the thermal capacity of the material is quite low. These expectations are fully confirmed in the full integrations (Figure 2b). Convection never carries more than $\sim$25% of the flux at a given Σ; critical turning values of Σ are hardly changed.

2.4   The prospect of a local thermal limit cycle

The physics we have discussed suggests the possibility of a local thermal limit cycle as indicated in Figure 1. After a prior outburst, temperatures and viscosities are low in some appropriate region. Material builds up locally, on the marginally optically thin lower branch. Point A is reached as the disk begins to be optically thick locally. Steady-states do not exist around point A for slightly larger Σ. Physically, the disk has become opaque to the energy produced by shear motions. The converted shear energy causes local heating which in turn increases the opacity, insulating the disk still more. A runaway vertical thermal transition occurs until the representative point has effectively tunneled through the opacity barrier and emerged on the hot, decreasing opacity side. At this point temperatures, and thus viscosities (if something like the α-model prescription is used, as here) have increased so much that the material can indeed be transferred on a faster time-scale. The representative point backtracks along the upper branch towards point B. If the mass transfer rate from the companion, $\dot{M}_*$, corresponds to an excluded value for $\nu\Sigma$ (between points A and B for the particular value of radius examined), the disk becomes sufficiently transparent at point B for it to cool dramatically towards the lower branch.

Although this scenario is highly suggestive, it suffers from several potential problems. Will it operate collectively at different radii? Ionization soaks up a lot of energy. How will material advection and radial thermal contact affect the postulated instability? These questions are answered, briefly, below.

3.   GLOBAL RELAXATION CYCLES

We have demonstrated the existence and character of collective, global relaxation cycles at two main levels of approximation. The first, purely hydrodynamic, exhibits most of the general behavior found in the second, more physical thermodynamic model. The hydrodynamic model enables us to appreciate the sometimes complex disk behavior as a natural consequence of angular momentum conservation.

3.1 Avalanche and snow-plough effects

When a local region has achieved a transition to a high-viscosity
state, it behaves like an isolated high-viscosity annulus. The inner
edge spreads inwards, promoting viscosity transitions at successively
smaller radii; similarly, the outer edge spreads outwards. In the
cases we have examined, there is always a small qualitative difference
between the two transition waves. Relatively small density enhancements
occur in the inwardly travelling "avalanche," until the central-most
regions are reached. In contrast, as the outwardly travelling high-
viscosity edge butts up against low-viscosity material, a significant
density enhancement occurs. This density spike travels outwards at the
transition edge in a "snow-plough" effect.

These differences in transition details may be understood as a
natural consequence of angular momentum conservation and the asymmetry
between the two cases. When high-viscosity material lies outside an
edge (as in the inwardly travelling case), it is able to remove angular
momentum outwards efficiently. When low-viscosity material is outermost
(the other case), it cannot remove the angular momentum rapidly, so a
strong density enhancement forms. For computational convenience we held
the outermost radius fixed. In practice, the disk would spread out during the outburst, as observed for U Gem by Smak (1971).

3.2 Outbursts in the thermodynamic model

Figure 3 illustrates a number of features of the early phases of
outburst in a thermodynamic model. This particular outburst was triggered at $\sim 1/4$ of the disk radius, as can be seen from the initiating

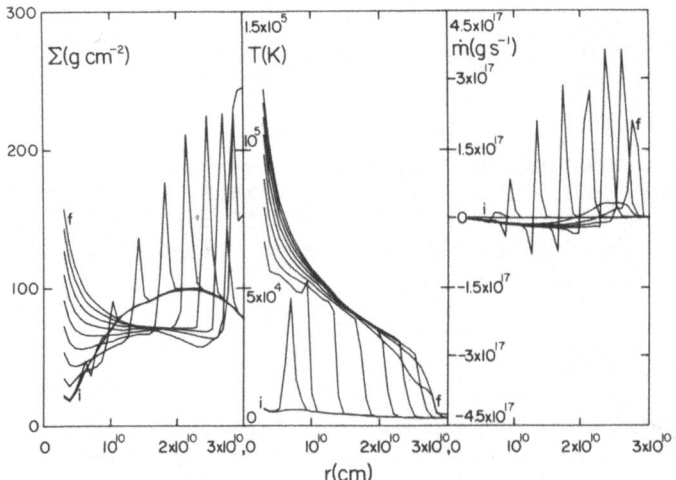

Figure 3. Onset and propagation of an upward thermal transition wave.
$\dot{M}_* = 4 \times 10^{-10}$ $M_\odot$ yr^{-1}, $\alpha = 0.1$ (standard case). Initial (i) and
final (f) states are identified; time-intervals $\simeq 2.3 \times 10^4$ s.

temperature jump. It should be noted that prior to this the entire
disk was in a low temperature ($\sim$4-5 x 10^3K), marginally optically thin
state, with a small unrelaxed thermal enhancement which served as a seed
for the outburst with the initial relatively smooth ṁ profile. We see
that the density transition waves are accompanied by marked ionization
transitions until almost the entire disk becomes hot and ionized. This
ionization wave travels predominantly outwards.

The character of this outburst dispels a number of myths which
have begun to appear in the literature. Most importantly, "cool" need
not be synonymous with "outermost regions." Our outbursts necessarily
start at a "cool" position, since they involve transitions from un-
ionized to ionized states. However, they can start quite deep within
the disk and move outwards. The tendency for the bluer parts of the
spectrum to react following the redder parts by $\sim$12-24^h (e.g. Hassall
et al., 1982; Hassall, 1985; Wu and Panek, 1982) is a natural consequence of the time delay for the initiating transition to communicate
with other parts of the disk.

We also note that true hysteresis effects occur. While successive
outbursts can be essentially identical, for other mass input ranges a
multiply repetitive pattern may emerge. The "standard model," with
major outbursts at $\sim$40^d intervals (see Figure 4) has a doubled secondary
interpulse, which may well relate to systems like SS Cyg. Here, the
nature of one outburst depends upon a memory retained of the previous
one. This example and others show that disks in outbursting systems may
never be in a truly or even approximately steady state. These implications agree with both the successes and failures obtained when steady-

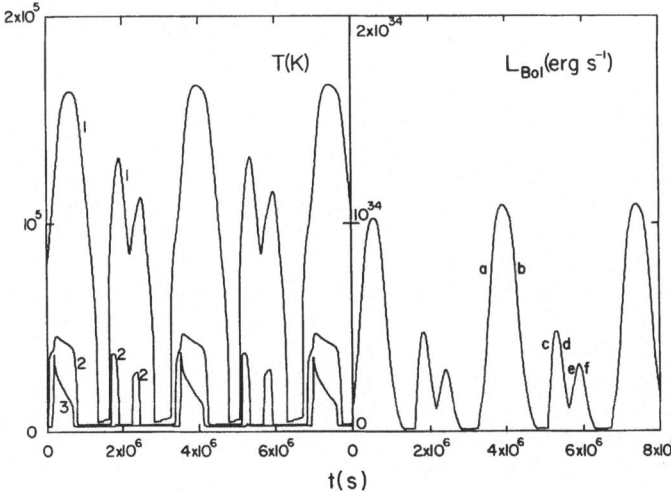

Figure 4. Limit cycles in temperature (at 3 chosen radii) and bolometric
luminosity for standard case. Note $\sim$ 40 day major outburst intervals
and similarity to SS Cyg.

state disk theory is applied to a broad spectrum of wavelengths (Hassall
et. al., 1982; Hassall, 1983; Szkody, 1982, 1983; Wu and Panek, 1982).
Our work opens up the prospect of many exciting applications to the observational data. We merely note here how ironic it is that the "steady-
state disk theory" has previously been applied with enthusiasm to the
more interesting, but necessarily non-steady, outbursting systems.

4.   CONCLUSIONS AND ACKNOWLEDGMENTS

We have demonstrated that global relaxation cycles are indeed triggered in DN accretion disks by the local hydrogen ionization thermal instability mechanism. Using reasonably realistic physics we have found
outburst characteristics and intervals ($\sim$ 20, 40, 60^d) in good accord
with observations for very reasonable mass-input rates (i.e., mass
transfer from the companion).

It is a pleasure to acknowledge the moral and even more enlightened
financial support of Robert P. Kraft, Director of Lick Observatory. In
the absence of specific external funding, this work was performed on the
Observatory's VAX computer, purchased with the aid of NSF grant AST-80-
17054. J.P. is grateful to Lick Observatory for hospitality during the
course of this work, which was unwittingly supported in part by the NSF
through grant no. AST-81-00163.

REFERENCES

Bath, G.T. and Pringle, J.E.: 1982, Mon. Not. R. astr. Soc. 199,
     p. 267 (BP).
Faulkner, J., Lin, D.N.C. and Papaloizou, J.C.B.: 1983, Mon. Not. R.
     astr. Soc. (FLP; in press).
Hassall, B.J.M.: 1985, this volume.
Hassall, B.J.M., Pringle, J.E., Wade, R.A. and Whelan, J.A.J.: 1982,
     Proc. Third IUE Conf., Madrid, p. 179.
Hōshi, R.: 1979, Prog. Theo. Phys., 61, p. 1307.
Meyer, F. and Meyer-Hofmeister, E.: 1981, Astron. Astrophys. 104, p. L10.
Papaloizou, J.C.B., Faulkner, J. and Lin, D.N.C.: 1983, Mon. Not. R.
     astr. Soc. (PFL; in press).
Pringle, J.E.: 1981, talk presented at UC Santa Cruz Summer Workshop.
Shakura, N.I. and Sunyaev, R.A.: 1973, Astron. Astrophys. 24, p. 337.
Smak, J.: 1971, Acta Astron. 21, p. 15.
Smak, J.: 1982, Acta Astron. (in press; delayed).
Szkody, P.: 1982, Astrophys. J. 261, p. 200.
Szkody, P.: 1983, Space Research 24 (in press).
Wu, C-C. and Panek, R.J.: 1982, preprint.

GENERAL DISCUSSION ON DWARF NOVAE ERUPTIONS

STARRFIELD: We're now to have a round table discussion with the four previous speakers. If the four gentlemen care to step forward, we can all take potshots at them. Icko?

IBEN: I'm sorry, I was going to ask if we could have some coffee first. [Sustained laughter.]

STARRFIELD: No!

FAULKNER: Bravo, Icko!

LONDON: John and Craig, what's the difference in the importance of convection in your two models. Craig, you said it was very important.

WHEELER: I think we all agree, Joe can pitch in on this too, that the opacity peak makes a kink in the $\Sigma(T)$ curve, and that's going to produce a thermal instability. If you do a mixing length calculation, it turns out that, no, the convection isn't very important. But I confess, I have tended to look at it and say, hey! convection is what's doing it. John has said, and I think Joe has too, well, it really is the opacities.

MANTLE: In our calculations, it is the opacity that produces the outburst. Convection...

WHEELER: But there is a legitimate difference. I'm working in a different regime, down in the very low part of the diagram, and arguing that convection is sometimes important. When convection turns on, it changes the viscosity.

F. LAMB: Why are you down there, and they're up there?

STARRFIELD: Fred's question is, why is Craig so far down in temperature from everybody else? Your're looking at molecular dissociation, are you not? Joe, do you want to show your transparency?

SMAK: I want to show a picture that I hope will partly answer your question. I think it's obvious that a very critical thing, for all time-dependent calculations, is the shape of the log Σ - log T relation. What John Faulkner was talking about, fundamentally, or mostly, is the relatively simple one-α model, with one parameter. What Craig and his colleagues are doing, is taking two α's, and it is of course nice to have two degrees of freedom. What you get at the extreme is what I got in my early calculations, where I assumed - very artificially, I admit - that there is no viscosity in the radiative zone. You get more complex, more complicated curves extending into the low temperature domain. Also, at low temperatures it becomes dramatically important what opacity you use, as usual. In particular, do we include grains or not? I'm very uncertain about the lower curves, although I tend to agree with Craig that there is something there.

D. Q. Lamb and J. Patterson (eds.), Cataclysmic Variables and Low-Mass X-Ray Binaries, 323–329.
© *1985 by D. Reidel Publishing Company.*

WHEELER: I think I need to finish answering Fred's question, if I may. In our picture, we were going to start off with nothing, and then bring matter over from the companion star. So you start at low densities and low temperatures, and watch them go up. That's sort of the way we were doing it. Then when you do the steady-state calculations self-consistently, you find out that convection breaks out down there at two or three thousand degrees. So, fundamentally that's why we were led to look there.

F. LAMB: You put in cold matter?

WHEELER: We certainly are putting in cold matter; if it's heated by energy transfer in the disk, that would be a problem. We're putting it in cold. Let me say one more thing, which comes back to what I think is the important difference. Having identified the point at low temperatures and densities as potentially unstable, we then considered the point at higher temperatures and densities, since we agree that an instability exists there. We decided that at the higher point, which corresponds to a higher accretion rate, you can't store enough matter to account for the size of the outburst, because the viscosity is higher and matter tends to move in toward the white dwarf instead of being stored.

So we looked at the instability at higher temperatures, and said that it doesn't account for the observations. We then went back down to the cold instability, because we can store more matter and get a brighter outburst. I remind you again that our models for SS Cyg are 10^{35} ergs s^{-1} at maximum light, while John's are less than 10^{34} ergs s^{-1}. I think that's a significant difference. We are working in the lower portion because it seems to account for the observations.

STARRFIELD: Thank you Craig. For people who are not afficionados of this model, I'd like to point out that the three authors actually seem to be in very basic agreement about the physics of what's going on.

SMAK: We don't know about the physics, only α. [Laughter.]

STARRFIELD: The physical origin of the instability is hydrogen ionization, which I think most people know a lot about in the context of stellar pulsations.

WHEELER: Well, I'm disagreeing. That's why I'm being so strident. I'm disagreeing.

FAULKNER: What we've all done is to proceed on the basis that "alph" a disk is better than none. [Groans, followed by laughter.]

LEIBERT: How fast does the disk go away when mass transfer from the secondary stops? TT Arietis and MV Lyrae seem to show that the disk can disappear on a timescale of days.

FAULKNER: Pringle pointed out some time ago that you get a time scale of days when α is about one, and it just scales.

MANTLE: Yes, it should be days, for an α of about one.

SMAK: Let me run a commercial for UX UMa and RW Tri, where I believe we are seeing a modulation of the mass transfer rate on a timescale of weeks, with the disk re-adjusting on virtually the same timescale. The commercial is, please observe these stars! They show very clearly how the luminosity and the disk radius change. They're a good testing ground for all of these models, because these stars are presumably in the high-temperature domain, where there are no instabilities, so we can check on everything else.

SHAVIV: You're doing a one-dimensional calculation. When you have a bump in Σ, your basic assumption is that the thermal time scale is very much shorter than the viscous time scale. On the other hand, when this bump moves to smaller radii, you actually tacitly assume that it has adjusted itself to the Keplerian velocity at that place. Isn't this inconsistent with the fact that you have assumed that the time scales are different?

FAULKNER: Yeah, there is an inconsistency of sorts there.

SHAVIV: Okay, my next question concerns convection. We all know the problems in stellar convection with the mixing length and so on. Now try to imagine convection in a highly sheared disk, where the radial scale is very large compared to the height in the z-direction. I would expect that convection takes a completely different form from what we are used to thinking about, in terms of the canonical mixing-length theory.

WHEELER: But convection in stars undoubtedly does too. [Laughter.]

FAULKNER: I'd just like to say what our philosophy was. The problem is a horrendous two-dimensional calculation with all the difficulties that you suggest. We wanted to see what the implications were of putting in the best treatment of local cooling we could do into a global analysis. We felt that the global problem hadn't been tackled yet, and there were interesting questions associated with the fact that, even though a hysteresis-type curve might exist at every radius, if it were to pop off independently of any other radius, we'd never see an outburst. So we wanted to see in what way the behavior might become collective.

SHAVIV: The intention of my remarks was to caution that we should not try to fit every small detail of the observations with such a crude model.

FAULKNER: Well, you'll notice that I didn't actually show any observations, except for SS Cyg.

VERBUNT: As to the vertical structure of the disk, if the viscosity is
magnetic, then you'd expect magnetic cells of about the same size as
the thickness of the disk. So that emphasizes again what Giora said.
I also want to urge you to calculate the optical luminosity versus
time, not just the bolometric luminosity, so we can compare with observations. And in this respect I'd like to ask you if you can make
outbursts in which the ultraviolet maximum precedes the optical maximum? That could be a major discriminant between your model and the
others, since the others can never produce such an outburst.

FAULKNER: I wouldn't like to say until we've actually done it, but I
would like to make an appeal to observers, as ever, that simultaneous
observations of a system going into outburst - preferably an eclipsing
system - would be extremely important here, because our work does suggest that there can be outbursts which start in the inner third of the
disk, or even lower.

PACZYNSKI: The standard disk instability model starts from outside, so
there's superficially very little difference between a disk instability
model and an accretion instability model. However, there's one very
spectacular difference at the beginning of an eruption, which has been
briefly mentioned but not emphasized here. In a disk instability model,
as the viscosity increases at the beginning of an eruption, the disk
should spread out, whereas in the other model, it should be the other way
around: the disk is flooded with a huge amount of matter with less
specific angular momentum than the outer part of the disk, so the disk
should momentarily shrink. Later on it may spread out again, because of
viscous effects, but at the very beginning of an eruption, these two
effects have the opposite sign. There's one object, U Geminorum,
which is both erupting and eclipsing, and two decades ago Krzeminski
and Mumford made the observations. Already at that stage, most people,
starting with Joe Smak, concluded that the disk size in fact increases
at the beginning of the eruption. That was one of the main reasons why
I've thought for a decade now that the issue has been observationally
settled in favor of the disk instability. It would be nice if someone
could repeat the observations for other objects...

MANTLE: In my mass-transfer burst models, if we use a reasonable value
of this beta parameter, say 0.5, which corresponds to an equipartition
of energy between the stream and disk, the disk doesn't shrink at all
during an outburst, even when we transfer a huge amount of mass. We
can't have the disk expanding, because our outer boundary is fixed, as
in your work. But the disk certainly doesn't shrink.

PACZYNSKY: I don't understand it, you dump...

MANTLE: Well, the stream penetrates into the disk, you see...

SHAVIV: What's the boundary condition at the outside?

MANTLE: The boundary condition we use is a fixed edge, so that you can't have any mass expanding outwards...

[The remainder of the exchange is drowned out in a cacophony of voices.]

FAULKNER: Do you get pulses of mass being forced outwards, because that's why I mentioned Joe's work, and Bep's comment is relevant. In our model we use a constant α, because there are already enough horrendous things about the calculation without changing this condition. But when the outburst starts, we definitely get mass barrelling outwards, owing to the increased viscosity and its effect upon the outer parts of the disk. I think it's clear that in practice our model would result in a fairly substantial, if temporary, increase in the radius of the disk.

MANTLE: In ours, you certainly get an increase in the size of the disk, because the viscosity contributed by the material that's being added forces the disk to spread outward. But it will happen on a viscous time scale, rather than a thermal one.

WHEELER: What sets the time scale of the outburst? Do you select the rate at which the matter comes over?

MANTLE: We just choose the mass to come over within a time that's less than that of the observed outbursts, but that's all. It's not really sensitive to the form of our mass transfer burst.

D. LAMB: I have a question for John, to sharpen the distinction between what Craig was saying and what you're saying. Do you think that the luminosity of SS Cyg during outburst is a difficulty for your model?

FAULKNER: Well, there's quite a large range to explore. And, as we've found, there are significant changes in detail as we change α or $\dot{M}$. The reason that α and $\dot{M}$ affect things like that is that there is always this lurking critical condition that the α somewhere determines whether material can, in fact, get down to a place where it will then be able to go unstable. So there's an intimate relationship there. I'm not very perturbed about the peak luminosities. I think that going up to something like four or five solar luminosities isn't bad. By increasing $\dot{M}$ and having some corresponding change in α, I'm sure that we'll be able to match that kind of luminosity. But it's a two-parameter set, and it's not linearly related, exactly, because of all the effects that go together.

Q: This question is for you, John. Do you view the mass which is moving outward as being ejected from the system, or just waiting around for the next cycle?

FAULKNER: What happens is that this mass temporarily increases the radius of the disk, but it also comes up against increased tidal torques from the companion, which help to limit the size of the disk. So as time goes on, the disk will relax back to its previous size.

PACZYNSKI: It was found by Krzmenski and Smak that the eclipses in U Gem get narrower and narrower as time goes on after the eruption. This has been interpreted as a gradual decrease in the disk size. This implies that the viscosity in the disk is small, so that it gets smaller and smaller even though you keep adding matter at a constant rate. If so, there should also be a secular change in the separation of the two emission line peaks, which reflect the rotational velocity of the disk. Again, twenty years ago, photometric evidence indicated that the emission line strengths, as indicate by the photometric U-B color, declined with time over 60 or 80 days in U Gem. This was seen against essentially constant hot spot brightness. Presumably, it implies that the white dwarf cools off on that time scale. I hope we can see that variation in the optical, without going to IUE.

SHAVIV: Would the outer layers of the white dwarf cool on that time scale?

PACZYNSKI: In any model, the white dwarf has accreted quite a lot of matter during the eruption and it takes a finite amount of time for it to cool. Theory can be very ambiguous about the time scale, so it would be very nice to see observational limits on it. I think the strength of the emission lines is the best test for it.

WADE: Have any of you included any energy deposition at the edge of the disk from tidal torques?

WHEELER: In our case we've not even included the energy from the accretion stream at the outer edge, let alone that.

MANTLE: We've included the energy of the impact, but nothing to do with tides.

STOCKMAN: In the AM Her stars we see high and low states, but nothing like dwarf nova outbursts. Since it's unlikely that the white dwarf's magnetic field significantly affects the secondary, this suggests that dwarf nova outbursts come from something which the AM Her systems don't have, namely a disk.

SMAK: May I just show one figure? Here the horizontal axis is the binary period (in seconds) and the vertical axis is the mass ratio (more exactly, the mass fraction of the secondary: 0.5 means a mass ratio of one). The red dots are dwarf novae, the green dots are novae or nova-like, or rather steady-state accretion cases. The data were taken from Ritter's catalogue, so there is no bias on my part. [Laughter.] The evidence is marginal that there is a separation, actually with the two Z Cam systems being somewhere in between. It implies that for higher mass ratios, we get higher mass transfer rates and stationary accretion, while for lower mass ratios we get lower mass transfer rates and nonstationary accretion.

STOCKMAN: Well, the AM Her stars cover that whole area, and they don't show these events. So, I think it argues against the secondary going through some unstable phase.

FIRST DETECTION OF RADIO EMISSION FROM A DWARF NOVA
(Reprinted from Nature)

A.O. Benz[1], E. Furst[2], and A.L. Kiplinger[3,4]
[1]Institut of Astronomy, ETH, Zurich, Switzerland,
[2]Max-Planck-Institut fur Radioastronomie Bonn, FRG,
[3]NASA/Goddard Space Flight Center Greenbelt, MD,
and [4]Applied Research Corp., Landover, MD, USA

ABSTRACT

Although radiation has been detected from dwarf novae from infrared through x-ray energies, radio emission has never been reported from these objects. We describe here a search for radio emission at 4.75 GHz from dwarf novae that has been carried out with the 100-m telescope at Effelsbrg, FGR. We have searched for radio emission from six dwarf novae and a source was discovered at the position of SU UMa. The source could only be detected during optical outburst and was below the threshold during quiescence. We suggest here that the radio emission arises from a non-thermal process.

1. OBSERVATIONS

SU UMa was observed with a double horn receiver system. System parameters and times of observation are listed in Table 1. The observations consist of scans in elevation, 15 arc-min long, that are centered on the optical position of SU UMa, with the offset horn situated 8 arc-min East. In fair weather conditions, effects of clouds are mostly cancelled out with the double horn system, but the scans suffer from source confusion by neighboring sources. At the galactic latitude of SU UMa (+30°) most confusion sources can be expected to be of extragalactic origin. Source counts by Maslowski et al. (1981) near 5 GHz suggest a probability of 16 ± 9% for finding a source stronger than 1 mJy in the Effelsberg beam. To smooth the effects of source confusion we scanned the optical position with different parallactic angles. Due to the siderial time periods available for observations, the directions of the scans varied by ± 35 degrees with respect to the celestial meridian. After rejecting poor scans obtained in unacceptable weather, 123 scans of SU UMa were found to be suitable for further analysis. Each scan consists of 31 three second integrations seperated by 30 arc-sec. A linear baseline was subtracted from all scans by using a mean of 11 data points at both ends of the scans. This subtraction would be expected

331

D. Q. Lamb and J. Patterson (eds.), Cataclysmic Variables and Low-Mass X-Ray Binaries, 331–335.

to fail if strong confusion sources exist; however, inspections of the individual scans do not show such sources around SU UMa. The scans were then combined by averaging data points as a function of distance from the opitcal position, thus minimizing source confusion.

TABLE 1

System Parameters and Observing Periods

Frequency.................4.75 GHz
System temperature.......65 K
Bandwidth.................600 MHz
Half power beam width....2.4 arcmin
Observing periods........22-23 April...outburst(decline)..36 scans
(1982).................13 June.........outburst(decline)..18 scans
.................25-27 June..........quiescence.........69 scans

The average flux per beam position as a function of distance from SU UMa is shown in Fig. 1. For comparison the measured antenna

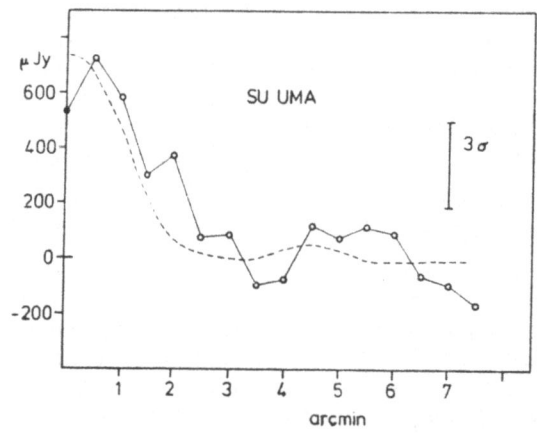

Figure 1: Radio flux vs. distance from optical source SU UMa. The antenna response of a point source (3C84) is shown for comparison.

response of a point source is also given. A source was detected at the optical position with a flux corresponding to about 6 standard deviations of the background noise at distances greater than 3 arcmin from SU UMa. The noise is close to the expected level from the receiver. Background sources seem to be weak and effectively smoothed out, hence, the field is apparently not rich in extragalactic sources. The half power width of the detected source is close to the actual beam size. No corresponding peak in the polarized intensity could be detected, suggesting that linear polarization is less than 30%.

In a second analysis we divided the observations according to the optical activity of SU UMa. The first two observing periods were obtained one or two days after the peak of optical brightness

as determined from the visual light curve of SU UMa supplied by the
American Association of Variable Star Observers. Similar to Fig. 1,
Fig. 2 shows two radial intensity profiles of SU UMa corresponding
to the outburst and quiescence observations. Although the noise
properties at distances beyond 2 arc-min are comparable during the
two states, a peak flux of 1.3 mJy is found at the position of SU
UMa only during outburst. The proability for a chance coincidence
with a source of 1.3 mJy or more is about 10%. However, the radio
flux dependence on optical activity suggested by Fig. 2 strongly
supports the idea that the detected emission during peak optical
brightness can be attributed to SU UMa.

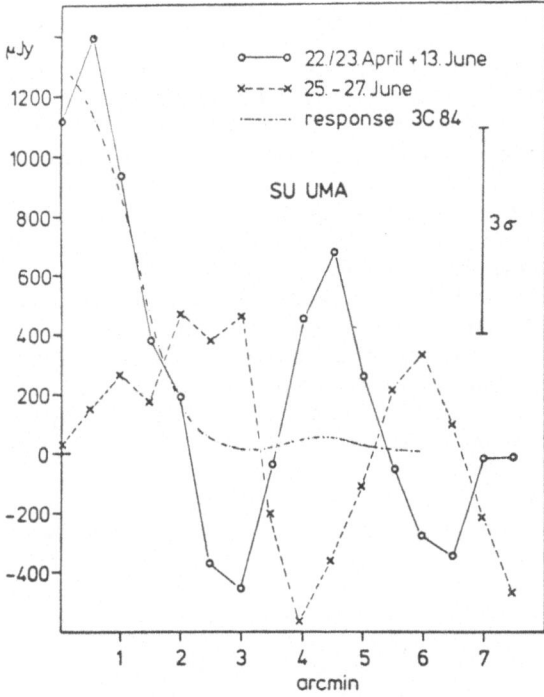

Figure 2: Radio flux
for outburst (solid
curve) and quiescent
(dased curve) periods
of SU UMa as a function
of distance from the
optical source.

2. DISCUSSION OF EMISSION MECHANISMS

X-ray observations of SU UMa in quiescence by Cordova and Mason
(1980) yield an emission measure of $7.6 \ 10^{54}$ cm^{-3} in the 0.1 –
4.5 keV band assuming a temperature of 10^8 K and a distance of 220
pc (cf. Bode et al. 1982). Independently of this hot plasma being
optically thick or thin to radio waves, the estimated free-free
emission turns out to be more than 3 orders of magnitude lower than
the observed flux. Alternatively, the mass loss observed in UV
lines of dwarf novae in outburst may be expected to emit free-free
emissions at a temperature of a few 10^4 K. The radio flux was
calculated following the derivation of Wright and Barlow (1975).
For an outflow velocity of 3000 km/s and a mass loss rate of 10^{-11}

$M_{\odot}$/y observed in similar systems (cf. Klare et al. 1982) the calculated flux is 7 orders of magnitude below the observed value.
It is therefore more reasonable to propose suprathermal electrons
as the origin of the radio emission of SU UMa.

Synchrotron emission has recently been considered for the radio
emission of the magnetic variable AM her (Chanmugan and Dulk, 1982).
The white dwarf of a dwarf nova system is expected to have lower
magnetic fields than in AM Her like binaries. If we apply this
mechanism to SU UMa, the source dimension turns out to be large
compared to the binary distance suggesting that magnetic confinement
is unlikely. However, a more detailed analysis is required to show
whether a synchrotron model is possible or not.

As an alternative to synchrotron radiation, a coherent process
may be considered. In this case, the ratio of plasma frequency to
gyrofrequency is an important parameter. There is very little
known at present about this source parameter, and it is not reasonable to propose a detailed model. However, the cyclotron maser
instability seems to be an attractive possibility. For a maser
efficiency of 0.1%, the same order as in a well-known model of
terrestrial auroral kilometric radiation (Wu and Lee, 1979), the
total energy requirement for suprathermal electrons would be only
10^{28} erg/s. This is 7 orders of magnitude below optical luminosity
during outburst. Maser spectra are expected to have small bandwidths
(a few octaves), and their emission is predicted to be circularly
polarized. Future observations may decide whether the radio emission originates from an expanding sychrotron source or a cyclotron
maser near dwarf.

We thank the A.A.V.S.O. for their prompt notification of dwarf
novae activities and Dr. D. Wills for providing an accurate optical
source position of SU UMa.

References

Maslowski, J., Pauliny-Toth, I.I.K., Witzel, A., and Kuhr, H.,
    Astron. Astrophys., 95, 285-294 (1981).
Cordova, F.A., and Mason, K.O., Nature, 287, 25-27 (1980).
Bode, M.F., Evans, A., and Bruch, A., Binaries and Multiple Stars
    as Tracers of Stellar Evolution, Proceedings of the 69th
    Colloqu., IAU, 475 (1982).
Wright, A.E., and Barlow, M.J., M.N.R.A.S., 170, 41 (1975).
Klare, G., Krautter, J., Wolf, B., Stahl, O., Vogt, N., Wargau,
    W., and Rahe, J., Astron. Astrophys., 113, 70-84 (1982).
Chanmugan, G., and Dulk, G.A., Astrophys. J., 255, L107-L110 (1982).
Wu, C.S., and Lee, L.C., Astrophys. J., 230, 621-626 (1979).

DISCUSSION

PACZYNSKI: Can this be thermal emission from a wind produced during the eruption?

KIPLINGER: I don't think so. We looked into that, but for a temperature of 10^4 $^{\circ}$K, the emitting region has to be $\sim 10^{13}$ cm in radius.

WHEELER: Was this a super-outburst or a normal outburst?

KIPLINGER: Normal. We detected it on two occasions, during two separate outbursts. Both were near the end of the decline.

MATTEI: I just wanted to bring YZ Cancri's friend, SU UMa, to everyone's attemtion. It used to have supermaxima every 250 days or so. Since March 1980, the supermaxima have stopped. And since June 1980, it has stopped having very frequent short outbursts. It's now having short outbursts every couple of months or so, compared to about 13 days previously. So something very peculiar is happening to the system.

WHEELER: Is the mean luminosity higher?

[Ed: Assorted chuckles.]

KIPLINGER: That's a good question. I think it's up a little.

Anonymous: John knows it's a good question.

FAULKNER: Yes, it came in stereo.

MATTEI: Not really. It's about the same in quiescence.

SMAK: Watch for a big event now.

STARRFIELD: Is that a prediction?

PHOTOIONIZATION MODELS FOR THE WIND FROM TW VIR

Timothy Kallman
NASA/Goddard Space Flight Center, Greenbelt, MD   20771

I. Introduction

One of the outstanding features of the UV spectra from dwarf
novae when viewed during outburst, and from many other cataclysmic
variable (CV) stars, is the existence of P-Cygni type profiles in the
resonance lines of C IV, Si IV, and N V (e.g. Heap et al. 1978;
Szkody 1981, 1982). Profiles of this type are familiar from the
study of O and B stars; in these stars P-Cygni profiles are due to
scattering by material in a stellar wind. If the line profiles from
dwarf novae are compared to those from O and B stars, it is found
that the line shapes are similar in the two classes of objects, but
that the line strengths differ. The line strengths depend on the
abundances of the scattering ions, and hence on the physical
conditions in the wind. For example, if the abundances of the ions
in the CV winds are determined by the rate of photoionization by the
local continuum radiation, then the observed UV line strengths can be
used to constrain the flux in the radiation field.

II. Line Profiles

The observed line profiles can be used to derive the density of
scattering ions in the CV wind as a function of outflow velocity by
fitting to simple model profiles. The lines we consider are the
resonance doublets of the ions C IV, Si IV, and N V in the spectrum
from the dwarf nova TW Vir as observed by Cordova and Mason (1982a;
see Figure 1 of their paper). The assumption that the TW Vir Wind is
analogous to an OB star wind has been used by Cordova and Mason to
fit the model OB star line profiles of Castor and Lamers (1979) and
Olson (1978) to the observations of TW Vir. These models are
conveniently parametrized in terms of the line optical depth at the
wavelength corresponding to a wind velocity of half the terminal
velocity, and yield values of 0.8, 0.4, and 0.8 for the optical
depths of C IV, Si IV, and N V, respectively. These fits also reveal
that the dependence of the optical depths on outflow velocity is
roughly the same for all three ions.

D. Q. Lamb and J. Patterson (eds.), Cataclysmic Variables and Low-Mass X-Ray Binaries, 337–342.
© *1985 by D. Reidel Publishing Company.*

The derivation of ion abundances from the observed line optical depths is simplified by the large outlow velocities implied by the line widths, so that the line optical depths depend only on the velocity gradient in the wind, the ion abundances, elemental abundances, and atomic physics parameters. Thus, the ratios of the line optical depths at a given velocity are proportional to the ion abundances there. If the element abundances have cosmic (Allen, 1976) values, then the ion abundance ratios are g(Si IV)/g(C IV) $\approx 2 \pm 1$, and g(N V)/g(C IV) $\approx 6 \pm 3$ at half the terminal velocity

III. Ion Abundances

The observed ion abundance ratios can be compared with models for the wind ionization in order to infer the conditions in the wind. In order to make such comparisons we assume that the ionization in the wind is supplied by radiation. We also assume that the ionization and recombination timescales are smaller than the wind flow timescale so that all ionization and recombination processes are in a steady state. Under these conditions the ion abundances and temperature at a given point in the wind depend only on the shape of the local ionizing spectrum and on the ionization parameter, the ratio of the total ionizing flux F to the gas density n, denoted $\xi = 4\pi F/n$ (Tarter, Tucker, and Salpeter, 1969). In the remainder of this discussion we explore the range of spectral shapes and ionization parameters which are consistent with the steady state

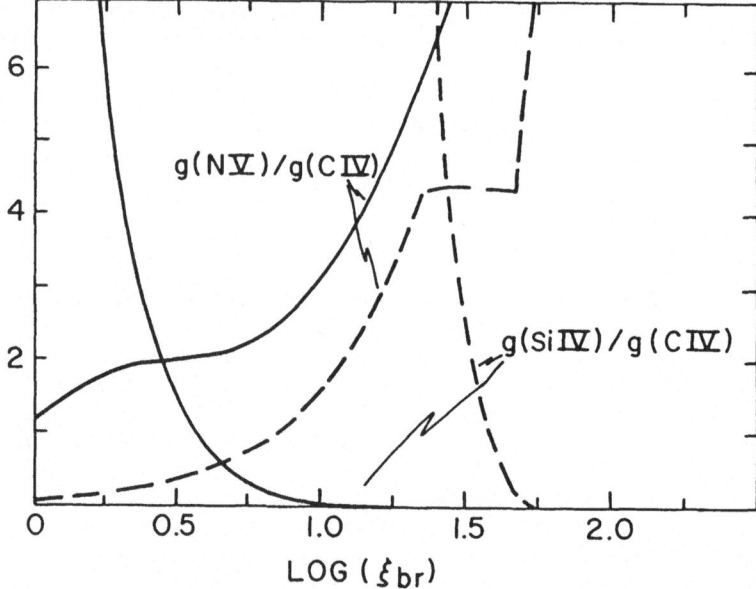

Figure 1:The ratios g(Si IV)/g(C IV) and the g(N V)/g(C IV) produced by a thermal bremsstrahlung spectrum with temperature kT_{br}=10 KeV, plotted as a function of the ionization parameter ξ_{br}. Solid curve represents spectrum which extends down to 1 eV, dashed curve represents spectrum attenuated by 3 x 10^{18} cm^{-2} of neutral matter.

assumption and with the observed values of the g(Si IV)/g(C IV) and g(N V)/g(C IV) ratios. We consider combinations of simple spectral shapes, such as bremsstrahlung or blackbody spectra, and explore values of the abundance ratios for a wide range of ionization parameters and spectral temperatures.

The simplest choice of ionizing flux distribution is one that is flat in the EUV spectral range, such as an optically thin thermal bremsstrahlung spectrum. The distribution of values of g(Si IV)/g(C IV) produced by 10 KeV bremmsstrahlung spectrum is shown as the solid curve in Figure 1 as a function of the ionization parameter corresponding to the total flux in the spectrum, ξ_{br}. These results were calculated using the photoionization equilibrium code described in Kallman and McCray (1982). At small values of ξ_{br} the g(Si IV)/g(C IV) ratio achieves large values, and at large values of ξ_{br} the g(Si IV)/g(C IV) ratio decreases to zero. The g(N V)/g(C IV) ratio, also shown in Figure 1, is a monotonically increasing function of ξ_{br}. The g(Si IV)/g(C IV) and g(N V)/g(C IV) ratios are seen to satisfy a roughly inverse relationship as a function of ξ_{br}, due to the fact that increases in ξ_{br} required to ionize N IV to N V also act to suppress Si IV relative to C IV. The response of the observed abundance ratios to a bremsstrahlung spectrum is only very weakly dependent on the bremsstrahlung spectral temperature due to the dominance of the valence shell cross sections in determining the photionization rate.

The inverse relationship between the two abundance ratios is less pronounced when the ionizing spectrum is deficient in EUV flux, relative to the soft X-ray flux. In this case the ionization of Si IV is suppressed relative to the ionization of N IV and C IV, and the g(Si IV)/g(C IV) and g(N V)/g(C IV) ratios can have values consistent with the observations in the same region of parameter space. The dashed curve in Figure 1 shows the distribution of abundance ratio values produced by a spectrum which consists of a 10 KeV bremsstrahlung spectrum which is cut off at energies below about 77 eV by absorption due to neutral materaial with a column density of 3×10^{18} cm^{-2} (Brown and Gould, 1970). These results show that values of the abundance ratios which are in accord with the observations are produced for values of the ionization parameter in the range $\xi_{br} \sim 30$-50.

Blackbody spectra do not provide a sufficiently steep decrease between the Si IV and the C IV and N IV thresholds in order to match the observed abundance ratios. In the blackbody case the g(Si IV)/g(C IV) and the g(N V)/g(C IV) rations satisfy the same rough anti-correlation as in the unattenuated bremsstrahlung case.

A combination of bremsstrahlung and blackbody spectra allows the parameters affecting the g(Si IV)/g(C IV) and the g(N V)/g(C IV) ratios to be adjusted independently, and therefore can simultaneously produce values of both ratios consistent with the observations. However, the blackbody component in this case must have a luminosity much less than that of the bremsstrahlung component (see Kallman, 1983, for details).

IV. Discussion

Our results show that the observed UV line ratios from TW Vir cannot be produced by photionization by either an unattenuated single bremsstrahlung or by a single blackbody spectrum. Satisfactory values of the abundance ratios can be produced by bremsstrahlung spectra extending into the X-ray region which have low energy cutoffs near the N IV threshold energy at 77 eV and have fluxes corresponding to a ionization paramter in the range $\xi_{br} \sim 10$. Further insight into the wind conditions can be gained if we assume that the observed UV and X-ray emission from CV's represents the continuation of our derived EUV spectra into these spectral regions. Typical CV spectra consist of an X-ray continuum with luminosity $L_x \sim 10^{31}$ erg sec^{-1}, and a UV continuum with luminosity $L_{UV} \sim 10^{33}$ erg sec^{-1} (Cordova and Mason, 1982b). The existence of such a strong UV flux will not affect our model results if the spectrum in the EUV is dominated by the blackbody component that we have considered. For example, a 150000K EUV blackbody will dominate the flux at the Si IV threshold energy produced by a 27000K UV blackbody unless the total flux in the UV blackbody exceeds the total flux in the EUV blackbody by a factor of $\sim 10^4$.

Our finding that the flux in the soft x-ray bremsstrahlung component must exceed the EUV blackbody flux by at least an order of magnitude has implications for theories of the continuum emission from cataclysmic variables. The X-ray luminosities typically observed from CV's fall short of the luminosities predicted by disk theories by a factor of 10-100 (Ferland, et al. (1982). Our models show that if the missing luminosity is hidden in the EUV region, it can't be illuminating the gas responsible for the UV lines in TW Vir.

Given the total observed CV X-ray luminosity, the ratio of the X-ray luminosity to the X-ray ionization parameter derived in Section III can be used to derive the value of the quantity nR^2, the product of the gas density and the mean square distance from the continuum source to the line-forming gas. This quantity takes on the value $nR^2 \sim 10^{30}$ cm^{-1} for $\xi_{br} \sim 10$. If we assume that the lines are formed in a spherically outflowing wind emanating from the continuum source, then the continuity equation requires that

$$nR^2 = \dot{M}/(4\pi v \mu)$$

In this equation $\dot{M}$ is the wind mass loss rate, μ is the mean molecular weight, and v is the outflow velocity. This equation implies that $\dot{M} \sim 10^{-10} M_\odot$ yr^{-1} for $v = 2400$ Km sec^{-1}. This value is smaller than the mass flux due to accretion required to produce the X-ray emission observed from the more luminous CV's by a factor of $10^2 - 10^3$.

References

Allen C. W. 1976, Astrophysical Quantities, 3d ed. (London: Athlone).

Brown, R. L., and Gould, R. J. 1970, Phys. Rev. D., 1, 2252.
Castro, J. I., and Lamers, H. G. L. M., 1979, Ap. J. Suppl., 39, 481.
Cordova, F. A., and Mason, K. O. 1982a, Ap. J., 260, 716, (CM)
                                  1982b, in IAU Transactions, Vol. XVIII A,
   Part II: Reports on Astronomy, 1982, ed. P. A. Wayman
   (Dordrecht: Reidel).
Ferland, G. J., et al. 1982, Ap. J. (Lett.), 262, L53.
Heap, S. et al. 1978, Nature, 275, 385.
Kallman, T. R. 1983 Ap. J. in press.
Kallman, T. R. and McCray, R. 1982, Ap. J. Suppl., 50, 263.
Olson, G. L. 1978, Ap. J., 226, 124.
Szkody, P., 1981, Ap. J. 247, 577.
Szkody, P., 1982, Ap. J. 261, 200.

DISCUSSION

vanPARADIJS: This whole interpretation of the line profiles comes from fitting models which have been calculated for spherically symmetric systems, where you get your emission from a large volume, and absorption from a column which is projected against the star. Now we have a completely different geometry in these flat systems, and the combination of emission and absorption depends very much on the angular dependence of emission. Isn't it very dangerous, then, to interpret the line profiles with a spherically symmetric model?

KALLMAN: Yes, you're right. If you assume it's plane-parallel and you're looking straight down it, at a jet directed at you, then you can also very easily figure out what the optical depth in the lines should be. You don't need to worry about scattering, etc. The ratio of those optical depths turns out to be very much the same as with the spherically symmetric assumptions. So it's my feeling that the results are not extremely sensitive to this. I've also included rather liberal error estimates on the fits. I think the results are fairly secure.

JENSEN: Are you saying that because of the line ratios, there has to be a larger photon flux at 77 eV than at 44 eV?

KALLMAN: Yes. I worked with energy flux, but that would still be true.

JENSEN: How do you do that with bremsstrahlung?

KALLMAN: You don't. A bremsstrahlung spectrum doesn't work, unless it is cut off. Then it will work.

WADE: To what extent does your conclusion about the ionizing spectrum depend on N V? I recall that a popular notion these days is that you go directly from N III to N V via the Auger process and soft X-rays, without needing the intermediate ionization state.

KALLMAN; According to my models, that's competitive, but not dominant.

It seems to be about fifty-fifty between the two processes. This is
true in O stars also. It also turns out that there are L thresholds
of Si right near the N V K edge, so both ratios are coupled. While it
may be true that if you had a spectrum that didn't extend out beyond,
say, 200 eV you might get a different answer, it would also affect the
Si IV/C IV ratio.

Q: Could you comment on any relation between line profile and binary
inclination?

KALLMAN: As I understand the observational situation, the absorption
component seems to be more pronounced in low inclination systems--that
is, when you're looking more down the axis of orbital motion. So, that
suggests that the wind may be more or less a plane-parallel thing.

HASSALL: Could I comment on that? If you follow a particular dwarf
nova through outburst, the spectrum changes from mixed absorption and
emission to pure absorption to pure emission, when it comes down again.
So I think it's a gross oversimplification to say that it's purely an
inclination effect. There may be some marginal correlation between low
inclination angles and the appearance of the P Cyg profiles in general,
but the statistics are rather poor.

ORBITAL PERIODS OF NOVAE BEFORE ERUPTION

Bradley E. Schaefer
Center for Space Research, Massachusetts Institute of
Technology

Joseph Patterson
Harvard-Smithsonian Center for Astrophysics

Since significant mass loss occurs in classical nova eruptions, it
is very likely that the orbital period of any classical nova will change
when it experiences an eruption. After the eruption, the orbital period
can be measured by conventional photometric and spectroscopic
techniques. However, we still do not know any pre-eruption orbital
periods, since no individual nova was known prior to its eruption. For
eclipsing systems, one possible way to measure the prenova orbital
period is to look for eclipses on archival photographic plates.

This method can be applied in only two cases, DQ Her (=Nova Her
1934) and BT Mon (=Nova Mon 1939). We have examined the archival plates
at Harvard College Observatory for prenova images of these stars. Our
results are presented in detail in the 15 May 1983 Astrophysical
Journal. A brief summary of our results is that we find:
- BT Mon's orbital period <u>increased</u> by forty parts per million as a
result of the eruption (see the figure).
- The matter ejected from the system has a mass certainly greater than
$10^{-5}M_\odot$ and probably equal to $3 \times 10^{-5}M_\odot$.
- The ejected matter is not extremely rich in angular momentum.
- BT Mon shows eclipses while it is within several magnitudes of its
peak observed brightness.
- DQ Her shows no evidence of eclipses in all available prenova data.

Below, we list various implications and lines of future research
suggested by our observations:
- For one eruption of one classical nova, the orbital period was
observed to <u>increase</u>. If we extrapolate this fact to nova outbursts in
general, we see that over the long term, nova events will increase both
the orbital period and the separation of the components. Since we
believe that classical novae are evolving to shorter periods, this shows
that the novae events themselves do not drive the evolution of the
system. On the contrary, the nova eruptions must tend to combat binary
evolution, although not very effectively. A rough calculation shows
that the long term rate of increase in the orbital period due to
eruptions is equal to the rate of decrease in the orbital period due to
gravitational radiation. So for the evolution of the system to proceed,

343

D. Q. Lamb and J. Patterson (eds.), Cataclysmic Variables and Low-Mass X-Ray Binaries, 343–346.
© *1985 by D. Reidel Publishing Company.*

there must be some other mechanism besides gravitational radiation which will decrease the orbital period. Angular momentum loss due to magnetic breaking is one such mechanism, extensively discussed by others during this workshop.

- If we ignore the small effect due to changes in the secondary's mass, the fractional change in the radius of the secondary's Roche lobe will be two thirds of the fractional change in orbital period. Due to the eruption, the radius of the secondary's Roche lobe increased by roughly 30km. However, the optical brightness (which comes primarily from the accretion disc) was identical before and after the eruption. This implies that either the secondary expanded roughly 30km in a several year period or that the radial thickness of the zone where matter is peeled off the secondary greatly exceeds 30km.

- Wachmann (1968) records many magnitude estimates for BT Mon during its 1939 outbust. Eclipses with the modern period are certainly visible in his data after 1941.0, at which time BT Mon had faded by roughly six magnitudes to 13^m. Of the 39 observations between 1939.9 and 1941.0, five of the six faintest measurements (after long term trends are subtracted) occur at the eclipse phase. It is difficult to assess whether these measurements are real detections of the eclipse phenomenon. The earliest of these possible eclipses occurs in 1940.0. The observations of the amplitude of eclipse during the nova outburst can be compared in detail to the models of the distribution of luminosity in the system. It appears that over one quarter of the light was being eclipsed while BT Mon was two to four magnitudes below the observed maximum. This implies a somewhat smaller photospheric radius than allowed for in the model of Bath (1978).

- Our observation of the period change in BT Mon provides a <u>dynamical</u> measure of the mass ejected from the system. Happily, our preferred value is not far from the "standard" value of $5 \times 10^{-5} M_o$ which has been determined for many novae using spectroscopic techniques. Unlike the spectroscopic techniques, our dynamical measure is independent of the shell's ionization, composition, and distance. It would be nice to compare the dynamical and spectroscopic measurements of the mass ejected from BT Mon. Such a comparison is now being carried out by Ferland, Schaefer, and Patterson.

It is unfortunate that BT Mon appears to be the only star for which the archival plate collections can yield an orbital period change. However, this one observed case can teach us a lot about nova eruptions.

References

Bath, G.J. 1978, MNRAS <u>182</u>, 35.
Wachmann, A.A. 1968, Astron. Abhand. Hamburger Stern., <u>7</u>, 387.

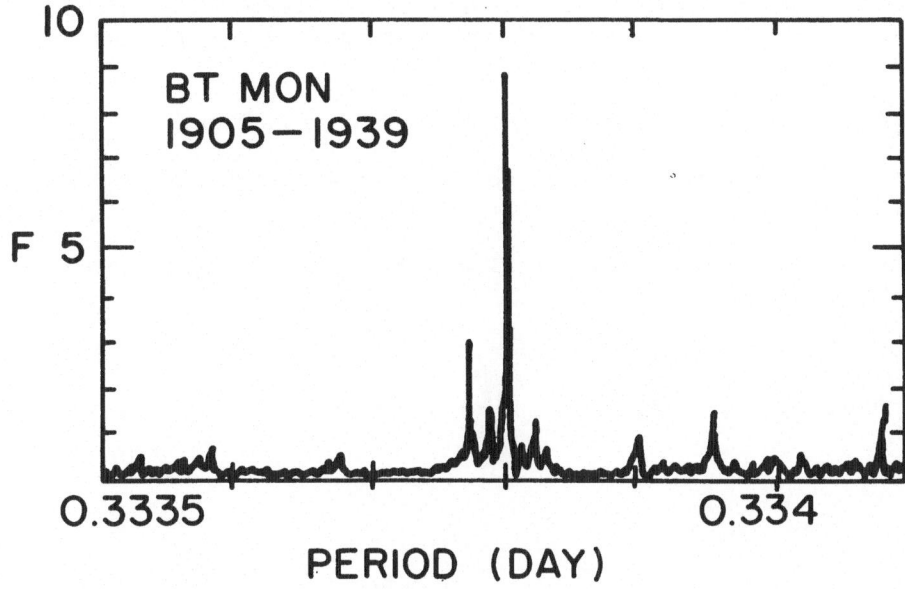

Figure. Periodogram for prenova data of BT Mon.
The sharp peak at 0.3338010^{d} indicates that before its nova eruption in
1939, BT Mon had an orbital period which is shorter than the modern
value. Along the vertical axis is plotted an F statistic for a given
trial period. The F test is applied to the two hypotheses that the
light curve is constant and that it is eclipsing with similar shape as
the the modern light curve.

DISCUSSION

JOSS: For a given mass of the ejected shell, the final period will be
different--even with spherical symmetry--depending on whether the mass
is lost on a timescale long, or short, compared to the orbital period.
Which of these assumptions did you make in deriving the ejected mass?

SCHAEFER: It doesn't really matter, as long as the orbit remains
circular. And even with an eccentricity, it won't be important. The
fractional change of the orbital period due to an induced eccentricity,
e, will be $-1.5 \ln (1 - e^2) = -1.5e^2$. On the assumption that the matter
is not ejected in a jet, the eccentricity induced by the eruption will
be roughly 3×10^{-5}. The observed approximate symmetry of the nova shell
tells us that jet ejection did not occur. Therefore, the induced
eccentricity should cause a negligible fractional period change of 10^{-9}.

PRIEDHORSKY: Why didn't you see eclipses of DQ Her before the eruption?
Is it consistent with the existence of an eclipse?

SCHAEFER: Because DQ Her has a relatively short orbital period, the photographic exposure times of 1-2 hours cause a large smearing of the true light curve. So we think that the data quality was just not sufficient.

PRIEDHORSKY: Which data was the previous result based on?

SCHAEFER: On 27 archival plates.   We're up to 40 plates, including their 27.

PACZYNSKI: Is it usual for such a slow nova to have so high an ejection velocity (2000 km/sec)?

STARRFIELD: No, it's not.

SHAVIV: There are no usual novae.

CATACLYSMIC VARIABLES AS PROBES OF X-RAY PROPERTIES OF INTERSTELLAR GRAINS

M. F. Bode
Los Alamos National Laboratory, Los Alamos, NM 87545, USA

A. Evans and G. A. Norwell
University of Keele, Staffordshire ST5 5BG, UK

ABSTRACT

Interstellar grain properties have previously been probed at wavelengths ranging from the infrared to the ultraviolet. Recent work by other authors has shown that we may also observe the effects of scattering by such grains at X-ray wavelengths. In this paper we suggest that investigations of the X-ray properties of interstellar grains may profitably be conducted in sight lines to variable sources. Particular emphasis is given in this context to cataclysmic variables and related objects.

I. INTRODUCTION

The prediction that scattering by interstellar grains of X-rays from celestial sources would tend to extend the images of such sources was first made by Overbeck (1965). Hayakawa (1970) and Martin (1970) both stressed that observations of these effects would lead to a fuller understanding of dust grain types, sizes and number density distributions. Trümper and Schönfelder (1973) suggested that by observing time variable scattered haloes around variable sources, extinction independent source distances could be derived. As grains preferentially scatter soft X-rays it was in this domain that it was suggested the search should be made.

EINSTEIN with its imaging capabilities gave the first real opportunity to look for scattered haloes. Indeed, for any plausible interstellar grain population these effects must be detectable at some level. However, X-ray telescope optics themselves introduce an instrumental scattering of X-rays from celestial sources (the point response function-PRF) making the detection of relatively faint haloes all the more difficult. Despite this and other problems, recent work by Rolf (1983) and Catura (1983) has uncovered detections of haloes in EINSTEIN IPC and HRI images of five distant, bright sources.

347

D. Q. Lamb and J. Patterson (eds.), Cataclysmic Variables and Low-Mass X-Ray Binaries, 347–353.
© 1985 by D. Reidel Publishing Company.

Perhaps a more promising method of detecting haloes and one which would potentially yield more information has been suggested by Norwell et al. (1983) and is outlined below. In essence if a variable source suddenly enters a low state (due to the end of an outburst for example) the scattered halo will persist due to finite light travel times. Thus it would be more easily visible against any intrinsic halo from the low state source formed by the effects of the telescope optics. In this short paper we will briefly outline the theory appertaining to the potential detection of transient haloes from variable sources. Particular attention will be paid to the detection of such effects around cataclysmic variables.

II. MODEL

At X-ray wavelengths interstellar grains of radius $a \lesssim 0.2$ μm are assumed to scatter according to the Rayleigh-Gans approximation (Hayakawa, 1970). The differential scattering cross-section (a measure of the fractional energy scattered into angle $\emptyset$ from the source-grain vector per unit solid angle) is thus given by

$$\frac{d\sigma}{d\Omega} = 2a^2 \left(\frac{2\pi a}{\lambda}\right)^4 |m-1|^2 \left(\frac{j_1(x)}{x}\right)^2 (1 + \cos^2\emptyset) \tag{1}$$

In equation (1) λ is the wavelength of observation; m is the wavelength dependent complex refractive index of the grain material and $j_1(x)$ is the first order Bessel function where

$$x = \left(\frac{4\pi a}{\lambda}\right) \sin\left(\frac{\emptyset}{2}\right).$$

The halo surface brightness can be calculated from

$$I_\lambda(\alpha,t) = L_\lambda(t) n_g d \int_{\Theta_A(\alpha,t)}^{\Theta_B(\alpha,t)} \frac{1}{4\pi r^2} \frac{d\sigma}{d\Omega} \exp\left(-(\tau_1+\tau_2)\right) \frac{\sin\alpha}{\sin^2(\alpha+\theta)} d\Theta \tag{2}$$

where L_λ is central source luminosity per unit wavelength; n_g is the

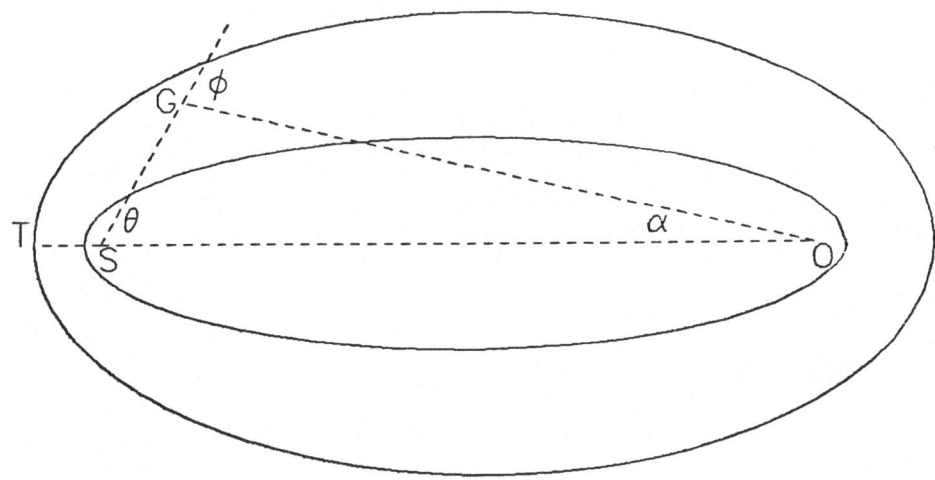

Figure (1). Scattered halo geometry in the case of an intrinsically variable source. S, O and G are positions of source, observer and grain respectively. Scattered X rays are seen only from grains lying between two ellipses defined by 'echo' of the source switching on and off. Distance ST = ct/2 where t is time from observed source switch on.

interstellar grain number density; d is the source observer distance (SO in figure (1)) and r is the source-grain distance (SG in figure (1)).

Angles α and Θ are as defined in figure (1). Optical depths τ_1 and τ_2 are those for paths SG and GO respectively. These include wavelength dependent extinction contributions from the interstellar grains and absorption from the interstellar gas, the latter as given by equation (3.2) of Hayakawa (1970).

For a constant central source $\Theta_A=0$ and $\Theta_B = \pi-\alpha$. However, if the source suddenly "switches on" at observer time t=0 and then "switches off" at time t=Δt then, due to light travel time effects, an observer will see scattered X-rays arising from scatterings occurring between two confocal ellipsoids of revolution with the source and observer lying at the respective foci. For any given azimuthal angle α from the central source the integration limits Θ_A and Θ_B will thus be time dependent.

The central source X-ray behavior has been chosen to approximate that which might be expected from certain dwarf novae at outburst. Observations by Rappaport et al. (1974) and Mason et al. (1978) and others have for example shown that U Gem and SS Cyg have increased in soft X-ray luminosity by factors of up to 100 at optical outburst. In these calculations, we have assumed L_{max}=100 L_{min} = 10^{33} ergs s^{-1} and

Δt = 2 days. The emission is assumed to be black body with T=10^6K (cf. Cordova et al., 1980).

In order to compare the expected halo brightness with the possibilities of observation, equation (2) was integrated numerically over Θ and then with respect to λ, over the energy range 0.23-2.48 keV. This energy range is that for which refractive indices of grain materials at X-ray wavelengths are available (Rolf, 1980) and approximates to part of the EINSTEIN IPC band (0.1-4.5 keV).

III. RESULTS

Figure (2) shows the temporal evolution of the halo surface brightness as a function of α for 0.1 μm silicate (olivine) grains with interstellar number density 2.10^{-12} cm^{-3}. Superimposed are approximate illustrative levels of 3000s IPC background and the PRF for the low state source.

It can immediately be seen that the resulting halo takes the form of a transient ring of soft X-ray emission about the central source. The ring may be visible above background for several days after outburst. The angular radius of the peak of the ring emission increases with time from ~2.5 arcmin at t=2.5 days to ~9 arcmin at t=4.5 days. At the distance of the source (200 pc) the expansion appears superluminal.

IV. DISCUSSION

The foregoing results indicate that the detection of transient haloes arising from scattering of soft X-rays by silicate grains may indeed be feasible. Further calculations have been performed using grains of varying sizes and compositions (e.g. graphite), and the evolution of the halo is found to be sensitive to both these variables. Halo evolution will also depend on soft X-ray light curves and source distances. If we can constrain some of these otherwise free parameters by using related observations (for example determination of the optical and ultraviolet extinction laws; monitoring the X-ray outburst light curve, etc.) we should in principle be better able to derive grain types and source distances. Of more than 30 dwarf novae observed by EINSTEIN, seven were observed <10 days after an optical outburst. It might be instructive to re-examine these IPC fields in the soft X-ray bands to look for possible extended, time variable structure.

Application of this model can of course be made much more widely to include any suitably variable source of soft X-rays. Extrinsically variable sources (i.e. those showing eclipses) would also be worth considering, though the variable halo model would of course be somewhat different in detail.

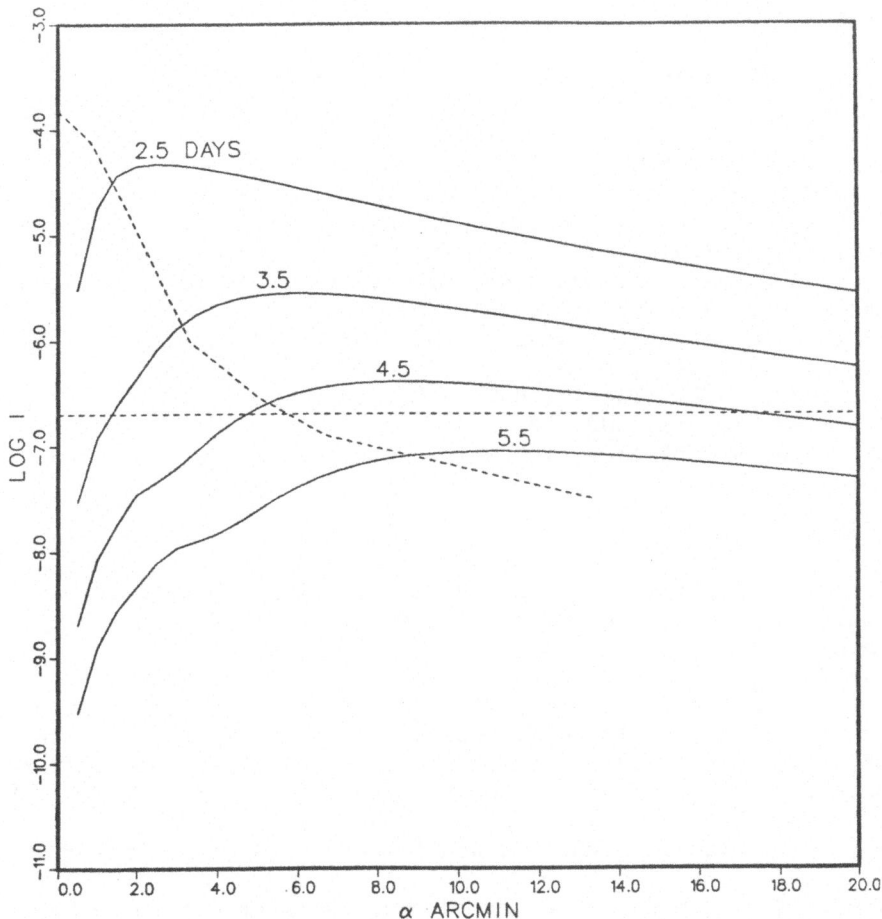

Figure (2). Temporal evolution of scattered halo intensity, I (in ergs s^{-1} cm^{-2} sterad^{-1}) as a function of azimuthal angle from source, in the energy band 0.23–2.48 keV. Results shown are for 0.1 μm silicate grains and model parameters as given in text. Approximate 3000s IPC background and PRF of low state source (Rolf 1980) are shown as dashed lines.

V. CONCLUSION

We have shown that suitably timed imaging X-ray observations of cataclysmic variables and related objects should in certain cases show evidence of transient scattered soft X-ray haloes. The study of such haloes will undoubtedly lead to a greater understanding of interstellar dust grain properties.

Work completed, in part, under the auspices of the U.S. Department of Energy.

REFERENCES

Catura, R., 1983, BAAS, 14, 893.
Cordova, F. A., Nugent, J. J., Klein, S. R. and Gamire, G. P., 1980,
     Mon. Not. R. Astr. Soc., 190, 87.
Hayakawa, S., 1970, Prog. Th. Phys., 43, 1224.
Martin, P. G., 1970, Mon. Not. R. Astr. Soc., 149, 221.
Mason, K. O., Lampton, M., Charles, P. and Bowyer, S., 1978, Ap. J.
     Lett., 226, L129.
Norwell, G. A., Evans, A. and Bode, M. F., 1983, Nature, submitted.
Overbeck, J. W., 1965, Ap. J., 141, 864.
Rappaport, S., Cash, W., Doxsey, R., McClintock, J. and Moore, G., 1974,
     Ap. J. Lett., 187, L5.
Rolf, D., 1980, Ph.D. Thesis, Univ. Leicester.
Rolf, D., 1983, Nature, 302, 46.
Trümper, J. and Schönfelder, V., 1973, Astr. Ap., 25, 445.

DISCUSSION

PRIEDHORSKY: I might mention that France Cordova has an observation of
SU UMa which appears to show a very faint ring at a distance of, I think,
11 arc-minutes.

BODE: It's about 14 arcminutes, I've seen that. I didn't mention it,
but that fits in extremely well, because it's soft X-ray, it's the right
distance, it's about 9 days after an optical outburst. We don't know
what the X-ray outbursts of that object are like.

FABBIANO: It's also true that it wasn't seen in a longer observation.

BODE: Yes. That was during a period when the star ceased to have outbursts. You wouldn't expect to see it then in this model. It's too
early to propose this as a definite model, but if the effect is real, if
it isn't something instrumental, then I think it's the first positive
detection of an X-ray halo.

JENSEN: It would be nice to have an Exosat observation during outburst.

BODE: We've got five 10^3 second observations planned.

CARROLL: Can you get information about the orientation of the grains?

BODE: Yes, you probably could, because the scattering will depend on
the orientation of the grains.

JENSEN: What's the high energy cutoff you'd expect?

BODE: Well, there isn't one. It just tails off to invisibility. I've
used a band pass from about 0.5 keV longwards. The hard X-rays aren't
important, because the scattering is so low.

Q: Can you say anthing about the chemical composition of the grains?

BODE: Yes. My collaborators have used graphite grains, and you get a halo with different properties. If you could find a halo bright enough to get a spectrum, you could find, say, oxygen in the grains.

ON MEASURING THE RADIAL VELOCITY OF WHITE DWARFS IN CATACLYSMIC
BINARIES

Allen W. Shafter
Department of Astronomy,
University of California, Los Angeles

It is notoriously difficult to determine the semiamplitude of the
white dwarf in a cataclysmic binary (Smak 1970). This is a result of
the fact that spectral features arising in the photosphere of the white
dwarf are rarely, if ever, seen in the spectra of these systems. The
motion of the white dwarf must be inferred from radial velocity variations of the broad emission lines which originate in the surrounding
accretion disk. This would not propose any particular problem if the
disk emissivity was axially symmetric. Unfortunately, this is usually
not the case, primarily because of enhanced emission in the vicinity of
the shock front (hot spot) where the interstar mass transfer stream im-
pacts the disk. The challenge, then, is to measure the velocity of the
emission lines using a method which is insensitive to contamination
from the hot spot. The purpose of this paper is twofold: (1) to review a simple method of measuring the velocity of emission lines in
cataclysmic binaries and (2) to remind the reader of the disastrous
consequences which can arise if care is not exercised when measuring
the emission line velocities.

It is generally agreed that the best method to determine the radial velocity of the white dwarf is to employ a line measuring technique which is mainly sensitive to the motion of the line wings. The
high velocity emission line wings are presumably formed in the inner
parts of the accretion disk near the white dwarf and thus should reflect its motion with the highest reliability. One such method was
outlined by Schneider and Young (1980). Here, I discuss a preliminary
analysis of this method and conclude by applying it to the dwarf nova
T Leo.

As pointed out by Schneider and Young, the wavelength, λ, of an
emission line in a spectrum $S(\Lambda)$ can be found by solving the equation:

$$\int_{-\infty}^{\infty} S(\Lambda) \ G(\lambda - \Lambda) \ d\Lambda = 0, \qquad (1)$$

where $G(x) = \exp[-(x-a)^2/2\sigma^2] - \exp[-(x+a)^2/2\sigma^2]$.

355

D. Q. Lamb and J. Patterson (eds.), Cataclysmic Variables and Low-Mass X-Ray Binaries, 355–358.
© 1985 by D. Reidel Publishing Company.

Generally speaking, this method consists of convolving S(Λ) with two identical Gaussian bandpasses whose centroids have a separation of $2a$. Equation 1 is satisfied and the wavelength of the spectral feature is determined when the counts in each bandpass are equal. The choice of the parameters a and σ can be chosen to suit the characteristics of the spectra being analyzed (i.e., the emission line width and the signal-to-noise ratio of the data). This method has the advantage that the entire velocity profile of the line can be mapped out by varying a.

In general, I have found that the value of the semiamplitude (K_1), which is derived from a fit of the resulting velocity points to a sinusoid, is a strong function of the parameter a (Shafter 1983; Shafter and Szkody 1983). This is easy to understand when the emission line profiles are asymmetrical *and* this asymmetry is a function of orbital phase. In most cases, the line asymmetry can be attributed to emission from the hot spot and/or the interstar mass transfer stream. As pointed out by Smak (1970), contamination by the hot spot results in a measured semiamplitude which is spurious; usually *smaller* than the true value of K_1. In the simplest model, the maximum velocity amplitude of the emission from the hot spot is equal to the rotational velocity at the outer edge of the accretion disk. This velocity is much smaller than the Keplerian velocity at the inner edge of the disk where the extreme line wings are formed. Consequently, K_1 should increase as one measures further out in the line wings, where the contamination from the hot spot is less pronounced (i.e. $K_1(a)$ should be an increasing function of a).

As an example, I have employed the above analysis in a radial velocity study of the dwarf nova T Leo (Shafter and Szkody 1983). I have co-added 121 individual 8 minute spectra of T Leo synchronously with the 84.7 minute orbital period and summed them into 10 phase bins. I then measured the velocities of the 10 Hα emission lines using 11 different values of the parameter a. The result is shown in Figure 1. As expected, the value of K_1 increases with increasing a. I argue that the best estimate of the semiamplitude is determined by increasing the value of a until the fractional error (σ_K/K_1) in the resulting semiamplitude begins to sharply increase. At this point, the velocity measurements are beginning to be

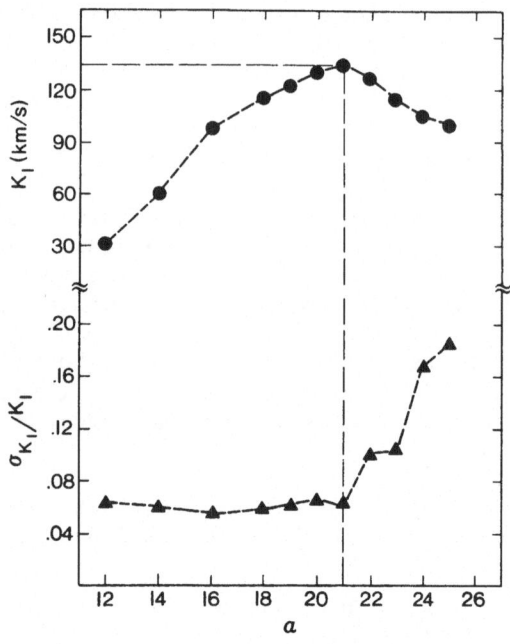

Fig. 1.   The diagnostic diagram for the dwarf nova T Leo.

dominated by noise in the continuum rather than by flux in the line
wings. In some, but not all, cases, this point will correspond to the
maximum value of K_1. The best estimate of K_1 for T Leo is indicated by
the dashed line in Figure 1. The resulting radial velocity curve is
shown in Figure 2.

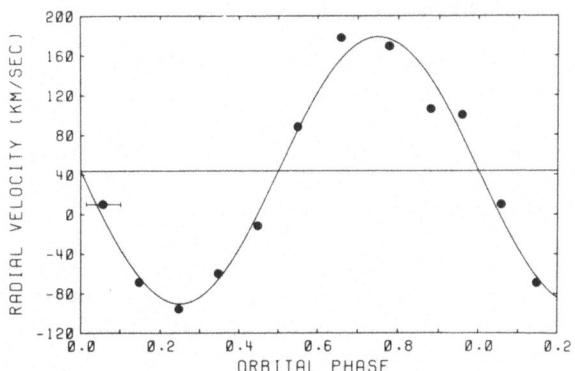

Fig. 2. The radial velocity curve for T Leo. The velocities were mea-
         sured using equation 1 with a = 21 Å and σ = 1 Å.

An example which dramatically illustrates the importance of re-
stricting the velocity measurements to the extreme line wings is provided by a comparison of the radial velocity curves presented in Figures 2 and 3. As described above, the curve presented in Figure 2
was derived using equation 1, with a = 21 Å. While, for purposes of
comparison, the curve presented in Figure 3 was derived from measurements of the centroid of the entire Hα emission line from the 10 co-
added spectra. By simply measuring the centroid of the line, I have

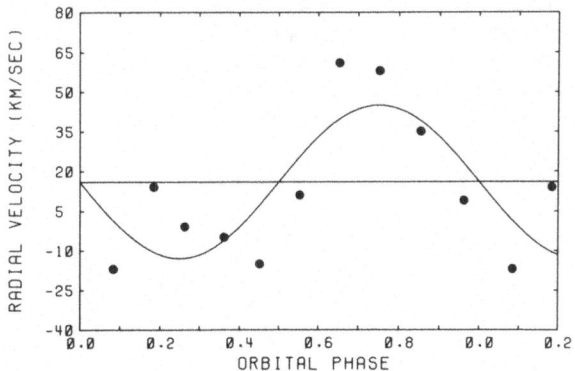

Fig. 3. The radial velocity curve for T Leo. The velocities were mea-
         sured from the centroid of the entire Hα emission line. Note
         the poor fit and low semiamplitude compared with Figure 2.

deliberately made no attempt to avoid contamination from the hot spot
or from any other non-orbital mass motions.   Although the curves shown
in Figures 2 and 3 were derived from measurements from the *same* 10
spectra, the difference is striking.   It is clear that failing to restrict the measurements to the extreme line wings can result in not
only a poor fit of the velocity points to a circular orbit, but also a
serious underestimation of the semiamplitude.

        Although I believe that the line measuring technique outlined in
equation 1 is particularly effective, I have not intended to imply that
it is the only method, nor that it is necessarily the best method of
estimating the radial velocity of the white dwarf in a cataclysmic binary. For example, other line measuring techniques which are frequently employed, such as fitting a Lorentzian profile (Gilliland 1982) or a
high order symmetric polynomial (Stover, Robinson, and Nather 1981) to
the line wings, may potentially be as effective as the method presented
here.   However, the important point is that, regardless of the line
measuring technique used, a special effort should be made to assure
that one is measuring as far out in the line wings as is practically
possible.   I believe that the best way to accomplish this is to construct a diagnostic diagram analogous to the one presented in Figure 1.
The diagnostic diagram is a useful tool for assessing the degree of hot
spot contamination and for determining the best estimate of K_1 which
can be extracted from the available data.

        I thank R.K. Ulrich for financial support through a grant from the
National Science Foundation (AST78-20236).

REFERENCES

Gilliland, R.L.: 1982, Astrophys. J. 254, pp.653-657.
Schneider, D.P. and Young, P.: 1980, Astrophys. J. 238, pp.946-954.
Shafter, A.W.: 1983, Astrophys. J., (in press).
Shafter, A.W. and Szkody, P.: 1983, (submitted to the Astrophys. J.).
Smak, J.: 1970, Acta Astronomica 20, pp.311-326.
Stover, R.J., Robinson, E.L., and Nather, R.E.: 1981, Astrophys. J.
    248, pp.696-704.

ACCRETION DISCS

J. I. Katz
Department of Physics and McDonnell Center
for the Space Sciences, Washington University
St. Louis, Missouri 63130

Accretion discs are believed to be important in a wide variety of astronomical objects, ranging from quasars to cataclysmic variables. Attempts to construct their theory from first principles founder on our ignorance of the processes of angular momentum transport. Phenomenological theories and empirical constraints, although model dependent, are useful. Empirical evidence and constraints are found from the study of systems as diverse as Her X-1, SS433, and the SU UMa stars.

INTRODUCTION

This talk is a deliberately personal and idiosyncratic view of the present understanding of accretion discs. It is not meant to be complete or balanced, but rather to concentrate on those problems which are presently interesting in the study of mass transfer binary stars. It therefore reflects my own prejudices, and I apologize to those whose work and views are slighted or ignored. I will be particularly concerned with the dynamics of accretion discs, largely as inferred from the comparison of phenomenological models with observations.

The theory of accretion discs is based on the assumption that matter orbiting a central object may lose its thermal energy by radiation much faster than it may lose angular momentum. This assumption is sensible, because radiation is generally a rapid process, but carries a negligible amount of angular momentum. Therefore, matter may be expected to settle to its state of lowest energy consistent with conservation of angular momentum. For any given fluid element the state of lowest energy is a circular orbit. Orbits of the same radius will collide and radiation will reduce them to a single circular ring whose size and orientation are determined by the mean angular momentum. This process will be chaotic but brief. If the surviving rings are coplanar (friction between them will tend to bring this about) and extend over a range of radii the resulting object will resemble a disc, with a thickness determined by the residual thermal energy.

D. Q. Lamb and J. Patterson (eds.), Cataclysmic Variables and Low-Mass X-Ray Binaries, 359–377.
© *1985 by D. Reidel Publishing Company.*

This sketch of the reason why we expect accretion discs to exist was implicit 200 years ago in Laplace's model for the origin of the solar system. The concept was revived about 40 years ago in the study of interacting binary stars, during which mass transfer streams and accretion rings and discs were observed spectroscopically by Struve and Joy. They were studied extensively in subsequent decades, initially in binaries containing only nondegenerate stars, but later by Kraft in cataclysmic variables. The recognition of the enormous luminosities and peculiar properties of quasars in the mid-1960's led to the consideration of supermassive accretion discs, or accretion discs surrounding supermassive objects. Following the blossoming of X-ray astronomy in 1971 the possibility and importance of accretion discs surrounding neutron stars and black holes of stellar mass were recognized. In the last decade accretion discs surrounding compact objects of all varieties have been the focus of energetic research and lively controversy, as cataclysmic variable and X-ray astronomers have found their interests converging.

The early theoretical work on accretion discs was principally concerned with their place in models of quasars and active galactic nuclei. Lynden-Bell (1969) formulated the important physical processes in this context. Novikov and Thorne (1973) extended this theory to include relativistic effects and elaborated it in great detail in order to apply it to accretion discs surrounding neutron stars and black holes in stellar X-ray sources. The theory consists of the equations of gravitation and hydrodynamics (conservation of mass, momentum, and energy), together with material constitutive relations (equation of state and opacity), thermodynamic relations and a theory of energy transport (usually either the diffusion of radiation or some assumption about convective equilibrium). In addition the viscous torque and energy sources must be described, as must sufficient boundary conditions. The problem resembles that of stellar structure, but is much complicated by the lack of spherical symmetry. The equations are written explicitly by Novikov and Thorne.

This very complex problem is simplified by the neglect of relativistic effects and self-gravitation (considering only the attraction of a central mass), and the assumption of azimuthal symmetry. Additional simplifications usually made are the assumptions that the disc is geometrically flat and thin and that the fluid follows nearby Keplerian orbits. Then the two-dimensional problem separates into two one-dimensional problems: the structure of the disc normal to its midplane (which resembles a one-dimensional atmosphere in an imposed gravitational potential) and the dependence of this structure on radius from the central object. These problems are tractable, and even admit approximate simple analytic solutions. Figure 1 shows a qualitative picture of such a solution.

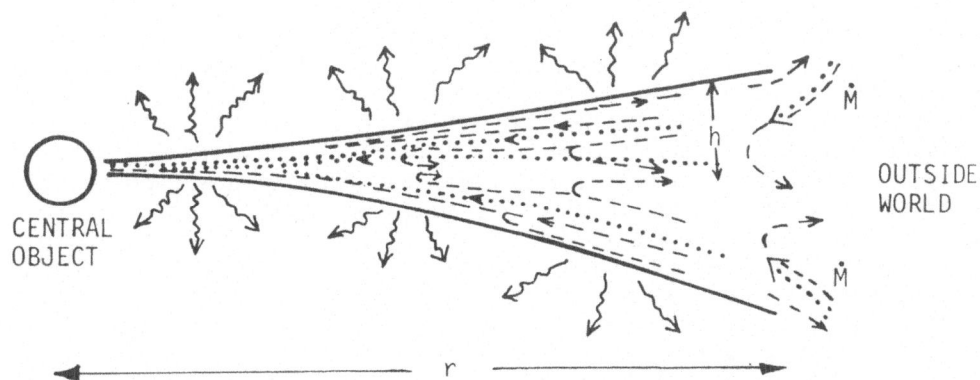

Figure 1: Qualitative simple disc model showing radiation from
           surface, mass flow $\dot{M}$ (dotted lines), and angular momentum
           flow (dashed lines), radius r and thickness h. Note that
           angular momentum flows in by advection and out by viscous
           torques, which need not balance exactly; circulatory flows
           are possible, as shown. Energy is generated by viscous
           heating, and radiated near its source; thermal pressure
           gives the disc its finite thickness.

     In a stellar interior the force of gravity is opposed by a pressure
gradient, while in a disc it is opposed by both pressure and angular
momentum. In a thin disc pressure is by far the less important ($p \ll \rho v^2$,
where p, ρ, and v represent pressure, density, and fluid velocity), and
its loss would only result in a collapse to zero thickness in the disc
midplane. Angular momentum prevents a sudden collapse onto the
central object. If the analogy between stars and discs is continued,
the stellar nuclear energy sources are insignificant in discs, and their
place is taken by viscous heating, for a Keplerian disc rotates differentially.

     The goal of disc theory is to calculate all the physical parameters
of discs — the distributions of density, pressure, temperature, and
velocity within them. In practice most of these quantities are unobservable, so that the more modest goals of calculating the emergent
spectrum and geometrical shape of the disc are sufficient. These are
hard enough problems, as is shown by observations of real discs whose
spectra and shapes are much more complicated than simple models predict.
The data are scarce, but they show that many (if not all) discs are not
stationary. Unlike most stars, a mean model does not capture the
essence of the problem.

VISCOSITY

     Viscosity is the central problem of disc theory because the value
and physical origin of the viscous stress are unknown. It is not
believed to be large enough to directly affect the disc equilibrium,

but its indirect effect is essential, for it supplies the energy which
determines the disc structure and the emitted radiation, and it is the
mechanism by which angular momentum flows outward in the disc, so that
mass may flow inward. Ultimately, energy is released by the descent
of matter into the gravitational potential, but this descent would not
be possible without viscous torques, and it is by these torques that
energy is thermalized and made available for radiation and observation.
Our present understanding of discs is analogous to that of stellar
interiors when Eddington wrote his book The Internal Constitution of
the Stars in 1926, before there was any understanding of the processes
of nuclear energy generation. Disc models have been calculated using
a variety of assumptions about the viscosity. Some of these may
resemble reality, just as many real stars are well represented by
polytropes, but we do not know which these are. The problem is most
serious when we wish to consider the transient and dynamical response
of discs, just as the evolution of stars is more critically dependent
on the understanding of nuclear processes than is the calculation of
equilibrium models. Fortunately, this serious problem is also an
opportunity, for empirical observation of these processes may illuminate
the problem of viscosity.

All fluids have viscosity, and for plasmas with and without radiation its value is known. Unfortunately, under most circumstances it
is quantitatively inadequate to produce the mass flow rates (10^{-10}-10^{-8}
$M_\odot$/yr) required to supply binary X-ray sources and cataclysmic variables.
Two possible exceptions are worth mention. If the plasma ion temperature
$T_i \gtrsim 10^{10}$°K (obtainable only if the plasma is nonequilibrium and electron
and radiation temperatures are much lower than this value) the viscosity
may become large and the accretion rate may be significant. Alternatively,
radiation viscosity may be significant in discs which are supported by
radiation pressure, but a simple estimate using the equations of hydrostatic equilibrium and radiation diffusion, estimating the viscous
torque from the accretion rate, shows that a self-consistent solution
is not possible for geometrically thin ($h \ll r$) discs. It is not easy
to determine whether self-consistent solutions are possible for geometrically thick discs, but the thin disc estimates suggest that they
are not.

In order to explain the accretion discs actually observed we must
assume the existence of a source of viscosity in addition to the microscopic fluid viscosity. This is usually called "turbulent viscosity",
although that phrase is a label rather than a description. Its origin
is unknown. Although Keplerian rotation implies large shear rates,
and Reynolds numbers are large, discs are expected to be stable to the
simplest forms of shearing instability because their specific angular
momentum increases outward (a stably stratified distribution). Discs
may be unstable to more complex (and weaker) double-diffusive instabilities, or may have turbulence or circulation driven by other forces
which incidentally mimics viscosity. Magnetohydrodynamic stress may
also transport angular momentum and have the effect of a viscosity.
Unfortunately, it is not possible to calculate any of these processes
from first principles.

It is useful to parametrize the unknown viscosity, reducing a complex problem to a simple model, described by one or more parameters. This has the advantage of radical simplification, and the concomitant disadvantage of hiding the physical problem behind a simple numerical value. The usual parametrization describes the $r\phi$ component of the viscous stress tensor $\sigma_{r\phi}$ (the most important component in a thin Keplerian disc) as a multiple of the fluid pressure p:

$$\sigma_{r\phi} = \alpha\, p \qquad\qquad\qquad\qquad (1)$$

This is widely used (often termed the "α model" or the "α disc") but it must be remembered that there is no reason to expect that α be a universal number, or even be single-valued at any place in a disc. Two problems face the astrophysicist who chooses to use Eq. (1): to determine if this is a useful description of disc viscosity, and if so to estimate the numerical value of α.

Theory, beginning with first principles, has little to say about the first question. Some work has been done on the second, assuming the first to be answered affirmatively. Most attention has been paid to the possibility of magnetohydrodynamic viscosity because of the ubiquity of magnetic fields in astrophysical plasmas, and because estimates are possible within the assumption of Keplerian orbits. Purely hydrodynamic turbulent viscosity is even harder to estimate.

Differential rotation will wind up any magnetic field frozen into a disc, producing a steadily and rapidly increasing B_ϕ from any B_r present, thus amplifying the magnetic stress. The stress is limited because if it exceeds the pressure an instability is expected to expel the field from the disc entirely. Then we expect $B_r \ll B_\phi$ and

$$\frac{B_r B_\phi}{8\pi} \ll \frac{B_r^2 + B_\phi^2 + B_z^2}{8\pi} \lesssim p \qquad\qquad (2)$$

which implies

$$\alpha \lesssim \frac{B_r}{B_\phi} \ll 1 \quad . \qquad\qquad\qquad (3)$$

The magnitude of α depends on the initial value of B_r with which the winding-up process began; it could be very small.

Coroniti (1981) has argued that a magnetized accretion disc will break up into regions within which the magnetic field suppresses differential rotation. In such a region it is possible to have $B_r \approx B_\phi$. He attempts to estimate the dissipation between adjacent regions which results from magnetic flux reconnection (enhanced resistivity produced by microscopic plasma instability in these thin sheets of high current density), and concludes that $\alpha \approx 0.1$ may be possible.

Unfortunately, it is not possible to evaluate critically arguments like these. A determined skeptic might argue that they amount only to dimensional analysis, and must inevitably yield $\alpha \sim 1$. Even if this skeptic were justified in disbelieving the argument α could be as large as 1.

Still larger values of $\alpha \gtrsim 1$ of either hydrodynamic or magnetohydodynamic origin are generally regarded as unlikely because such large fluid shear stresses are likely to produce comparable fluid pressures if the fluid motions become random and thermalize. This is analogous to the argument, familiar in the study of stellar interiors, that turbulent convective velocities are unlikely to be supersonic. The argument may be restated by considering the rate at which a viscous stress does mechanical work. This leads to an approximate equation for the energy density ε:

$$\frac{\partial \varepsilon}{\partial t} \sim \sigma \frac{\partial v}{\partial r} \sim \sigma \frac{v}{r} \sim \sigma \omega \sim \alpha \varepsilon \omega \qquad (4)$$

where ω is the Keplerian angular orbital frequency. The hydrostatic relaxation time (normal to its midplane) of a disc is about $1/\omega$ so that a disc with $\alpha \gtrsim 1$ need not be in hydrostatic equilibrium; it will probably also be violently unstable. Such a disc may not be physically impossible, but the usual theory would not apply to it.

Very small values of α cannot be ruled out on theoretical grounds, and disc models with $\alpha \sim 10^{-10}$ (for example) have been proposed and should be taken seriously. Most opinion favors larger values, but the rings of Saturn should be remembered; if they were once fluid or embedded in a fluid disc its lifetime was long and α low.

Finally, the confident theorist should not forget that the assumption of a numerical value for α may hide the most interesting accretion disc phenomena. The distribution and variability of viscous stress may be complex beyond the description of a scalar constant.

EVIDENCE — RADIATIVE

Most compact X-ray sources are expected to contain accretion discs. If the central object is a black hole then its luminosity is expected to be produced in a disc if the matter supplied by the binary companion has significant angular momentum. Thus accreting black holes are the natural place to look for a comparison of the observed and predicted radiation of accretion discs. Cyg X-1 is the best known candidate for such an object, and it has been studied carefully in order to test theoretical models and predictions. Most of the interest in Cyg X-1 has been motivated by the desire to verify the existence of a black hole and to understand better its properties, rather than to study accretion discs, but the results are equally applicable to both problems.

Cyg X-1 turned out to be unexpectedly complex. Its X-ray flux varies on a wide variety of time scales, including fractions of a

second (Rothschild, et al. 1974). This is not explicable within
ordinary theoretical disc models, which generally assume a stationary
solution, although Lightman and Eardley (1974) have pointed out that
within certain versions of the α disc model an instability may account
for it. On time scales of months and years two discrete states are
seen, with relatively rapid transitions between them. The X-ray
spectrum contains a very hard component (Tananbaum, et al. 1972) not
explicable in simple models in which the disc surface radiates as an
optically thick photosphere. Thorne and Price (1975) and Shapiro,
Lightman, and Eardley (1976) have suggested explanations of some of
these phenomena in more complex disc models which may have hot coronae
or reprocess radiation from one part of the disc in another part.
These explanations may be correct, but they were only developed after
the complexity of the data became apparent and do not alter the conclusion that disc theory failed its first predictive test.

More recently time-dependent models of the outer portions of
accretion discs in cataclysmic variables have been computed in order
to describe dwarf nova outbursts. Because this promising work has
been described by other speakers at this meeting I will only express
optimism that it will lead to a better understanding of discs in
general.

EVIDENCE — DYNAMICAL

Fortunately there are other sources of useful observational information about accretion discs. At least three astronomical objects
(or classes of objects) have yielded information about the geometry
or dynamics of accretion discs. This information is not obtained from
the spectrum or intensity of the accretion disc's intrinsic radiation,
but rather is found in the modulation the disc imposes on other sources
of radiation. In each case the observations led to the conclusion that
real accretion discs are more complex than was expected, and in each
case the interpretation of the data was or remains controversial.
Although the disc viscosity and radiation are not directly measured
in some cases it is possible to interpret the data to estimate α.

Her X-1. Her X-1 is the best studied pulsating binary X-ray
source. Its X-ray radiation has a 1.24 second period, attributed to
the spin of an accreting neutron star, and a 1.7 day orbital eclipse
period. In addition there is a large (1.5 magnitudes) modulation of
the optical intensity at the orbital period, attributed to an X-ray
reflection effect. The orbital parameters are unusually well determined
(Bahcall 1978). The intensity is modulated with a period of 35 days,
during about 12 days of which X-rays are seen, as shown in Figure 2.
During the remaining 23 days of the cycle X-rays are undetectable,
or nearly so.

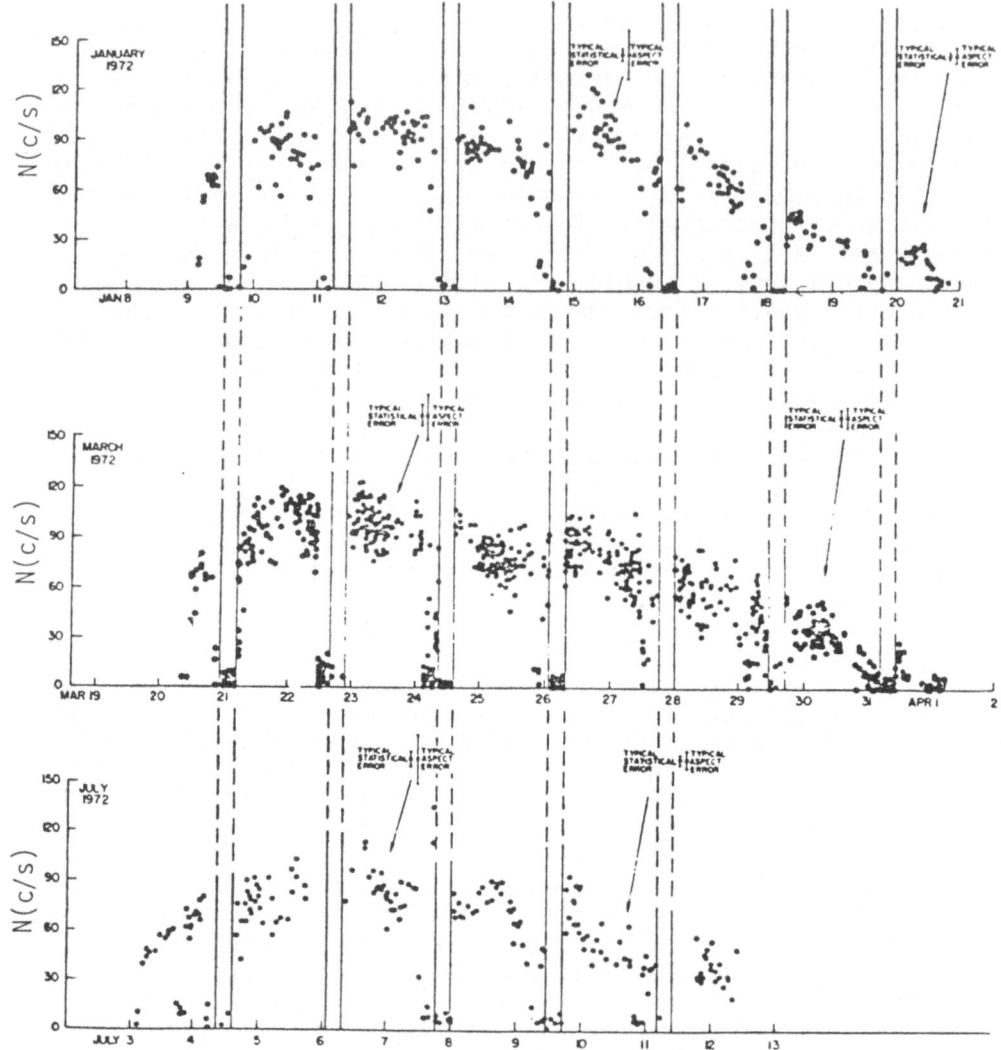

Figure 2:   The "on" portion of the 35 day cycle of Her X-1 (from
            Giacconi, et al. 1973). Abrupt eclipses are apparent,
            as is the smooth modulation envelope.

        The key to understanding this system was the observation that the
optical modulation is not strongly affected by the 35 day X-ray cycle;
the companion star sees X-rays even when we do not. The natural interpretation of this fact is that the radiation pattern is partly occulted
by a twisted accretion disc, as shown in Figure 3. The 35 day period
is then interpreted as the precession period of the outer portion of
the disc, whose varying orientation produces the observed X-ray modulation.

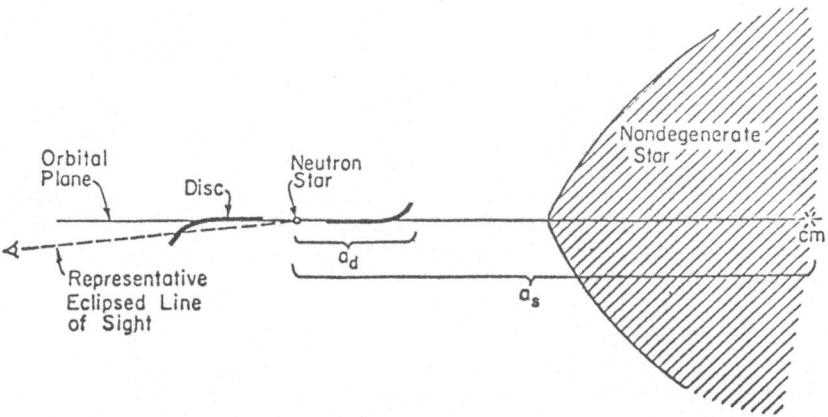

Figure 3:   Sketch of the Her X-1 system, showing how twisted
            accretion disc produces 35 day cycle (from Katz 1973).

        Several predictions of this model have been confirmed, but the
mechanism of disc precession remains controversial, as do the implications for disc mechanics. Katz (1973) proposed that the precession of
the outer disc is driven by the gravitational torque of the companion
star, while Roberts (1974) proposed that the precessing disc is flat
with an orientation which follows (is slaved to) the companion's
orientation, precessing under the neutron star's gravitational torque.
Each mechanism is capable of explaining the observed precession period.
Both models predict that the sense of disc precession should be opposite
to that of orbital motion. Careful photometric studies of the reflection effect (Deeter, et al. 1976) have verified this. Driven precession
requires $\alpha \ll 10^{-2}$ while slaved precession requires $\alpha \gtrsim 10^{-2}$ (Katz
1980a). The actual disc shape will be complex. Its theory has
been developed by Hatchett, Begelman, and Sarazin (1981) and for each
mechanism is qualitatively consistent with the X-ray modulation pattern
shown in Figure 2. The companion's gravitational torque contains a
component periodic at twice the synodic precession frequency (slightly
more than twice the sidereal orbital frequency). This induces (for
either precession mechanism) a modulation of the disc rim which can
explain (Katz, et al. 1982, Levine and Jernigan 1982) a peculiar
correlation of the onset of the 12 day "on" state with orbital phase
observed by Giacconi, et al. (1973). An analogous effect is observed
in greater detail in SS433.

        Each precession mechanism has problems. Slaved precession requires
that the companion's spin be misaligned with the orbital angular momentum.
This is not expected to be the case because the accurately circular orbit
implies that tidal dissipation has been effective. Dissipation is
expected to align and synchronize spins even faster than it circularizes
orbits (Papaloizou and Pringle 1982). Driven precession requires a
mechanism of excitation. There is no theory of this, though there
may be self-excited processes in which a tilted disc leads to asymmetric

illumination of the companion star, which leads to asymmetric mass
transfer and in turn to a tilted disc ... . Boynton, Crosa, and Deeter
(1980) have suggested that neither of these mechanisms is correct. A
hybrid driven-slaved-feedback model may be possible in which the precession occurs in the mass transfer stream, and the precession rate
is the ratio of the angle of angular momentum rotation during the
flow through the stream to the duration of the flow. Mass flows
rapidly through the disc, so that it is slaved to the angular momentum
delivered by the stream. This possibility has not yet been quantitatively investigated.

Her X-1 is probably not unique. It is a very favorable object
for the discovery of a precessing disc, partly because of the extensive
body of data available, partly because we have the good fortune to be
located close to its orbital plane, where precession may manifest itself in the 35 day modulation of X-ray intensity, and partly because
its unusually large reflection effect has made optical studies fruitful.
If it were oriented at an inclination of 70° or less we probably would
have no inkling of its many wonders. There have been several suggestions
of similar X-ray modulation phenomena in other binary X-ray sources, so
that precessing discs may be quite frequent. The mechanism of excitation
of disc precession should be widespread or universal, which suggests
some form of self-excitation or feedback, which can operate in any
system with significant X-ray luminosity.

SS433. SS433 contains the accretion disc with the most accurately
observed geometry. It is a binary with a 13 day period, and produces
high velocity jets whose direction traces out a cone of half-angle 20°
with a period of 164 days. In the most successful model for this
system the jets are produced at the center of an accretion disc and are
directed normal to its surface (Katz 1980b, van den Heuvel, Ostriker,
and Petterson 1980). The variation of jet direction is attributed to
precession of the disc. Such a disc will also show a shorter period
(6 days) nodding motion produced by the periodic terms in the companion's
gravitational torque. This effect is observed in great detail (Katz,
et al. 1982), confirming the model. The theory and a portion of the
data are shown in Figure 4. The jitter in the observations has been
shown statistically (Margon 1982) to be the consequence of fluctuations
in the jet orientation, rather than speed, and therefore to contain
additional information about the disc orientation; some implications
have been discussed by Katz and Piran (1982).

If the model just described is accepted SS433 is a good laboratory
for the study of disc dynamics because easy and accurate measurements
of jet Doppler shifts may be interpreted as measurements of the disc
orientation. There is much more quantitative information in these data
than in observations of Her X-1 because information is continually
obtained, rather than only when the disc rim cuts our line of sight,
and is not dependent on uncertain hypotheses about the thickness and
shape of the rim. The most important conclusion which may be drawn
comes from the fact that the precession and nodding motions are observed

with significant amplitude at the disc center, where the jets are
produced. In order to avoid attenuation by phase mixing mass and
angular momentum must rapidly flow from the source of these motions
at the rim in to the center. This rules out the driven precession
model considered for Her X-1 (in which the precession is observed at
the rim), but is consistent with either slaved precession or hybrid
feedback models.

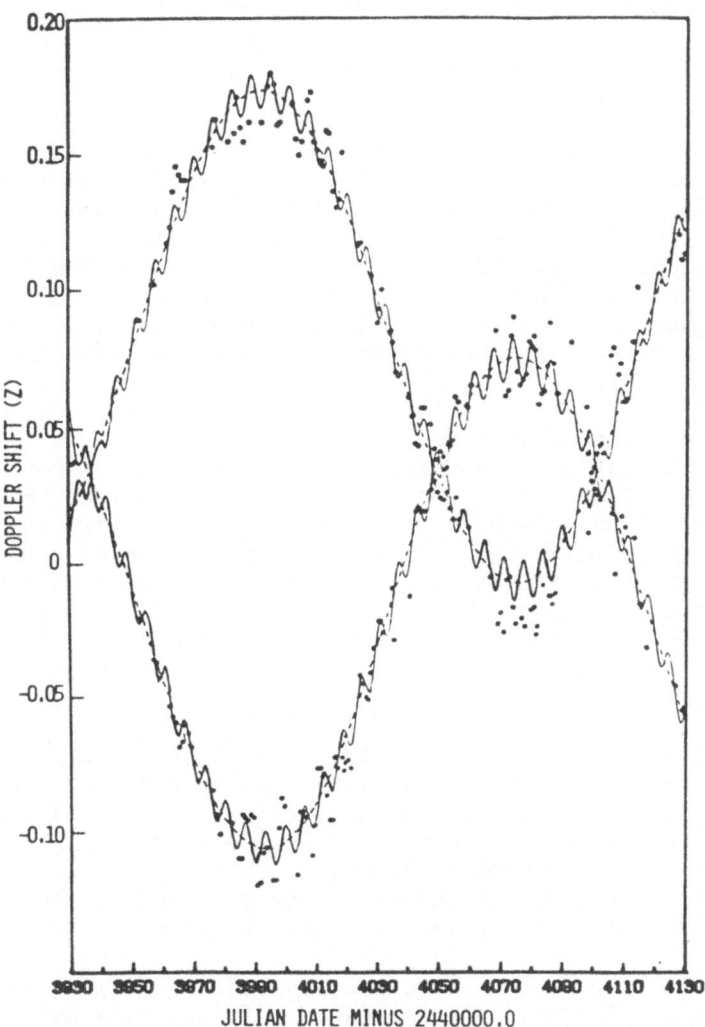

Figure 4: SS433 data (dots) with mean precession model (dashed) and
model with precession plus nodding motion (solid) for one
cycle of precession. The nodding motion shows up clearly
in Fourier transforms, but here is partly hidden by the
significant residual jitter (figure courtesy of S. Anderson
and B. Margon).

The nodding motion is particularly significant because of its short
period. In an α disc model the mass flow time t_f is approximately given
by

$$t_f \simeq \frac{r^2}{h^2 \alpha} \sqrt{\frac{r^3}{GM}} \qquad (5)$$

where M is the mass of the compact object at the disc center. Then the
observation of nodding with a significant amplitude in SS433 implies
$t_f \lesssim 1$ day (one radian of the nodding cycle), or

$$\frac{r^2}{h^2 \alpha} \lesssim 6 \qquad (6)$$

where I have made the conservative but plausible estimate of 1 day for
the orbital period of the disc rim (Katz, et al. 1982). Anderson,
Margon, and Grandi (1983) have estimated $r/h \approx 1.5$ which implies
$\alpha \gtrsim 0.3$. This result depends on an uncertain but plausible estimates
for the disc dimensions, and on a rough approximation to the process
of angular momentum flow through the disc. Its numerical value may
be revised somewhat by a different or more careful analysis, but the
conclusion that α is near the upper range of values considered reasonable
will remain.

The interpretation of this bound on α requires care. It is derived
from the rapid propagation of angular momentum, and therefore may
reflect the fastest propagating elements of a complex disc. For
example, magnetic fields embedded within the body of a disc may exert
stresses on its upper layers which far exceed the material pressure
there, yet the mean magnetic stress in the disc (and mean α) may be
quite small. In addition, most of the propagation time is expected
to be in the outer part of the disc, where the Keplerian motion is
slow and angular momentum high; bounds set on α in the inner region
are very much weaker.

The most fundamental implication of SS433 is just that its
accurately observed disc precesses with an amplitude of 20°. Each
of the two well observed accretion discs (Her X-1 and SS433) precesses
with significant amplitude. As mentioned in the previous section,
there is evidence for phenomena analogous to those of Her X-1 in
other X-ray sources. Precession appears to be a common, if not
ubiquitous, feature of discs in mass transfer binaries.

SU UMa Stars. The SU UMa stars are a subclass of dwarf novae.
Their characteristic feature is the presence of "supermaxima" in
addition to (in one case apparently in place of) ordinary maxima.
During these supermaxima there are also   periodic variations of
intensity known as superhumps. When both periods are known the
superhump period is invariably observed to be 1% to 7% longer than
the orbital period.

A number of models have been suggested, but one seems to me clearly superior. The near equality between superhump and orbital periods indicate that the superhump clock is a sideband between the orbital motion and another periodic phenomenon with a period 14-100 times longer. The sense of the inequality in the periods implies that the two phenomena have the same sense of rotation. Disc precession has provides of the right magnitude, but the wrong sense of rotation (the nodes of the Moon's orbit recede, rather than advance). The apsidal motion of an eccentric accretion ring has both the right magnitude and sense (the Moon's apse advances). This hypothesis was confirmed by Vogt's (1982) discovery of an eccentric accretion ring in Z Cha from his spectroscopic data. The photometric modulation is explained if much of the optical luminosity during supermaximum is produced in the "hot spot" where the mass transfer stream collides with the disc, because both the relative speed and angle of collision are modulated at the apsidal sideband of the orbital frequency. A quantitative investigation is underway of this modulation and of the constraints imposed on the dimensions of the accretion ring by the observed rate of apsidal advance (Katz and Trainoff 1983).

This interpretation of the superhump phenomenon has several interesting implications. It demonstrates, yet again, how complex accretion discs (or rings) may be. The discovery of eccentricity should not have been completely unexpected, because standard calculations of binary mass transfer (Flannery 1975, for example) show that transferred matter goes into an eccentric orbit about the accreting star. (Note that the large apsidal advance apparent in figures in these papers is mostly the consequence of showing the orbits in a frame corotating with the binary orbit.) If mass transfer is steady over a binary orbital period then all orientations of the apse should be present, and collisions should rapidly circularize the orbits. Therefore the observation of significant eccentricity implies that mass transfer was impulsive, as pointed out by Vogt, or that it is correlated with superhump phase (perhaps by a radiative feedback mechanism?). This is a major problem for the model, and is not understood.

During the duration of the superhump phenomenon, typically dozens of binary orbits, the eccentricity remains. This constrains dissipation processes in eccentric rings. If the ring broadened into a disc differential apsidal motion might be expected to circularize the orbit, or to change the apsidal period. The persistence of the superhumps rules out circularization; it should be possible to extract some information about the evolution of the ring size and eccentricity from the variation of the apsidal period and of the amplitude of the superhumps.

The creation of an eccentric ring during supermaximum implies that the pre-existing (presumably circular) accretion disc was rarefied enough not to interfere with its formation.

It may be possible to infer the binary mass ratio from the ratio of the apsidal to the orbital periods, but we have not yet succeeded. This would be very useful, for determinations of masses of cataclysmic variables are few and controversial.

The association of supermaxima with superhumps, occurring in the outermost portion of the disc, and involving or following impulsive mass transfer, suggests that the ordinary outbursts of dwarf novae are a different phenomenon. Those theorists who presented at this meeting outburst models involving instabilities within accretion discs will probably agree. This may offer a partial explanation of why only a few cataclysmic variables show supermaxima--only those which have impulsive (or suitably modulated) mass transfer do.

This work was supported in part by NSF AST 81-21704.

REFERENCES

Anderson, S., Margon, B., and Grandi, S.: 1983, Ap. J. 269, in press.
Bahcall, J. N.: 1978, Ann. Rev. Astr. Ap. 16, pp. 241-264.
Boynton, P. E., Crosa, L. M., and Deeter, J. E.: 1980, Ap. J. 237, pp. 169-174.
Coroniti, F. C.: 1981, Ap. J. 244, pp. 587-599.
Deeter, J., Crosa, L., Gerend, D., and Boynton, P. E.: 1976, Ap. J. 206, pp. 861-868.
Flannery, B. P.: 1975, M.N.R.A.S. 170, pp. 325-331.
Giacconi, R., Gursky, H., Kellogg, E., Levinson, R., Schreier, E., and Tananbaum, H.: 1973, Ap. J. 184, pp. 227-236.
Hatchett, S. P., Begelman, M. C., and Sarazin, C. L.: 1981, Ap. J. 247, pp. 677-685.
Katz, J. I.: 1973, Nature Phys. Sci. 246, pp. 87-89.
Katz, J. I.: 1980a, Ap. Lett. 20, pp. 135-136.
Katz, J. I.: 1980b, Ap. J. (Letters) 236, pp. L127-L130.
Katz, J. I., Anderson, S., Margon, B., and Grandi, S.: 1982, Ap. J. 260, pp. 780-793.
Katz, J. I., and Piran, T.: 1982, Ap. Lett. 23, pp. 11-15.
Katz, J. I., and Trainoff, M. A.: 1983, in preparation.
Levine, A. M., and Jernigan, J. G.: 1982, Ap. J. 262, pp. 294-300.
Lightman, A. P., and Eardley, D. M.: 1974, Ap. J. 187, pp. L1-L3.
Lynden-Bell, D.: 1969, Nature 223, pp. 690-694.
Margon, B.: 1982, Ann. N.Y. Acad. Sci. 275, pp. 403-
Novikov, I. D., and Thorne, K. S.: 1973, in Black Holes, ed. C. DeWitt and B. S. DeWitt, pp. 343-450 (Gordon and Breach, New York).
Papaloizou, J., and Pringle, J. E.: 1982, M.N.R.A.S. 200, pp. 49-69.
Roberts, W. J.: 1974, Ap. J. 187, pp. 575-584.
Rothschild, R. E. Boldt, E. A., Holt, S. S., and Serlemitsos, P. J.: 1974, Ap. J. (Letters) 189, pp. L13-L16.
Shapiro, S. L., Lightman, A. P., and Eardley, D. M.: 1976, Ap. J. 204, pp. 187-199.

Tananbaum, H., Gursky, H., Kellogg, E., Giacconi, R., and Jones, C.:
    1972, Ap. J. (Letters) 177, pp. L5-L10.
Thorne, K. S., and Price, R. H.: 1975, Ap. J. (Letters) 195,
    pp. L101-105.
van den Heuvel, E. P. J., Ostriker, J. P., and Petterson, J. A.:
    1980, Astron. Ap. 81, pp. L7-L10.
Vogt, N.: 1982, Ap. J. 252, pp. 653-667.

DISCUSSION

FAULKNER: Jonathan, I have a problem with your model. Isn't it the
case that the superhump period is longer than the orbital period for a
sufficient time that you get the equivalent of four or five rotations
of your "blimp" or "blump", or whatever you want to call it?

KATZ: Yes.

FAULKNER: It looks to me as though in your model you would always have
to have your stream impact the disk in about the 180 degree region, so
I would expect it not to be phase coherent. I would expect your
intersection to have to jump back in phase and then go forward a bit,
and then jump back again. I don't think there is any observational
indication of that.

KATZ: The intersection will travel around the eccentric ring as the
orientation of the star varies, but there is no reason why there should
be a jump. What you have to be careful of is that, in fact, the ring
is not so eccentric that the mass transfer stream misses it entirely.
But if the ring circularizes even the slightest amount, as it will
coming to a first collision here, than it will be continuous. Its
periastron will be larger then the periastron of the mass transfer
stream and, hence, there must always be an intersection.

FAULKNER: Yeah, there'll be intersections, but they're always going to
occur in the upper half of that diagram. You're not going to have the
intersection migrating five times around the whole system.

KATZ: They will occur in the upper half, but what you need is to have
the angle between the intersection and the line of apsides migrate
through $5 \times 2\pi$. That's all you need. The absolute location of the
intersection in the corotating frame is of no relevance; all you
observe is the bolometric luminosity, and that will vary with the
apsidal period in the corotating frame, which is a sideband period for
us inertial observers.

PATTERSON: Can I put that another way that, if I understood you
correctly, might be a little easier to understand? In your model,
there are two possible periods at which you can get a modulation. One
is the orbital period. If you have a spot there and the disk is optically thick, then as you wheel around it, you get the orbital period, in
principle. The amplitude might be only 1%, of course, but there it is.

And the other period you get is a different period having to do with
the fact that the distance the falling material goes down the gravitational potential well varies with the rather long period of 4 days, or
whatever it is, corresponding to the beat between the two periods. And
so both periods can be produced. Precisely which one will have a higher
amplitude is left open.

KATZ: If the emitting region emits isotropically, then it's only the
sideband period that you'll see. If there's an aspect effect, then
you can get all combinations of frequencies.

WEBBINK: Vogt makes a cryptic remark in his paper that he observes a
displacement in the times of eclipse, which implies that the hot part
of the disk responsible for the superhumps does, in fact, oscillate
about the position of the center of the hot star. As I understand your
model, one is essentially modulating the depth of the potential at
which the hot spot occurs.

KATZ: Well, it's not quite that simple because what's also relevant is
the angle of collision. It's a combination of the two.

WEBBINK: But my interpretation of this remark, which he didn't illustrate or justify, is that the hot spot responsible for the superhumps
sometimes lagged and sometimes preceded the accreting star in the orbit.

KATZ: It can move around quite a bit, depending both on the mass ratio
and on what size the ring has actually come to.

F. LAMB: Can you place constraints on the value of α?

SHAVIV: Me, or him...?

F. LAMB: I'm asking Jonathan. [Laughter.]

KATZ: I haven't tried to do it quantitatively. You'd really have to
do it numerically. To do so, you'd have to make some assumption about
the pressure in the ring, and so on. Clearly, it's going to be low,
compared to the other values we've been discussing.

PACZYNSKI: Two α questions: One, can you see any meaning to α larger
than one?

KATZ: By meaning, I assume you mean, is it at all plausible? The
meaning is just a number. I'd be a little surprised if α were greater
than one, but on the other hand it doesn't violate any fundamental law
of physics, like conservation of momentum. It tends to violate people's
intuition about magnetohydrodynamic instabilities. An extreme example
of a very large α is an AM Herculis star, where the companion star is
threaded by a permanent magnetic field and the magnetic stress is the
largest stress in the system.

PACZYNSKI: The second question is, has anybody estimated α for the sun, which is a differentially rotating object ?

KATZ: It's extremely low. Off the top of my head, I don't know the number, but if you tell me how much differential rotation you want, in five minutes we can turn that into a value for α. If you're talking about the surface layers, one presumes that the differential rotation is driven by the convective flow of heat, as is also the case in the major planets.

PACZYNSKI: You can place some limit because the shear produces extra heat.

KATZ: I haven't done that numerically. It's certainly worth doing.

MANTLE: Have you considered that these superhumps might be caused by a bright spot on the secondary star that's asynchronously rotating with respect to the binary?

KATZ: Yes, there have been about a half dozen other models...

MANTLE: If you imagine it is, every time it passes the inner Lagrangian point, the mass transfer rate might be enhanced. That could explain the modulation of the light curve.

KATZ: There are a half a dozen other models. I find them unaesthetic, but I can't prove them wrong. The problem with the bright spot idea is, why should these companion stars all be rotating nonsynchronously in the same sense by a few percent?

MANTLE: They could fluctuate in size as they transferred mass. Their shrinkage or enlargement won't conserve angular momentum unless the rate of spin changes. And a couple of percent is enough.

KATZ: I can't prove it wrong. On the other hand, Vogt did spectroscopically observe an eccentric ring.

GRAUER: I was wondering about that also. What would you see if you're looking at a system like this nearly perpendicular to its orbital plane?

KATZ: If you look perpendicular to the orbital plane, you will see a modulation. In fact, the point is, as I believe Joe Patterson was careful to point out to me when we discussed this some time ago, the modulation is bolometric, not an aspect effect. Joe, you probably remember more specifically your evidence for that.

PATTERSON: Actually, I don't think the conversation ever occurred. [Laughter.]

KATZ: I'm very sorry...[more laughter.] It's my impression that the variation is bolometric.

FAULKNER: Can I come back again to my earlier point? Isn't it
the case that the intersection of the stream and the disk, where
you want to have the light produced, simply slops sometimes
forward, sometimes back, sometimes forward, sometimes back. So
on average, I would expect the hump, if it's just a light hump,
to have on average the orbital period. Sometimes its period
will be somewhat longer, sometimes it will be somewhat shorter.

KATZ: If you could measure the phase in the orbit...

FAULKNER: It's not even a question of measuring the phase.
It's the peak of the hump.

KATZ: It oscillates back and forth... Its position in that
corotating picture oscillates with the sideband period, and that
produces a bolometric oscillation of the luminosity at the sideband period.

PATTERSON: Can I illustrate that? Look at that orientation up
there. Well, look at that...there it is! That blue spot is
good enough, the one way up to the upper right. The well there
is not deep, right? Matter does not have much of a distance to
fall, and so you don't get much of a modulation. Three days
later, the spot has wheeled 180 degrees around. Now the gravitational well is quite deep, and so now we get a greater infall.
There is a prediction that the actual intensity, the overall
intensity of the system, should go up and down with this period.
It needn't be a tremendous effect because, of course, most of
the light is coming from the inner disk. Do you follow what I'm
saying? Just look up there, the intensity is low at that point.
Flip it around 180 degrees, and the intensity is high. I think
this addresses your question.

BOND: There have been several cases where superhump periods
have been observed to decrease.

KATZ: That's associated with the reduction in the eccentricity
of the ring, as it gradually circularizes. We ought to be able
to quantitatively interpret that and turn it into an α, but we
haven't yet.

BOND: So you can't accommodate a period that always decreases?

KATZ: Yes, I think so.

F. LAMB: A more general question - I wondered if you would comment on the series of papers by Jim Pringle and John Papaloizou?
They address, first of all, the question of the precession of
the companion star in HZ Herculis and in other systems. They've
done a very careful calculation of the rotation and normal modes

of the system, and find that the precession gets all scrambled
up in a very short time. The other work that they've just published concerns precession in thin accretion disks. They show
that you have to have a certain ordering of the parameters (the
tilt angle, the value of α, and the parameter γ) in order for
the calculations to be valid. This ordering is not satisfied by
any of the calculations that have been done so far.

KATZ: My comment on the first is the following. For Hercules
X-1, I have never believed in "slaved" precession. I have other
pet models for this system. For SS433, I have to admit that I
did put my name down for slaved precession. My personal guess
is that these systems are not slaved precession, but are probably apparent precession as the result of radiative feedback. I'm
not surprised to hear that a quantitative investigation agrees
with our intuition, that trying to make a fluid star precess is
a problem. As for your second, I haven't seen the paper, so I
don't know what's in it. I don't think there's too much doubt
from the phenomenology that these disks precess. This might
just be telling us that our favorite papers on the theory of
disk precession haven't treated quite the right parameter regime,
and maybe some of us theorists should do something useful and do
it. I plead guilty, I do too much phenomenology.

THE CATACLYSMIC VARIABLE GD552

R. J. Stover
Lick Observatory, University of California

ABSTRACT

Time-resolved spectroscopic observations of the proper motion star GD552 were made at Lick Observatory using the image-tube scanner of the Shane 120-inch telescope. Radial velocity variations were measured for the Hβ emission line. No velocity variations were detected, with an upper limit of 35 kilometers per second on the semi-amplitude of the K-velocity of the accretion disk. A strong 'S-wave' component was observed in the emission lines, and, as a result, the orbital period could be determined. The most likely orbital period is 1 hour 36 minutes, with a night-to-night cycle count uncertainty of ± 1.

1. INTRODUCTION

As a suspected white dwarf, the proper-motion star GD552 (Giclas, Burnham, and Thomas 1970) was observed spectroscopically by Greenstein and Giclas, (1978). Their observations showed an emission-line spectrum similar to a U Geminorum star at minimum light, with strong, double-peaked Balmer emission lines. However, because they made only a few observations of GD552, the close-binary nature of the system could not be confirmed.

In this paper I present new, time-resolved spectroscopic observations of GD552. From these observations I establish that GD552 is an interacting close-binary system, and that the orbital period is 1 hour 36 minutes.

2. OBSERVATIONS AND ANALYSIS

Time-resolved spectroscopic observations of GD552 were made at Lick Observatory using the image-tube scanner and automated Cassegrain spectrograph of the Shane 120-inch telescope. A total of 56 spectra of GD552 were obtained over three consecutive nights beginning on U. T.

D. Q. Lamb and J. Patterson (eds.), Cataclysmic Variables and Low-Mass X-Ray Binaries, 379–384.
© 1985 by D. Reidel Publishing Company.

1982 August, 21. The observations consist a series of either four- or eight-minute observations with wavelength calibration spectra interspersed approximately every 50 minutes. Each spectrum covers a wavelength range from about 4200 Å to 5000 Å and shows strong Balmer and weak He I emission lines. All of the emission lines have the same broad, double-peaked profiles typical of an accretion disk. Absorption lines from the secondary star are not detectable in any of the spectra. Unfortunately, the Hγ emission line is badly contaminated by mercury lines from the nearby city of San Jose. Hβ was relatively uncontaminated and was used in the analysis presented here. A cursory examination of the emission lines immediately confirmed the close-binary nature of GD552. A very strong S-wave is clearly visible in both the Balmer emission lines and He I 4471, with the relative strength of the S-wave strongest in He I.

The apparent radial velocity of the Hβ emission line was measured by fitting a symmetric 6th order polynomial to the emission line wings. This procedure, described previously by Stover et. al. (1980), minimizes the effects of the S-wave which distorts the central region of the line. The resulting measurements showed no detectable systematic radial velocity variations. Using an orbital period of 0.0666 days (see below), and assuming a circular orbit, a 2σ upper limit of 35 kilometers per second can be placed on the semi-amplitude of any radial velocity variations.

The material in the area of localized emission which gives rise to the S-wave component of the observed emission lines is near the accretion stream shock front and has a velocity nearly equal to the velocity of the material at the outer edge of the accretion disk. Therefore, as the binary system rotates the S-wave exhibits radial velocity variations with an amplitude similar to the velocity separation of the peaks of the double-peaked emission line. This behavior is illustrated in Figure 1 where several profiles of Hβ are shown as they appear at different orbital phases. The variation of the profile created by the S-wave can be used to measure the orbital period. This was done by first forming the quantity $D=(V-R)/(V+R)$ for each spectrum, where V and R are computed by summing the counts in 20-angstrom wide bins centered on the violet and red peaks of Hβ. The resulting values were then subjected to periodogram analysis which revealed the most likely orbital period to be 0.0666 days (= 1 hour 36 minutes = 15 cycles per day), with an uncertainty of $\pm$ 1 cycle per day.

3. DISCUSSION

As the profiles in Figure 1 show, the double peaks of the Hβ profile are separated by about 1000 kilometers per second, and the wings of the line extend to at least ±1600 kilometers per second either side of the line center. Such broad lines suggested that the binary system may have a relatively high inclination and that a photometric search for eclipses might be worthwhile. At the author's request, Howard Bond and

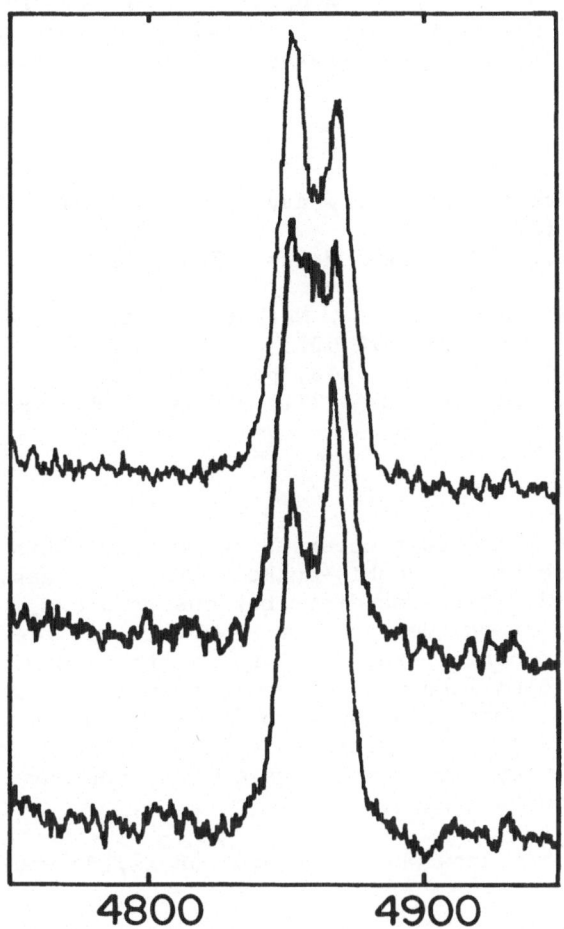

Figure 1. The profile of Hβ shown at three different orbital phases.

Al Grauer made high-speed photometric observations of GD552 at Kitt
Peak.   Also, E. L.   Robinson made similar observations at the
University of Texas McDonald Observatory.   In   both   cases   the
observations lasted for more than the orbital period, and although rapid
photometric activity was observed, no eclipses were detected.   In
addition, there was no obvious orbitally related photometric 'hump' in
the light-curve, as is observed in U Gem (Warner and Nather, 1971).
Since nothing is known about the outburst activity (if any) in GD552 a
search of existing plate collections might be worth while.

      In U Gem the strength of the S-wave varies widely relative to the
doubled profile, and the variations are a function of orbital phase
(Stover, 1981). In GD552 this is not the case.   The strength of the
S-wave is relatively constant throughout the orbit. As a consequence of

the simple behavior of the S-wave, the lack of eclipses, and the broad
nature of the double-peaked component, the observed emission line
profiles in GD552 may be especially useful in studies which attempt to
physically model the emission arising from an accretion disk.

REFERENCES

Giclas, H. L., Burnham, R., Jr., and Thomas, N. G. 1970, Lowell Obs.
    Bull. No. 153.
Greenstein, J. L., and Giclas, Henry 1978, Publ. A. S. P., 90, pp.
    460-462.
Stover, R. J., Robinson, E. L., Nather, R. E., and Montemayor, T. J.
    1980, Ap. J., 240, pp. 597-607.
Stover, R. J. 1981, Ap. J., 248, pp. 684-695.
Warner, B., and Nather, R. E. 1971, M. N. R. A. S., 152, pp.
    219-229.

DISCUSSION

STOVER: One nice thing that occurred to me about this system is that
it shows very nice broad, double-peaked emission lines. For people who
want to try to model emission distributions in disks, this is a very
nice system to do it on, because the S-wave is visible essentially all
the way around the orbit. You don't see a strong orbital hump in the
photometry. It looks like the hot spot, even though we can see it very
clearly in the photometry, doesn't go out of view, so you don't have
lots of problems due to geometric shielding of the disk and to other
effects. It's a very nice, clean system, and you have very good velocity resolution in the lines.

WADE: But you don't know where conjunction is, do you?

STOVER: No.

PATTERSON: Did you give an upper limit of 40 km s^{-1}?

STOVER: That's right.

PATTERSON: And that's the upper limit on the semiamplitude?

STOVER: Yes.

PATTERSON: Can you try to push that farther, because this is a very
high mass-ratio system. Just how high it really is would be interesting.

PACZYNSKI: What's the period again?

STOVER: One hour thirty-six minutes.

PACZYNSKI: So this could finally be one of the degenerates?

PATTERSON: Yeah. So, are you trying to push the limits further, or is anybody else?

STOVER: I'm not currently.

LIEBERT: There seem to be two kinds of eclipsing systems edge on like this, those that show double peaks and those that don't. Is the difference between the two that the former are high mass-ratio systems like WZ Sge?

STOVER: I don't know.

PATTERSON: DQ Her has strongly doubled lines and its mass ratio is near unity, so probably not.

LIEBERT: But that's an old nova.

PATTERSON: There go all my long orbital period systems!

PACZYNSKI: Is it thought to be a dwarf nova, or what?

STOVER: It has never been seen to have an eruption.

MCCLINTOCK: Can you mention some more characteristics of the system, like does it have absorption lines?

STOVER: I didn't see any absorption lines. I didn't see any He II lines.

MCCLINTOCK: So it's just hydrogen?

STOVER: Yeah, just hydrogen Balmer lines.

MCCLINTOCK: And there's no photometric data on it, as far as whether there's a bright spot?

STOVER: I got about two hours at McDonald. It shows photometric variations of about a quarter of a magnitude. But the variations didn't look like they were particularly correlated with the orbital period.

STOCKMAN: Is it unusual to see the hot spot during the whole orbital period?

STOVER: I don't know of any other case. In U Gem, for instance, the spot goes behind the disk and you don't see it, and the appearance of the spectrum around the orbit is quite different.

PACZYNSKI: This system is not eclipsing?

STOVER: It is not eclipsing.

SMAK: I would like to comment on the S-wave and the hot spot.
They're close together, but they are not the same. Frankly, I
don't know what is the region where the S-wave is produced, and
if anyone knows, please tell me. But my feeling, for the time
being, is that somewhere near the hot spot, say above it,
conditions are just right for the atoms to ionize and then
recombine, which is what we observe. Depending on what are the
densities of the incoming disk material and the incoming stream
material, we can get any effective velocity for those atoms.
As far as I know, there are cases when we see a velocity almost
identical with the stream velocity, and cases where the velocity
of the S-wave is closer to the Keplerian disk velocity. It's a
subject that hasn't even been touched upon so far during this
workshop, so I thought it's worth bringing up.

SHAVIV: Would anyone like to comment about that?

PATTERSON: I would have thought that the stream, which you
didn't mention but which is the other possibility, is disfavored
as a place to situate the S-wave emission region because of the
4π steradian diminution of the photons coming from the hotspot.
I would have thought that the dissipation in the hot spot itself
would make it quite a bit more favored as a location for the
emission lines.

SMAK: As I said, I don't know.

WADE: It depends on whether you think the spot is optically
thick or not.

SMAK: That's right.

PATTERSON: The statement that "the spot is optically thick" is
probably not right. There's an outer part and it's not optically
thick. The outer part of everything is optically thin.
[Laughter.]

PACZYNSKI: Can you see the absorption lines of the white dwarf,
as in WZ Sge?

STOVER: No.

A SUMMARY OF THE UV, OPTICAL AND IR PROPERTIES OF DISKS IN CV'S

Paula Szkody
Dept. of Astronomy, University of Washington

ABSTRACT
The available observations of dwarf novae and novalike systems that are
relevant to the study of accretion in cataclysmic variables are reviewed.
This includes UV data from IUE and optical spectroscopy and infrared
photometry. The flux distributions are compared to available steady
state models to obtain an estimate of the mass accretion and the range
of this parameter among the dwarf novae (U Gem, Z Cam and SU UMa types)
and the novalikes (UX UMa, AM Her, DQ Her types). The normal and abnor-
mal range of variation in the continuum and line fluxes is discussed.

I. INTRODUCTION

        Since the disks in cataclysmic variables involve a large
temperature range, the observed emission spans infrared to ultraviolet
wavelengths (and the X-ray if the boundary layer is included). Thus,
to construct an adequate data base with which to compare theoretical
disk models requires observations covering this entire wavelength range.
This review will attempt to summarize the observations that have been
accomplished in the ultraviolet (with IUE) and in the optical and infrared (with ground based telescopes) which allow the determination of good
flux distributions. Comparison of the slopes of these flux distributions
versus wavelength with predictions from steady state disk models can
lead to an estimate of the mass accretion rate. However, these rates
are dependent on the type of model used, i.e. summing black bodies within a temperature range (Bath, Pringle and Whelan, 1980), summing model
atmospheres over a temperature range (Kiplinger, 1979, Mayo, Wickramasinghe and Whelan, 1980), or inclusion of the optically thin emission
areas with the black bodies (Tylenda, 1981, Williams and Ferguson, 1982).
In general, the black bodies give the highest mass accretion rates and
the largest disks. The model atmospheres result in a steeper flux distribution for a lower temperature range and smaller disk size, yielding
accretion rates an order of magnitude smaller, but they cannot reproduce
the Balmer jump and line emission correctly. Inclusion of the optically
thin areas can reproduce the line emission, but is highly dependent on
the value of the viscosity and the abundance.

D. Q. Lamb and J. Patterson (eds.), Cataclysmic Variables and Low-Mass X-Ray Binaries, 385–401.
© 1985 by D. Reidel Publishing Company.

       In view of the most recent explanation for the outburst
mechanism for dwarf novae, i.e. the hot dam instability of Faulkner et al.
(see paper this volume) and the cold dam instability of Wheeler et al.
(see paper this volume), the disk is not in a stable situation between
outbursts and so the classic steady state models are not appropriate.
However, these new models have not yet reached the point of computing
flux distributions or making a detailed match with observations. Thus,
throughout this review, the emphasis will be on observed properties,
i.e. the slopes of the flux distributions vs. the derived quantities
like $\dot{M}$. Within the available models, a steeper slope will, in general,
correspond to a higher mass accretion whatever the zero point may be.

       Besides the contribution to the total flux distribution, each
wavelength region has an important value of its own in terms of interpretation of the disk and system parameters. The infrared can provide
the identification of the secondary star and lead to a distance estimate
for the system. The amount of IR flux also provides an estimate of the
extent of the outer disk and the temperature of the optically thick
boundary (Frank and King 1981). The optical line emission ratios and
strengths are also indicators of the conditions in the outer disk (Williams 1980). Through study of eclipsing systems, the origins of the
various lines have been identified, i.e. H, HeI with the outer disk and
HeII with the inner disk (Williams and Ferguson 1982). The ultraviolet
provides information on mass loss through analysis of P Cygni features
seen in the UV resonance lines of CIV1550 and SiIV1400 of systems near
outburst (Krautter et al. 1981, Szkody, 1982, Cordova and Mason, 1982,
Klare et al. 1982, Guinan and Sion 1982). The C, N and O lines evident
in the UV can also be used to study abundance and evolutionary scenarios
(Williams and Ferguson, 1983). Finally, the effects of suspected magnetic fields on the disk can be explored since the UV arises from the
inner regions which should be disrupted if the magnetic field is strong
enough to channel the inner parts of the disk to the poles of the white
dwarf. Thus, the disks in these systems would be expected to exist as
a cool outer ring.

       For the purpose of comparing the disk conditions among the
cataclysmic variables, the objects are first grouped into the dwarf novae
and novalikes. The dwarf novae are further divided into the U Gem
("normal" dwarf novae), the Z Cam systems (with standstills) and the
SU UMa category (with supermaxima and superhumps). The novalike systems
are divided into the UX UMa types, the AM Her types (with circular
polarization indicating high magnetic fields) and the non-synchronized
rotators (DQ Her systems). Some of these novalikes show high-low states
with different parameters at each state. Even with this breakdown, there
are many CV's which do not fit into these categories, e.g. V442 Oph,
V794 Aql, KR Aur, CM Del, V425 Cas, Lanning 10, LX Ser, etc.

       The primary emphasis will be on the ultraviolet, where in
general, more than half of the luminosity of the system emerges. Thus,
the sample of individual objects reviewed in terms of UV, optical and
infrared properties was chosen on the basis of availability of IUE

spectra. Whereas the optical and infrared are severely "contaminated" (in terms of studying the disk) by the presence of the hot spot and the secondary, the short wavelength UV ($\lambda \lesssim$ 2500 Å) is free of this problem.

II. ULTRAVIOLET

In dealing with the flux distribution from IUE data, it is important to keep in mind several potential problems:
1) The UV is very sensitive to reddening corrections. In IUE data, the long wavelength camera allows a determination of the reddening from the 2200 Å feature, but in some cases, only short wavelength exposures are available. In general, E(B-V) $\lesssim$ 0.1, but values as high as 0.25 are known.
2) There may be some "bumps" in the flux distribution in the spectra of faint objects due to the breakdown of an accurate calibration at low light levels.
3) Since half of the IUE shifts occur in the daytime and weather can (and in general, does) conspire against optical observing, simultaneous observations in the UV and other wavelengths are very rare. This is the largest problem for very active systems. On the other hand, a V magnitude (good to ±0.1 mag. down to about 15th mag.) is always available from the Fine Error Sensor (FES) on board the satellite. Also, the exposures are generally several hours so that any short time variations like flickering will average out. Thus, the joining of UV with optical and infrared data taken at comparable outburst states is possible.
4) The spot may represent an important contribution at the end of the long wavelength UV. About half the observed systems at decline and quiescence show an upturn in flux at wavelengths $\gtrsim$ 2500 Å corresponding to a temperature near 10000°K (Szkody, 1981, 1982). From ANS data, Wu and Panek (1982) have obtained a spot temperature for U Gem on this order and the flux distribution for this system put together by Cordova and Mason (1982b) shows that this spot is dominant from the long wavelength UV through the optical. Thus, this spot can result in the flattening of the observed flux distribution and the raising of the optical fluxes above those expected from the disk alone. It is also possible that this temperature component near 10,000°K represents the ionization zone of the entire disk (D. Lin, private communication).

Table 1 summarizes the available UV observations of the dwarf novae and novalike systems (see the papers by B. Hassall, E. Sion and J. Liebert, this volume, for a more detailed treatment of the dwarf novae, UX UMa and AM Her systems). The given α is the slope of the flux distribution where $F_\lambda \alpha \lambda^{-\alpha}$. The listed $\dot{M}$'s (mass accretion rates) are from a comparison of the α values with the steady state models of Williams and Ferguson (1982), which use a $1M_\odot$ white dwarf with radius = 6×10^8cm and a disk radius of 6×10^{10}cm. Their computed flux distributions in the UV (1000 - 3000 Å) for α = 0 to 2.3 are shown in Figure 1 along with 3 examples of the fits to the models. Near $\dot{M} \sim 10^{-10}M_\odot$/yr, the slope does not fit a power law due to the turnover near the Lyman limit.

TABLE 1. IUE Results

A. Dwarf Novae

U Gem	α	$\dot{M}$	Ref.
U Gem m[a]	4.1	+LWR comp.	1
SS Cyg M	2.3	10^{-7}	2
m	2.2	10^{-7}	3
EX Hya m	2.0	10^{-8}	4
RU Peg m	1.7	10^{-8}	5
UZ Ser M	4 + LWR comp.		6
TW Vir M	2.3	10^{-7}	1
m	1.0	10^{-9}	7
AR And M	2.4	10^{-7}	5
BV Cen m	0	10^{-10}	4
1E0643 m	0.8	10^{-10}	8
T Leo m	0	10^{-10}	7

Z Cam	α	$\dot{M}$	Ref.
Z Cam s	2.3	10^{-7}	9,5
m	2.3	+LWR comp.	7
M	2.0	10^{-8}	10
RX And r	1.0	10^{-9}	9
s	1.7	10^{-8}	5
SY Cnc M	2.3	10^{-7}	9
AB Dra mid	1.4	5×10^{-9}	9
AH Her m	2.3 + LWR comp.		9
WW Cet M	1.4	5×10^{-9}	9
TZ Per s	2.0	10^{-8}	5
EM Cyg m	2.3 + LWR comp.		9

SU UMa	α	$\dot{M}$	Ref.
SU UMa m	2 + LWR comp.	10^{-8}	9
YZ Cnc m	2 + LWR comp.	10^{-8}	9
SM	2.3	10^{-7}	
AY Lyr M	2.3	10^{-7}	10
m	2-4		
VW Hyi mid-m	2.3	10^{-7}	4,7
TU Men SM	2.4	10^{-7}	5
IR Gem m	1	10^{-9}	7
TY Psc m	1	10^{-9}	7

B. Novalikes

UX UMa	α	$\dot{M}$	Ref.
UX UMa	2.3	10^{-7}	11
TT Ari H	2.3	10^{-7}	12,13
L	4 + LWR comp.		7
RW Sex	2.0	10^{-8}	14
V3885 Sgr	2.0	10^{-8}	15
LSI+55°8	2.3	10^{-7}	16
KQ Mon	1.8	10^{-8}	17

AM Her	α	$\dot{M}$	Ref.
AM Her H	1+4		3
L	4		18
VV Pup H	2		19
AN UMa H	2		19
E1405-451 H	1.9		20
EF Eri H	4		19

DQ Her	α	$\dot{M}$	Ref.
DQ Her	4+0		19, 21
AE Aqr	2 + LWR comp.	10^{-8}	7
H2252	1.2	10^{-9}	22
H2215	0	10^{-10}	23
3A0729	1.8	10^{-8}	7
V1223 Sgr	2.0	10^{-8}	8
2A0526	2.3+LWR comp.	10^{-9}	24

a SM=supermax M=outburst m=quiescence s=standstill mid=mid state H=high state L=low state

1 Cordova & Mason 1982	14 Greenstein & Oke 1982
2 Heap et al 1978	15 Guinan & Sion 1982
3 Fabbiano et al 1981	16 Guinan & Sion 1982b
4 Bath, Pringle & Whelan 1980	17 Sion & Guinan 1982
5 Klare et al 1982	18 Szkody, Raymond & Capps 1982
6 Echeverra et al 1981	19 Hartmann & Raymond 1980
7 Szkody, in preparation	20 Nousek & Pravdo 1982
8 Bonnet-Bidaud, Mouchet & Motch 1982	21 Lambert & Slovak 1981
9 Szkody 1981	22 Hassall et al 1981
10 Szkody 1982	23 Szkody 1982b
11 Holm, Panek & Schiffer 1982	24 Mouchet et al 1981
12 Krautter et al 1981	
13 Jameson et al 1982	

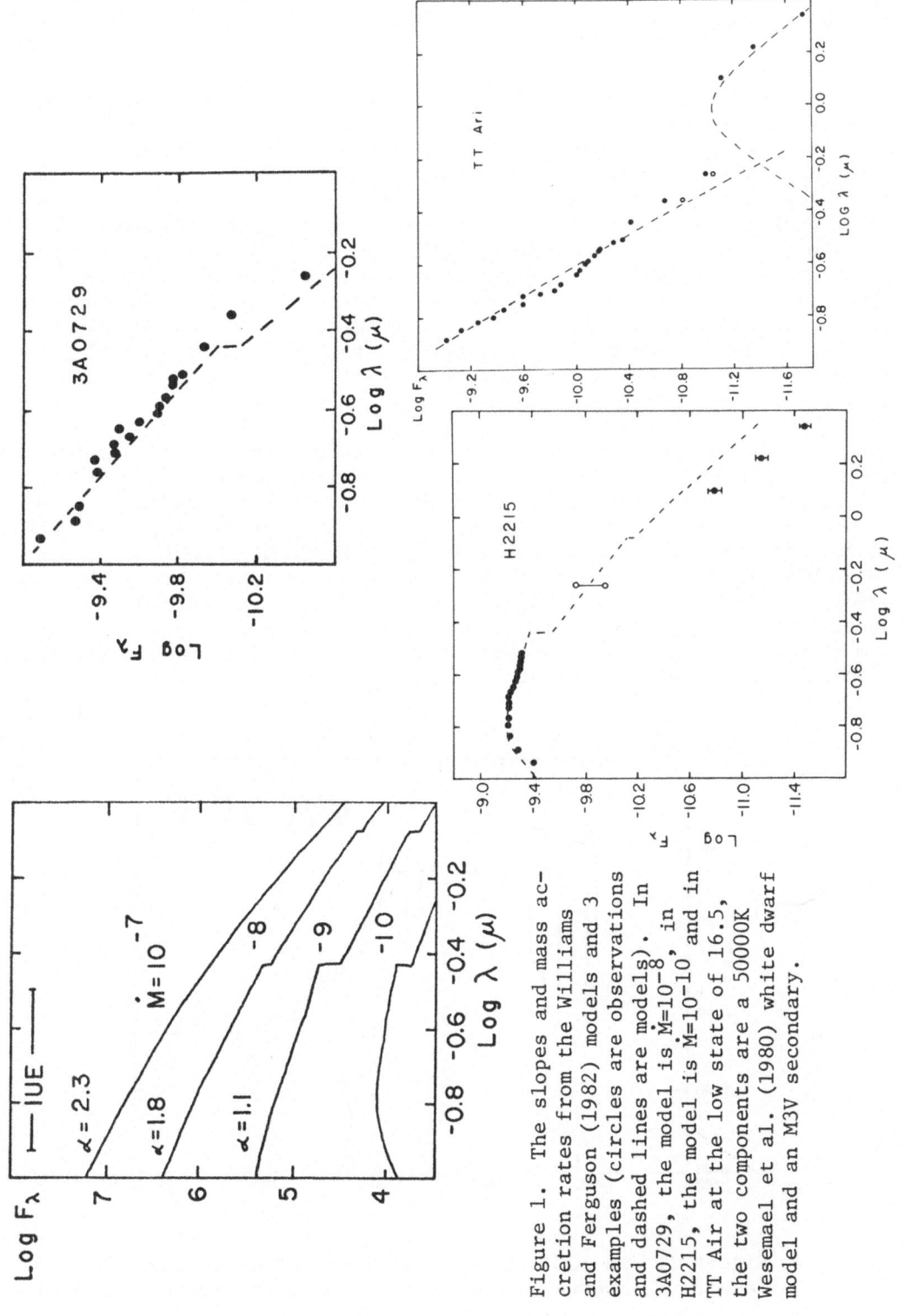

Figure 1. The slopes and mass accretion rates from the Williams
and Ferguson (1982) models and 3
examples (circles are observations
and dashed lines are models). In
3A0729, the model is $\dot{M}=10^{-8}$, in
H2215, the model is $\dot{M}=10^{-10}$, and in
TT Air at the low state of 16.5,
the two components are a 50000K
Wesemael et al. (1980) white dwarf
model and an M3V secondary.

A. Dwarf Novae

        With the exception of U Gem and UZ Ser which have $\alpha = 4$
(probably due to viewing a hot white dwarf), most of the U Gem stars
at quiescence have flat slopes ($\alpha \lesssim 2$), implying accretion rates varying
over a range of 100 ($\dot{M}$ from 10^{-8} to $10^{-10} M_\odot$/yr using Williams and
Ferguson 1982). At maximum, the systems are close to a steady state
disk slope of $\alpha = 2.3$. The Z Cam systems have slopes close to the steady
state disk at maximum and at standstill. The 3 systems observed at
minimum (Z Cam, AH Her and EM Cyg) all show the same slope at minimum
as at maximum (implying a higher accretion rate than in the U Gem
systems). In addition, a prominent spot (or cooler disk) component is
evident in the long wavelength UV in all three. Among the sample of
SU UMa's, there is no basic difference in the UV from the normal U Gem
stars, i.e. the slopes are $\alpha \lesssim 2$ at quiescence and close to $\alpha = 2.3$ at
maximum and supermaximum.

        Among the dwarf novae, VW Hyi presents one of the best examples
of the difficulties encountered in the determination of the characteristics of the accretion disk. This dwarf nova of the SU UMa type has a
UV and optical flux distribution with $\alpha = 2.3$. Bath, Pringle and Whelan
(1980), fit this continuum with a hot ($T > 10^5$K) high mass accretion
($\dot{M} > 10^{-8} M_\odot$/yr) disk. However, VW Hyi has an extremely broad Lyman α
absorption line (Mateo and Szkody, in prep.) with the turnover beginning
about 1350 Å. A good signal/noise spectrum obtained by the United King-
dom at minimum light has been binned and is shown in Figure 2 (other
spectra at minimum and decline from outburst show this same feature).

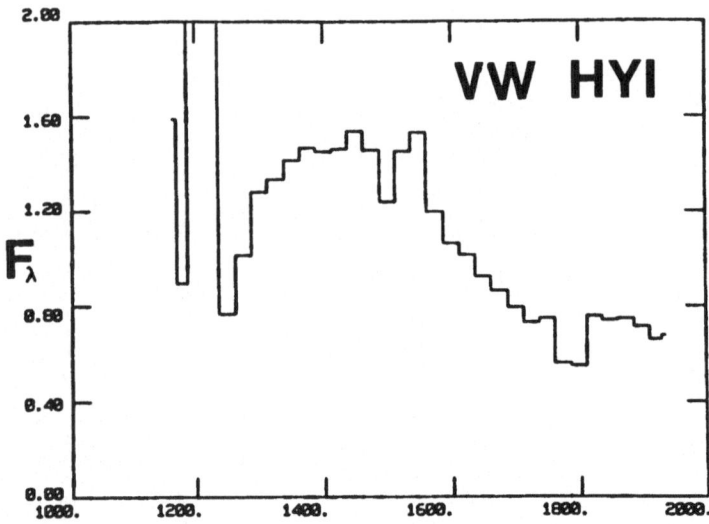

Figure 2. An IUE spectrum of VW Hyi at quiescence (binned
into 25 Å intervals) that shows the Lα absorption turnover.

The turnover is too sharp to be due to the inner edge boundary of a steady state disk (and the continuum appears to come up on the other side of Lα). It is too wide (by a factor of 10) to be due to Doppler motions in the disk. A high magnetic field could cause some broadening, but it is broader than highly magnetic single white dwarfs (Liebert, private communication) and the required magnetic field of 10^8G would not allow the formation of any disk in this close binary. The last, and best possibility for the broadening is a cool white dwarf. The best, although not perfect, fit is obtained with a Wesemael et al. (1980) log g = 9, T = 20,000K model (lower g or higher T gives a narrower line). Such a cool white dwarf would imply a very low accretion rate (10^{-11} M$_e$/yr) for steady state models. The exact solution for this system needs better models of the disk to combine with the white dwarf. U Gem and AM Her (low state) and TT Ari (low state) also show Lα absorption, but in these cases, the narrow absorption is consistent with a hot white dwarf. The flux distribution is $F_\lambda \alpha \lambda^{-4}$, also consistent with viewing a white dwarf, rather than a dominent disk.

B. Novalikes

The UX UMa type systems are all close to $F_\lambda \alpha \lambda^{-2.3}$ (steady state disk slopes), consistent with continual high mass accretion rates. The AM Her systems show an accretion column component that is near α = 2 (with sometimes an addition from a hot black-body) but the magnetic fields are too strong to enable much of a disk to be present. The non-synchronized rotators in general show flat slopes that indicate lower accretion rates than the average dwarf nova. This is consistent with a disk situation more like an outer ring, with the inner area disrupted and channeled by a magnetic field.

The most dramatic changes in the accretion areas of novalike systems occur in those objects which undergo occasional very low states during which most of the accretion turns off. This includes MV Lyr, TT Ari, AM Her, AN UMa and VV Pup (only the first 3 have the low states observed in the UV). In these systems, the "normal" flux distribution at the high state changes over to α = 4 at the low state, which fits with the flux expected from a hot (T $\gtrsim$ 50,000K) white dwarf. At the long wavelength end of the UV, there remains a small disk or column remnant contribution, which implies that the accretion has not stopped entirely. These rare states provide the best opportunity for studying the underlying stars in the system.

C. Spectral and Continuum Variability

With the exception of U Gem, all of the dwarf novae show absorption lines near outburst or standstill and increasing emission during the decline to quiescent light. The UX UMa's show the absorption lines characteristic of a thick disk. The AM Her's and non-synchronized rotators are all emission line systems. The largest difference among the lines present occurs in HeII1640, which is strong in the latter 2

classes but generally absent in the dwarf novae and UX UMa's. The other
lines (NV, SiIV, CIV, MgII) are normal for all types. The striking
feature that has become commonplace in all of the UX UMa's and in the
dwarf novae at outburst is the presence of P Cygni profiles (Krautter
et al. 1981, Szkody, 1982, Cordova and Mason, 1982, Klare et al. 1982,
Guinan and Sion 1982). The blue shifted absorption is indicative of
mass loss at rates $\sim 10^{-11} M_\odot$/yr, a small fraction of the mass accretion
rate ($\sim 10^{-8} M_\odot$/yr). The profile is enhanced in low inclination systems,
implying a conical shaped wind emanating from the inner disk. The
presence of this mass loss is short lived at the outburst situation
($< \sim$ 3 days - Szkody 1982) and intermittent at standstill (Szkody, 1981,
Klare et al. 1982).

Besides the outburst changes in dwarf novae, spectral changes
are also evident in novalike systems. In MV Lyr and AM Her, the normal
emission line spectrum present at the high state essentially disappears
at the low (Szkody and Downes, 1982, Chiapetti et al. 1982, Szkody,
Raymond and Capps, 1982) consistent with the absence of an extensive
accretion area at the low state. TT Ari, which shows the absorption
lines of a thick disk at its normal high state, reveals an emission
spectrum at the low, also consistent with the thinning out of the disk
(see Figure 1). UX UMa shows a surprising changeover from an absorption
to an emission line spectrum with no correlation with the continuum
intensity or the optical state (Holm, Panek and Schiffer 1982). This
could be related to the location of the emission lines in this system.
Holm, Panek and Schiffer found that the UV lines are not eclipsed,
whereas the optical and UV continuum does undergo eclipse. This implies
an origin for the UV lines in an extended region, larger than the Roche
lobe. Other eclipsing systems (RW Tri, EX Hya, - F. Cordova, private
communication) do show eclipses of the lines, implying an origin closer
to the white dwarf. Ongoing studies of the variations throughout an
orbit should help to further identify the line emission regions.

Sometimes, the variability that is seen in the UV is totally
unexpected. One example of this is TV Col = 2A0526-328, which has been
known to be a novalike at V $\sim$ 14th mag. since 1979 (Charles et al. 1979).
A recent spectroscopic study by Hutchings et al. (1981) showed a
difference between the spectroscopic and photometric periods which tied
into an observed 4 day beat period, implying that TV Col is a member of
the non-synchronous rotators. Previous IUE observations (Mouchet et al.
1981) showed typical UV spectra for a cataclysmic. However, recent
simultaneous IUE and optical observations to track the 4 day period
(Szkody and Mateo, in prep.) showed a 2 mag. mini-outburst which sets
some constraints on the timescale of an accretion event. In the better
time resolution optical data, the flare began with a 0.5 mag. increase
in one hour, followed by a 1.5 mag. increase in UBV in 20 minutes. In
this interval, the UV continuum increased by 1.5-2 mag. The spectra
before and after the outburst are shown in Figure 3. Unlike a dwarf
nova outburst, the lines are strongly in emission during the flare
(fluxes 10-60 times the pre-outburst values) and the highest excitation
lines (NV and HeII) show the largest increase. P Cygni absorption

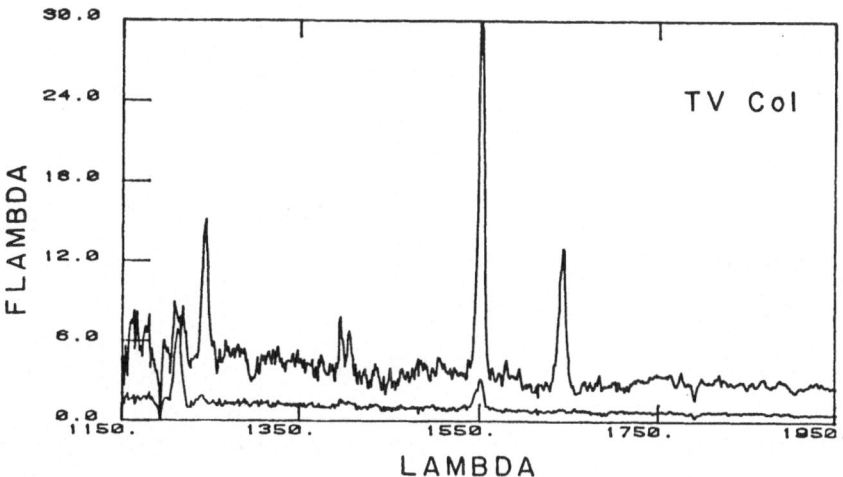

Figure 3. Two IUE spectra obtained before (bottom spectrum) and after (top spectrum) the two mag. flare.

profiles with terminal velocities of 3870 km/sec appeared and disappeared on timescales of one hour. The situation appears to be a small scale disk outburst which, in contrast to dwarf novae, lasts for hours rather than days and ejects mass only for minutes. The disk apparently does not establish the steady state situation seen in dwarf novae at outburst and in the UX UMa's. Any successful model of the outburst mechanism in a disk will thus have to take into account the normal outbursts in dwarf novae, standstills in Z Cams, supermaxima in the SU UMa's, infrequent transitions to a low state as in TT Ari and MV Lyr, and infrequent periods of short lived activity as in TV Col!

III. INFRARED

Table 2 lists the V-J and J-K colors that are available for the sample of CV's with UV data. For reference, the colors of an M star are V-J $\sim$ 3.7 and J-K $\sim$ 0.8, the colors of a steady state disk $(F_\lambda \alpha \lambda^{-2.3})$ are V-J $\sim$ 0.7, J-K $\sim$ 0.9, and a Rayleigh-Jean's distribution $(F_\lambda \alpha \lambda^{-4})$ in the infrared would have colors near 0. For a disk that is optically thick, it is expected there will be a change in the flux distribution at a temperature corresponding to the outer edge. For most systems, this change comes near one micron, where the distribution becomes Rayleigh-Jeans unless a late secondary is visible to override the turn-down of the flux distribution. Unfortunately, when the secondary does not dominate, it is difficult to sort out the exact contribution of the secondary and the disk, so that only limits to the magnitude of the secondary are possible in most cases. The best determination of the spectral type of the secondary comes from spectroscopic measurement of the strength of TiO bands.

TABLE 2.   Infrared Results

A. Dwarf Novae

U Gem	V–J	J–K	Ref.	d(pc)[b]
U Gem m[a]	2.7	0.9	1,2	78
M	0	0.3	3	
SS Cyg m	2.0	0.6		95
EX Hya m	0.7	0.8	4	
m	0.8	0.3	5	
TW Vir m	2.0	0.6	6	
AR And m	1.6	0.6	6	
T Leo m	1.4	0.7	6	

Z Cam	V–J	J–K	Ref.	d(pc)
Z Cam s	0.6	0.6	3	
m	1.1	0.5	6	
TX And m	1.6	0.6	3	
SY Cnc m	1.8	0.3	5	
AB Dra M	0	0.2	3	
m	0	0.2		
AH Her m	1.8	0.8	3	
TZ Per s	0.7	0.4	6	
m	1.7	0.4	6	
EM Cyg m	2.1	0.5	5,7	352

SU UMa	V–J	J–K	Ref.	d
SU UMa m	0	0.6	6	
YZ Cnc m	0.1	0.7	5	
AY Lyr m	0		6	
VW Hyi m	1.1	0.8	4	
IR Gem m	0		6	

B. Novalikes

UX UMa	V–J	J–K	Ref.	d(pc)
UX UMa H	0.6	0.5	3,8	
L	0	0	3	
TT Ari L	2.4	0.8	6	250

AM Her	V–J	J–K	Ref.	d(pc)
AM Her H	1.3	0.6		
L	3	0.9	9	60
VV Pup H	1	1	10	
L	3	0.9	11	144
AN UMa L	1	0.7	11	

DQ Her	V–J	J–K	Ref.	d
AE Aqr	1.8	0.87	3	84
H2252	0.4	0.2	12	
H2215	0	0.4	6	
DQ Her	0		6	
3A0729	<0.9		6	

[a] m = quiescence   M = outburst   s = standstill   H = high state   L = low state

[b] d from Bailey 1981 method

1 Berriman et al., 1983
2 Panek, 1982
3 Szkody, 1977
4 Sherrington et al., 1980
5 Berriman and Szkody, 1983
6 Szkody, in preparation
7 Jameson, King and Sherrington, 1980
8 Frank et al., 1980
9 Szkody, Raymond and Capps, 1982
10 Szkody, Bailey and Hough, 1983
11 Szkody and Capps, 1979
12 Hassall et al., 1981

From Table 2, it is apparent that if V-J $\gtrsim$ 2, the secondary
may be identified and classified and the distance to the system determined. Among the 42 objects in Table 1 there are only 7 with well
determined secondaries and distances. These systems provide the best
hope for determining the contribution of the disk. Several candidates
for detection of spectral features from the secondary in the optical and
near infrared are TW Vir, AR And, SY Cnc and AH Her, since these systems
have V-J colors close to 2, indicating a larger than normal contribution
from the secondary to the infrared.

With the exception of VW Hyi, all of the SU UMa stars have
Rayleigh-Jeans distributions, consistent with short orbital periods
which imply small mass (faint) secondaries and small Roche Lobe sizes
(small disks). This also appears to hold for H2215 and H2252, which
may be due to the existence of the disk as a thin ring because of the
field strength of the white dwarf (hence the optically thick portion
does not extend to long wavelengths or does not exist at all).

IV. OPTICAL

Optical wavelengths are certainly the oldest area of observational data in the field of CV's. In the past few years, the advent of
flux calibrated spectra from detectors like the Wampler scanner, SIT and
Shectograph have greatly extended the data base by permitting the rapid
acquisition of spectra of a large number of objects. Surveys have been
accomplished by H. Bond (unpublished), G. Williams (1982) and R. Wade
(Oke and Wade 1982). These instruments have also allowed a large number
of spectra per time interval to provide radial velocities and periods
(see review by Robinson 1982) which yield estimates of the masses of
the white dwarfs (see paper by A. Shafter, this volume). The mass is
an important parameter for the determination of the mass accretion and
disk structure.

Table 3 lists the objects from the UV sample which have available flux calibrated emission line spectra, along with their orbital
periods where known. These spectra have been compared using the width
of the Hβ line in Å at the continuum, the strength of Hβ (where strong
(s) means the Hβ peak flux is at least twice the continuum) and the
Balmer decrement (whether it is flat (f) or recombination (r)). The
width of the line can provide information on the extent of the disk
(complicated by inclination effects). If the inner parts are disrupted
(thin ring case), the lines should be narrower. The strength of the
lines and the Balmer decrement are measures of the conditions in the
outer disk regions (Williams 1980).

From Table 3, the differences among the different classes of
dwarf novae and the non-synchronous rotators are not drastic. There is
a large range in line widths even among systems of the same type and
inclination (see Figure 4). The non-synchronous systems are on the
narrow end of the width distribution for the dwarf novae and 4 out of the

TABLE 3. Optical Spectra of Emission Line Objects

A. Dwarf Novae at Quiescence

U Gem	W[a]	D	S	Ref	P[b](hr)	Z Cam	W	D	S	Ref	P(hr)	SU UMa	W	D	S	Ref	P(hr)
SS Cyg	75	f	s	1	6.6	RX And	70	f	s	1	5.08	SU UMa	80	f	s	2	1.9
EX Hya	100	f	s	1	1.64	SY Cnc	60	f	w	1	-	YZ Cnc	100	f	s	1	2.2
TW Vir	75	f	s	1	-	AB Dra	75	f	s	1	-	AY Lyr	30	f	s	2	1.8
AR And	40	f	s	2	-	AH Her	75	f	s	1	6.2	IR Gem	40	f	s	2	1.6
BV Cen	33	-	-	3	14.66	WW Cet	75	f	s	1	3.8	TY Psc	30	f	s	2	1.7
1E0643	75	f	s	1	4.3	EM Cyg	50	f	w	2	6.98						
T Leo	60	f	s	1	1.42												

B. Novalikes

UX UMa		W	D	S	Ref	P(hr)	AM Her		W	D	S	Ref	P(hr)	Rotators	W	D	S	Ref	P(hr)
TT Ari	H	40	f	s	1,2	3.2	AM Her	H	70	f	s	2	3.09	DQ Her	50	f	w	1	4.65
	L	13	r	s	4			L	20	r	s	5		AE Aqr	40	f	w	2	9.88
							VV Pup	H	40	f	s	2	1.67	H2252	30	f	w	2	3.59
								L	none			6		H2215	40	f	s	2	4.85
							AN UMa	H	40	f	s	2	1.91	3A0729	40	f	s	2	3.24
								L	none			7		V1223 Sgr	50	f	w	9	-
							E1405-451		45	f	s	8	1.69	2A0526	50	f	w	2	5.49
							EF Eri		40	f	s	2	1.35						

[a] W = width in Å   D = decrement (flat or recombination)   S = strength
[b] Robinson, 1982 review

1 Williams, G., 1982.
2 Szkody, in preparation.
3 Vogt and Breysacher, 1982.
4 Shafter et al., in preparation.
5 Young, Schneider & Shectman, 1981.
6 Liebert et al., 1978.
7 Liebert et al., 1982.
8 Mason et al., 1983.
9 Steiner et al., 1981.

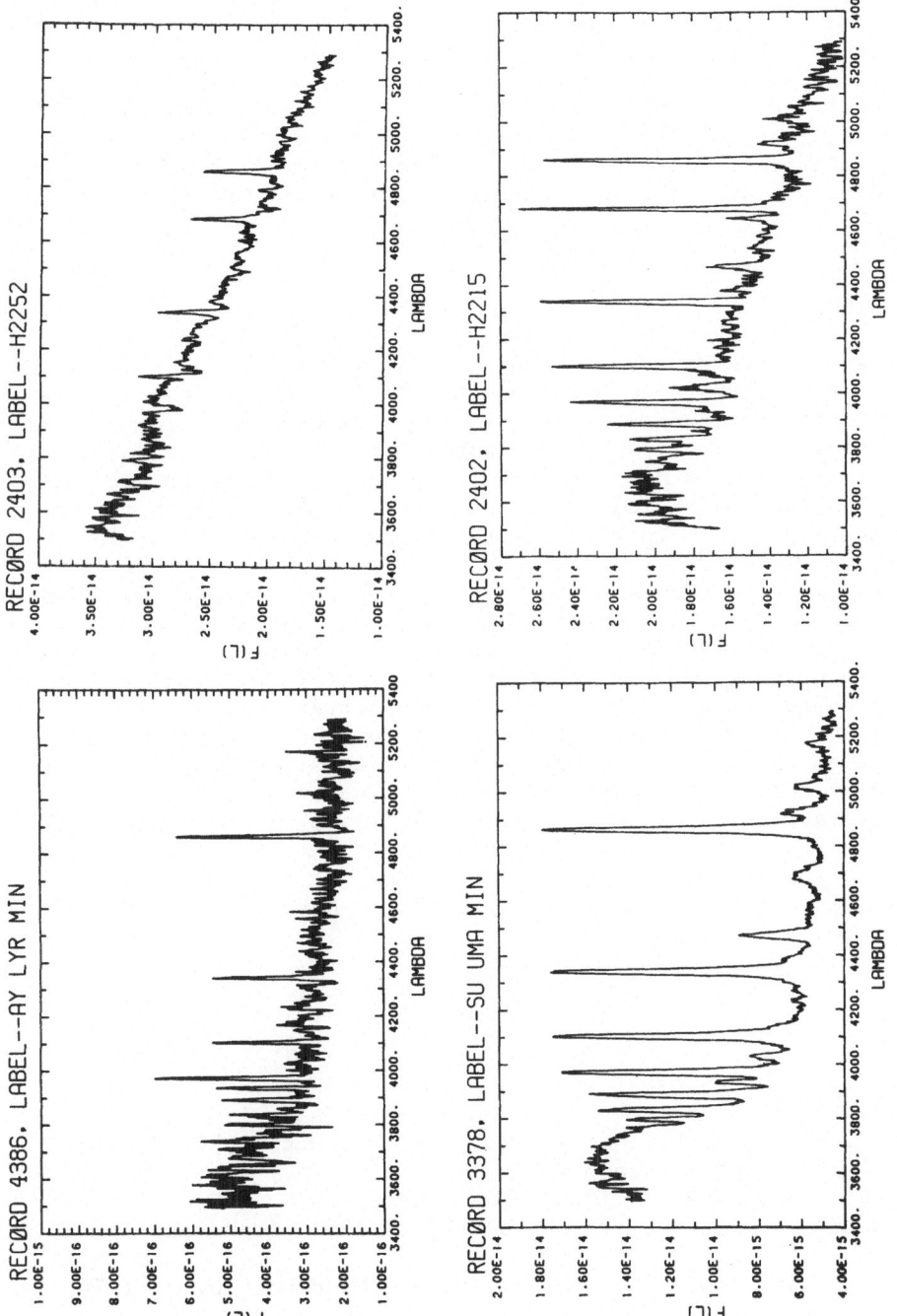

Figure 4. The optical spectra of 2 low inclination SU UMa type dwarf novae (AY Lyr and SU UMa) and two DQ Her type systems (H2252 and H2215). Note the differences in the line strengths, widths, Balmer decrement and He/H line ratios.

6 have weak line emission (see Figure 4), whereas the weak line systems among the dwarf novae occur when there is known or suspected contribution by the secondary star to dilute the emission. The most dramatic differences occur for the nova like systems which undergo the low states. At this time, the width decreases (or the lines disappear entirely) and the decrement steepends to recombination. In AM Her and MV Lyr (Young, Schneider and Shectman, 1981; Schneider, Young and Shectman, 1981), the low state emission is ascribed to the secondary rather than the disk, but this does not appear to be the case in TT Ari (A. Shafter, private communication).

V. CONCLUSIONS

At the present time there is an available data set covering most types of CV's that includes wavelengths from the UV through the IR. This total information on a system provides the necessary input to sort out the actual contribution of the disk light from the secondary star and the hot spot. This data set anxiously awaits the opportunity to constrain realistic disk models of the future. These models must explain the large range in flux distributions for dwarf novae in quiescence, the steeper slopes of the Z Cam distributions, the cause of the supermaxima situation when the disks of SU UMa stars appear similar to other dwarf novae, the mass loss mechanism during the dwarf novae outburst and the UX UMa states, the large UV variability that is evident in some systems and the cause of the low state phenomenon. While these models are being developed, further observational work is needed on systems of high inclination to pin down the location of the optical and UV line emission and the mass outflow regions. Also, continued monitoring of the novalike systems should be pursued to determine the range of variability - UX UMa and TV Col are just two examples to guard against thinking that the emission is constant in any system.

This work was partially supported by NSF grant AST 82-04488 and NASA grant NSG 5395.

REFERENCES

Bailey, J.:1981, MNRAS, 197, p.31.
Bath, G., Pringle, J. and Whelan, J. A.:1980, MNRAS, 190, p.185.
Berriman, G., Beattie, D., Gatley, I., Lee, T., Mochnocki, S. and
  Szkody, P.:1983, MNRAS, in press.
Berriman, G. and Szkody, P.:1983, MNRAS, in press.
Bonnet-Bidaud, J., Mouchet, M. and Motch, C.:1982, Astr. and Ap. 112,
  p.355.
Charles, P., Thorstensen, J., Bowyer, S. and Middleditch, J.:1979,
  Ap. J. 231, p.L131.
Chiapetti, L., Maraschi, L., Tanzi, E. and Treves, A.:1982, preprint.
Cordova, F. and Mason, K.:1982, Ap. J. 260, p.716.
Cordova, F. and Mason, K.:1982b, Accretion Driven Stellar X-ray Sources,
  eds. Lewin, W. and van den Heuvel, E.:Cambridge Univ. Press.

Echevarria, J., Jones, D., Wallis, R., Mayo, S., Hassall, B., Pringle,
    J. and Whelan, J.:1981, MNRAS, 197, p.565.
Fabbiano, G., Hartmann, L., Raymond, J., Steiner, J., Branduardi-Ray-
    mont, G. and Matilsky, T.:1981, Ap. J. 243, p.911.
Frank, J. and King, A.:1981, MNRAS, 196, p.507.
Frank, J., King, A., Sherrington, M., Jameson, R., Axon, D.:1981, MNRAS,
    195, p.505
Greenstein, J. and Oke, J.:1982, Ap. J. 258, p.209.
Guinan, E. and Sion, E.:1982, Ap. J. 258, p.217.
Guinan, E. and Sion, E.:1982b, Adv. in UV Astr., NASA, 465.
Hartmann, L. and Raymond, J.:The Universe at UV Wavelengths, NASA, 495.
Hassall, B., et al.:1981, MNRAS, 197, p.275.
Heap, S., et al.:1978, Nature, 275, p.385.
Holm, A., Panek, R. and Schiffer, F.:1982, Ap. J. 252, p.L35.
Hutchings, J., Crampton, D., Dowley, A., Thorstensen, R. and Charles,
    P.:1981, Ap. J. 249, p.680.
Jameson, R., King, A. and Sherrington, M.:1980, MNRAS.
Jameson, R., Sherrington, M., King, A. and Frank, J.:1982, Nature, 300,
    p.152.
Kiplinger, A.:1979, Ap. J. 234, p.97.
Klare, G., Krautter, J., Wolf, B., Stahl, O., Vogt, N., Wargau, W. and
    Rahe, J.:1982, Astr. and Ap. 113, p.76.
Krautter, J., Klare, G., Wolf, B., Duerbeck, H., Rahe, J., Vogt, N. and
    Wargau, W.:1981, Ast. and Ap. 102, p.337.
Lambert, D. and Slovak, M.:1981, PASP, 93, p.477.
Liebert, J., Stockman, H., Angel, J., Woolf, N., Hege, K. and Margon,
    B.:1978, Ap. J. 225, p.201.
Liebert, J., Tapia, S., Bond, H. and Grauer, A.:1982, Ap. J. 254, p.232.
Mason, K., Middleditch, J., Cordova, F., Jensen, K., Reichert, G., Mur-
    din, P., Clark, D. and Bowyer, S.:1983, Ap. J. 264, p.575.
Mayo, S., Wickramasinghe, D. and Whelan, J.:1980, MNRAS, 193, p.793.
Mouchet, M., Bonnet-Bidaud, J., Ilovaisky, S. and Chevalier, C.:1981,
    Astr. and Ap., 102, p.31.
Nousek, J. and Pravdo, S.:1982, preprint.
Oke, J. and Wade, R.:1982, A. J. 87, p.670.
Panek, R. and Eaton, J.:1982, Ap. J. 258, p.572.
Robinson, E., Barker, E., Cochran, A., Cochran, W. and Nather, R.:1981,
    Ap. J. 251, p.611.
Robinson, E.:1983, in IAU Coll. No. 72, p. 1.
Schneider, D., Young, P. and Shectman, S.:1981, Ap. J. 245, p.644.
Sherrington, M., Lawson, P., King, A. and Jameson, R.:1980, MNRAS, 191,
    p.185.
Sion, E. and Guinan, E.:1982, Adv. in UV Astr., NASA, 460.
Steiner, J., Schwartz, D., Jablonski, F., Busko, I., Watson, M., Pye,
    J. and McHardy, I.:1981, Ap. J. 249, p.L21.
Szkody, P.:1977, Ap. J. 217, p.140.
Szkody, P.:1981, Ap. J. 247, p.577.
Szkody, P.:1982, Ap. J. 261, p.200.
Szkody, P.:1982b, Adv. in UV Astr., NASA, 474.
Szkody, P. and Downes, R.:1982, PASP, 94, p.328.
Szkody, P., Raymond, J. and Capps, R.:1982, Ap. J. 257, p.686.

Szkody, P. and Capps, R.:1980, A. J. 85, p.882.

Szkody, P., Bailey, J. and Hough, J.:1983, MNRAS, in press.

Tylenda, R.:1981, Acta. Astr. 31, p.267.

Vogt, N. and Breysacher:1982, preprint.

Wesemael, F., Auer, L., Van Horn, H. and Savadoff, M.:1980, Ap. J. Suppl.
   43, p.159.

Williams, R.:1980, Ap. J. 235, p.939.

Williams, G.:1982, preprint.

Williams, R. and Ferguson, D.:1982, Ap. J. 257, p.672.

Williams, R. and Ferguson, D.:1983, preprint.

Wu, C. and Panek, R.:1982, Ap. J. 262, p.244.

Young, P., Schneider, D. and Shectman, S.:1981, Ap. J. 245, p.1043.

DISCUSSION

SHAFTER: Do your estimates of the mass transfer rate confirm Joe
Patterson's period versus mass transfer rate diagrams? It looked to me
like the SU UMa stars had quite a high mass transfer rate, or at least
as high as...

SZKODY: As Barb pointed out, the estimates of the mass transfer rate
you get depend on how you did the fitting. I did the fitting based on
Williams's models, which will give you a higher number. On the other
hand, there is a range. The spectral slopes range from one to two.
So there is no one set mass transfer rate that you can assign to an SU
UMa star. On the other hand, the slopes are less than 2.3, which you
find for some of the Z Cam's. Generally, I can say that mass transfer
rates in the SU UMa's, as far as we know, are less than the rates in
the Z Cam's. But I haven't sorted the mass transfer rate out according
to period.

WADE: This is just to muddy the waters some more about how we should
interpret the spectra that we get. I have here three disk models, if
you like, all of which have a slope $F(\nu) \sim \nu^{1/3}$ in the IUE range.
The bottom one is simply an 18,000 K stellar atmosphere from the Kurucz
grid. I think France Cordova has also pointed out that temperatures
around 18,000 K look very much like what we used to think disks ought
to look like. The next one up is a steady-state disk with a maximum
temperature of 24,000 K and a minimum temperature of 12,000 K. Of
course that also brackets 18,000 K. Based on IUE data alone, you
cannot say which of these models is the right one. The difference in
this case is in the Balmer jump and the level of the visual continuum.
Of course, the inferred mass transfer rate through a steady-state disk
of this kind is going to vary as T_{max}^4. So this is just one more
caution about deriving mass transfer rates from the data, without
worrying a little bit about how well the model actually fits throughout
the entire observable spectrum.

BODE: I saw some work recently which suggested that in one or two of
these CV's the 2200 Å feature was actually variable. That's rather
worrying, because you're going to have to form and destroy grains over

very short time scales. I was wondering whether the minimum between the hot spot spectrum and the disk spectrum might occur around 2200 Å in some of these objects, and whether it might be variability in these two spectra that makes people think they've seen a variable 2200 Å feature. What is the wavelength of the minimum between the hot spot spectrum and the disk spectrum?

SZKODY: In most cases, it's around 2500 Å.

BODE: So it's longward of the 2200 Å feature, and you wouldn't confuse them?

SZKODY: Right. The only system in which I've had some real problems with the 2200 Å feature is Stepanian's object. I was dealing with a low exposure and there was a bump at 2200 Å, but it didn't fit the interstellar feature at all.

BODE: These people actually suggested that the E(B-V) was changing, since the depth of the 2200 Å feature, which looked quite normal, changed significantly.

SZKODY: In most cases, there is very little reddening.

WHEELER: I have a comment about this object that changes its emission spectrum in the space of an hour or so. An hour is short, even compared to the thermal time scales, never mind a viscous time scale. I'd be tempted to think that it's entirely different physics than what John is modeling.

SZKODY: I didn't mean to refer just to John, but to whoever is going to model something like this. The magnitude is about what you would expect for a dwarf nova, but the time scale is not right. There is mass outflow like that during a dwarf nova eruption, so in that sense they are similar. And it's happening in the inner part of the disk because the lines we see have high excitation energies. I might also mention that there's no lag between the optical and the UV. They seem to go up together.

WHITE DWARF/ACCRETION DISK INTERACTIONS: INSTABILITIES OF INFINITE,
DIFFERENTIALLY ROTATING CYLINDERS

Bradley W. Carroll and Hugh M. Van Horn
Department of Physics and Astronomy, University of Rochester

ABSTRACT

We have initiated an investigation of the coherent and quasi-
periodic oscillations observed in some dwarf novae. The periods and
relatively high Q associated with the coherent oscillations suggest a
nonradial pulsation of the white dwarf primary for these variations,
while the longer periods and lower Q of the quasi-periodic oscillations
indicate a pulsation of the accretion disk. Guided by a terrestrial
analogy, we have made a preliminary exploration of the shear layer
instability which may drive the observed oscillations.

Our initial, highly idealized model consists of an infinitely long
cylinder, rotating as a solid body, which is surrounded by an infinite
non-self-gravitating sheath. The density of each component is assumed
constant, and the rotation of the sheath about the central cylinder
obeys a Keplerian velocity law. The pulsation of this model is
analyzed both analytically and numerically, using the linearized
adiabatic pulsation equations. The oscillations are taken to be
independent of z (the direction of the axis of the cylinder), with the
eigenfunctions varying as $\exp(im\phi)$.

For incompressible oscillations, the only allowed pulsation mode
is the f-mode. We have found that although the (unphysical) case of
$|m| = 1$ is stable, instability does occur for sufficiently large $|m|$.
The stable $|m| = 1$ mode is retrograde, while the unstable large-$|m|$
modes are prograde.

When compressible oscillations are considered, this simple picture
becomes more complicated: an infinite spectrum of p and g-modes
accompanies the f-mode. Preliminary results suggest that instability
does not occur for large $|m|$ for the compressible case; all modes
examined thus far have been stable (and retrograde). However, further
work using different models and intermediate values of $|m|$ must be
completed before a reliable conclusion can be reached concerning the
stability of the compressible oscillations.

D. Q. Lamb and J. Patterson (eds.), Cataclysmic Variables and Low-Mass X-Ray Binaries, 403–405.
© *1985 by D. Reidel Publishing Company.*

DISCUSSION

PACZYNSKI: Is there any viscosity between your rotating cylinder and the outside world?

CARROLL: No, none at all.

PACZYNSKI: How does the outside world, the rotating fluid, know the angular velocity of the cylinder?

CARROLL: It enters through pressure differences. If you imagine a small pump, the pressure difference depends upon the relative rotation rates of the fluid. I think it finds it out that way.

PACZYNSKI: If there is no viscosity, what is the difference between a wall which is stationary and one which moves?

CARROLL: It comes in through Bernoulli's...

PACZYNSKI: It's a rigid cylinder isn't it, or can it oscillate?

CARROLL: Both the cylinder and the fluid oscillate.

SHAVIV: This is a high-m instability. So you don't actually have to have the disk very high, it can be narrow.

CARROLL: Yes. On the other hand, you don't really want to look at a high-m instability, because you're not going to see it, it's going to wash out. But you can lower the value of m by speeding up the rotation rate or by decreasing the density difference.

SHAVIV: How do you expect the instability to change if you plug in, say, a more realistic or self-consistent solution for the unperturbed state?

CARROLL: It's hard to predict. What I expect is that as the model is refined to make it more realistic, the tendency toward instability will decrease. I expect it to be more stable. That's not the direction I want it to go, obviously. But I still think we'll find an instability; the organ pipe analogy is very strong.

SHAVIV: The radiation in the boundary layer is very important. This kind of adiabatic stuff ignores everything about the radiation, so from this point of view, the chances are very high that everything will be damped.

D. LAMB: I would like to follow up on Bep's question. If you don't have any viscosity, coupling the star and the fluid outside, I'm worried that whatever results you do find will not apply to the real situation, because the boundary layer between the inner part of the disk and the surface of the star is a complicated structure. It's bound to affect this enormously.

CARROLL: I agree with you absolutely. I make no pretense at trying to apply this to a real system. I can tell you that in the organ pipe analogy, if the jet in an organ pipe is fully turbulent, it makes the calculations simpler. I hope that I can introduce something like turbulence in the boundary layer here, which would be more realistic and easily treated. We'll just have to see what happens.

BOUNDARY LAYERS AND CORONAE IN CATACLYSMIC VARIABLES

Kenneth A. Jensen[1]
Laboratory for High Energy Astrophysics
NASA/Goddard Space Flight Center
Greenbelt, MD 20771, U.S.A.

ABSTRACT   X-ray observations of novae, nova-like variables, and dwarf novae in outburst show them to be weaker X-ray sources than predicted by boundary layer models. The discrepancy can be resolved if these systems have optically thick boundary layers with temperatures less than 30 eV, provided the neutral hydrogen column densities of circumstellar gas are greater than 10^{21} cm^{-2}. The absorption of greater than 99 percent of the flux from the boundary layer must occur in a region nearer the white dwarf than the region in which spectral lines are formed. The weak hard X-ray fluxes observed from these systems probably do not come from the boundary layer but may be produced in an accretion disk corona.

1. INTRODUCTION

Accretion models for the radiation from cataclysmic variables (CVs) predict the formation of a boundary layer at the interface between the white dwarf surface and an accretion disk. The boundary layer is expected to radiate a soft X-ray flux if it is optically thick [13] or a hard X-ray flux if it is optically thin [14,15]. Approximately half the gravitational energy released by accretion should be released by dissipation of rotational energy in the boundary layer so that boundary layer luminosities greater than 10^{33} ergs s^{-1} are expected. X-ray observations of CVs have shown them to be weaker X-ray sources, with X-ray luminosities less than 10^{32} ergs s^{-1} [4]. A discrepancy between the observed low X-ray luminosities and the luminosities expected from the boundary layer has been noted particularly for the classical nova subclass of CVs [2,3,6]. It is difficult to explain the absence of a boundary layer if the standard accretion model for CVs is correct. A resolution of this discrepancy is important for an understanding of the accretion process in CVs.

[1]NAS/NRC Research Associate

D. Q. Lamb and J. Patterson (eds.), Cataclysmic Variables and Low-Mass X-Ray Binaries, 407–416.
© *1985 by D. Reidel Publishing Company.*

2. THE PROBLEM

HEAO-1 soft X-ray observations of ~ 200 novae, dwarf novae, and
nova-like variables [3] failed to detect X-ray emission in the 0.18-
2.8 keV energy band, with the notable exception of two dwarf novae in
outburst, U Geminorum and SS Cygni. These CVs, therefore, generally
have X-ray luminosities less than 10^{31} $(d/100pc)^2$ ergs s^{-1}.
Detections of magnetic variables, such as AM Herculis, and intermediate polars, such as EX Hydrae, were made. These types of CVs
probably do not have boundary layers, however, and are not relevant to
a consideration of the boundary layer problem. Einstein Imaging
Proportional Counter (IPC) observations of ~ 50 novae, dwarf novae,
and nova-like variables have been reviewed [4]. The conclusions from
the IPC observations are that these CVs are weak X-ray sources
with Lx < 10^{32} ergs s^{-1} and they are hard X-ray sources, with
temperatures greater than 0.5 keV. The very soft X-ray fluxes
observed from U Gem and SS Cyg in outburst have not been observed in
any other dwarf nova, nova, or nova-like variable. If the hard X-ray
fluxes observed are produced in an optically thin boundary layer, it
is difficult to explain why the ratios of hard X-ray luminosity to
optical luminosity are much lower than unity for dwarf novae in
outburst, novae, and nova-like variables.

Even tighter constraints on X-ray luminosity can be determined
for the classical nova subclass. It is probable that the novae
examined by HEAO-1 and Einstein constitute only a small fraction of
the entire class of novae. Older novae, those whose last outbursts
occurred more than 100 years ago, are probably much more numerous,
since the recurrence time scale for nova outbursts is believed to be
much greater than 100 years [1]. These novae have not been
identified, but some of them should appear in X-ray sky surveys as
serendipitous sources if their X-ray luminosities are sufficiently
high. From the lack of detections of older novae in the HEAO-1 soft
X-ray survey, the luminosities of novae in the 0.18-2.8 keV energy
band are less than $10^{30}(n_{-4})^{-2/3}$ ergs s^{-1}, where n_{-4} is the number
density of novae in units of 10^{-4} pc^{-3} [3]. From the low detecta-
bility of older novae by the Einstein IPC, the luminosities of novae
in the 0.1-4.0 keV band are less than 5 x 10^{29} $(n_{-4})^{-2/3}$ ergs s^{-1} [2].

Dwarf novae, nova-like variables, and classical novae therefore
appear to have X-ray luminosities much lower than expected if the
boundary layer models for the X-ray emission from CVs are correct.

3. THE SOLUTION ?

The constraints on X-ray luminosity described above are a result
of analyses which may be too limited to adequately address the
problem. The X-ray luminosities derived from Einstein IPC
observations apply only to a hard X-ray spectrum [2,4]. HEAO-1
constraints apply to a soft X-ray spectrum, but only for a column

density $N_H = 10^{20}$ cm^{-2} [3]. This choice for N_H was due to spectral fits of the soft X-ray spectra from U Gem and SS Cyg, which indicated $N_H \lesssim 10^{20}$ cm^{-2} for these systems. We cannot rule out the possibility that U Gem and SS Cyg have anomalously low column densities. The need to consider a wider range of source temperatures and column densities is critical because the sensitivity of the X-ray detectors often depends strongly on these parameters.

Another limitation is that constraints on the X-ray luminosity can only be used to test boundary layer models if a total accretion luminosity, hence a mass accretion rate, is assumed. Since boundary layer models predict that one quarter of the total accretion luminosity should be observable as a boundary layer component, the ratio of the boundary layer luminosity to the total accretion luminosity should be ~ 0.25, independent of the mass accretion rate [9]. Constraints on this ratio are more useful than constraints on the X-ray luminosity.

3.1 Classical Novae

HEAO-1 limits on the X-ray emission from novae can be used to derive constraints on the ratio of the boundary layer luminosity to the total accretion luminosity for a wide range of source temperatures and column densities. The fraction of the total accretion luminosity that can be radiated as an X-ray component in novae can be constrained solely as a function of the source spectral parameters. By applying these constraints to the predictions of boundary layer models, the lack of detections of novae by HEAO-1 can be explained if novae have optically thick boundary layers with temperatures $\lesssim 30$ eV and column densities $\gtrsim 10^{21}$ cm^{-2}. In that case, we do not observe a strong hard X-ray flux from novae because their boundary layers are optically thick, and we do not observe a strong soft X-ray flux from novae, even though it is produced in the boundary layer, because more than 99 percent of this flux is absorbed by an intervening column of gas with a column density $\gtrsim 10^{21}$ cm^{-2} [9].

The amount of absorption of the soft X-ray flux from a 30 eV blackbody is very sensitive to column densities in the range 10^{20}–10^{21} cm^{-2} because the cross section for photoelectric absorption ranges from 3–10x10^{-21} cm^2 for the 30 eV blackbody flux accessible to the HEAO-1 soft X-ray detector. For a column density of 10^{20} cm^{-2} less than half of this soft X-ray flux is absorbed, while for a column density of 10^{21} cm^{-2} more than 99 percent of the flux is absorbed.

3.2 Dwarf Novae and Nova-like Variables

The analysis of HEAO-1 constraints on the X-ray emission from novae [9] cannot be repeated for other CV subclasses. The nova subclass is uniquely suitable for an analysis constraining the fraction of the accretion luminosity contained in an X-ray component because the number density of the novae is a function of the white dwarf mass and mass accretion rate, through its dependence on the outburst recurrence time scale.

The Einstein IPC observations of selected novae, nova-like variables, and dwarf novae may help us understand the nature of boundary layers in dwarf novae and nova-like variables, when considered in the context of the constraints on the boundary layers in novae. Particularly useful is the ratio of the received flux from the hard X-ray component in the 0.1-4.0 keV band, f_x, to the received flux in the visual band, f_v. The dwarf novae in quiescence have larger f_x/f_v ratios than the dwarf novae in outburst. Although there is more scatter in f_x/f_v for the novae and nova-like variables, as a group they tend to have f_x/f_v ratios closer to those of the dwarf novae in outburst than the dwarf novae in quiescence [12].

It is more difficult to use f_x/f_v ratios to test boundary layer models than it is to use ratios of the X-ray component luminosity to the total accretion luminosity, because the ratio of visual band flux to total bolometric flux may depend on poorly determined factors such as reprocessed fluxes from the white dwarf, the accretion disk, and the secondary star. It is tempting though to infer from the similarity in f_x/f_v between the novae, the nova-like variables, and the dwarf novae in outburst that nova-like variables and dwarf novae in outburst have boundary layers similar to nova boundary layers. Since the dwarf novae in quiescence are believed to have lower mass accretion rates, it may be that they produce hard X-rays in an optically thin boundary layer while the boundary layers in novae, nova-like variables, and dwarf novae in outburst are optically thick soft X-ray producers.

3.3   The Circumstellar Gas

Since the nearest novae, nova-like variables, and dwarf novae are at distances $\lesssim$ 100 pc, it is unlikely that interstellar column densities are greater than 10^{20} cm^{-2}. If these systems have optically thick boundary layers, the required large absorption of the soft X-ray boundary layer flux means that these must be enough absorbing gas in the system to produce column densities $\gtrsim 10^{21}$ cm^{-2}. The flux from the boundary layer should have an effect on the ionization of this gas. The application of photoionization models to fit the ultraviolet and optical line spectra of the nova V603 Aquilae results in better fits for models in which the line forming region is not exposed to an EUV/soft X-ray flux expected from an optically thick boundary layer [7]. A photoionization analysis of the abundances of Si IV, C IV, and N V in the ultraviolet line forming region of the dwarf nova TW Virginis in outburst also results in the conclusion than an EUV/soft X-ray boundary layer flux cannot be illuminating the gas responsible for the ultraviolet lines [11].

It therefore appears that not only must greater than 99 percent of the boundary layer flux be absorbed by gas in the system, but also this absorption must occur in a region nearer the white dwarf than the region in which the ultraviolet and optical lines are formed.

3.4 The Hard X-Ray Source

If novae, nova-like variables, and dwarf novae in outburst typically have optically thick boundary layers, what then is the source of the weak hard X-ray flux observed from them? The HEAO-1 constraints cannot directly identify this source, but they do constrain the fraction of the accretion luminosity which can be radiated by this source for the novae and, by inference, for the nova-like variables and dwarf novae in outburst. If we adopt a column density of 10^{21} cm^{-2} as typical, less than 1 percent of the total accretion luminosity can be radiated by a 1-10 keV optically thin plasma [9].

3.5 Accretion Disk Coronae

Where in a CV system can we find a hot thin plasma which radiates only a small fraction of the gravitational energy released by accretion? It has been suggested that the weak hard X-ray flux observed from the nova-like variable TT Arietis is produced in an accretion disk corona [10]. If only a small fraction of the energy dissipated in the accretion disk and boundary layer is transported to the corona, it can explain why only a small fraction of the accretion luminosity is radiated by the hard X-ray source. This explanation may apply to the weak hard X-ray fluxes generally observed from novae, nova-like variables, and dwarf novae in outburst.

A coronal model may also provide an explanation for the curious time delay of 60 ± 20 seconds in the X-ray flickering with respect to the optical flickering in TT Ari [10]. If this delay reflects a time scale for radiative cooling of the X-ray corona, coronal densities of $\sim 10^{13}$ cm^{-3} are required. These densities are lower than densities expected in the inner accretion disk [14] and are higher than densities inferred for the region of line formation in V603 Aql [7]. The emission measure constraints on an X-ray corona with this density require a coronal volume $\lesssim 10^{28}$ cm^3, which is a reasonable volume for a corona formed about the inner accretion disk of a white dwarf.

The presence of short wavelength shifted absorption lines in the ultraviolet spectra of novae, nova-like variables, and dwarf novae in outburst indicates that these systems have accretion disk winds as well as hard X-ray emission [5]. This suggests that magnetic fields may play a role in the formation of a structured corona [8] in which the hard X-ray emission is produced in closed structures (coronal loops) while the wind flows through open structures (coronal holes).

If CVs are producing hard X-rays in accretion disk coronae, the analysis of their X-ray emission may be considered in the general context of stellar coronal X-ray emission [16]. They may therefore extend the study of coronal formation and heating to smaller spatial scales, faster rotational time scales, and higher gravitational potentials than are available for stellar coronae.

REFERENCES

1. Bath, G. T., and Shaviv, G.: 1978, M.N.R.A.S., 183, pp. 515-522.
2. Becker, R. H.: 1981, Astrophys. J., 251, pp. 626-629.
3. Cordova, F. A., Jensen, K. A., and Nugent, J. J.: 1981, M.N.R.A.S., 196, pp. 1-12.
4. Cordova, F. A., and Mason, K. O.: 1983, in Accretion Driven Stellar X-Ray Sources, ed. W. H. G. Lewin and E. P. J. v.d. Heuvel.   pɪ
5. Cordova, F. A., and Mason, K. O.: 1982, Astrophys. J., 260, pp. 716-721.
6. Ferland, G. J., Langer, S. H., MacDonald, J., Pepper, G. H., Shaviv, G., and Truran, J. W.: 1982, Astrophys. J., 262, pp. L53-L58.
7. Ferland, G. J., Lambert, D. L., McCall, M. L., Shields, G. A. and Slovak, M. H.: 1982, Astrophys. J., 260, pp. 794-806.
8. Galeev, A. A., Rosner, R., and Vaiana, G. S.: 1979, Astrophys. J., 229, pp. 318-326.
9. Jensen, K. A.: 1983, Astrophys. J., submitted.
10. Jensen, K. A., Cordova, F. A. Middleditch, J., Mason, K. O., Grauer, A. D., Horne, K., and Gomer, R.: 1983, Astrophys. J., 270, in press.
11. Kallman, T. R.: 1983, Astrophys. J., in press.
12. Mason, K. O., and Cordova, F. A.: 1982, in COSPAR, Advances in Space Exploration (Pergamon Press, Ltd.).
13. Pringle, J. E.: 1977, M.N.R.A.S., 178, pp. 195-202.
14. Pringle, J. E., and Savonije, G. J.: 1979, M.N.R.A.S, 187, p. 177.
15. Tylenda, R.: 1981, Acta Astr., 31, p. 267.
16. Vaiana, G. S. et al.: 1981, Astrophys. J., 244, p. 163.

DISCUSSION

PACZYNSKI: There were reports some years ago that when a dwarf nova goes into eruption, the hard X-ray flux goes down. Is this true for most of them, or is it not known?

JENSEN: Well, let's take SS Cyg as an example. The Einstein observations are usually snapshots and don't cover an outburst. I don't know of any pointing observations which do. If anybody knows of some, I'd like to hear about it. I haven't seen any confirmation, but my idea of what's happening in SS Cyg during outburst is that the boundary layer becomes optically thick and suppresses the hard X-rays, and all you see then is the residual coronal emission.

LANGER: We looked at the problem of the ratio of the X-ray to optical fluxes, and what not. From what I recollect, if you have an optically thick boundary layer with a temperature of 30 eV, and try to square that with the bolometric luminosity that you expect from the mass accretion rates that people quote for these systems, you find out that it implies a rather large boundary layer. It doesn't have anything like the kind of thickness that a disk would have, or that one would expect for a simple boundary layer. Factor of ten uncertainties in determining accretion rates could have a considerable bearing on this problem. I would like to know if there is a possibility of developing better disk models and getting better fits to these systems, so we can have a better idea of what the accretion rate, and therefore the bolometric luminosity, is?

JENSEN: That's a question for somebody else to answer. What I would emphasize is that my results apply for mass accretion rates ranging from $10^{-7} - 10^{-11}$ $M_\odot$ yr^{-1}. What they show is that the disks can't be any hotter; they can be cooler.

LANGER: The bigger the boundary layer, the worse the theory is doing. So, either somebody's got to fix up the bolometric luminosity so that we can tolerate a boundary layer like what's being discussed, or some theorist has to sit down and come up with a rather different kind of theory for the boundary layer itself.

JENSEN: As an observer, I would simply say that 30 eV is about as high a temperature as you can get.

LANGER: I accept that result. It's just we don't understand why it happens.

WHEELER: There's been a question kicking around of whether, when you come into the boundary layer, it expands and "wastes" some energy in PdV work, eventually turning it into a wind. Ed Sion and I were trying to do some numbers at the coffee break. It seemed that the kinetic energy of the wind is down by a couple of orders of magnitude from what you're trying to account for. Can you or anybody verify that this is a crummy idea?

JENSEN: I can say that it is only a few percent because, if we believe the mass flux in the wind derived by Cordova and Mason, the amount of momentum and energy flux in the wind is on the order of the amount of X radiation itself. Since the rotational energy in the boundary layer is the same as the escape energy, to order of magnitude, we're talking about a few percent. So all that does is say we can double the amount of X-rays or the energy that can go into the boundary layer.

SWANK: Do you have to assume an area?

JENSEN: No.

STOCKMAN: This result holds for novae or nova-like systems, not
dwarf novae, right?

JENSEN: This result holds for novae as a class, which means
objects like the ones we've seen having outbursts. Whether the
nova-like variables are the same class or not, I don't know.
This result holds for objects that have recurrent hydrogen
flashes on some time scale, like those modelled by Nariai and
Nomoto.

HERTZ: Just to let you know that some more data is coming,
we've accumulated about three hundred serendipitous Einstein
sources in the galactic plane, and we're searching for their
optical counterparts. As of yet, we haven't discovered any new
cataclysmic variables. We've searched down to a limiting magnitude of about 16.5 for a sample of about a hundred sources.
That limits the X-ray/optical flux ratio to something greater
than about 0.1 for these sources, as a class.

JENSEN: Greater than 0.1?

HERTZ: Right. We've searched all the brightest optical sources
in the error boxes.

JENSEN: I'm glad you mentioned that, because Einstein is a lot
more sensitive. The problem, of course, is finding the optical
counterparts before these things novae. But if some enterpris-
ing person could do that, could repeat my LED analysis with
Einstein, we could tell whether or not novae fade in X-ray
luminosity as they get older, for example. This is a question
which can only be marginally answered by the LED.

HERTZ: Our survey should give a pretty good limit on the contribution of non-optically selected CV's.

LIEBERT: Maybe this is a good place to put in an advertisement.
In the first medium survey sample done with Einstein by the
Harvard group in collaboration with us at Steward, one entire CV
was discovered out of about - well, there were many more than 50
extragalactic fields - something like 30 stars. And in the second
sample that's being prepared for publication now, there are thus
far zero CV's. That one CV is written up in a preprint that's
now available by Stockman et al. We didn't write a separate
paper for it; it's, you know, the "ninety-fourth" X-ray - selected CV by now. The point is, it's just an ordinary-looking
dwarf nova at high galactic latitude.

JENSEN: There's no question about it, they are not as strong X-ray emitters as you would believe from AM Her and SS Cyg. As a class, they are much weaker. They might be the high X-ray flux end of the stellar sources; rapid rotation of the inner accretion disk might make their coronae more active. It might be important to consider cataclysmic variable X-ray emission in the context of stellar coronal emission, rather than in the context of the more active systems, the neutron star binaries.

KIPLINGER: Do you have an estimate of the hard X-ray flux, or total X-ray flux from boundary layers of dwarf novae in outburst?

JENSEN: I don't, and I want to tell you why I don't. It's because the distances aren't known that well. That's why I deal with flux ratios. You can make an estimate, if you want to pick a distance. Luminosities of around $10^{31} - 10^{32}$ ergs s^{-1} are as much as you get.

GRINDLAY: What do you think the flickering mechanism might be then? Why should they show flickering if the emission is, in fact, coronal?

JENSEN: That's for theorists to work on. [Laughter.] I don't know, it's not my bailiwick. I would say that the most active systems show P-Cygni profiles, which change around the orbit. It looks like these objects have winds. It looks like vertical angular momentum transport definitely occurs in the active systems with thick accretion disks. In analogy with the sun, for example, where there is hard X-ray emission from the surface and a wind further out, one wonders whether the flickering time scale is somehow related to magnetic activity.

SHAVIV: First of all I would like to thank you, and then I would like to add a small contribution of my own to the problem of the boundary layer. Papers on the boundary layer, such as those by Pringle, Pringle and Savonije, and Tylenda, assume that the boundary layer is optically thin, and that the radiation really comes out. If you assume that the boundary layer is optically thin, you get a first-order ordinary differential for the temperature. This equation has a singularity at the place where the derivative of the rotational velocity versus radius vanishes because it is Keplerian outside and must then go down. This is a singular point of the equation. So, in order to save the model, you choose the one free parameter you have in this equation so that it takes care of the singularity. If you do that, everything is solved and you have a model. But! the model fails to fit the boundary layer to the star itself. It's a boundary layer which is somehow lighter, but it does not fit the conditions on the white dwarf itself. This is an internal inconsistency that the boundary layer has, if you assume it to

be optically thin. If you want to take into account radiation
that comes out of the boundary layer, you find out that the net
flux in it is much larger than both the flux in the accretion
disk above it and in the white dwarf below it. So, not only
does it not fit the boundary conditions of the unperturbed white
dwarf, it perturbs what is below it. There are theoretical
problems with thin boundary layers that we cannot cope with in a
simple way. One obvious solution is to assume the boundary
layer is optically thick, rather than optically thin, and turn
the first-order differential equation into a second-order one
by means of diffusive phenomena. But that's not the whole problem. The other one is that the only way to solve the problem at
that singular point is to assume that the temperature vanishes.
Then you don't have the proper disk that comes to that place.
So there are still problems...

JENSEN: Perhaps I should switch roles with Giora, and be the
session chairman and say, maybe that's enough till later.

ACCRETION DISKS IN SYMBIOTIC STARS AND THEIR RELATIONSHIP TO
CATACLYSMIC BINARIES

Scott J. Kenyon
University of Illinois

I. INTRODUCTION

The symbiotic stars were discovered in the early 1900's by Cannon
and Merrill as late-type giants with surprisingly strong emission
lines from H I, He I and He II. Infrared photometry has confirmed the
presence of K or M giants in most symbiotics (Allen 1982), while IUE
spectra show intense emission lines superimposed on a strong continuous source which is generally associated with a hot companion to the
cool giant (Slovak 1982). Accurate binary periods are known for only a
handful of symbiotics (cf. Kenyon 1983), and thus the physical parameters of the two components are not well-determined. This lack of basic
observational data has slowed progress in understanding the hot component in these systems. Indirect methods involving emission line
fluxes and/or fitting the optical continuum have been employed to estimate the luminosity and effective temperature of this component, but
these have provided little information concerning the physical nature
(white dwarf, subdwarf or main sequence star) of the underlying star.

The launch of the IUE satellite has allowed the first direct observations of the continuum emitted by the hot component. A few attempts have been made to fit the observed continua with stellar atmosphere models, but the results have not been entirely convincing: a
few symbiotics are almost certainly hot stars (e.g. AG Peg; Keyes and
Plavec 1980), while other systems are not so simply understood (e.g.
YY Her; Michalitsianos et al 1982). Kenyon and Webbink (1983; Paper I)
explored a wide variety of binary models, and devised a simple set of
reddening-free, diagnostic criteria to classify symbiotics according
to the nature of their hot components. Applying these criteria to a
small sample of systems, they found many of these objects appear to be
main sequence stars accreting at rates of 10^{-5} M_o/yr. Other systems
were more easily explained as hot stellar sources with effective temperatures of 25-100,000 K.

D. Q. Lamb and J. Patterson (eds.), Cataclysmic Variables and Low-Mass X-Ray Binaries, 417–427.

II. CALCULATIONS

Our aim in this paper is to expand the sample of systems discussed in Paper I, and to present a different analysis of symbiotic UV continua. We consider a two-component model for the UV continuum (cf. Boyarchuk 1967), consisting of:

(i)   a hot component, being either an accretion disk surrounding a low mass dwarf star, or a hot, compact star similar to the central star of a planetary nebula, and

(ii) a surrounding gaseous nebula ionized by the hot component.

The observed flux from this combination can be written as:

$$\log F_{obs}(\lambda) = \log [F_H(\lambda) + F_N(\lambda)] + 2 \log R_1/d - 0.4 A(\lambda) \qquad (1)$$

where $F_H(\lambda)$ and $F_N(\lambda)$ are the emitted flux from the hot component and the nebula (in erg $cm^{-2}s^{-1}\mathring{A}^{-1}$), R_1 is the radius of the hot component, d is its distance and $A(\lambda)$ is the extinction as a function of wavelength (Savage and Mathis 1979).

The observed spectrum of the hot component, F_H, is naturally dependent on whether the hot component is a star or an accretion disk. We follow Bath et al (1974), and write the temperature in a disk as:

$$T(x) = T_* x^{-3/4}(1 - x^{-1/2})^{1/4} \qquad (2)$$

where $x = R/R_1$ and $T_* = (3GM_1\dot{M}/8\pi\sigma R_1^3)$. The mass of the accreting star and the accretion rate are given by M_1 and $\dot{M}$, respectively. We also adopt the Lynden-Bell and Pringle (1974) formula for the temperature of the boundary layer, namely $T_{bl} = 2.4 \cdot T_*$. For this investigation, we assume the hot component (whether it be a disk or a stellar source) radiates locally as a blackbody. This is a fairly reasonable assumption for stars with effective temperatures greater than 50,000 K, but is a poor representation for lower temperatures (see Paper I). At the accretion rates of interest here, the disk is optically thick and radiates much like a blackbody (Williams 1980). Thus, we follow Tylenda (1977) and assume the flux from the disk is simply a superposition of blackbodies with the temperature law given in equation 2. Since our major interest lies in determining the general nature of the hot components in the symbiotics (rather than in a detailed model for a specific symbiotic star), it seems most natural to begin with these simple approximations and develop more rigorous models as detailed observations become available.

In the IUE wavelength range, the nebular flux is predominantly free-bound and free-free emission from H I and He II. We assume the nebula is optically thick to all photons below the Lyman limit, and optically thin to radiation above the Lyman limit. The nebular flux is then proportional to the number of ionizing photons, which is a unique function of the temperature of the ionizing source (T_h for a hot star or T_* for a disk). Thus the total emitted flux, $F_H + F_N$, is simply a

function of T_* or T_h, depending on the model assumed, and the electron
temperature. Electron temperatures of 10-20,000 K appear to be typical
of symbiotics (Nussbaumer 1982), and both these values were considered
for hot star and accreting white dwarf models. The nebular contribution to the continuum of model main sequence accretors is never more
than a few percent, and thus the nebular component has been neglected
for these cases.

Our procedure to determine the nature of the hot component is to
minimize the quantity δ as a function of E(B-V), $\log(R_1/d)$ and either
T_* or T_h, where:

$$\delta = \sum_\lambda [\log F_{obs}(\lambda) - \log(F_H(\lambda) + F_N(\lambda)) - 2 \log R_1/d + 0.4A(\lambda)]^2. \quad (3)$$

Typically, we pick 20 well-spaced wavelength points on SWP and LWR IUE
spectra and determine the continuum, F_{obs}, at each point. Since F_H and
F_N are non-linear functions of temperature, a simple least-squares
procedure is not adequate to minimize δ. The Levenberg-Marquardt algo-
rithm was chosen to minimize this function, since it offers rapid con-
vergence and high accuracy (Bard 1974). A disadvantage of non-linear
least-squares algorithms is that they generally find local minima,
rather than absolute minima. Multiple calculations must then be made
to adequately explore parameter space.

III. RESULTS

The calculations presented here were made on the University of Illinois Cyber 175 computer, and a few of the graphical results are displayed in Figure 1. The observed continuum fluxes have been plotted as
filled circles, while the derived reddened continua are drawn as solid
lines. The average r.m.s. deviation of the derived continuum from the
observed fluxes, $\langle \delta^2 \rangle^{1/2}$ has been adopted as a "goodness of fit" parameter. Ideally, this parameter should be smaller than, or at least
roughly comparable to, the errors in the observed continuum points.
The position of the continuum can probably be determined to ± 0.05
magnitude for the brighter symbiotics such as AG Peg, with a possible
error of at least 0.2 magnitude for the fainter systems Y CrA and
V1329 Cyg. For this investigation, a fit was accepted only if $\langle \delta^2 \rangle^{1/2}$
was less than 0.07.

Satisfactory fits were obtained for 11 spectra, and these are
listed in Table 1 (the numbers in parentheses refer to the number of
continuum points measured for each system). Two marginally good fits
for Z And have been included in the Table, since Z And is the proto-
typical symbiotic star. Acceptable solutions could not be found for
Y CrA and V1016 Cyg. Y CrA has an extremely weak far-UV continuum; our
inability to find a good solution is probably caused by the difficulty
of accurately locating the continuum. V1016 Cyg has a rather bright UV
continuum: a marginal, but not good, solution was found for an effect-

ive temperature of roughly 130,000 K. Both V1016 Cyg and HM Sge possess circumstellar dust shells, and our inability to obtain a good solution for V1016 Cyg may be a result of substantial circumstellar reddening, which may not follow the standard interstellar extinction law
(Paper I; Kenyon 1983). Multiple solutions were found for Z And,
YY Her and V 443 Her, and thus the nature of their hot components
could not be determined solely from UV continuum measurements. Unique
fits were found for EG And, CI Cyg, AG Peg, AX Per and HM Sge. Two of
these systems, CI Cyg and AG Peg, were considered in Paper I, and the
solutions presented here agree with those reported in Paper I to within 5-10%. It is encouraging the two methods developed to determine the
nature of the hot components yield similar results.

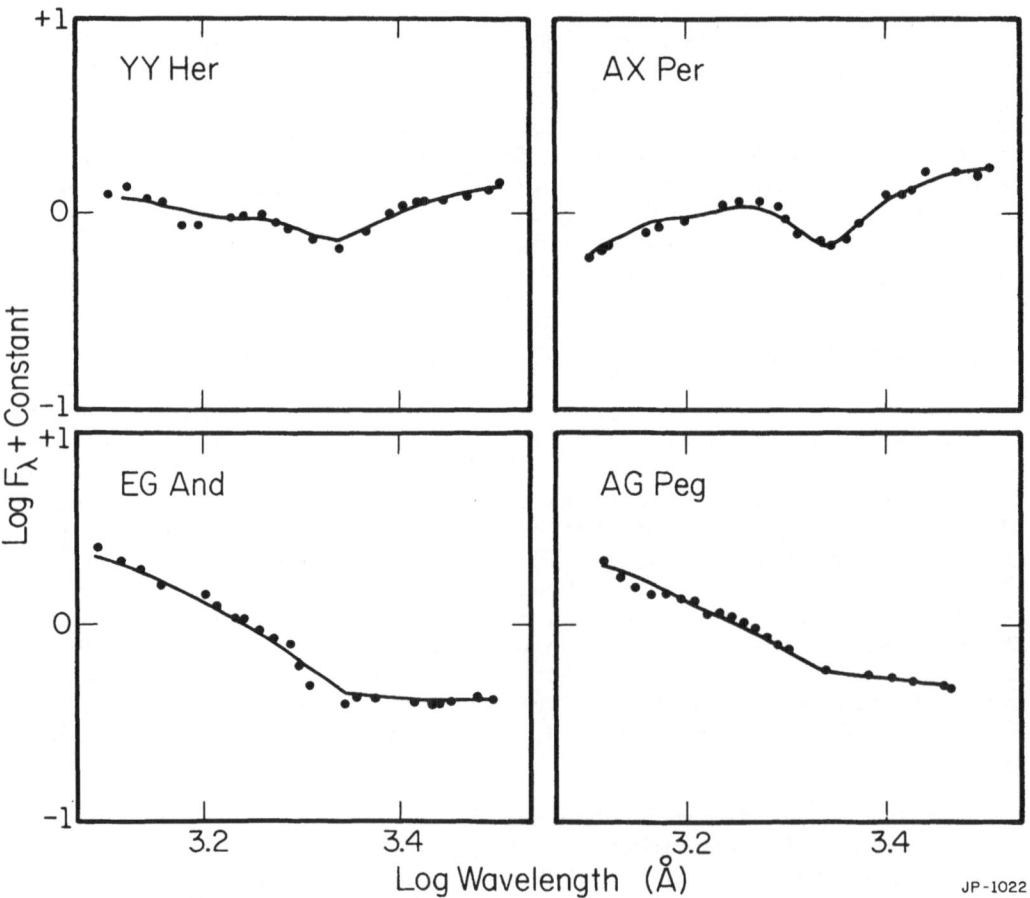

Figure 1 - Least squares fits to the continua of four symbiotic stars.
           Accretion disk models are shown for YY Her and AX Per,
           while hot stars best represent the continua of EG And and
           AG Peg.

Table 1
Symbiotic Stars with Acceptable Continuum Fits
Hot Star Models

symbiotic	JD	$\langle \delta^2 \rangle^{1/2}$	$\log T_h$	$\log R_1/d$	E(B-V)
	2,443,000+				
Z And(26)	1613	0.06	4.99±0.03	-11.69±0.09	0.35±0.04
EG And(23)	873	0.03	4.73±0.03	-11.68±0.04	0.08±0.02
EG And(23)	1613	0.05	4.86±0.04	-11.81±0.05	0.05±0.04
YY Her(22)	1422	0.05	5.12±0.03	-12.47±0.08	0.20±0.03
V443 Her(22)	1345	0.03	4.84±0.03	-11.65±0.07	0.25±0.03
AG Peg(23)	1613	0.02	4.77±0.02	-11.13±0.05	0.08±0.02
HM Sge(21)	1686	0.06	4.41±0.03	-10.78±0.05	0.61±0.04

Accretion Models

symbiotic	JD	$\langle \delta^2 \rangle^{1/2}$	$\log T_*$	$\log R_1/d$	E(B-V)
Z And (26)	1613	0.06	4.28±0.12	-10.58±0.21	0.31±0.05
CI Cyg(23)	881	0.05	4.58±0.04	-11.18±0.06	0.45±0.03
CI Cyg(19)	1054	0.05	4.51±0.04	-11.01±0.06	0.48±0.03
YY Her(22)	1422	0.03	4.38+0.04	-11.45±0.08	0.15±0.03
V443 Her(22)	1345	0.04	4.20±0.04	-10.62±0.08	0.19±0.04
AX Per(23)	1613	0.02	4.56±0.02	-11.21±0.06	0.28±0.02

Three systems have unique solutions corresponding to hot stellar sources with effective temperatures of 25-75,000 K. Best solutions for EG And and HM Sge were obtained with electron temperatures of 10,000 K, while the best model for AG Peg had an electron temperature of 20,000 K. Rough distance estimates can be made using the observed K magnitudes for these objects (Kenyon and Gallagher 1983), and these may be combined with the derived R_1/d values to estimate luminosities of 6-16 L_o for EG And, 500-3000 L_o for AG Peg and 300-2000 L_o for HM Sge. These are consistent with previous estimates, as are the derived extinctions (Slovak 1982, Kenyon 1983). Since these other estimates rely on optical data as well as UV data, this agreement is quite encouraging. The two observations of EG And yield slightly different results for the hot star effective temperature. The observation obtained on JD 2444613 has a significantly stronger far UV continuum, which suggests the hot star in this system may be intrinsically variable.

CI Cyg and AX Per are the only symbiotics uniquely identified as accreting systems in this study. Reliable distance estimates of 2.1 kpc for CI Cyg and 2.8 kpc for AX Per (Kenyon and Gallagher 1983) imply central star radii of roughly 1 R_o for each system! Thus, both CI Cyg and AX Per appear not to be accreting white dwarfs but rather accreting main sequence stars. The implied accretion rates for these

systems are 1-4 x 10^{-5} $M_{\odot}$/yr if a solar-type object is assumed to be
the accreting star. The derived extinctions are in excellent agreement
with estimates based on optical data (Kenyon 1983 and references
therein), which encourages us to believe these accretion rates are
fairly accurate. Two observations for CI Cyg were analyzed in this
study, and these indicate some variability in the accretion rate,
which is probably not significant given the overall faintness of the
UV continuum near 1300 Å. The inclination of the disk, i, has been
assumed to be 0° in the results presented here. Although the observed
flux of the disk depends on cos i, the relatively small wavelength
range of IUE observations does not allow sufficient constraint on the
value of cos i. Once EUV observations of symbiotics become available,
it may be possible to provide constraints on the inclinations of the
disks in these systems. Much better boundary layer models are also required.

IV. DISCUSSION

     The important result of this study is that many symbiotic stars
appear to contain low mass main sequence stars accreting material from
their red giant companions at roughly 10^{-5} $M_{\odot}$/yr. Thus, these systems
provide an opportunity to examine disk accretion under far more extreme conditions than those found in a typical cataclysmic binary.
Furthermore, a rapidly evolving red giant with a low mass main sequence companion is precisely the combination imagined as the progenitor for the typical cataclysmic binary, as Dr. Paczynski discussed
earlier in these Proceedings. If low mass main sequence stars are present in symbiotics, then each of these systems could give birth to a
cataclysmic binary!

     The evolution of a binary containing a Roche-lobe filling red gi-
ant and a main sequence star is a complicated process, involving expansion into contact, the formation of a common envelope and the ejection of this common envelope as the two compact stars spiral-in to
form a close binary. The symbiotic stars CI Cyg and AX Per appear to
be at the onset of this process, although very little information concerning the giants in these systems is currently available. Kenyon and
Gallagher (1983) attempted to determine the luminosity class of the
giants in 11 symbiotics, and only the giant in CI Cyg definitely fills
its Roche Lobe. This red giant appears to be on the asymptotic giant
branch, and should lose mass more rapidly as it evolves up the AGB.
The recent eruptions of the hot component appear to be accretion-
powered (Webbink and Kenyon, in preparation), and may indicate the giant in CI Cyg is approaching a phase of extremely rapid mass loss
which could lead to the formation of a common envelope binary. Other
symbiotics may also be in or approaching this phase of evolution, and
may provide important information regarding the formation of cataclysmic binaries.

The author thanks Drs. J.S. Gallagher, I. Iben, J.W. Truran, A.V. Tutukov and R.F. Webbink for providing advice, comments and encouragement during the course of this project. Drs. A.G. Michalitsianos and R.E. Stencel and the National Space Data Center graciously provided the IUE data used in this analysis. Computing funds and travel support were generously provided by the University of Illinois. Additional support for this research was provided by the National Science Foundation through grants AST 80-18859 and AST 80-18198 to the University of Illinois.

REFERENCES

Allen, D.A. 1982, In The Nature of Symbiotic Stars, IAU Colloquium No. 70, edited by M. Friedjung and R. Viotti (Reidel:Dordrecht), p.27.
Bard, Y. 1974, Nonlinear Parameter Estimation (Academic:New York)
Bath, G.T., Evans, W.D., Papaloizou, J. and Pringle, J.E. 1974, Mon. Not. Roy. astr. Soc.,169,447.
Boyarchuk, A.A. 1967, Soviet Astron. J.,11,8.
Kenyon, S.J. 1983, Ph. D. thesis, Univ. of Illinois.
Kenyon, S.J. and Gallagher, J.S. 1983, Astron. J., in press.
Kenyon, S.J. and Webbink, R.F. 1983, Astrophys. J., submitted.
Keyes, C.D. and Plavec, M.J. 1980, In The Universe at Ultraviolet Wavelengths, edited by R.D. Chapman, (NASA CP-2171), p. 443.
Lynden-Bell, D. and Pringle, J.E. 1974, Mon. Not. Roy. astr. Soc., 168,603.
Michalitsianos, A.G., Kafatos, M., Feibelman, W.A. and Hobbs, R.W. 1982, Astrophys. J.,253,735.
Nussbaumer, H. 1982, In The Nature of Symbiotic Stars, IAU Colloquium No. 70, edited by M. Friedjung and R. Viotti (Reidel,Dordrecht) p.85.
Savage, B.D. and Mathis, J.S. 1979, Ann. Rev. Astron. Astrophys.,17, 73.
Slovak, M.H. 1982, Ph. D. thesis, Univ. of Texas.
Tylenda, R. 1977, Acta Astron.,27,3.
Williams, R.E. 1980, Astrophys. J.,235,939.

DISCUSSION

PACZYNSKI: On purely observational grounds, there are two categories of symbiotic stars. In one type, the strength of emission lines, or the degree of ionization, is positively correlated with luminosity. These are usually argued to be symbiotics, novae, slow novae, or something like that.

KENYON: Right, like V1016 Cyg.

PACZYNSKI: Yeah! Then there's another class where the strength of the emission lines and the excitation temperature indicated by the emission lines are inversely correlated with luminosity. These are the classical symbiotic stars described by Boyarchuk. I am curious to which category your model is supposed to apply.

KENYON: The model of accreting main sequence stars applies to the classical symbiotics.

PACZYNSKI: How do you explain that the luminosity in the visual goes up at the same time the emission lines go to lower and lower states of ionization as they disappear?

KENYON: In an outburst, the high excitation emission lines vanish. The Balmer lines are still strong.

PACZYNSKI: But clearly the far UV goes down at the same time the visual light goes up.

KENYON: Right. It's possible that with a main sequence star, when you have an outburst and you go to the Eddington limit, the boundary layer becomes optically thick and then expands. So you get a reduction in the number of ionizing photons. Whereas in the case of a nova, you won't get that because you just have an increase of luminosity in a wind. The hot star solutions are more applicable to novae or nova-like symbiotics, in which the strength of the emission lines is correlated with the luminosity. But the accreting systems, and even the cataclysmic variables, show the opposite behavior. For example, dwarf novae have emission lines at minimum and absorption lines at maximum.

PACZYNSKI: Oh, but that's a different story, because the absolute strength of the emission lines may stay just the same. It's not known.

KENYON: Right. But it's not known in symbiotics either.

PACZYNSKI: No, no! There's one big difference. The excitation temperature goes down gradually. It's very well inversely correlated with the observed luminosity. It's not like you shut off the UV. If the emission lines are a good measure of the far UV, the far UV goes down simultaneously as the optical goes up. That's exactly what would be expected, as Boyachuk pointed out fifteen years ago, if a star had a constant bolometric luminosity and just changed its radius.

KENYON: That's true, but the observed spectra of these systems in the UV are not consistent with the spectrum you get from a hot star. You just can't fit the observed IUE spectra with a hot star, or a hot star plus a nebula, unless you are willing to distort the extinction curve and make it very peculiar. They are very flat in the ultraviolet; from 1200 to 3000 Å they tend to be very flat. That can only be obtained with a main sequence accretor, which tends to have a Planck peak in the UV.

PACZYNSKI: Someone, I don't remember who, was showing us model spectra of single stars. They are flat in the far UV, all of them.

KENYON: Not as flat as the symbiotics.

PACZYNSKI: Well, they were as flat as the power law indicated on the figures.

KENYON: That was $\nu^{1/3}$.

PACZYNSKI: Yeah!

KENYON: These are flatter; F_λ is approximately constant with λ.

WEBBINGK: The problem is that the stars that give relatively flat spectra are not hot enough to ionize the nebula.

KENYON: Right.

PACZYNSKI: Oh, I see.

SHAVIV: What is the mass of the nebula?

KENYON: In the accreting symbiotics, I can estimate the mass of the ionized nebula. It's around 10^{-7} or 10^{-8} $M_\odot$.

WEBBINK: If one assumes a hot star and has to have temperatures like 120,000 or 150,000 K, there's no escaping having an intrinsically steeply rising continuum, at least at the short end of the IUE range. That is, from 1700 to 1300 Å the continuum should be rising quite rapidly. Either that or, if you want to fill it in with nebular emission, you ought to see incredibly large Balmer jumps.

PACZYNSKI: Do IUE observations extend through the end of the outburst or the brightening?

WEBBINK: I'm convinced SY Mus is an eclipsing system. The first time it was observed, it was at phase 0.95 of the photometric ephemeris. It appeared, if anything, to have a very cool "hot star" but still had lots of emission lines. It was re-observed later and had a far stronger UV continuum, characteristic of a hot star with a temperature of about 120,000 to 150,000K. AG Dra was observed before the optical eruption and once during the decline.

PACZYNSKI: Are they correlated? Do people know how the IUE EUV behaves?

WEBBINK: That's a crucial test, because the accretor models and the hot star solutions predict quite different developments in the near ultraviolet during outburst.

SMAK: Is it correct to say that the terms "symbiotic stars: or "symbiotic binaries" were introduced by the spectroscopists? The physical situation seems to be that if we have a red giant trans-ferring material at a very fast rate to a main sequence star, or something not much smaller, it's a symbiotic binary. If another red giant transfers material at the same fast rate to a white dwarf, it isn't a symbiotic binary, it's something much more complicated that we won't even see as a binary. Am I correct?

KENYON: I would imagine that if you had a white dwarf that was accreting at the rate that I'm giving for the accreting main sequence stars, which is near the Eddington limit for a white dwarf, and it would just expand to engulf the system.

SMAK: That's what I mean: we wouldn't see it as a binary.

KENYON: I believe that's correct. The hot stars that I discussed could be accreting at very low rates, like 10^{-10} or 10^{-9} $M_\odot$ yr^{-1}, via a wind or some similar interaction. You'd never know it, because the intrinsic luminosity of the hot star dominates everything.

PACZYNSKI: What are the masses of your main sequence stars?

KENYON: I chose stars of 1 $M_\odot$ and 1 $R_\odot$.

PAZYNSKI: Kippenhahn's calculations, and lots of other people's calculations, indicate that when you accrete at such a high rate, a main sequence star swells up just as much.

KENYON: But on a time scale of 10^5 years.

PACZYNSKI: So you think these stars are just turning on?

KENYON: I think that the time scale for the symbiotic phenomenon is not that different from the time scale for the red giant. The time scale for the main sequence star to expand to fill its Roche lobe is within a factor of ten, maybe, of the time scale for the red giant to go up the asymptotic giant branch, which is roughly 10^6 years.

WADE: Two questions about the spectroscopy. You mentioned that you have a hard time predicting the observed H_β flux. How

much of the flux at 4860 Å is from He II? Do you see the
Pickering series in your spectra?

KENYON: No, generally not. H_α is slightly broader than it
should be, but it's not a strong component.

WADE: The second question is, for your main sequence systems you
infer high mass transfer rates. Does that mean that you require
that the red giant should be filling its Roche lobe?

KENYON: Yes.

WADE: That leads me to an information question. What are the
spectroscopic distinctions between luminosity class II and
luminosity class III?

KENYON: They are generally in the near infrared. The TiO bands
are a good temperature indicator, and for luminosity class, the
CN bands and other singly ionized or neutral metals are good
indicators. That's been generally unexplored. I've tried to
use CO at 2.2 microns as a luminosity indicator. It indicates
that CI Cyg, for example, is much more luminous than a normal
everyday giant but is not quite a supergiant, sort of a luminosity
class II giant.

WADE: So a fair statement is that the present ambiguity could be
resolved if someone wanted to do it?

KENYON: Yes. I'm planning to do that, to observe as many as
possible in CO in the near infrared to try to determine it. For
the two eclipsing systems, you can get an estimate of the size.
For CI Cyg, it again works out. CI Cyg, of all the symbiotics,
appears to be filling its Roche lobe.

ACCRETION DISKS

Warren M. Sparks
Los Alamos National Laboratory
G. Siegfried Kutter
The Evergreen State College

Derivations are made for the mass and the mass turnover time scale of an accretion disk as a function of the accretion rate, the observed disk radius, the non-viscous disk radius and two parameters. These parameters depend on the effectiveness of viscosity and tidal angular momentum loss. Application is made to DQ Herculis.

1. INTRODUCTION

Recently, Sulkanen, Brasure, and Patterson (1981, hereafter called SBP) have calculated disk radii in cataclysmic variables using eclipse observations. These disk radii are about 2-3 times larger than the radii of Flannery's (1975) non-viscous disks. The non-viscous disks are calculated by assuming that the angular momentum of the outer material of the disk is equal to the angular momentum of the infalling stream of material from the secondary. The fact that the observed disks are larger than the non-viscous disks should not be surprising since the same viscosity which reduces the angular momentum of the material that reaches the surface of the white dwarf increases the angular momentum of the outer disk material (Kruszewski 1967).

Following Paczynski (1977), we will assume that angular momentum is transferred from the disk to the binary system via tidal interaction. Assuming an equilibrium disk, the balance between angular momentum input and removal determines the disk radius. Thus, we are able to use the disk radius to determine something about these mechanisms and the structure of the disk.

2. ASSUMPTIONS

We make the following assumptions.
1) The pressure gradient in the disk is negligible and the disk material is in circular Keplerian orbits. Thus the orbital velocity and the specific angular momentum at a distance R from the center of the

D. Q. Lamb and J. Patterson (eds.), Cataclysmic Variables and Low-Mass X-Ray Binaries, 429–433.
© *1985 by D. Reidel Publishing Company.*

primary star is given by

$$v_O = \sqrt{\frac{GM_1}{R}} \quad , \tag{1}$$

and

$$j = \sqrt{(GM_1 R)} \quad , \tag{2}$$

where M_1 is the mass of the primary. Differentation of Equation (2) gives

$$\frac{dj}{dt} = \frac{1}{2} \sqrt{\frac{GM_1}{R}} \, v_{tot}, \tag{3}$$

where v_{tot} is the average total radial velocity of the material. Actually we do not strictly require the assumption of circular Keplerian orbits, but only that the material at radius R has an average angular momentum given by Equation (2).

2) The inward radial velocity of the material <u>due to viscosity</u> is proportional to the orbital velocity (Pringle and Rees 1972),

$$v_V = -\beta v_O \quad . \tag{4}$$

In analogy to Equation (3), the rate of change of angular momentum due to viscosity is

$$\frac{dj_V}{dt} = \frac{1}{2} \sqrt{\frac{GM_1}{R}} \, v_V \quad . \tag{5}$$

This represents an outward flow of angular momentum. Combining Equations. (1), (4) and (5) gives

$$\frac{dj_V}{dt} = -\frac{1}{2} \beta \frac{GM_1}{R} \quad . \tag{6}$$

3) Angular momentum is lost from the disk only by tidal interactions with the secondary and by accretion of material onto the primary. Therefore, there is no mass loss after the material is on the disk, other than onto the primary.

4) Furthermore, we estimate the tidal angular momentum loss by the expression

$$\frac{dj_T}{dt} = -K \frac{G(3M_2 + M_1)R^2}{D^3} \quad , \tag{7}$$

where M_2 is the mass of the secondary and D is the binary separation. The proportionality constant K is a measure of the efficiency of tidal frictional breaking and is always much less than 1.

5) The total angular momentum loss rate is the sum of the viscous term and the tidal term,

$$\frac{dj}{dt} = \frac{dj_V}{dt} + \frac{dj_T}{dt} \quad . \tag{8}$$

6) The disk is in steady state. This means that the mass flow across any spherical surface at R is constant and given by

$$\dot{M} \equiv v_{tot}\, A\rho \quad ,$$
(9)

where ρ is the density and A is the surface perpendicular to R.

7) The area, A, is given by

$$A \equiv \gamma R^2 \quad ,$$
(10)

where γ is a constant.

3. DERIVATIONS

We define R_E as the radius where the two rates of angular momentum loss are equal and obtain

$$-\frac{1}{2}\,\beta\,\frac{GM_1}{R_E} = -K\,\frac{G(3M_2+M_1)R_E^2}{D^3} \quad ,$$
(11)

or

$$R_E = D\,\sqrt[3]{\frac{\beta M_1}{K2(3M_2+M_1)}} \quad .$$
(12)

From the steady state condition of angular momentum, we derive

$$\dot{M}\,\sqrt{(GM_1 R_N)} = \int_{disk}\frac{dj_T}{dt}\,dm + \dot{M}\,\sqrt{(GM_1 R_1)} \quad ,$$
(13)

where R_N is the non-viscous radius that the input material would have after it is circularized, and R_1 is the radius of the primary star. The two terms on the right hand side of Equation (13) are the mass-integrated loss of angular momentum by tidal interaction and the angular momentum accreted onto the primary. Note that the viscosity term does not remove angular momentum from the disk but only transports it outward. Changing the integral from over mass to over radius and integrating, we derive

$$\sqrt{R_N} = \sqrt{R_D} - \sqrt{R_E}\left\{\frac{1}{4\sqrt{3}}\ln\left[\frac{R_D + \sqrt{(3R_E R_D)} + R_E}{R_D - \sqrt{(3R_E R_D)} + R_E}\right]\right.$$

$$\left. + \frac{1}{6}\arctan\frac{\sqrt{(R_E R_D)}}{R_E - R_D} + \frac{\pi}{6} + \frac{1}{3}\arctan\sqrt{\frac{R_D}{R_E}}\right\},$$
(14)

under the conditions $R_1 \ll R_D$ (the disk radius) and $R_1 \ll R_E$. An integration of the time derivative of the angular momentum flow due to viscosity gives

$$\dot{J}_V \equiv \int_{disk}\frac{dj_V}{dt} = \dot{M}\,\sqrt{(GM_1 R_D)} - \dot{M}\,\sqrt{(GM_1 R_N)} \quad ,$$
(15)

which shows that the viscosity term is responsible for the extension of the disk beyond R_N.

Integrating over the mass of the disk gives

$$M_{disk} = \frac{2}{3} \dot{M} \sqrt{\left(\frac{R_E^3}{\beta^2 M_1 G}\right)} \left[\text{arc tan} \sqrt{\frac{R_D^3}{R_E^3}} - \text{arc tan} \sqrt{\frac{R_1^3}{R_E^3}} \right]. \quad (16)$$

The time scale, $\tau = M_{disk}/\dot{M}$, for a particle to move through the disk is

$$\tau = \frac{2}{3} \sqrt{\left(\frac{R_E^3}{\beta^2 M_1 G}\right)} \left[\text{arc tan} \sqrt{\frac{R_D^3}{R_E^3}} - \text{arc tan} \sqrt{\frac{R_1^3}{R_E^3}} \right]. \quad (17)$$

4. APPLICATION TO DQ HERCULIS

DQ Her probably has the best determined parameters needed for this analysis. From eclipse observations, SBP have derived a disk radius of $R_D = 0.78 \, R_{R1}$ where R_{R1} is the Roche lobe radius of the primary. Using a primary mass of $0.83 \, M_\odot$, and a secondary mass of $0.60 \, M_\odot$ (Warner 1976), and a binary period of 1.67×10^4 sec (Walker 1954), we derive a binary separation, $D = 1.10 \times 10^{11}$ cm and a disk radius, $R_D = 0.32D$. The non-viscous disk radius has been calculated by Flannery (1975) as a function of the stellar masses and their separation. Interpolating in his tables we find $R_N = 0.10 \, D$.

Using the above values of R_D and R_N in Equation (14), we iterate to find $R_E = 0.056D$. Substituting this value into Equation (12) gives $(\beta/K) = 1.13 \times 10^{-3}$. Thus β must be much smaller than 1.13×10^{-3}. The disk time scale from Equation (17) gives $\tau = 3.74 \times 10^4$ sec/K. If we estimate $K \simeq 10^{-3}$, then $\beta \simeq 10^{-6}$ and $\tau \simeq 1$ year. The mass accretion rate for DQ Her is not known, but if we use a typical accretion rate found for other cataclysmic variables of $\sim 10^{-8} \, M_\odot/yr$, we obtain a disk mass of $1 \times 10^{-8} \, M_\odot$.

We thank Clara DeMaria for typing this paper.

REFERENCES

Flannery, B.: 1975, Monthly Notices Roy. Astron. Soc. 170, pp. 325-331.
Kruszewski, A.: 1967, Acta Astr. 17, pp. 297-310.
Paczynski, B.: 1977, Astrophys. J. 216, pp. 822-826.
Sulkanen, M. E., Brasure, L. W., and Patterson, J.: 1981, Astrophys. J. 244, pp. 579-581 (SBP).
Walker, M.F.: 1954, Publ. Astron. Soc. Pacific 66, pp. 230-232.
Warner, B.: 1976, in "Structure and Evolution of Close Binary Systems", eds. P. Eggleton, S. Mitton, and J. Whelan, Reidel, Dordrecht, Holland, pp. 85-140.

DISCUSSION

STOCKMAN: Could you just briefly go over the choice of "K", and why you chose such a small value?

SPARKS: If you have all the disk material along a line connecting the two stars, instead of at 90 degrees to the line of centers, then the tidal torque is a maximum. In a realistic situation, you have material distributed in a disk through a range of angles. It is certainly not a line, and I would think "K" would be a small number.

STOCKMAN: I agree it is probably small, but it's not intuitive to me why it couldn't be $10^{-1.5}$, in which case it would bring your numbers into better agreement with what other people have found.

PACZYNSKI: I'm somewhat misled by the picture you have drawn, because it shows the angular distortion appropriate to a star, which is pressure supported. In the case of a disk, which is not pressure supported, the tidal effect makes it elongated perpendicular to the line joining the two stars.

SPARKS: Okay. But as long as it is symmetric about the line of centers joining the two stars, you don't get any net torque.

WADE: On your last transparency, you had a time scale of a year. What does that time scale correspond to?

SPARKS: It corresponds to the time scale for material to flow through the disk. It should not be related to rise times of the disk luminosity.

WADE: I'm wondering if that has any connection with the observation that the disk in DQ Her has been getting fainter over the last couple of decades?

PACZYNSKI: I think the most obvious explanation of that is that the white dwarf is cooling off after the eruption. Most of the light is being reprocessed, as observations clearly show. That gives us the cooling time scale of the white dwarf. It actually could be compared with theory, but this has never been, as far as I know.

WADE: If you adopted rather lower values for the masses of the stars, as has been suggested by Young and Schneider, for example, what numbers would change and by how much?

SPARKS: I have not calculated that. It shouldn't change them by very much.

DISTANCES, ACCRETION RATES, AND SPACE DENSITIES OF CATACLYSMIC VARIABLES: PIECE O' CAKE, AND OTHER HERESIES

Joseph Patterson
Harvard-Smithsonian Center for Astrophysics

"I have no special regard for Satan; but I can at
least claim that I have no prejudice against him.
It may even be that I lean a little his way, on
account of his not having a fair show. All religions
issue bibles against him, and say the most injurious
things about him, but we never hear his side . . . .
To my mind, this is irregular. It is un-English; it
is un-American; it is French . . . . Of course Satan
has some kind of a case, it goes without saying. It
may be a poor one, but that is nothing; that can be
said about any of us."

                                        -- Twain (1894)

Over the last few years, it has become virtually compulsory for
every review article, and generally stylish for research papers, to
include at least a little hand-wringing about how poorly we know the
fundamental parameters of cataclysmic variables (hereafter CVs).
Probably all of us remember the time when we thought that CVs contained
main sequence stars transferring matter to 1 $M_\odot$ white dwarfs via α-model
accretion disks. Such views are now considered naive, and cause dozens
of eyes to roll heavenward in disgust. But, listening to the relentless
criticism of yesterday's wisdom at meetings such as this one, I get
confused as to who or what is being criticized, since it seems to me
that there are no defenders of the "standard models" left--that virtually
all of us have embraced today's conventional, pessimistic wisdom. Most
of you know this litany of pessimism: standard disk models are too
naive; the lobe-filling secondaries are frequently not on the main
sequence; the distances are very poorly known; the accretion rates are
uncertain by orders of magnitude; so are the space densities; etc., etc.
The unanimity with which these discouraging views are professed strikes
me as unhealthy. So I will step up in defense, *mutatis mutandis*, of
yesterday's optimistic views--partly because I do not share the current
pessimism, and partly because we are in need of a whipping boy on which
to focus our criticisms. Most of these arguments are spelled out in
much greater detail in a bloated paper (Patterson 1983, hereafter P83)
which I am not so sure anyone will actually read!

D. Q. Lamb and J. Patterson (eds.), Cataclysmic Variables and Low-Mass X-Ray Binaries, 435–444.
© *1985 by D. Reidel Publishing Company.*

1. DISTANCES

Until a few years ago, most of us felt that the secondaries in CVs were main-sequence stars, because that is the natural way to obtain a long-lived phase of mass transfer, and because the mass-radius determinations for the secondary in EM Cyg and Z Cam indicated a main-sequence structure (Robinson 1976). But recent observational results have been interpreted as indicating that some secondaries are not on the main sequence (e.g., Warner 1978, Wade 1979, Stover 1981, Robinson *et al*. 1982), and this has led many of us to doubt or reject altogether that very useful and plausible assumption of main-sequence membership.

However, the observers have frequently not allowed for uncertainties in the main-sequence mass-radius law, which is not precisely known. They have compared their results to theoretical models of zero-age main sequence stars, and there is substantial evidence (Hoxie 1973, Lacy 1977) that these models do not correctly predict the mass-radius-luminosity relations for any stars. We can derive an empirical mass-radius law-- approximately $R/R_\odot = (M/M_\odot)^{0.88}$--for ZAMS stars from Lacy's extensive data, and we have shown that if this law is used, the empirical data on CV secondaries in systems with an orbital period ≤ 9 hours require a ZAMS membership (P83, see also Ritter 1982). The importance of the mass-radius law can be seen in Figure 1, where we show the dependence of the secondary's spectral type and absolute visual magnitude on orbital period, for both theoretical and empirical mass-radius laws. We have also super-imposed the observed spectral types of CV secondaries, and it is clear that all of them are too cool to be theoretical ZAMS stars, but those with $P \leq 9$ hours agree very well with the empirical ZAMS. Secondaries in long-period systems are too cool for either mass-radius law, but they too are "on the main sequence", as they have expanded by $\lesssim 30\%$ over their ZAMS radii (P83).

If the secondaries are on the main sequence, then we can use their apparent magnitudes (or limits thereon) to derive distance information for any system of known orbital period. This yields 22 good-quality distances, plus another ~ 60 good-quality limits. We can add another 7 good-quality distances by using the "expansion parallax" method on classical nova shells. Finally, we can derive additional, rougher distance constraints by a variety of methods: proper motions, inter-stellar reddening, position in the galaxy, "standard candle" arguments, and the empirical relation between absolute magnitude and the equivalent width of $H\beta$ emission. By such means we can give a distance estimate for virtually every CV of known orbital period, and I have done so in P83. The uncertainties vary from star to star, but for the ~ 50 best-observed stars, I estimate a typical $\pm 1\sigma$ uncertainty of $\Delta \log d = \pm 0.1$-0.3.

There is an additional constraint available from the space density argument of Section 3. At the very least, we should check that the space density of known CVs does not exceed the total CV space density revealed by surveys which are thought to contain a complete sample. In P83, these numbers are found to be $\sim 2 \times 10^{-6}$ pc^{-3} and 7×10^{-6} pc^{-3}, respectively.

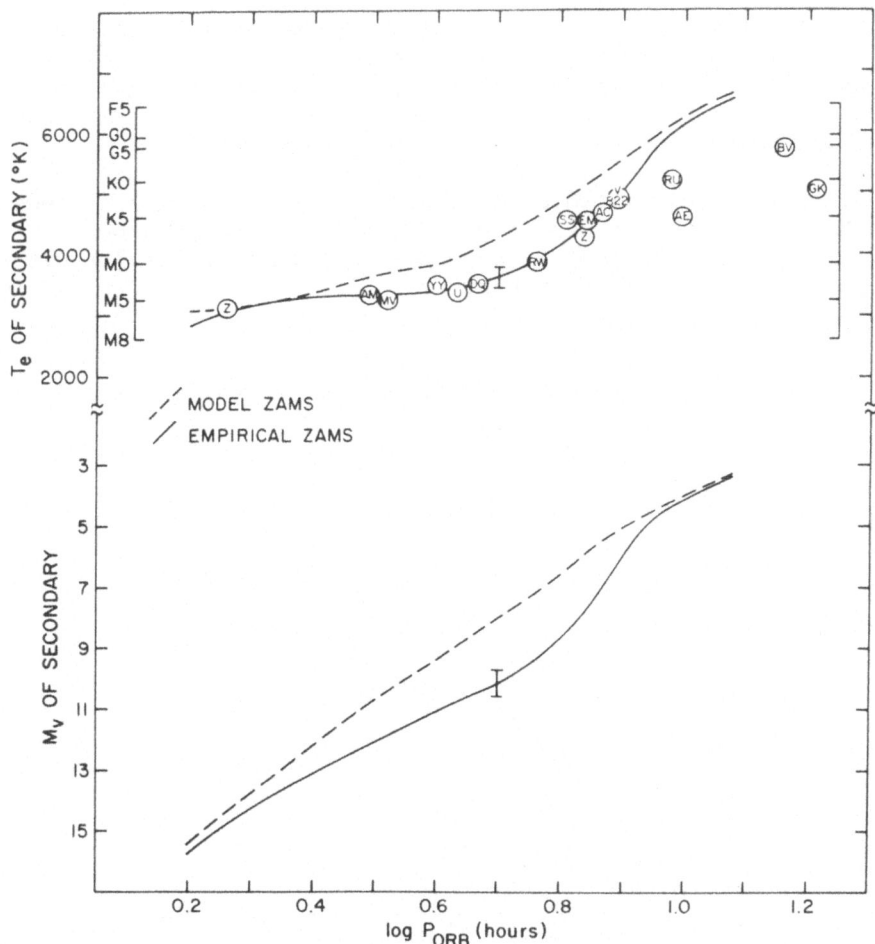

Figure 1. Dependence of the secondary's effective temperature (or
           spectral type, given by the inset scale) and absolute visual
           magnitude on orbital period for two choices of mass-radius
           relation: the theoretical ZAMS and the empirical ZAMS.

To take this argument a bit further, we could require that the ratio of
the space density of known CVs to the space density of all CVs should be
approximately equal to the fraction of previously known CVs in the complete sample (11/43, from the surveys cited below). As you can see, these
numbers check reasonably well, i.e., 2/7 ≈ 11/43. This is not a decisive
argument, but provides an order-of-magnitude check for consistency.

2. MASS TRANSFER RATES

   Over the years, several methods have been used to derive accretion
rates, but I believe that the best general method available today matches
the observed absolute visual magnitude of the accretion disk with
<u>theoretical</u> <u>disk</u> <u>models</u>. Williams and Ferguson (1982) and Tylenda (1981)

have published disk models which calculate the optical depth and radiation self-consistently from physics (although the amount of viscous
heating is still obtained from α-disk theory).   This is still not widely
appreciated; as recent criticisms continue to focus on the inadequacy of
assuming high optical depth and/or blackbody emission.

In P83 we have calculated accretion rates in individual systems by
matching M_V with Tylenda's disk models.   The results are shown in
Figure 2, plotted versus orbital period.   A strong correlation with
orbital period ($\dot{M} \propto P^{3.3}$) is evident.

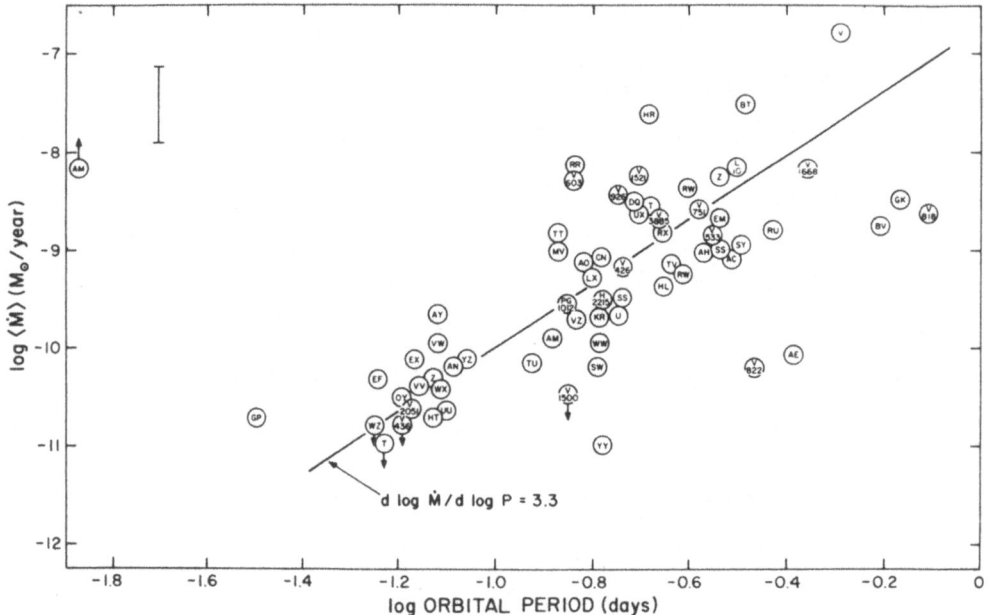

Figure 2.   Estimated accretion rate versus orbital period.   The straight
            line shows the best linear fit to the data.

How reliable are such estimates of $\dot{M}$?   One cannot give a simple
answer to so oversimplified a question, but I would be surprised if the
models are wrong by as much as a factor of 10.   For a given $\dot{M}$, the bolometric accretion disk luminosity L_{disk} is just $GM_1\dot{M}/2R_1$.   The models tell
us at what wavelengths the radiation emerges.   Because the disk temperature varies with radius ($T \propto R^{-3/4}$ in the optically thick portions),
the luminous contribution in the visual band is relatively insensitive
to the assumptions made about the disk, unless all of the disk temperatures are substantially too low (< 4000 °K) or too high (> 30 000 °K) to
contribute significantly in the visual band.   But the observed colors of
CVs indicate effective temperatures in the range 8000–25000 °K, and the
bolometric corrections for such temperatures are fairly well-determined
and small (0.5–3 magnitudes).   As long as the light comes from the
viscous heating of matter spiralling through the disk, it is hard to see

how the models can be seriously wrong in predicting the visual luminosity (although the infrared and ultraviolet luminosities can be quite sensitive to the details of the outer and inner disk structure, respectively).

3. SPACE DENSITIES

We also hear many doleful proclamations of our ignorance of the space density of CVs. In the last decade, only one serious attempt to deduce a space density has been made: Bath and Shaviv (1978) argued that the classical novae <u>alone</u> have a space density $D_{CN} \sim 10^{-4}$ pc^{-3}, and this number has generally been adopted when needed. However, I believe that the space density is not so intractable a problem--and, in particular, that this estimate is much, <u>much</u> too high.

There are two ways of arriving at this conclusion. First, one can adopt typical values of M_V and L_X for classical novae, and ask how many should be detected down to some convenient flux limit. Presently available data suggests that mean values for classical novae are: $M_V = +4.2$ (Payne-Gaposchkin 1957, Duerbeck 1981), $L_X(0.1-4.0$ keV$) = 3 \times 10^{31}$ erg s^{-1} (Becker 1981). Using these values and a galactic scale height of 150 pc (the result is not sensitive to this number), we show in Table 1 how many old novae should be seen in the sky at various familiar flux levels, together with how many <u>are</u> seen.

TABLE I. CONSTRAINTS ON THE NOVA SPACE DENSITY

Flux Limit	Predicted Number of Stars*		Observation
	$D_{CN} = 10^{-4}$ pc^{-3}	$D_{CN} = 3 \times 10^{-6}$ pc^{-3}	
$m_V < 6.0$ (naked eye star)	5	0.15	none!
$m_V < 9.5$ ($\sim$ HD star)	400	13	1
$m_V < 11.0$	2400	70	5
$F_X(2-10$ keV$) > 3 \times 10^{-11}$ erg cm^{-2} s^{-1} (HEAO-1 limit)	450	14	13[†]
$F_X(0.1-4.0$ keV$) > 6 \times 10^{-13}$ erg cm^{-2} s^{-1} (HEAO-2 IPC limit)	25000 or 0.5 deg^{-2} (all sky)	750 or 0.016 deg^{-2} (all sky)	$\lesssim 0.02$ deg^{-2} (high-latitude) $\lesssim 0.10$ deg^{-2} (low-latitude)

*Assuming $L_X(0.1-4.0$ keV$) = 3 \times 10^{31}$ erg s^{-1}, $M_V = +4.2$.

[†]Including all CVs <u>other</u> than AM Her stars, but not including the $\sim$ 3-10 stars which are probably yet to be found among the $\sim$ 100 unidentified HEAO-1 sources.

While there are certainly grounds for worrying about how complete our
census of "observed" novae is, I think Table 1 demonstrates that a space
density of 10^{-4} pc^{-3} is completely ruled out by observation. I trust
that every X-ray and optical observer in the room will agree that there
are not 5 old novae among the naked-eye stars, 400 among the HD stars,
450 among $HEAO-1$ X-ray sources, and 25,000 within $Einstein$ IPC limits.
Even our favored estimate of 3×10^{-6} pc^{-3} may be a bit high, as suggested
by the optical data. There is a well-known difficulty in discovering
quiescent old novae at optical wavelengths, because such stars typically
lack the means (eruptions, bright emission lines) to call attention to
themselves. This could account for the order-of-magnitude discrepancy
in the optical data, but it is also possible that we have overestimated
D_{CN} somewhat because of the statistical uncertainty implicit in the
number of novae (12) that have occurred within 1000 pc in the last 100
years.

This argument would change dramatically, yielding a much higher
space density, if one were to adopt much lower mean visual and X-ray
luminosities for classical novae (e.g., $M_v = +9$, $L_x = 10^{30}$ erg s^{-1}).
But this can be ruled out with great confidence since: (1) it is inconsistent with observations of known classical novae; and (2) it would
flood the high-latitude sky with faint blue stars that are not observed
(see below). For example, for $M_v = +9$ and $D_{CN} = 10^{-4}$ pc^{-3}, approximately
600 CVs should have been seen by the Palomar-Green survey--about twenty
times greater than observed.

Secondly, we can study the space density of all CVs from the results
of three recent surveys of the high galactic latitude sky: the Palomar-Green
survey for objects with blue U-B colors (Green $et\,al.$ 1982); the various
$Einstein$ surveys for serendipitous X-ray sources (e.g., Stocke $et\,al.$ 1983);
and the Michigan-Tololo survey for emission-line objects (e.g., MacAlpine and
Williams 1981). The first two of these were sufficiently sensitive to detect
any CV in the fields, given the mean empirical values for M_v, L_x, and scale
height. The emission-line survey required a minimum equivalent width of about
30 Å, and something like half of known CVs would have been caught. Because the
detection thresholds were sufficient to identify essentially any CV in the
field, and because the scale height is approximately known, it is simple to
convert the number per square degree to a number per cubic parsec. The result is
that the total space density in the solar neighborhood is $\sim 7 \times 10^{-6}$ pc^{-3}, and
various arguments suggest that about 30-50% of these are classical novae.

If there were some strong reason for favoring a high space density,
we could still escape these arguments by further adjustments to the
assumed M_v and L_x. Of course, it is formally true that if one does not
place any constraints on the radiation coming from an object, then any
results from any survey are consistent with any space density. But this
is not a very useful viewpoint! Let us state more carefully the conclusion that we draw from these arguments: stars with properties similar to
those of known classical novae have a space density of $\sim 3 \times 10^{-6}$ pc^{-3},
and stars with properties similar to those of known CVs have a total
space density of $\sim 7 \times 10^{-6}$ pc^{-3}. The second figure is likely to be

accurate to $\sim 30\%$, while the first is probably uncertain to a factor of 2 or so.

4. WHERE ARE THE DEAD NOVAE?

The high mass transfer rates ($\sim 10^{-8}$ $M_\odot$/year) deduced for classical novae, and for long-period systems generally, imply very short lifetimes, approximately 10^8 years. Unless we live at a very special time in the history of the galaxy, this implies that the number of active systems we see today is only 1-2% of the total number that have ever existed. Where are the descendants of the classical novae which presumably lit up the skies of yesteryear? One's first guess is that they could now be short-period systems, which have $<\dot{M}> \sim 5 \times 10^{-11}$ $M_\odot$/year and can therefore survive for a time comparable to the age of the galactic disk. But this would predict that the space density of short-period systems should be $\sim 2 \times 10^{-4}$ pc^{-3}, and this is completely excluded by the high-latitude surveys.

I have discussed this problem in more detail in P83, but have not found any particularly desirable solution. A reasonably trouble-free solution would be provided by a plausible way to destroy the binary systems while the orbital period is still $\gtrsim 3$ hours. In principle, the secondary can be whittled down to planetary size if a sufficiently high rate of angular momentum loss is inflicted on the system *sine fine* (Paczynski 1980).

5. EXHORTATION

This has been a quick foray into the demography of CVs, and I have had to present the arguments in skeletal form. Strictly on the basis of what I have been able to present here, I hope that no one will con-clude that what I have said is definitely wrong--or definitely right. But perhaps this will stir up the waters a bit, or induce a few aimless souls to take a peek at P83. My hope is that some of you will be sufficiently interested to explore these issues in greater detail--and I think that you will find yourself staring at uncertainties considerably smaller than you probably expected. The observers must do this if we are ever to get down to the important and exciting business of deciding just which numbers we are going to ask theories of CV evolution to reproduce.

OK, you've patiently endured my whimsical remarks. Now it is time for the mudslinging to begin.

REFERENCES

Bath, G. T., and Shaviv, G.: 1978, *Mon. Not. R. Astron. Soc.* **183**, 515.
Becker, R. H.: 1981, *Astrophys. J.* **251**, 626.
Duerbeck, H. W.: 1981, *Publ. Astron. Soc. Pac.* **93**, 165.
Ferguson, D. H. 1982, private communication.

Green, R. F., Ferguson, D. H., Liebert, J., and Schmidt, M.: 1982, *Publ. Astron. Soc. Pac.* **94**, 560.
Hoxie, D. T.: 1973, *Astron. Astrophys.* **26**, 437.
Lacy, C.: 1977, *Astrophys. J. (Suppl.)* **34**, 479.
MacAlpine, G. M., and Williams, G. A.: 1981, *Astrophys. J. (Suppl.)* **45**, 113.
Paczynski, B.: 1981, *Acta Astron.* **31**, 1.
Patterson, J.: 1983, *Astrophys. J. (Suppl.),* in press (P83).
Payne-Gaposchkin, C.: 1957, *The Galactic Novae*, Dover, New York.
Ritter, H.: 1982, *Workshop on High Energy Astrophysics* (Nanking).
Robinson, E. L.: 1976, *Ann. Rev. Astron. Astrophys.* **14**, 119.
Robinson, E. L., Nather, R. E., and Kepler, S. O.: 1982, *Astrophys. J.* **254**, 646.
Stocke, J., *et al.*: 1983, *Astrophys. J.,* submitted.
Stover, R. J.: 1981, *Astrophys. J.* **248**, 684.
Twain, M.: 1894, *In Defense of Harriet Shelley and Other Essays*, Harper & Bros, London.
Tylenda, R.: 1981, *Acta Astron.* **31**, 267.
Wade, R.: 1979, *Astrophys. J.* **84**, 562.
Warner, B.: 1978, *Acta Astron.* **28**, 303.

DISCUSSION

PACZYNSKI: I find it hard to believe that you can claim to know the space density to a factor of two. This requires knowing the distance to 25%, and you cannot know that.

PATTERSON: No, I don't claim to know the distance to any individual system to better than 25%. It's the PG and *Einstein* surveys that tell us the number of CVs in a particular number of square degrees of sky (since their thresholds for detection were such that essentially all known CVs would be seen). With some rough distance, you can get the scale height for CVs. Then you can convert from square degrees to pc^3, and you have the space density. The cumulative error, in my opinion, is no worse than a factor of two.

GARRISON: I'd like to challenge your statement that you know all the CVs within 100 pc. I think that some pole-on systems may still be hidden. A pole-on system would mimic a Be star, and I for one am going to go back and inspect my spectra of Be stars more closely.

PATTERSON: When I say that they're all known, I'm making an approximate statement. Surely there are still some to be found--in fact, I guess you just found one recently. In fact, probably less than half are known. But the number known is an appreciable fraction of the total. We're not just observing the tail of the elephant.

HORNE: Don't you depend on the assumption that classical novae remain throughout quiescence in a state similar to the classical novae that we see?

PATTERSON: Yes. That assumption is abundantly supported by photographic patrols which show that if you chop out the nova eruptions, classical novae essentially have the same brightness before and after eruption.

McCLINTOCK: But it's not clear their X-rays hold up.

PATTERSON: Yes, and if you're worried about that, you can delete half of the argument from X-rays.

FAULKNER: What fraction is 100 years of the timescale we're talking about?

PATTERSON: I don't understand. One hundred is 1% of ten thousand.

PACZYNSKI: How atypical is DQ Herculis? It declines in brightness for decades. It doesn't show any X-rays. . . . It is an old nova.

PATTERSON: Gosh, I don't know. That's such a many-sided question.

SMAK: I have a comment concerning two-object statistics. Only two cataclysmic binaries were discovered as eclipsing binaries without realizing what they were: UX UMa and RW Tri. All other eclipsing systems were discovered as cataclysmic binaries first. It doesn't prove anything, but. . . .

SHAVIV: Joe, do you want to comment?

PATTERSON: Yes, I would. I haven't thought of that argument, but I believe that if one needed any extra proof that these numbers are not completely out to lunch, then one could say: Yes, although old novae at quiescence are hard to find, they're not hard to find if they're at eighth magnitude and eclipsing. One certainly can't have such things as that lurking in the sky. And that's your point, right?

SMAK: Just 10% of that 400.

PATTERSON: Forty eclipsing systems brighter than 9.5! Let me at them!

WADE: On the correlation between equivalent width and Hβ emission: Did you make any corrections for binary inclination?

PATTERSON: No. I thought about it, but decided not to.

WADE: Another point: If, as we seem to be feeling these days, irradiation is an important aspect of disk structure, then the idea of a physical understanding of the absolute magnitude-equivalent width relation goes out the window. In particular, for the brighter systems, the higher $\dot{M}$ is, the more irradiation matters and, if you adopt a chromospheric model, you might expect the equivalent width to increase.

PATTERSON: OK. A very quick answer: Irradiation will not be important if the energy source is the gravitational energy in the disk. It seems to me that only with nuclear burning can you make the middle of the disk sufficiently luminous to cause important heating effects. I would love to hear a theoretical row on this issue of nuclear burning. Observations show no sign of it, but maybe I can look a little harder.

LIEBERT: I have a comment and a question about the database. I'm worried that the 39 PG objects may contain some stars that are not truly cataclysmic variables, and also that there may be additional cataclysmic binaries among the stars for which our spectra were too poor to classify them. So, I'm worried about using the numbers to better than 30-40%. Finally, my question: Do you think that weak emission-line objects could be found in the Michigan-Tololo survey?

PATTERSON: No, certainly not. It's only sensitive to equivalent widths above about 30 Å. I didn't count the results from this survey for this reason. And I'll give you 30% on these numbers any day.

PACZYNSKI: I'm totally confused about the brightness of novae before eruption, because we heard a claim that at least some novae are declining over 50 or 100 years.

PATTERSON: Somewhat.

PACZYNSKI: Yes, but they do. Now there is another claim that before the eruptions, novae are just as bright. Is this true for every nova which is examined?

PATTERSON: Basically, every nova which goes off is inspected on photographic patrols for its pre-outburst brightness. The published numbers are about 30 out of 30, and I think the unpublished numbers are probably greater.

PACZYNSKI: And in every case, the brightness is just about the same as after the eruption?

PATTERSON: Yes, within a magnitude or so. Of course, "after the eruption" might require a decade, or even more. The observed "tailing off" of DQ Her's brightness probably just reflects the very late stages of the eruption, since it is now about as bright as before the eruption. I just think that you don't go far wrong if you adopt the simple statement Robinson adopted 9 years ago: novae before eruption are the same as they are after eruption.

FAULKNER: I thought that he showed evidence for precursor activity.

PATTERSON: Well, it all depends on how large an effect you're interested in. There are some effects of a magnitude or so. But we're concerned with the possibility of much larger effects here, aren't we? And the point is, all the evidence suggests that they do not exist.

AUTHOR INDEX

STAR INDEX